W9-ATO-649

STRUCTURE AND DEVELOPMENT OF MEAT ANIMALS

H.J. Swatland

Ontario Agricultural College
University of Guelph

PRENTICE-HALL, INC., Englewood Cliffs, New Jersey 07632

Library of Congress Cataloging in Publication Data

Swatland, H. J., (date).
Structure and development of meat animals.

Includes bibliographies and index.
1. Livestock—Carcasses. 2. Livestock—Anatomy.
3. Meat. 4. Muscle. 5. Muscles. I. Title.
SF140.C37S9 1984 636.089'2 83-13938
ISBN 0-13-854398-4

Editorial/production supervision
and interior design: Paula Martinac
Cover design: Edsal Enterprises
Manufacturing buyer: John B. Hall

Printed in the United States of America

10 9 8 7 6 5 4 3 2 1

ISBN 0-13-854398-4

Prentice-Hall International, Inc., *London*
Prentice-Hall of Australia Pty. Limited, *Sydney*
Editora Prentice-Hall do Brasil, Ltda., *Rio de Janeiro*
Prentice-Hall Canada Inc., *Toronto*
Prentice-Hall of India Private Limited, *New Delhi*
Prentice-Hall of Japan, Inc., *Tokyo*
Prentice-Hall of Southeast Asia Pte. Ltd., *Singapore*
Whitehall Books Limited, *Wellington, New Zealand*

Contents

Preface

The meat industry is one of the world's oldest and most important industries. In the national balance sheets of the industrialized nations, the entries for the production and consumption of meat represent vast sums of money. The recognition that scientific research and development are essential for competitive productivity, however, came rather late for the meat industry. In some countries, research institutes dedicated to the scientific study of meat proved their worth decades ago. In other countries, the mortar between the bricks is still damp.

Perhaps the most difficult problem in an applied science such as meat science is finding the right balance between fundamental and applied research. To outsiders, it might appear strange that in-house fundamental research is necessary at all, since it is popularly believed that applied scientists merely apply the discoveries made in the fundamental sciences. There is, however, a two-way flow of ideas between scientists who work on general problems and those who apply themselves to specific technical objectives, and meat scientists have contributed their fair share of fundamental discoveries. One of the most important was probably the discovery by Marsh and Bendall that calcium ions are responsible for the biochemical regulation of muscle contraction. More recently, meat scientists have made some major contributions with their work on the cytoskeleton and the calcium-activated proteases of skeletal muscle.

A gap exists between the mainstream physiological study of animal growth and development and the study of the growth and development of meat animals because of the types of animals used for experimentation. For obvious economic and practical reasons, most fundamental research on animal growth and development is performed on small laboratory animals. As Chapter 7 of this book demonstrates, however, the cellular growth patterns of laboratory rats and mice provide a poor model for muscle growth in meat animals. Therefore, there exists a whole area of fundamental research on muscle growth in large animals that must be undertaken by meat scientists. Much of this research has some relevance to human muscle growth.

A balance between fundamental and applied knowledge must also be achieved when developing an undergraduate curriculum in animal science. Meat science is an academic subject that has emerged slowly from its antecedent disciplines: animal husbandry, microbiology, biochemistry, physiology, and veterinary anatomy. In many universities this has happened within what used to be called departments of animal husbandry. Most of these departments are now called departments of animal science. Our colleagues within such departments—geneticists, nutritional biochemists, and reproductive physiologists—have wasted no time in upgrading their scientific endeavors. All too often, however, meat scientists involved in the study of animal structure and development have inherited the simplistic tools and techniques of animal husbandry. In reality, the structure and development of meat animals is a complex scientific subject, and at the university level, there is no need to pretend that it is not.

This book is intended for anyone with a serious academic interest in the structure and development of meat animals and in the special problems of muscle growth and development in large animals. Particularly in mind are students who would like one volume to span the complete range from the commercial structure of the meat carcass to the physiology of animal growth. This book includes poultry; even though they may not often be called meat animals, they certainly produce a lot of meat. Practically all of the line drawings in this book were drawn by my own shaky hand. I hope that the introduction of new illustrative material relating directly to meat animals will make up for their lack of artistic merit.

Chapter 1 is a general introduction to the visceral anatomy and physiology of meat animals. It briefly describes the operations involved in animal slaughtering and then demonstrates some of the wealth of scientific information that can be gleaned by careful observation in the abattoir. Important animal byproducts such as leather are also included. Chapter 2 deals with the connective tissues of the carcass, that is, with bones, gristle, and fat. The range in structure is from the gross anatomy of the skeleton to the biochemistry of collagen. The developmental topics include cellular growth and metabolism in adipose tissue and in the skeleton.

Chapters 3, 4, and 5 are dedicated to three major aspects of the meat of the carcass. Chapter 3 describes the commercial structure of the carcass, in particular, cuts of meat and meat grading. The major emphasis is on the United States and Canada, with a brief introductory look at meat cutting in other countries. Chapter 3 starts with a simple introduction to the arrangement of the major muscles in the carcasses of meat animals and poultry. This is necessary to understand the anatomical principles of meat cutting and, for many readers, will be all the information on this topic that is required. All the details of carcass myology, both trivial and important, have been packed into Chapter 4 so that this chapter can be bypassed by the general reader. For the devotee of carcass structure, however, Chapter 4 provides step-by-step instructions for the identification of all the muscles in commercial carcasses of beef, pork, and lamb. The general properties of meat are outlined in Chapter 5. The major features of the microstructure and chemistry of skeletal muscles are described and are then used to explain, as far as is possible, the everyday properties of meat such as tenderness, color, and wetness.

After Chapters 1 to 5, with or without a diversion through Chapter 4, the reader should have a balanced understanding of the anatomical and commercial aspects of the structure of meat animals and poultry. Chapters 6, 7, and 8 will then show that animal structure is dynamic. In other words, as an animal increases in age it develops or changes in structure. The nervous system (Chapter 6) is in command of the muscular system, both for the immediate control of muscle contraction and for the long-term regulation of muscle development. The cellular structure of skeletal muscle (Chapter 7) dictates the manner in which muscles can grow and the ways in which an animal's meat yield can be improved. Muscle development, however, is only one part of the overall growth and development of the body (Chapter 8). Chapters 6, 7, and 8 cover topics such as muscle fiber histochemistry, double-muscling, the analysis of growth curves, and the hormonal regulation of muscle growth.

Last but not least, Chapter 9 takes us back to the abattoir to show how meat quality is related to animal structure and development. Chapter 9 covers topics such as the production of lactic acid in meat, stress-susceptibility of pigs, electrical stimulation of meat, and reflex muscle activity. Both commercially and scientifically, the factors that affect the quality of meat are as important as the factors that affect the yield of meat.

I am grateful to Thayne Dutson, Chris Haworth, Steve Jones, Barbara Lewars, and J-P Schoch for their help and criticism in gathering the material for this book. The staff at Prentice-Hall has also played a vital role: Paul McInnis first prompted me to write this book, and then Dudley Kay and Paula Martinac provided valuable help and advice in getting the book into print. I apologize to those foreign authors whose names now lack their proper accents in my bibliography. I typed the text using a computer word processor, and my printer did not have any foreign punctuation. I do not profess to have an expert knowledge in all the various disciplines that I have brought together in this book, and I hope the mistakes that I have made will not cause too much grief.

H. J. Swatland

1

Offals and Byproducts

INTRODUCTION

Apart from the present chapter, the remainder of this book is concerned with the structure and development of muscle, fat, and bone in meat carcasses. The animal byproduct industry, however, is a vital part of the overall economic system of animal agriculture, and must be included in any consideration of animal growth or meat science. In many countries of the world, the byproducts of animal production and slaughtering are an important source of fuel and fertilizers, and a source of materials for clothing, bedding, and building (McDowell, 1977). From a scientific viewpoint as well, it is not possible to ignore all other body systems except those that contribute directly to the commercial carcass. *Offals* are the parts of the meat animal that are removed in the *abattoir* or slaughterhouse to leave a *dressed carcass*. This chapter briefly reviews the structure, function, and economic importance of the offals which form the life-support system for the commercial carcass during its growth and development.

SOME DESCRIPTIVE TERMS

A number of anatomical terms are needed to describe the relative positions of structures within the body.

Anterior toward the head
Posterior toward the tail

Dorsal	toward the upper part or back of the standing animal
Ventral	toward the lower part or belly of the standing animal
Medial	toward the midline plane that separates right and left sides of the body
Lateral	toward the sides of a standing animal
Proximal	toward the body in a limb of the animal
Distal	away from the body in a limb of the animal

The names for different types of farm animals may also be unfamiliar to some readers. The adjectives that relate to cattle, sheep, and pigs are bovine, ovine, and porcine, respectively. The first of these may be used elliptically so that bovine may stand for bovine animal. For cattle, sheep, pigs, and poultry, the sire or father is called a bull, a ram, a boar, or a cock (tom in turkeys), respectively, while the dam or mother is called a cow, a ewe, a sow, or a hen, respectively. A heifer is an immature female bovine, and a gilt is an immature female pig. A hogget is a yearling sheep. Neonates, or newborn cattle, sheep, and pigs, are called calves, lambs, or piglets, respectively. For pigs, the process of birth or parturition is called farrowing. Newly hatched chickens, turkeys, ducks, and geese are called chicks, poults, ducklings, or goslings, respectively. For cattle, sheep, pigs, and poultry, a castrated male is called a steer, a wether, a barrow, or a capon, respectively.

ABATTOIR METHODS

Strictly speaking, an abattoir is a place where cattle are slaughtered, but other species of meat animals are often slaughtered in the same building. The methods used to slaughter meat animals have a profound effect on meat quality. The scientific basis of this relationship is considered in detail in Chapter 9. At this early point in a long and complicated story, only the general principles of slaughter methods need to be introduced.

Preparation for Slaughter

The optimum amount of rest required by meat animals before they are slaughtered depends on the climate, the distance that they have traveled, their method of transport, and their general health. In some countries, where animals are auctioned at stockyards before they are taken to an abattoir, the rest periods are sometimes inadequate. This creates a commercial problem that is difficult to evaluate. On one hand, animals lose weight during transport and in holding pens, and it is undesirable to use pens and labor to prolong a rest period that confers no immediately obvious commercial advantage. On the other hand, stressed or weary animals sometimes produce meat with an unacceptable appearance or water-holding capacity, and this may create economic losses further along the industrial system.

It is traditionally maintained that bleeding or *exsanguination* is less complete when animals are extremely fatigued at the time of slaughter, and that bacteria from the gut more readily enter the bloodstream and contaminate meat from exhausted animals. Any residual blood in meat is often regarded as a good medium for bacterial growth that may cause meat spoilage. In some situations, a rest period of 1 day for cattle and 2 or 3 days for pigs is considered to be optimum. However, such rest periods may be counterproductive if the animals fight among themselves. Animals are not fed in the 24-hour period prior to slaughter.

Stunning Methods

There are several criteria for a good slaughter method: (1) animals must not be treated cruelly; (2) animals must not be unnecessarily stressed; (3) exsanguination must be as rapid and as complete as possible; (4) damage to the carcass must be minimal; and the method of slaughter must be (5) hygienic; (6) economical; and (7) safe for abattoir workers.

To avoid the risk of cruelty, animals must be stunned or rendered unconscious before they are actually killed by exsanguination. When religious reasons do not allow stunning, extra care is needed to ensure that exsanguination causes the minimum of distress to the animal. In the kosher method of killing, conscious cattle are suspended with the head stretched back, and then the throat and its major blood vessels are severed. Drugs cannot be used in the meat industry to induce unconsciousness in animals for slaughter since unacceptable residues would remain in the meat.

Animals can be effectively stunned by *concussion*. Concussion can be induced by a bullet or a bolt that penetrates the cranium, or by the impact of a fast-moving *knocker* on the surface of the cranium. In modern abattoirs, the primitive poleax has been replaced by devices that use expanding gas, either from an air compressor or from a blank ammunition cartridge. First, the animal is restrained in a narrow pen or *knocking box* to minimize its head movements. The concussion instrument can then be accurately located at a point on the midline of the skull, above the level of the brow ridges of the eye sockets. Concussion stunning should not be applied on the neck or posterior part of the skull (Lambooy and Spanjaard, 1981). The knocker is a heavy instrument that is held with both hands. There is a safety catch on the handle, but the actual trigger protrudes from the head of the knocker and is activated as the knocker is tapped against the animal's head (Figure 1-1B). The *captive bolt pistol* (Figure 1-1A) resembles a heavy hand gun, but a blank cartridge is used to propel a cylindrical bolt rather than a bullet into the skull. After penetration, the bolt is withdrawn into the barrel of the pistol and the pistol is reloaded. Steers, heifers, and cows are normally stunned with a knocker or a heavy captive bolt pistol, but bulls and boars which have massive skulls are sometimes shot with a rifle bullet. Pigs and lambs can be stunned with a lightweight captive bolt pistol.

Pigs can be stunned by passing an *alternating electric current* through the brain. Unconsciousness can be induced by a wide range of voltages, from about 70 V to several hundred volts. The length of time that the current is passed through the brain can be reduced to 1 or 2 seconds if abattoir workers are waiting to *shackle* the pig's hind limb with a chain, and then to exsanguinate the animal immediately. The electric current is applied to the pig's head with two electrodes which protrude from an insulated handle. The electrodes must be cleaned at frequent intervals to ensure good electrical contact with the pig. The transformer that supplies the current is usually mounted on a nearby wall.

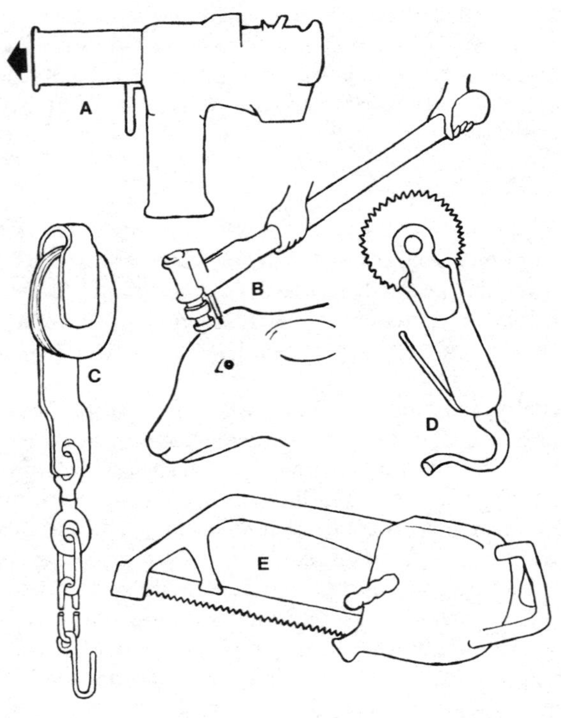

Figure 1-1 Typical abattoir tools: (A) heavy captive bolt pistol; (B) knocker; (C) roller and shackling chain; (D) air-powered dehider; (E) power saw.

In some abattoirs, pigs are stunned by placing them in an atmosphere that contains 65 per cent *carbon dioxide*. Carbon dioxide is heavier than air, so it can be trapped in a pit or deep tunnel into which the pigs are conveyed. After about 1 minute, the pigs are withdrawn in a cage or on a conveyer belt, and they are then exsanguinated as rapidly as possible.

Meat animals are usually stunned, shackled, and exsanguinated, in that order. However, poultry may be shackled or hooked by their feet as soon as they are unloaded from the crate. Getting poultry into crates prior to transport is a major commercial problem. Live birds are easily bruised or more seriously damaged; this causes suffering to the birds and creates carcasses with an unattractive appearance. Poultry are sometimes exsanguinated without first being stunned. However, electrical stunning is very effective, and it facilitates the subsequent removal of the feathers. Concussion from a hammer blow is commonly used to stun ducks.

Exsanguination

Cattle and pigs are usually exsanguinated by a puncture wound which opens the major blood vessels at the base of the neck, not far from the heart (Figure 1-2). The name used in the trade for this process is *sticking*. In sheep, lambs, and small calves, the major blood vessels may be severed by a transverse cut across the throat near the head. Poultry can be exsanguinated with a diagonal cut from the corner of the jaw toward the ear on the other side, or by a knife thrust through the roof of the mouth to sever the

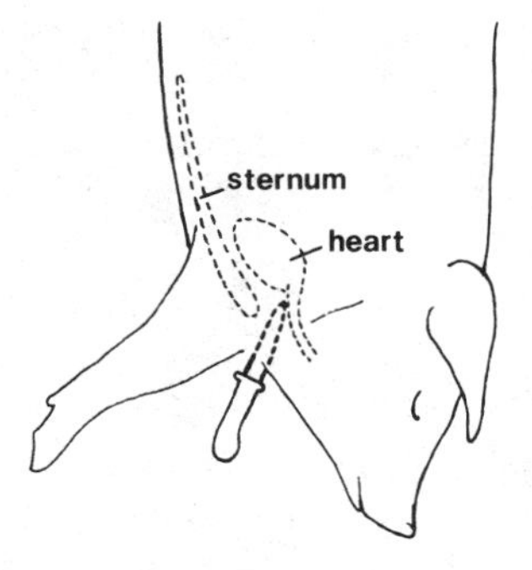

Figure 1-2 Exsanguination of a pig. The knife between the forelimbs is angled so that it cuts the major blood vessels from the heart near the anterior edge of the sternum.

brain and its major blood vessels. For poultry, the cut may be made on the side of the head if the head is later to be removed automatically by machine.

If the sticking wound is inaccurately placed, exsanguination may be too slow, and it may be almost halted by the formation of large blood clots. The formation of blood clots is accelerated when large areas of tissue are damaged by repeated inaccurate punctures. If the trachea is severed by the sticking wound, blood may be drawn into the lungs as the animal breathes. Later in the slaughtering procedure, this may necessitate the trimming of blood clots from the pleural membranes after they have been inspected. If the esophagus is severed, the vascular system may be contaminated by the entry of food particles into the venous system. If the connective tissues of the shoulder are opened, blood may seep into the shoulder region to form blood clots between the muscles.

Incomplete exsanguination increases the amount of residual blood in the carcass. The lean meat may then appear unduly dark and the fat may become streaked with blood. On the surface of incompletely exsanguinated poultry, the skin may appear dark and bloody over the breast, neck, shoulders, and wings. The microscopic tissue damage that may later be caused by the freezing and thawing of poultry enables residual blood to leak from skin capillaries. Thus the results of incomplete exsanguination are often more noticeable to the consumer than to the producer.

Carcass Dressing

Slaughter procedures are extremely variable, and common methods in one locality may be unheard of in another. The introduction of new equipment, such as a hide puller, may lead to a reorganization of the slaughter line. The main objectives of a slaughtering procedure are to get the job done neatly, hygienically, and quickly. One possible method for beef carcasses is described below. In this case, the carcass has already been suspended on an overhead rail in a manner that enables the removal of the distal parts of the hind limbs.

1. Skin the head and remove the skull and lower jaw, leaving the whole of the neck and the skin of the head hanging on the carcass.
2. Remove each foot and the distal part of each limb by cutting through the joint immediately proximal to the long cannon bone.

3. Make a long incision through the hide in the midline of the chest and abdomen, and continue the incision along the medial face of each limb.
4. Remove the hide completely if suitable equipment is available, or just remove it from the ventral part of the body and leave it temporarily hanging from the animal's back.
5. Open the thoracic cavity with a midventral saw cut through the breast bone or sternum.
6. Open the abdomen with a long midventral incision, and remove the penis or udder tissue, and any loose fat in the abdominal cavity.
7. Split the pelvic girdle with a midventral knife cut or saw cut through the cartilage that separates the pelvic bones in the midline.
8. Cut around the anus and close it off with a plastic bag.
9. Skin out the tail (if this was not done earlier).
10. Separate the esophagus (which takes food to the stomach) from the trachea (which takes air to the lungs), by pulling the esophagus through a metal ring; close off the esophagus by knotting it.
11. Eviscerate the carcass by pulling out the bladder (and uterus if present), intestines and mesenteries, rumen and other parts of the stomach, and liver; after cutting through the diaphragm, remove the *plucks* (heart, lungs, and trachea).
12. Separate the left and right sides of the carcass by sawing down the midline of the carcass, through the vertebral column.
13. Trim and weigh the carcass to obtain its *hot weight*.
14. Wash the carcass and pin a shroud over it to smooth the subcutaneous fat.

The other species of meat animals are treated in a corresponding manner, except for the head, feet, and hide. With calves, the skin may be left on until the eviscerated carcass has been chilled. Beef carcasses are first shackled with a chain around the foot, but before the feet are removed, the carcasses are resuspended from a hook under the Achilles tendon at each hock. However, the feet are usually left on pork carcasses. After being shackled during exsanguination, usually by one hind limb, pork carcasses are resuspended from a hooked bar or *gambrel*. This is inserted beneath tendons which have been freed underneath the hind feet. When pigs are shackled by one hind limb during exsanguination, differences in meat tenderness may be created between left and right hams (Cagle and Henrickson, 1970). In some abattoirs, carcasses are skinned while they are on a metal cradle which holds them off the floor.

Hairs and bristles must be removed from pork carcasses when the skin is to be left on the carcass. After exsanguination, otherwise intact pork carcasses are usually immersed in a *scalding tank* that contains water at about 60° C. Scalding at this temperature for longer than 6 minutes damages the skin (Mowafy and Cassens, 1975). Lime salts or a *depilator* such as sodium borohydride are added to the water to facilitate loosening of the hair. After 5 to 6 minutes, the carcass is lifted out of the tank and placed in a *dehairing machine*, where it is repeatedly slapped by strong rubber paddles with metal edges. Loosening of the hair by hot water can also be accomplished by the action of steam on carcasses hanging vertically from an overhead rail. Microbial contamination is minimized but costs due to energy and water are increased.

After removal of the hoof from each toe, the pork carcass is resuspended from the overhead rail. The carcass is then quickly singed with a gas flame which burns all the fine hairs that have escaped the dehairing machine. The carcass is shaved with a sharp knife until it is clean. However, this often damages the skin and it may spoil the skin for leather production. Sometimes it is almost impossible to remove the stumps of strong bristles from the skin, particularly in the early months of the winter.

In some abattoirs, pork carcasses are skinned like beef carcasses. This enables better quality leather to be made from the skin and, in the long run, is less expensive than scalding (Judge et al., 1978). In this procedure, the skin is manually detached in the ventral region of the head and body, and on the medial faces of the limbs. The skin can then be removed from the dorsal part of the carcass with an air knife (Figure 1-1) or with a hide puller. The hide puller is driven by a powerful motor or hydraulic piston and it simply rips the skin off the carcass. However, this usually displaces several kilograms of fat from the edible carcass to the inedible hide, with a consequent loss in revenue (Rust, 1974).

Poultry are usually scalded to facilitate the removal of their feathers. The ease with which feathers may be removed is related to the temperature and duration of scalding. However, high temperatures (over 58° C) cause the skin to become dark, sticky, and easily invaded by bacteria. Consequently, *hard scalding* (at 70 to 80° C) is used only for low-grade poultry that are destined for immediate use in processed products. For broilers, the appearance of the skin is unharmed by about 30 seconds of *semiscalding* in water at 50 to 54° C. Both temperature and duration are precisely controlled, depending on the age and condition of the birds. After the feathers have been loosened, they are removed by machines that have thousands of rubber fingers mounted on rotating drums. However, many of the strong *pinfeathers* on the tail and wings may survive this treatment, and these feathers have to be removed manually.

The feathers on the carcasses of ducks and geese are difficult to remove. Following scalding and the mechanical removal of as many feathers as possible, ducks and geese may be quickly dipped in hot wax. After the birds have been removed and cooled, the wax sets hard and can be pulled off together with large numbers of feathers. The wax is melted and recycled, and the birds are picked bare manually.

Methods for the evisceration of poultry are even more variable than those for meat animals. Many of the operations for poultry evisceration have been successfully automated. Poultry are usually suspended on some type of moving overhead rail. Sometimes they are suspended by their feet, sometimes by their heads, and sometimes by both, so that the vent or cloaca bulges downward. One possible method for the evisceration of poultry is as follows:

1. After stunning and exsanguination, the bird is suspended from its head, and the oil gland at the base of the tail is removed.
2. An incision is made through the skin along the back of the neck, from the head to the shoulders.
3. The crop and the trachea are removed.
4. The bird is resuspended by its feet, and an incision is made through the skin, around the cloaca and toward the sternum.
5. The viscera and the intact cloaca are pulled out and inspected for signs of disease.

6. The liver is removed and the green gallbladder is discarded, without contaminating the carcass with bile.
7. The muscular wall of the gizzard is slit open so that the inner lining and the contents can be discarded.
8. The heart is removed from the hanging viscera and trimmed.
9. The remaining viscera are removed and discarded, and the lungs, kidneys, and ovary or testes are removed from under the vertebral column with a suction tube.
10. The head, neck, and feet are removed.
11. The carcass is chilled in a mixture of ice and water.
12. After chilling, the *giblets* (neck, gizzard wall, liver, and heart) are packed into the carcass.

Although mass-produced poultry are now almost all eviscerated prior to distribution to retail outlets, intact poultry carcasses keep quite well if their viscera are left in place. Growth of intestinal bacteria is minimal below 7° C and, at temperatures below 4° C, uneviscerated carcasses can be stored for at least as long as eviscerated carcasses (Barnes and Impey, 1975).

Meat Inspection

Meat inspection involves the examination of live animals (*antemortem inspection*), carcasses and viscera (*postmortem inspection*), and finished products. The buildings and equipment of the abattoir must conform to a prescribed standard of hygiene, and abattoir workers must be properly trained. The main objectives of meat inspection are (1) to ensure that consumers receive only wholesome products for consumption; (2) to ensure that byproducts are properly treated so as to cause no direct hazard to health; and (3) to provide a warning of the presence of serious contagious diseases among farm livestock. The purpose of antemortem inspection is to identify injured animals that must be slaughtered before the others, and to identify sick animals which must be slaughtered separately or subjected to special postmortem examination.

Human beings have exhibited a fear and dislike of contaminated meat throughout recorded history. The Mosaic food laws, with a religious basis, have survived to the present day, while Greek and Roman civil laws have slowly evolved into modern civil legislation. Many European countries have a long history of legislation relating to meat hygiene and, by 1707, these laws had reached Canada (Heagerty, 1928). However, it was not until the early years of the present century that modern meat inspection started to develop with the science of veterinary microbiology as its basis (1906 in the United States, and 1907 in Canada).

One of the major factors in the design of a slaughter line is the minimization of the spread of salmonella (Anon., 1981a). There are well over 1000 species of this bacterium, and they are frequently found in the feces of meat animals and poultry. When the bacteria are transmitted to human food, they may infect the human digestive system and cause a *food-borne illness*. Although salmonellae on meat are killed by the heat of thorough cooking, they can cause illness by contaminating other foods which are eaten raw. For example, they may contaminate a salad which has been prepared on a cutting board previously used for contaminated poultry.

Most of the hygienic precautions which are taken in the abattoir are quite straightforward. For example, the knives used to remove animal hides often become severely contaminated. Thus they must not be used for later operations when the carcass meat has been exposed, and they must be decontaminated by a method such as dipping in hot water at 82° C for 10 seconds (Peel and Simmons, 1978). Contamination is not limited to knives, and relatively large numbers of salmonellae can be found on steel-mesh safety gloves, cutting boards, and stainless steel tables (Smeltzer et al., 1979). Salmonellae may also contaminate the mixtures of ice and water that are often used to chill poultry carcasses after evisceration.

Lymphatic System

Blood is brought to the body tissues in arteries, and it is removed by veins. However, *interstitial fluid* from between the cells of a tissue is also removed by the *lymphatic system*. Lymph vessels have extremely thin walls, and the *lymph* fluid that they contain is wafted along by any body movements which massage the lymph vessels. A system of flaplike valves prevents backward movement of the lymph. As well as fluids, the lymphatic system also recycles proteins that leak from the vascular system. This is an important factor in the determination of the osmotic balance between the interstitial fluid and the blood (Mayerson, 1963). In starved animals, the scarcity of blood proteins unbalances the system and leads to the accumulation of interstitial fluid (*edema*).

If body tissues are invaded by disease-forming bacteria, some of the bacteria will drift into the lymphatic system. The lymphatic system is arranged like a system of rivers leading to an estuary. The final opening of the system is called the *right thoracic duct*, and this returns the lymph to the vascular system at a point where the main veins of the body enter the heart. *Lymph nodes*, which have a glandlike appearance, are located at regular intervals throughout the lymphatic system (Figure 1-3). Their

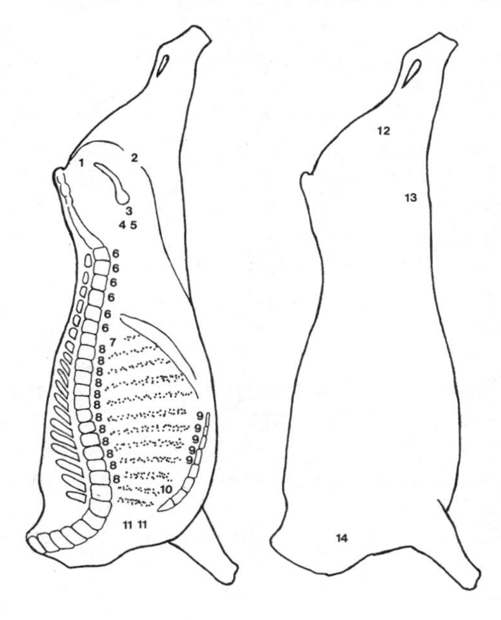

Figure 1-3 Approximate location of lymph nodes on a dressed steer carcass: (1) ischiatic; (2) scrotal; (3) deep inguinal; (4) internal iliac; (5) external iliac; (6) lumbar; (7) renal; (8) intercostal; (9) sternal; (10) anterior sternal; (11) prepectoral; (12) popliteal; (13) precrural; (14) prescapular.

function is to filter and destroy invading bacteria. When lymph nodes are successful, they prevent the spread of disease from the region of tissue that has been invaded. The activated *lymphocytes* of the lymphatic system may play a major role in attacking and destroying invading bacteria.

The lymph nodes that guard healthy tissues are compact in structure and pale brown in color. When they are activated by invading bacteria, they become swollen and discolored. The meat inspector systematically examines the lymph nodes of the viscera and the dressed carcass. Lymph nodes that appear to be abnormal are sliced open for inspection. Knives must be resterilized once they have been used to open an infected lymph node. Once alerted to the presence of diseased tissue, the inspector determines the type and severity of the disease. The whole of the carcass or just the diseased parts may be condemned. It is essential, therefore, that any offals that have already been removed from the carcass can all be traced back to the carcass from which they originated. This also includes any blood that may have been collected as an ingredient for processed meat products. Blood for human consumption is usually collected with a hollow knife in order to minimize contamination from the surface of the carcass.

Tuberculosis is a bacterial disease which can be transmitted from cattle to humans by the ingestion of milk or meat. Diseases may also be transmitted on byproducts such as hides or fleeces. The bacteria that cause *anthrax* require free oxygen to form spores. Workers who handle infected hides or wool may be infected by skin contact or by inhalation. Fortunately, these two serious diseases are rare in the industrialized countries, and the everyday work of the meat inspector is really part of the overall system for the quality control of meat products. Most industrialized nations have a complex system of legislation relating to the disposal of condemned meat. In many cases, condemned meat can be rendered safe for consumption by cooking or prolonged freezing prior to sale.

There are many parasites that can attack farm animals and retard their growth. In temperate climates, however, only a few types of parasite occur in the muscles of a dressed carcass. *Trichinella spiralis* is a small *nematode* worm that sometimes occurs within bundles of muscle fibers in pork carcasses. If the worms are not destroyed as the meat is cooked (at about 60° C), they will reproduce in the human intestine. The larvae burrow through the wall of the intestine and into the body tissues. This causes a disease known as *trichinosis*. Although mild cases are not serious, heavy infections can be fatal. Pigs may become infected when they eat uncooked garbage or the flesh of rodents that carry encysted worms in their own muscles. Once a pork carcass is infected, the encysted worms are most likely to be found in the muscles of the diaphragm, tongue, larynx, abdomen, or under the vertebral column. The great problem for the meat inspector is that the encysted worms in pork are too small to be seen without a microscope. In Germany, pork carcasses are examined by a simplified microscope technique, but the number of infected carcasses that are detected is very small (Ten Horn, 1973). Thus under typical commercial conditions, although no pork carcass can be guaranteed to be free from trichinella, it is not a serious problem provided that the incidence of the parasite is kept low. For this reason, pork producers have the responsibility to cook any waste food or garbage that is fed to pigs, and consumers have the responsibility to make sure that all pork products are thoroughly cooked.

Echinococcus granulosus is a tapeworm, a *cestode* parasite. The adult tapeworm is quite small (8 mm long) and lives in the intestine of a dog or fox. The eggs of the parasite leave the body of the host in its feces. If a sheep eats grass that is contaminated by these eggs, the eggs will hatch in the intestine of the sheep and the larvae may migrate into the bloodstream of the sheep. The parasite then becomes lodged in the body tissues and grows to form a large (10 cm) *hydatid cyst* that contains inactive worms. The life cycle of the parasite is completed if hydatid cysts from the flesh of a dead sheep or from abattoir waste are consumed by another carnivore. Any parts of a lamb or mutton carcass that contain hydatid cysts are condemned by the meat inspector following postmortem examination. The hazard to human health is in the possible contamination of human food by fecal material from dogs. A hydatid cyst may then develop in the human body. To prevent the completion of the parasite's life cycle through sheep, it is essential that dead sheep be disposed of properly and that dogs be prevented from gaining access to abattoirs or to abattoir waste. In areas where there is a high incidence of this parasite, dogs should be dewormed regularly. There are various subspecies of *E. granulosus* that involve other herbivores and carnivores (Thompson, 1979).

Taenia saginata and *T. solium* are tapeworms that live in the human intestine. Contamination of feed for cattle and pigs by eggs from human feces completes a life cycle that leads to the presence of tapeworm larvae in meat. *Cysticercus bovis* is the larval form of *T. saginata* in beef, and *Cysticercus cellulosae* is the larval form of *T. solium* in pork. Cysticerci in meat appear as oval vesicles, almost a centimeter in length and with a white, gray, or translucent appearance. Cysticerci are most commonly found in the heart and masticatory muscles. Cysticerci are detected during postmortem examination, after cuts have been made through these muscles. Once a cyst has been found, carcasses should be cooked or made safe by prolonged freezing (Juranek et al., 1976).

Carcass Refrigeration

Carcasses are chilled to reduce the microbial spoilage of their meat. The rate of chilling is determined by the temperature, the relative humidity, and the velocity of the air in the meat cooler. In addition to direct heat losses by conduction, convection, and radiation, heat is lost from a carcass when water evaporates on its surface. Carcasses cool rapidly if they have a large surface area relative to their mass, and if they have only a thin covering of subcutaneous fat for insulation.

The heat exchange units in meat coolers resemble automobile radiators filled with a refrigerant. The refrigerant is a gas that has been compressed to a liquid by a powerful compressor. Compression liberates heat, and the hot liquid is now pumped to another unit, usually outside on the roof of the meat cooler. The hot liquid, still under pressure in a pipe, is cooled as it passes its heat to the atmosphere or to a water fountain. The cold liquid is then pumped through a small orifice that resembles, in principle, the carburator of an automobile. The conversion of a liquid to gas absorbs heat, so that the resulting gas is very cold. This cold refrigerant passes through the heat exchange units inside the meat cooler, where it cools the air inside the meat cooler.

The air inside the meat cooler is kept moving by powerful fans. The arrangement of heat exchangers and their fans in the meat cooler is carefully planned so as to produce an even distribution of cold air. However, there is always a problem in overcrowded coolers. When hot carcasses are first placed in a cooler, a high air speed is maintained so as to accelerate initial cooling. Later, the air speed is reduced so that the surface of the carcass does not become desiccated. To minimize surface dehydration, the air should not blow directly on the carcasses.

Evaporation losses from pork carcasses can be reduced by rapid cooling after slaughter. Pork carcasses can be briefly prechilled by very cold air or by immersion in liquid nitrogen for about 30 seconds, sufficient to precool the skin without cracking it (Anon., 1981b). In bad conditions, evaporation losses from pork carcasses within 24 hours of slaughter can exceed 3 per cent of the initial hot carcass weight. With rapid chilling, evaporation losses can be held below 1 per cent in the first 24 hours after slaughter. In laboratory conditions, rapid chilling may enhance the quality of the pork by reducing the incidence of paleness and softness (Borchert, 1972), but there is also some risk of causing a deterioration in meat quality under commercial conditions. Evaporation losses from pork carcasses can also be minimized by the use of a spray-on cellulose film (Anon., 1982).

Large carcasses take a long time to cool to 0° C in an ordinary meat cooler, and it is not usually until the day after slaughter that the deep parts of a beef carcass reach the temperature of the surrounding air. Once carcasses have been chilled after slaughter, they can be stored just above 0° C at a relative humidity of 90 per cent with some slight air movement (0.3 m/sec). Higher relative humidities will reduce evaporation losses but will also encourage surface spoilage by microorganisms.

Meat must never be frozen before *rigor mortis* has occurred and the meat has become stiff and inextensible. When meat that has been frozen before the onset of rigor mortis is thawed prior to consumption, it undergoes *thaw shortening* and becomes very tough. Even if the meat is not frozen, cooling that is too rapid may also make the meat very tough. This is called *cold shortening*. Beef carcasses should not be subjected to air less than about 5° C with a velocity over 1 m/sec within 24 hours after slaughter (Cutting, 1974). The temperature of lamb carcasses should not be forced below 10° C within 10 hours of slaughter. These topics are considered in more detail in Chapter 9.

There is usually no reason why meat cannot be removed from the carcass while it is still warm (*hot boning* or *hot processing*). It is more difficult to handle floppy warm meat, but requirements for refrigerated storage space are reduced and there are favorable changes in the water-holding capacity of meat so that drip losses are reduced. The temperature, the hygiene, and the shape of isolated pieces of hot meat must be carefully controlled (Cuthbertson, 1980).

At present, the least expensive method of chilling poultry is by immersion in a mixture of water and ice. Carcasses have an 8% to 12% water uptake and microbial contamination is controlled by chlorination. Spray cooling wastes water, while air cooling may cause dehydration of the carcass (Lilliard, 1982).

DIGESTIVE SYSTEM

The alimentary tract of the digestive system is composed of the mouth, pharynx, esophagus, stomach, small and large intestines, rectum, and anus. Associated with the alimentary tract are the following accessory organs: teeth, tongue, salivary glands, liver, and pancreas.

Mouth, Pharynx, and Esophagus

The jaw and tongue muscles are an important source of low-grade meat from cattle, pigs, and lambs. The mouth is lined by a slippery mucous membrane which in cattle and sheep (ruminants) bears posteriorly directed tags of tissue known as *papillae*. Lips, cheeks, and teeth are, of course, absent in poultry. Whereas mammals have a secondary palate that separates the mouth from the nasal cavity, the nostrils of poultry open directly into the roof of the mouth. Thus when drinking, a bird must use its long neck to keep its head in a horizontal position.

After the mouth, the alimentary tract leads to the *pharynx*. The pharynx is a complicated junction since also opening into it are the *posterior nares* from the nasal cavity; the *eustachian tubes*, which balance the air pressure behind the ear drums; the *larynx*, which tops the windpipe from the lungs; and finally, the *esophagus*, which continues the alimentary canal. When a *bolus* of food is swallowed, the pharyngeal muscles contract to force it down the esophagus.

The *esophagus* is a long muscular tube that runs to the stomach. It is located dorsally to the trachea so that it is behind the trachea when the throat is opened up prior to carcass evisceration in the abattoir. At this point in the slaughter of ruminants it is desirable to tie off the esophagus so as to minimize the spread of ruminal contents onto the carcass. In the meat trade, the esophagus is known as the gullet or *weasand*. Weasands from beef and lamb carcasses may be used as *sausage casings* after they have been cleaned and scraped.

In poultry, just before the esophagus enters the thoracic cavity, there is a large saclike expansion on the right side known as the *crop*. The crop is a temporary storage area for feed. The muscular wall of the mammalian esophagus starts with two oblique or spiral layers, which then develop into inner circular, and outer longitudinal layers farther down. In ruminants, all the muscle tissue of the esophagus is striated. Smooth muscle replaces striated muscle at the level of the diaphragm in pigs. Smooth muscle occurs along the remainder of the alimentary tract and in other organs such as the uterus. Microscopically, smooth muscle is formed from thick layers of elongated cells, each with a single nucleus. The contractile elements of smooth muscle cells do not form microscopically striated fibrils as they do in heart and skeletal muscles.

Stomach

The stomach differs in structure between pigs, ruminants, and poultry. Pigs have a relatively simple, single-chambered stomach (*monogastric*). Cattle and sheep have three additional chambers before the true stomach. Poultry have a second chamber after the true stomach.

In pigs, the entrance of the esophagus into the stomach is controlled by a sphincter. This region is called the *cardia* (not to be confused with the cardiac gland region farther into the stomach). The *esophageal region* of the stomach receives incoming food and is lined by *stratified squamous epithelium* (Figure 1-4). An epithelium is a sheet of cells; squamous cells are flattened in shape; stratified tissues have more than one layer of cells. In the adjacent *cardiac gland region*, the epithelium is supplemented by simple and by compound tubular glands which produce mucus to keep the food sliding through. Tubular glands are formed from a layer of cells rolled into a tube. Each cell secretes its products into the lumen of the tube which then opens

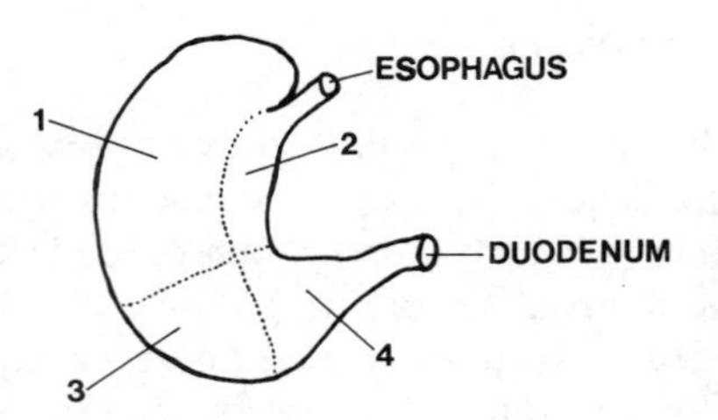

Figure 1-4 Regions of the pig stomach: (1) cardiac gland region; (2) esophageal gland region; (3) fundic gland region; (4) pyloric gland region.

into the stomach. Compound tubular glands are formed by branching tubes. In the *fundic gland region* of the porcine stomach, simple (unbranched) tubular glands open into pits in the stomach wall (Figure 1-5). Cells in the necks of these glands produce mucus. *Parietal* cells in the body of the gland produce hydrochloric acid. Other cells, called *chief* or *zymogen cells*, produce *pepsinogen*, which is split by hydrochloric acid to release the digestive enzyme *pepsin*. The zymogen cells of milk-fed young animals produce *rennin*, which initiates the digestion of milk. The *pyloric glands* are deeper and more branched, and they produce a small amount of *protease* and a lot of mucus.

In cattle and sheep, instead of opening directly into a glandular stomach where digestion begins, the esophagus leads to a series of three extra compartments: the *rumen*, the *reticulum*, and the *omasum*. These compartments are lined with stratified squamous epithelium. In young lambs and calves which are still drinking milk, the rumen and reticulum can be bypassed. The presence of the milk is detected by sensory nerve endings in the mouth and pharynx. Reflex activity causes heavy muscular folds in the walls of the rumen and reticulum to meet and form an *esophageal groove* that leads directly from the cardia to the omasum. The rumen or paunch is a very large muscular bag on the left side of the body, and it extends from the diaphragm back to the pelvis. The smooth muscle of the rumen wall consists of two layers: a superficial layer which runs from anterior to posterior, and an inner layer which runs transversely to form musclar pillars. The reticulum is lined by thin, wall-like ridges which are arranged in a honeycomb pattern. The reticulum is posterior to the heart and to the diaphragm (Figure 1-6). The rumen and reticulum contain countless microorganisms whose metabolic activity greatly enhances the nutritive value of low-grade feed.

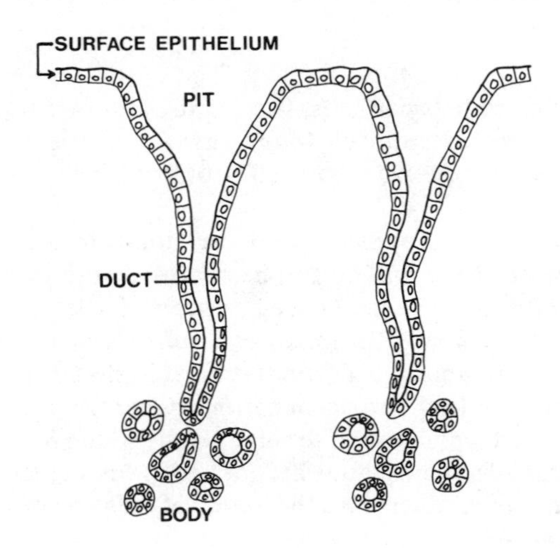

Figure 1-5 General microstructure of a stomach gland.

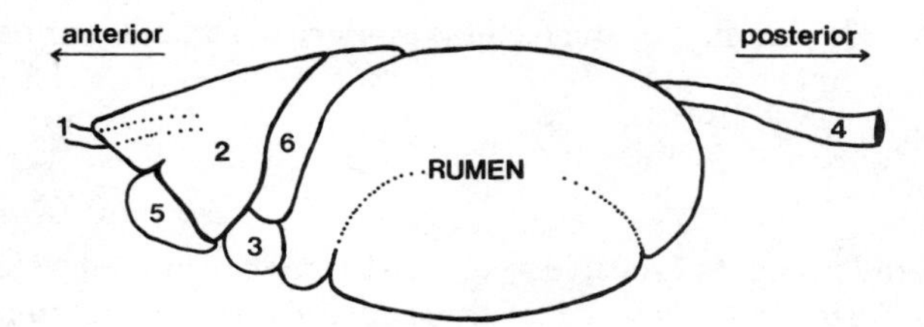

Figure 1-6 Relative positions of the viscera in beef as seen from the left side: (1) esophagus; (2) lungs; (3) reticulum; (4) rectum; (5) heart; (6) spleen.

The *omasum* is almost spherical in shape and is filled with muscular plates which hang from the dorsal roof. These plates or laminae are studded with short, blunt papillae whose function is to grind roughage. The name used for the omasum in the meat trade is *manyplies* or *book bag*. The true glandular stomach or *abomasum* is located ventrally to the omasum (Figure 1-7). The epithelium of the abomasum is glandular with many mucus cells. In a typical lean beef steer, the emptied weight of the rumen, reticulum, omasum, and abomasum comprises about 2.5 per cent of the live weight. The growth of the rumen and reticulum in calves is very rapid, while the abomasum grows more slowly (Godfrey, 1961). The gut fill is extremely variable, but is often around 15 per cent of the live weight.

In poultry, the stomach is divided into two chambers (Figure 1-8). The first chamber, the *proventriculus* or glandular stomach, secretes *pepsin* and hydrochloric acid. The second chamber, the *gizzard*, is thick and muscular with a horny internal epithelium. The gizzard grinds the feed and mixes it with the enzyme mixture from the proventriculus. This is the reverse of the sequence found in the omasum and

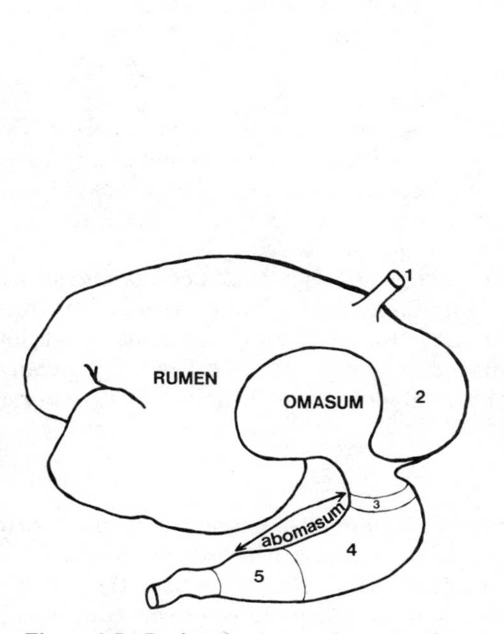

Figure 1-7 Beef stomach seen from the right side: (1) esophagus; (2) reticulum; (3) cardiac gland region; (4) fundic gland region; (5) pyloric gland region.

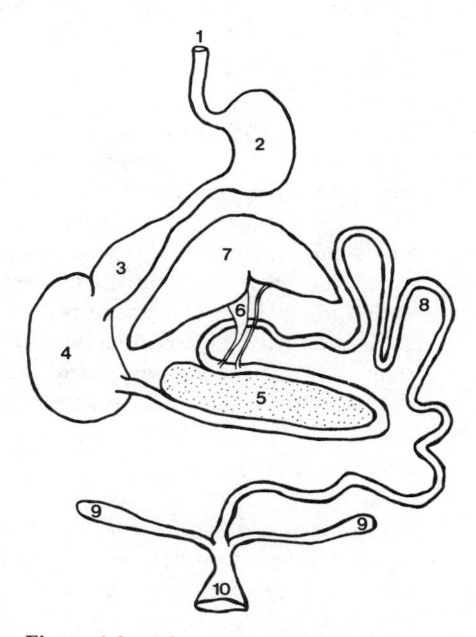

Figure 1-8 Alimentary tract of poultry: (1) esophagus; (2) crop; (3) proventriculus; (4) gizzard; (5) pancreas; (6) gallbladder; (7) liver; (8) small intestine; (9) cecum; (10) cloaca.

abomasum of ruminants, where grinding of the feed takes place before it is exposed to the enzymes from the true stomach.

Small Intestine

In beef animals, slightly over 2 per cent of the live weight is due to the emptied weight of the intestines. The *small intestine* is composed of three regions: the *duodenum*, the *jejunum*, and the *ileum*. The duodenum receives the *hepatic* and *pancreatic ducts* and is lined by a simple epithelium of columnar (tall) cells together with numerous *goblet* cells which produce mucus. The surface area of the epithelium is greatly expanded by two structural features. First, individual epithelial cells have a *brush border* which faces into the lumen of the duodenum. The brush border forms a surface like a carpet with short bristles; this is only just visible by light microscopy and is best seen by electron microscopy. Second, the epithelial surface area is expanded by large numbers of fingerlike inpushings or *villi* which extend into the lumen. Villi are quite large structures (approximately 0.5 to 1 mm long and 0.2 mm thick) which bear the columnar cells on their surface. In the axis of each villus is a tubular lymph vessel called a *lacteal*. The intestinal glands of *Lieberkühn* are tightly packed glands with a simple tubular structure. Their main products are (1) mucus; (2) enzymes, which attack peptides, fats, and carbohydrates; and (3) *enterokinase*, which activates *trypsinogen* from the pancreas to produce the digestive enzyme *trypsin*. *Duodenal* or *Brunner's* glands are tubuloalveolar glands (glands composed of tubes and thin-walled vesicles), which produce mucus to protect the duodenal mucosa. In the intestinal wall are nodules or aggregations of lymphatic cells known as *Peyer's patches*.

Large Intestine

The mammalian *large intestine* consists of the *cecum* and the *colon*. A cecum is a sac that opens into the alimentary tract. The colon is divided into ascending, transverse, and descending parts, and it terminates at the *rectum* and *anus*. Poultry have two ceca (Figure 1-8) just before the rectum. In poultry, but not in cattle, sheep, or pigs, the inner surface area of the large intestine is expanded by villi.

In poultry, the equivalent aperture to the anus is part of a compound structure called the *cloaca*. The cloaca is divided into three regions, but these are often difficult to distinguish. The rectum enters the cloaca at the *coprodeum*, the urinary and genital ducts enter at the *urodeum*, and the opening to the exterior is called the *proctodeum*. Dorsal to the proctodeum is a region of lymphoidal tissue called the *bursa of Fabricius*.

Byproducts

Some parts of the alimentary canal have a considerable commercial value as natural casings. After extensive cleaning and preparation, they can be used to contain different types of sausages and processed meat products. When first taken from the animal and emptied, natural casings typically have five layers. From lumen to exterior these are: (1) the *mucosa*, composed of epithelial, glandular, and vascular components; (2) the *submucosa*, composed of connective tissues which strengthen the gut wall; (3) a circular layer of smooth muscle cells; (4) a longitudinal layer of smooth muscle cells;

and (5) an irregular layer of visceral fat covering the outside. The fat is trimmed off manually and by machine brushing. The intestinal contents are squeezed out and washed away in a process called *stripping*. Finally, the layers of muscle and mucosa are removed as the intestine passes between a pair of rollers in a process called *sliming*. This leaves the strong connective tissues of the submucosa as the sausage casing.

The commercial properties of casings are due to the high *collagen* content of the submucosa, together with smaller amounts of *elastin*. Casings are often turned inside out to facilitate processing. Clean casings are preserved with dry sodium chloride prior to sale. With approximate lengths given in parentheses, the commonly used beef casings are the weasand (0.6 m), rounds from the small intestine (32 m), the bung or cecum (1.8 m), and middles from the large intestine (8 m). The commonly used hog casings are rounds or small casings from the small intestine (18 m), the cap from the cecum (0.4 m), middles from the middle part of the large intestine (1 m), and the bung from the terminal end of the large intestine (1 m). The small intestine from sheep provides 27 m of casing.

Teeth

The embryological formation of teeth starts by the cooperative action of two types of tissue. The *dental lamina* tucks downward from the surface epithelium of the mouth to meet a *dental papilla*, which is growing toward the surface. Tooth *enamel* is formed by the *ameloblast* cells of an *enamel organ* which is formed from the dental lamina. *Dentine* is formed by the outer cells of the dental papilla, and *pulp* is formed by the inner cells of the dental papilla. *Cementum* is a type of modified bone which develops around the roots of a tooth.

Teeth that have a simple structure and do not grow very tall are called *brachydont teeth*. The crown of a brachydont tooth is made of enamel, the inner core of the tooth is made of dentine, and the root is covered by cementum. All the teeth of humans and pigs, and the incisor teeth of cattle and sheep, are of this type. The premolar and molar teeth of pigs have a grinding surface covered by rounded bumps or tubercles; this is called a *bunodont* type of tooth (Figure 1-9). Bunodont teeth can be used for grinding and for crushing a variety of feeds. However, they would soon be

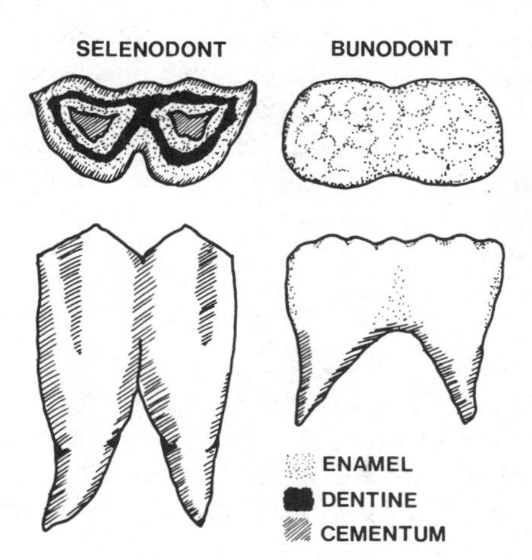

Figure 1-9 Views of the grinding surfaces (at the top of the figure) and the sides of a selenodont molar of a cow and a bunodont molar of a pig.

worn down by the constant grinding of tough fibrous feeds such as those consumed by foraging ruminants.

Ruminants have a special type of tooth which is distinguished by a greater height (*hypsodont*) and by complex curved ridges of enamel (*selenodont*). Hypsodont teeth last longer because they develop a greater depth to be worn down. Selenodont ridges are composed of alternating layers of enamel, dentine, and cementum, and they exert a powerful grinding action when top and bottom teeth are moved horizontally against each other (Figure 1-9).

Dental formulas may be used to describe the patterns of teeth in mammals. The four types of teeth are indicated by a letter notation: I, for incisors or biting teeth, C, for canine or tearing teeth, P, for premolars or anterior grinding teeth, and M, for molars or posterior grinding teeth. The numerator and denominator of a fraction are used to indicate upper and lower numbers of teeth, respectively. Left and right sides of the jaws are not written separately but are indicated by the initial factor, × 2. A prefix, D, is used to denote deciduous teeth which are present in the young animal but replaced in the older animal.

$$\text{Calf and lamb} = 2 \times (\text{DI } 0/4, \text{ DC } 0/0, \text{ DP } 3/3)$$

$$\text{Mature cattle and sheep} = 2 \times (\text{I } 0/4, \text{ C } 0/0, \text{ P } 3/3, \text{ M } 3/3)$$

$$\text{Young pig} = 2 \times (\text{DI } 3/3, \text{ DC } 1/1, \text{ DP } 4/4)$$

$$\text{Mature pig} = 2 \times (\text{I } 3/3, \text{ C } 1/1, \text{ P } 4/4, \text{ M } 3/3)$$

The transition from deciduous to permanent dentition follows a rather complex pattern between 1.5 to 4 years in cattle, 0.25 to 4 years in sheep, and 8 to 20 months in pigs. Most commercially reared meat animals will, therefore, be at an intermediate stage between deciduous and permanent dentition when slaughtered.

The fate of the missing canine teeth in ruminants is quite interesting. The teeth of the upper jaw are inserted into two bones, the *premaxilla* and the *maxilla.* Many anatomists define the upper canine tooth as the tooth that is immediately posterior to the suture between the premaxilla and the maxilla. The lower canine is then defined as the tooth that articulates immediately anterior to the upper canine. Thus the fourth incisor in the lower jaw may be claimed as a canine tooth (Andrews, 1981). In embryonic ruminants, the control system that causes the most anterior three pairs of teeth to become shaped as incisors appears to spread posteriorly and to take control over the developing canine tooth (Osborn, 1978). Although poultry do not have any teeth, the genetic information for tooth formation may still be present in an unexpressed form. When grafted onto jaw tissue from mice, the dental epithelium from chicks may develop ameloblasts which can synthesize enamel (Kollar and Fisher, 1980).

Salivary Glands

Salivary glands have a yellowish color and occur at three major locations. The *sublingual* glands are located under the tongue and between the lower jaw bones, and they have a multiple duct system which drains saliva into the mouth. The *submaxillary* glands are located at the angle of the lower jaw, and they have large ducts which open onto the floor of the mouth, beneath the tip of the tongue. Beneath each ear is a *parotid* salivary gland with a duct that opens into the mouth near the molar teeth.

Liver

The liver is a large organ, about 1.5 per cent of the live weight in beef cattle. In mammals, it is located in the anterior part of the visceral cavity, just posterior to the diaphragm. In pigs, the liver has four equally large lobes plus a small caudate lobe on the right side. Apart from a small caudate lobe on the right side, the bovine liver is not subdivided into lobes. The liver in sheep is similar to that in the bovine, but there is something of a fissure in the main lobe. The liver stores and processes newly digested nutrients which are brought to the liver by the blood vessels of the *hepatic portal system*. The liver receives oxygenated blood for its extensive metabolic activities from the *hepatic artery*. Processed blood is returned via the *hepatic vein* to the general circulation. The liver also produces *bile*, which is emptied into the intestine to aid in the digestion of fats. Histologically, the liver is divided into lobules (Figure 1-10). Livers are condemned if they are infected by trematode flukes such as *Fasciola hepatica* in ruminants, or by nematodes such as *Ascaris suum* in pigs.

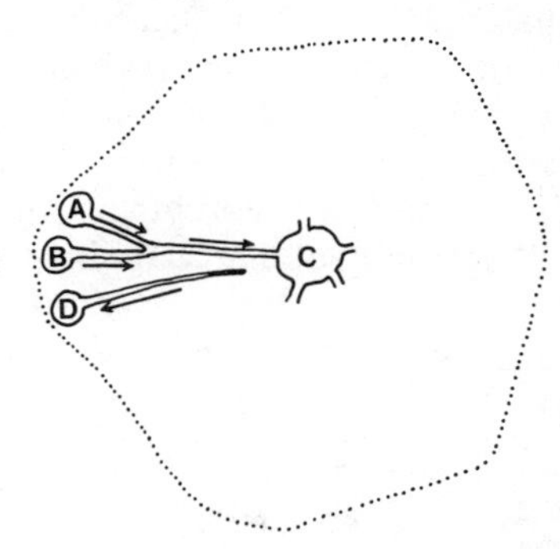

Figure 1-10 Generalized cross-sectional structure of a liver lobule. Oxygenated blood arrives in a branch of the hepatic artery (A) and mixes with blood which carries digested nutrients from a branch of the hepatic portal vein (B). Treated blood leaves via the central vein (C) and newly formed bile exits to the bile duct (D). The radiating pattern is repeated round the lobule.

Pancreas

The *pancreas* is a pale yellow gland located between the stomach and the small intestine in mammals, and in a loop of the duodenum in poultry. It has one or two ducts which convey pancreatic juice to the duodenum. The external secretions of the pancreas are controlled by the nervous system (*vagus nerve*) and the endocrine system (the hormone *secretin* from the duodenum). The three major constituents of the pancreatic juice are *trypsin* (for the hydrolysis of proteins when in conjunction with enterokinase present in the small intestine), *amylase* (for initial digestion of starch), and *lipase* (for the digestion of fats).

RESPIRATORY SYSTEM

The nasal cavity of the skull contains the *turbinate* bones. Left and right turbinate bones have a shape that resembles a loosely rolled sheet of paper. This creates a large surface area for the *nasal epithelium*. The nasal cavity opens into the *pharynx* (shared with the alimentary canal), and then opens into the *larynx*. The larynx has a cartilagenous skeleton with muscles that support and stretch the vocal cords. In poultry, however, sound is produced by a separate organ, the *syrinx*, which is located farther down the respiratory system. The *epiglottis* is a spout-shaped cartilage which protects the entrance to the larynx. The larynx leads to the *trachea* or windpipe.

The trachea is a flexible tube which is held open by rings of cartilage. The continuity of each ring of cartilage is broken by a small dorsal gap. The trachea divides into two *bronchi* at a "Y" fork (Figure 1-11). The bronchi connect with the right and left lungs, where they branch into progressively smaller ducts called *bronchioles*. The trachea, bronchi, and bronchioles are lined with ciliated epithelium and mucous glands. *Cilia* are extremely fine whiplike hairs on the lumenal surfaces of cells. A complex system of mobile protein strands along the length of each cilium provides the motive power for movements that appear whiplike. Millions of cilia beat in a coordinated manner so that they can propel a continuous stream of mucus from the lungs to the nasal cavity. Thus any small particles which have entered the lungs, despite the protective filtering of incoming air by the turbinate bones, can be removed. Gaseous exchange between inhaled air and the blood in the lungs takes place across the moist surfaces of *alveoli* or *alveolar sacs*. In mammals, the alveoli are the final blind-ending branches of the air duct system. Beneath the moist epithelium that lines each alveolus is an extensive meshwork of lung capillaries.

Oxygen is taken up by the blood in a loose combination with the *hemoglobin* of red blood cells or *erythrocytes*. There are three ways in which carbon dioxide may be carried in the blood: (1) in solution; (2) combined with blood proteins; or (3) as bicarbonate. Carbon dioxide is more soluble than oxygen. The ratio of bicarbonate

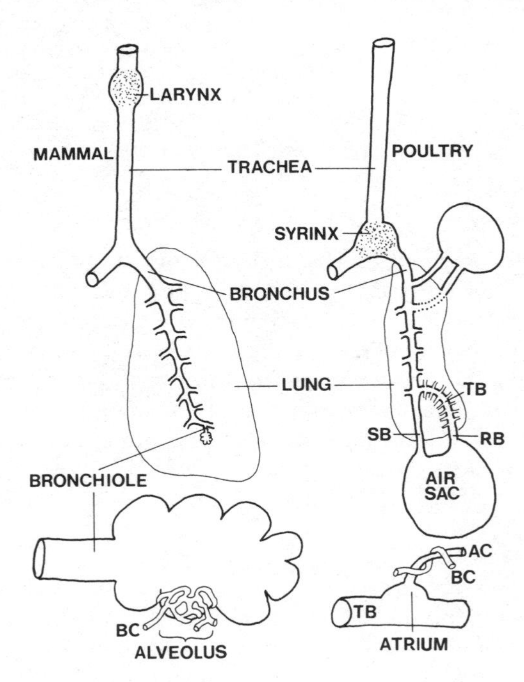

Figure 1-11 Comparison of the respiratory systems in farm mammals and in poultry: (AC) air capillary; (BC) blood capillary; (RB) recurrent bronchus; (SB) secondary bronchus; (TB) tertiary bronchus.

to carbonic acid determines the pH or acidity of the blood. This ratio can be regulated by the rate of escape of carbon dioxide from the blood in the lungs; loss of carbon dioxide increases pH (decreases acidity). Gaseous exchange does not occur across the walls of the major air ducts that lead into the lungs. Thus the last fraction of air that is inhaled becomes the first fraction to be exhaled, and the oxygen it contains is not utilized. Typical resting rates of respiration are 12 to 18 breaths per minute in cattle, 12 to 20 in sheep, and 10 to 18 in pigs.

When the lungs are removed from the body, slippery *pleural membranes* can be seen covering both the inner surface of the thoracic cavity and the lung surface. Pleural membranes prevent friction between the lungs and the body wall. Inspiration and expiration are largely due to movements of the *intercostal muscles*, the ribs, the *diaphragm*, and sometimes, the abdominal muscles. The diaphragm resembles a strong drumskin which divides the thoracic and abdominal cavities, but it is thickened by muscle where it joins the body wall. In a dressed carcass, the muscular part of the diaphragm remains as a flap of muscle running diagonally across the inside of the rib cage. The rate of respiration is controlled by the *medulla oblongata* in the posterior part of the brain. The medulla responds primarily to the pH and the carbon dioxide content of the blood; it increases the rate of respiration if the blood becomes acidic with more carbon dioxide.

The proteins of the lungs, together with those of the rumen and spleen, can be recovered by alkaline extraction followed by reacidification (Swingler and Lawrie, 1979). These proteins can be isolated as a powder (Levin, 1970) or they can be texturized to form fibers (Swingler and Lawrie, 1978). Lungs can be processed to isolate *heparin*, an anticoagulant for medical use (Levin, 1970).

The avian respiratory system is quite different from that found in mammals (Figure 1-11). There is no diaphragm separating thoracic from abdominal cavities. Instead of being drawn *into* the lungs and then exhaled, air is drawn *through* the lungs and into *air sacs* outside the lungs. On exhalation, the air passes back through the lungs to the exterior. In poultry, therefore, the gaseous exchange between air and blood takes place as the air is moving *through* the lungs. The lungs of poultry are much smaller than those of mammals. Instead of occupying almost the whole of the thoracic cavity, they are located under the vertebral column, where they are shaped to fit between the deep arches of the ribs where they meet the vertebral column. The lungs of poultry are usually removed with a suction tube during commercial slaughtering procedures. In meat animals, the lungs are removed together with the trachea, bronchi, and heart, as "plucks."

The extensive system of air sacs in poultry extends between many of the viscera and even into certain bones. The *interclavicular* air sac is a single structure in the midline but the other air sacs are paired (right and left). The *cervical* sac extends toward the neck, the *axillary* sac is within the body at the junction with the wing, and the *anterior thoracic, posterior thoracic*, and *abdominal* sacs are in the body cavity. The *humeral air sac* is located within the humerus and is connected to the axillary sac. Air sacs have extremely thin walls and, when poultry are dissected, they should be identified while the viscera are in a relatively undisturbed condition. Special techniques for the demonstration and dissection of air sacs are described by Goodchild (1970).

URINARY SYSTEM

The urinary system has two major functions: (1) to remove waste products from the bloodstream; and (2) to regulate the amount of water present in the body. In mammals, the *kidneys* are ventral to the vertebral column in the anterior lumbar region. The kidneys of pigs and sheep are oval in shape, while the kidneys of cattle are each divided into approximately 20 lobules. Pork kidneys are flatter and paler than lamb kidneys. In healthy, well-fed animals, the kidneys are usually surrounded by *perirenal fat*. This is called leaf fat in the pork carcass. Each kidney has a depression or *hilum* where the *renal artery* enters the kidney and where the *renal vein* and *ureter* leave the kidney (Figure 1-12). The ureter from each kidney carries urine to the *bladder*.

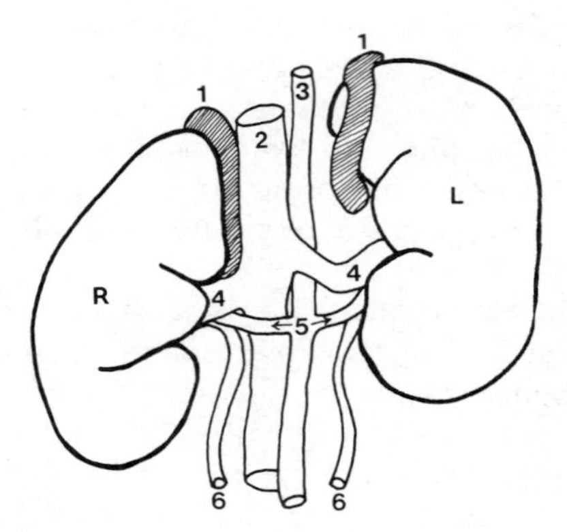

Figure 1-12 Vascular supply to the kidneys of a pork carcass, as seen when the surrounding fat is removed; (L) left kidney; (R) right kidney; (1) adrenal gland; (2) posterior vena cava; (3) posterior aorta; (4) renal veins; (5) renal arteries; (6) ureters.

When a kidney is cut open, a pale inner *medulla* is seen surrounded by a dark red *cortex* (Figure 1-13). The wide entrance to the ureter is called the *pelvis* of the kidney. Running through the medulla, toward the pelvis of the kidney, are many small *collecting tubules*. Each of these terminates at a small conical mound called the *pyramid*, so that the pyramids project into the pelvis of the kidney. Urine is produced from the blood by a functional unit of the kidney called a *nephron*. There are large numbers of nephrons in each kidney. Urine leaves the bladder in a single tube, the *urethra*, which runs to the *penis* or to the vagina.

The main parts of a nephron are shown in Figure 1-14. Ultrafiltration occurs in the *malpighian corpuscle*, so that blood plasma passes from the capillary *glomerulus* to the *Bowman's capsule*. Large molecules cannot leave the blood because the pores of

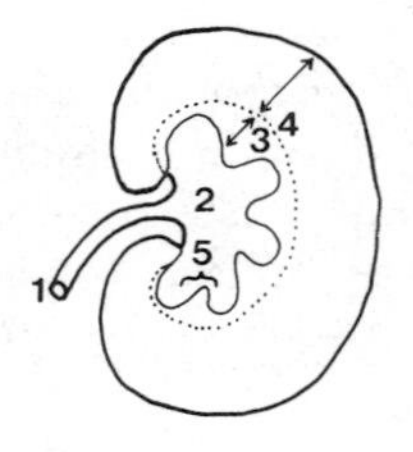

Figure 1-13 Longitudinal section through the kidney of a lamb: (1) ureter; (2) pelvis; (3) medulla; (4) cortex; (5) papilla.

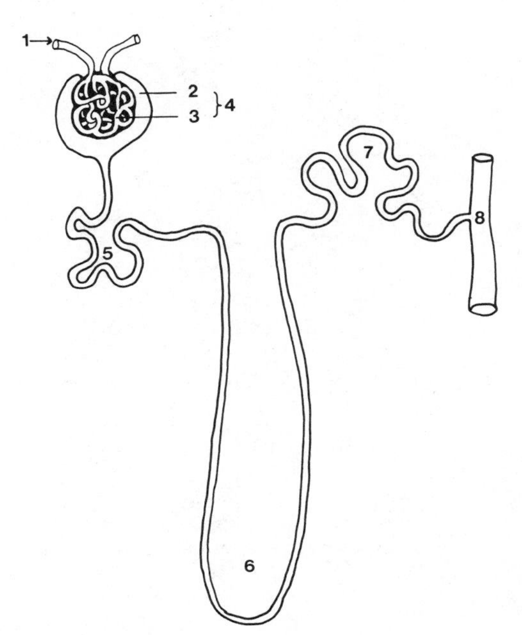

Figure 1-14 Diagram of the microstructure of a nephron: (1) renal capillary; (2) Bowman's capsule; (3) glomerulus; (4) Malpighian corpuscle; (5) proximal convoluted tubule; (6) loop of Henle; (7) distal convoluted tubule; (8) collecting tubule.

the filter are too small. In the *proximal convoluted tubule*, sodium chloride together with useful substances such as glucose, amino acids, proteins, and ascorbic acid may be absorbed back into the bloodstream. The main function of the *descending* and *ascending loops* of *Henle* is to create an osmotic gradient. The cells of the ascending loop pump out sodium ions. Sodium ions cannot get back into the ascending loop because its wall is impermeable. However, the wall of the descending loop is permeable, and water can diffuse out while sodium ions can diffuse in. Because of these movements, the osmotic pressure in the tissue surrounding the loops of Henle increases toward the bottom of the loop. Thus when urine eventually flows down the collecting tubule, water is lost from the urine in the collecting tubule by osmosis into the surrounding tissue (since the concentration is higher outside the tubule near the bottom of the loops). This forms an efficient countercurrent exchange system. The final concentration of the urine is controlled by *antidiuretic hormone* (ADH) from the posterior lobe of the pituitary gland beneath the brain. ADH makes the wall of the collecting tubule more permeable, and water can leave more readily by osmosis. Thirsty animals produce a lot of ADH and only a small volume of urine. Their urine is more concentrated. The *distal convoluted tubule* is the principal site of acidification of the urine, and this is where potassium, hydrogen, and ammonium ions enter the urine. Nitrogen is excreted from the body as *urea*.

In poultry, the kidneys are pressed closely against the ventral surface of the vertebral column, posterior to the lungs. Urine from each kidney leaves in a ureter but passes directly to the cloaca (Figure 1-15). Here the urine rapidly loses water and the main component of nitrogenous excretion, *uric acid*, is precipitated as a sludge.

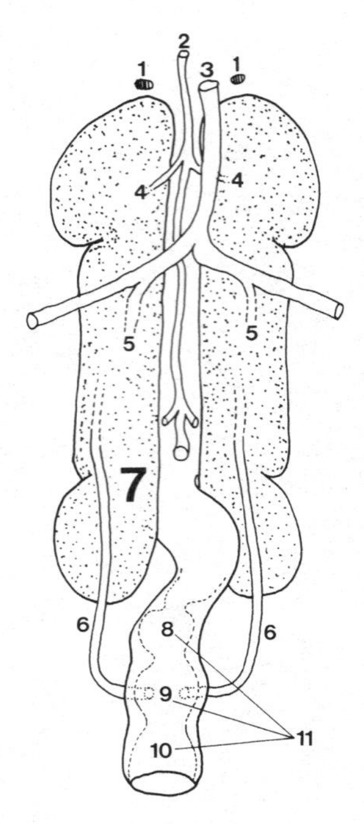

Figure 1-15 Ventral view of the kidneys in poultry: (1) adrenal gland; (2) aorta; (3) vena cava; (4) renal arteries; (5) renal veins; (6) ureters; (7) kidney (right side); (8) coprodeum; (9) urodeum; (10) proctodeum; (11) cloaca.

REPRODUCTIVE SYSTEM

Males

The paired *testicles* or *testes* of male mammals are located in the *scrotum*. Here they can be maintained several degrees below body temperature for the efficient production of *spermatozoa* or sperm. Each testis can be raised by a detached strip of the obliquus abdominis internus muscle (Chapter 4, Group 5 muscles) called the *cremaster externus*. Evidence of cremaster muscles may appear in carcasses of sows and gilts and so cannot be used to identify male carcasses. Spermatozoa are produced in *seminiferous tubules*, which are tightly packed into the oval shape of the testis. Immature spermatozoa pass to a further tubular system, the *epididymis*, on the surface of the testis (Figure 1-16). Spermatozoa become mature during storage in the epididymis. They are carried to the urethra during mating by peristalsis of the *vas deferens*. The urethra is located ventrally in the penis. Seminal fluid to carry the spermatozoa is produced by the paired *seminal vesicles*, by the *prostate gland*, and by the paired *bulbourethral glands* (Cowper's glands). The glands are located along the urethra near the bladder. In boars, the length of the bulbourethral glands is correlated with the *androstenone* content of the carcass (Forland et al., 1980). As discussed briefly toward the end of Chapter 2, androstenone is associated with the level of boar taint in the carcass.

The penis contains a *sigmoid flexure* or S-shaped bend along its length. The sigmoid flexure is straightened out when the penis is extended for mating. This occurs when a pair of muscles, the *ischiocavernosus* muscles, compress the veins which drain the blood from the penis. Arterial blood pressure then expands the volume of vascular

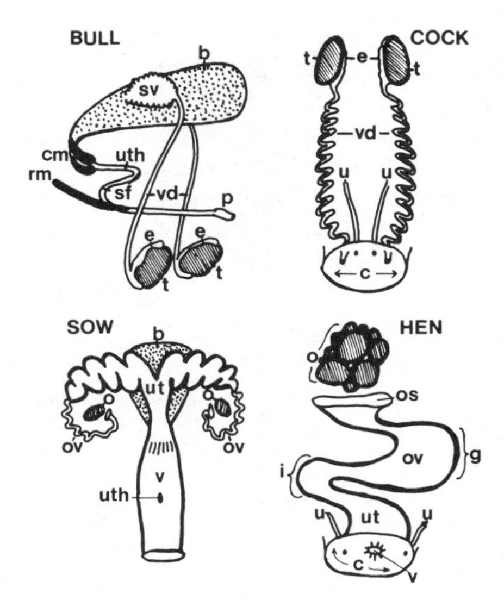

Figure 1-16 Comparison of the reproductive systems in the bull (lateral view), in the sow (dorsal view), and in male and female poultry (ventral views): (b) bladder; (c) cloaca; (cm) cavernosus muscles; (e) epididymis; (g) glandular region of oviduct; (i) isthmus of oviduct; (o) ovary; (os) ostium; (ov) oviduct; (p) penis; (rm) retractor muscle; (sf) sigmoid flexure; (sv) seminal vesicle; (t) testis; (u) ureter; (ut) uterus; (uth) urethra; (v) vagina; (vd) vas deferens.

tissue in the penis. The ischiocavernosus muscles are attached to the ischium (the posterior bone of the pelvis) and the trimmed stump of the muscle can be seen on dressed sides of beef as a *pizzle eye* (Figure 2-12). The pizzle eye is poorly developed in steer carcasses and is larger and darker in bull carcasses.

During the embryonic development of both mammals and birds, the testes are formed from tissue that is located near the kidneys. In male mammals, the testes normally move to the scrotum outside the body cavity, and they pass through the body wall in the *inguinal canal*. In *cryptorchid* pigs or ridgelings, movement of the testes is incomplete and they do not reach the scrotum. This abnormality causes infertility, but an older cryptorchid pig may still develop boar taint like a normal boar. In male poultry, the testes remain in their original position near the kidneys, and a highly coiled vas deferens links each testis separately to the urodeum of the cloaca. The testes of cockerels can be removed through an incision made posterior to the ribs and anterior to the pelvis. *Capons* produced in this way are less aggressive in their behavior and they tend to deposit fat more readily than entire males. Except for ducks, male poultry have no functional penis.

Females

Female mammals have a pair of *ovaries* located posteriorly and dorsally in the abdominal cavity. *Ova* develop in the cortex (outer layer) of the ovary. Each ripe ovum is enclosed in a fluid-filled *follicle*. At estrus, ova are released into a ciliated funnel or *infundibulum* at the end of each *oviduct* (*fallopian tube*). The oviduct on each side leads into a horn of the uterus where embryonic development takes place. At birth, the offspring emerge through the dilated *cervix* and *vagina*.

The mammary glands are derived from highly modified sweat glands of the skin. The *udders* of sheep and goats are divided into right and left halves, each with a teat. The cow's udder has four quarters, so that there are two teats on each side. Most sows have seven pairs of mammary glands and a total of 14 teats. Milk is produced in glandular *alveoli*, and it collects in the *cistern* of the teat. The bovine udder is supported by medial and lateral suspensory ligaments which are dominated by elastin and collagen fibers, respectively.

In female poultry there is only a single ovary since the ovary and oviduct of the right side do not normally develop. In poultry, the ovary usually contains a cluster of ova in different stages of development. The ova in the most advanced state of development appear as full-sized egg yolks. A large infundibulum (*ostium*) leads to a thick glandular region of the oviduct where egg albumen is formed, then to a narrower *isthmus* where shell membranes are added, and finally to a wide uterus where a calcareous shell is formed. The vagina opens into the cloaca and forms mucus to facilitate egg laying.

NERVOUS SYSTEM

Components

The nervous system is a single integrated system composed of distinct regions. The *central nervous system* is composed of the brain and spinal cord. The *peripheral nervous system* is composed of the nerves, which radiate from the central nervous system to all parts of the body. The peripheral nervous system includes the *cranial nerves*, which radiate from the base of the brain; the *spinal nerves*, which radiate from the spinal cord; and the *autonomic nervous system*. The autonomic nervous system is subdivided into the *sympathetic nervous system*, which originates from the thoracic and lumbar regions of the spinal cord, and the *parasympathetic nervous system*, which originates from the brain and the sacral part of the spinal cord. The salivary glands are innervated by the parasympathetic system. The heart, lungs, alimentary canal, and bladder receive dual innervation from both the parasympathetic and the sympathetic nervous systems.

A *ganglion* is a beadlike swelling along a nerve, and it contains the cell bodies of certain nerve cells (*neurons*). The autonomic nerve fibers, which radiate from the central nervous system to the ganglia of the autonomic nervous system, are called *preganglionic nerves*. The autonomic nerve fibers which continue on from the ganglia to the organs which they innervate are called *postganglionic fibers*. The ganglia of the sympathetic nervous system are mostly located under the vertebral column so that their preganglionic nerves are short and their postganglionic nerves are long. The ganglia of the parasympathetic nervous system are mostly located close to or within the innervated organs so that their preganglionic fibers are long and their postganglionic fibers are short. The peripheral nervous system innervates muscle and skin while the autonomic nervous system innervates glands and viscera.

Brain

The surface of the brain is covered by a delicate membrane (*pia mater*) which carries a network of small blood vessels supplying the brain. The pia mater is covered by another thin membrane called the *arachnoid membrane*. On top of this membrane is a layer of tough tissue, the *dura mater*, which adheres to the inner surface of the skull.

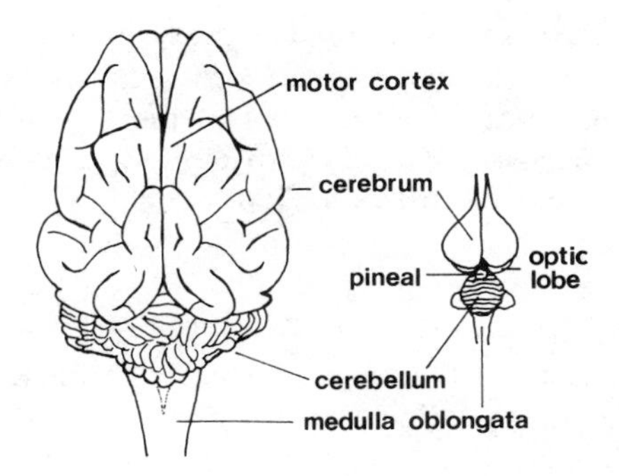

Figure 1-17 Dorsal views of the brain of the pig (left side of figure) and the brain of the chicken (right side of figure).

The surface of the brain in meat animals (Figure 1-17) is increased in area by folds (*gyri*) and grooves (*sulci*).

The *cerebrum* is composed of left and right *cerebral hemispheres* which are separated by a deep fissure. If the brain is exposed by cutting through the skull with a band saw, the outer layers of the brain appear gray in color while the inner parts appear white. The gray areas are dominated by nerve cell bodies while the white areas are dominated by *axons*. Axons are cablelike extensions of the nerve cell body (Figure 1-18), and they are electrically insulated by a sheath of *myelin*. The gyri and sulci on the surface of the brain allow large numbers of nerve cell bodies to connect with the bundles of axons which carry information within the central nervous system. The function of the cerebrum is to regulate higher forms of nervous activity, such as recognition, learning, communication, and behavior. The *cerebellum* is posterior to the cerebrum, and is formed from a middle lobe called the *vermis* and two lateral hemispheres. The cerebellum coordinates muscle movements during locomotion and in the maintenance of posture.

The *thalamus* is the region of the brain that is located ventrally to the cerebrum. It links the cerebrum to the rest of the central nervous system. The *hypothalamus* is located ventrally to the thalamus, and it connects the major regulatory gland of the

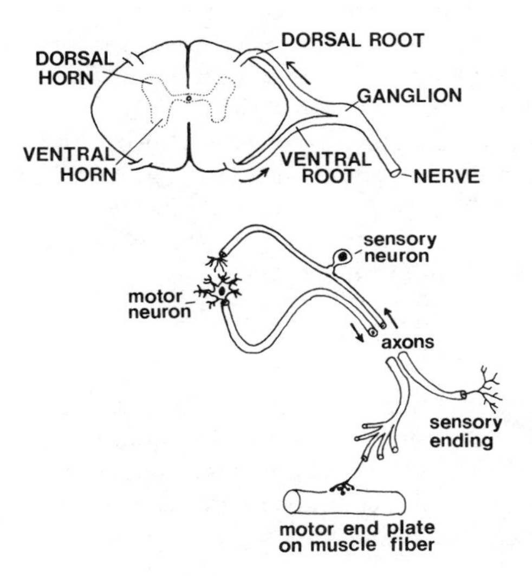

Figure 1-18 Gross components (top diagram) and microscopic components (bottom diagram) of a spinal reflex pathway. Arrows indicate the direction of nerve impulses. In the top diagram, the spinal cord is seen in transverse section.

endocrine system, the *pituitary*, to the brain. The most posterior region of the brain, where it tapers down to the diameter of the spinal cord, is called the *medulla oblongata*. This region controls the heart rate via elements of the autonomic nervous system described earlier. Cavities or *ventricles* filled with *cerebrospinal fluid* run through the brain and extend down the spinal cord as a small canal.

Spinal Cord

The spinal cord is an extension of the brain. It emerges from the skull through the *foramen magnum*, and extends posteriorly to the first (cattle and sheep) or third (pigs) sacral vertebrae in the sirloin region. At regular intervals, pairs of *dorsal* and *ventral roots* enter and leave the spinal cord (Figure 1-18). Dorsal and ventral roots unite outside the spinal cord to form the nerves of the peripheral nervous system. Sensory neurons with their cell bodies in the *dorsal root ganglia* at the side of the spinal cord carry incoming sensory information from the skin, muscles, and tendons. Sensory axons terminate on a variety of different types of neurons in the spinal cord. These neurons may relay information up the spinal cord to the brain, or to other regions of the spinal cord, or to nearby *motor neurons*. Motor neurons have their cell bodies in the central gray region of the spinal cord (in the *ventral horn;* Figure 1-18). Their axons leave the spinal cord in ventral roots and innervate muscle fibers in the carcass muscles. Motor axons do not branch much before they reach their muscles, but once inside a muscle, they branch extensively to innervate a large group of muscle fibers called a *motor unit*. Myelinated axons, which ascend and descend the spinal cord, are located outside the central gray areas, so that, unlike the situation in the brain, the white matter is placed outside the gray matter. The spinal cord is located within the vertebral column. It is separated from the inner bony surfaces of the vertebrae by an *epidural space*.

Neurons

Many different types of neurons contribute to the complex circuitry of the central nervous system. Two relatively simple types of neurons are sensory neurons, with their cell bodies located partway along their axons outside the spinal cord, and motor neurons, with their cell bodies located within the spinal cord. *Dendrites* are small rootlike branches which provide the input to motor neuron cell bodies and to the axons of sensory neurons. Information is communicated along axons by waves of ionic activity called *action potentials*. Information is communicated between neurons by the release and reception of chemical transmitters. A *synapse* is a junction between two neurons, or between a neuron and a muscle fiber.

Action Potentials

The cell bodies, dendrites, and axons of a neuron are bounded by a cell membrane which is able to pump sodium ions outward. This allows the concentration of potassium ions to build up inside the neuron. Because of the unequal distribution of these and other ions, the neuronal membrane carries an electrical charge of 50 to 70 mV, with the negative charge on the inner face of the membrane. If the membrane is briefly short-circuited by a change in its ionic permeability, sodium ions rush inward and potassium ions rush outward for a brief instant. This rapid movement of ions short circuits an adjacent region of the membrane so that the cycle is propagated along the membrane. This self-propagating ionic and electrical change is called an *action potential*. Once an action potential has passed a region of a membrane, an equilibrium is restored ready for the next action potential. During this brief restoration period, which is called the *refractory period*, the membrane does not respond to further stimuli. Action potentials are normally carried in only one direction, away from the origin of the action potential. Action potentials are all identical once they are under way, and information is coded by their number and frequency pattern.

CIRCULATORY SYSTEM

Heart and Blood Vessels

The *right ventricle* (Figure 1-19) pumps blood into the *pulmonary arteries* and then to the lungs. Oxygenated blood returns to the *left atrium* in the *pulmonary veins*, the atrium fills the *left ventricle*, and oxygenated blood is then pumped through the *aorta* to the body tissues. The aorta branches to form the major arteries. These branch again many times and eventually give rise to *arterioles* and, finally, to *capillaries*. Blood is collected from the body tissues by the venous system, and it eventually returns to the *right atrium* via the *anterior vena cava* or the *posterior vena cava* for another cycle through the lungs. Thus, relative to other arteries, the pulmonary artery is unusual because it contains deoxygenated blood. Relative to other veins, the pulmonary vein is unusual because it contains oxygenated blood.

In meat animals, the aorta bends to the left side of the body, and then runs posteriorly in the midline of the body, ventral to the vertebral column. The right forelimb is supplied from the *right brachial artery* which, like the *common carotid*

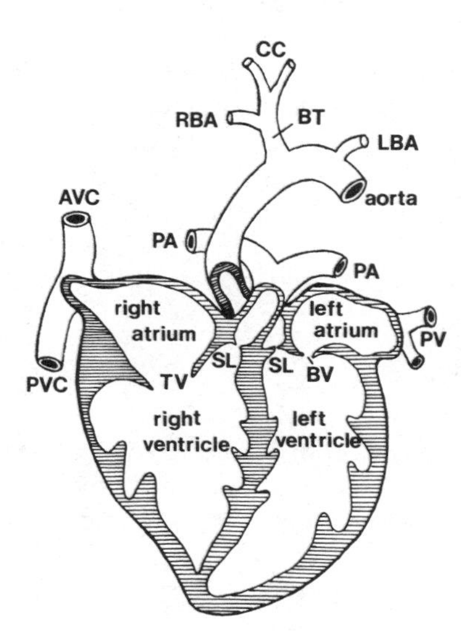

Figure 1-19 Structure of the heart in meat animals: (AVC) anterior vena cava; (BT) brachiocephalic trunk; (BV) bicuspid or mitral valve; (CC) common carotid arteries to right and left sides; (LBA) left brachial artery; (PA) pulmonary artery; (PV) pulmonary vein; (PVC) posterior vena cava; (RBA) right brachial artery; (SL) semilunar valve; (TV) tricuspid valve.

arteries to the head, branches from the *brachiocephalic trunk* (Figure 1-19). The *left brachial artery* to the left forelimb comes directly off the aorta. As it passes under the vertebral column in the ribcage, the aorta gives rise to a series of small *intercostal arteries* before the first main branch to the viscera, the *coeliac artery*. In cattle, the coeliac artery has five main branches: (1) the *hepatic artery* to the liver, pancreas, and nearby structures; (2) *left* and (3) *right ruminal arteries* to the rumen; (4) an *omasoabomasal artery* to the omasum and abomasum; and (5) the *splenic artery* to the spleen. Proceeding posteriorly, the remaining major branches from the aorta are the *anterior mesenteric, right* and *left renal arteries*, a *spermatic* or *utero-ovarian artery* (depending on the sex of the animal), and the *external* and *internal iliac arteries* to the hind limb and rump.

In poultry, the aorta swings to the right side of the body after leaving the heart. There are arteries to the head (*right* and *left common carotid arteries*), to the shoulder and wing (*brachial artery*), and to the large flight muscles covering the sternum (*pectoral artery*). These arteries originate symmetrically just after the aorta leaves the heart. The first major visceral branch of the aorta, the *coeliac artery*, supplies the proventriculus, gizzard, spleen, liver, duodenum, pancreas, ceca, and posterior intestine. The two major pairs of arteries to the leg (*external iliac* and *sciatic*) branch from the aorta just posterior to the *renal arteries;* they are anterior to the *posterior mesenteric artery*.

In meat animals, the anterior vena cava (Figure 1-19) receives blood from the *internal* and *exterior jugular veins*, which drain the head, and from the *brachial vein*, which drains the forelimb. The posterior vena cava passes through the liver, where it receives blood from the liver in the *hepatic vein*. The *mesenteric veins* collect blood from the intestines. This blood contains newly digested nutrients. It is brought to the liver by the *hepatic portal vein* (Figure 1-10) so that it can be biochemically processed before being returned to the general circulation of the body. The remaining major inputs to the posterior vena cava are the *renal veins* (Figure 1-12), the *ovarian* or *spermatic veins* (depending on the sex of the animal), the *external iliac veins* from the hind limb, and the *internal iliac veins* from the penis or udder.

In poultry, the anterior vena cava receives symmetrical pairs of *jugular veins* (from the head), and *pectoral, brachial*, and *subclavian veins*. Birds have a well-developed *jugular anastomosis* which links the left and right jugular veins. This allows an efficient venous return if the veins on one side or the other are compressed by the position of the neck. Poultry also have a hepatic portal system, which allows blood-borne nutrients from the gut to be processed by the liver before entering the general circulation. The venous return of blood from the leg and pelvic regions travels in a complex pattern of veins, part of which is shown in Figure 1-15.

Cardiovascular Function

Most cardiac muscle cells are mononucleate. They are arranged in rows to form branching fibers, but individual cells are separated by *intercalated disks*. Cardiac muscles cells have a striated appearance, which is due to the precise alignment of sliding filaments in their contractile fibrils.

Cardiac muscle cells are continuously pumping out sodium ions through their membranes. This causes the inside of the cell to have an electrical charge of approximately −90 mV with respect to the outside of the cell. This is called a *resting potential.* Extrinsic factors such as electrical activity (ionic movements) in adjacent cells may decrease the resting potential toward zero. When it reaches a value of approximately −65 mV, the *threshold potential*, the decrease in electrical potential accelerates, and it shoots past the zero value so that for a brief instant (about one-tenth of a second) the membrane potential is positive. This sudden reversal of electrical charges is called an *action potential.* Action potentials are propagated into the interior of cardiac muscle cells by *transverse tubules*. In each cell, the transverse tubular system is an extensive series of fingerlike indentations of the surface membrane.

The arrival of an action potential in the interior of the cardiac muscle cell causes the release of calcium ions from the *sarcoplasmic reticulum*. The sarcoplasmic reticulum is a series of membrane-bounded vesicles in the interior of the cell. Unlike the transverse tubular system, the sarcoplasmic reticulum does not open to the surface of the muscle cell. Units of the sarcoplasmic reticulum surround the contractile fibrils in the interior of cardiac muscle cells. The sarcoplasmic reticulum sequesters and stores calcium ions, but it releases them again when prompted to do so by the transverse tubular system. Calcium ions activate the system of sliding protein filaments which is responsible for muscle contraction (Chapter 5).

The intrinsic rhythm of heart contraction originates from a group of cells at the *sinuatrial node*. The membranes of these cells behave as if they had a sodium ion leak. Thus at regular intervals their resting potentials drop to their threshold values, and they initiate action potentials. Action potentials then spread in a coordinated wave through right and left atria. The atria then contract and pump blood into the ventricles. However, the ventricles are also capable of filling themselves as they expand after pumping out their previous fill of blood. Under normal conditions, atrial contraction contributes to the overall cardiovascular efficiency, but its contribution may become vital when the heart is weakened by disease. In the medial wall of the heart, at the junction between the atria and ventricles, is a sensitive group of cells which form the *atrioventricular node*. This node is connected to a conduction system called the *bundle of His*, which runs down the medial wall separating left and right ventricles. The atrioventricular node is activated by contraction of the atrial cells, and the bundle of His conducts the action potential wave to the base of the ventricles. From this point, a wave of contraction spreads upward through the ventricles so that the blood that has just filled the ventricles is now pumped out of the heart. The intrinsic heart rate is determined by the rate at which the sinuatrial cells "leak" or depolarize, by the value of their threshold potentials, and by their resting potential. The flow of blood through the heart is directed by the heart valves (Figure 1-19). Mitral and tricuspid valves make a "lub" sound and the semilunar valves make a "dup" sound.

Coordinated electrical activity of cardiac muscle cells generates an electrical signal which can be picked up on the surface of the fore flank as an *electrocardiogram* (Figure 1-20). The P wave is due to atrial excitation, PQ is the delay as the action potential passes down the bundle of His, QRS is due to ventricular contraction or *systole*, and T is caused by repolarization of the ventricles. Activity of the heart is greatly influenced by its ionic environment. Isotonic sodium chloride plus calcium ions tend to stop the heart in systole (contracted), while isotonic sodium chloride plus potassium ions tend to stop the heart in *diastole* (relaxed).

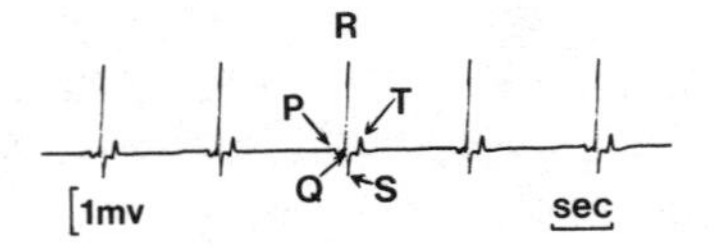

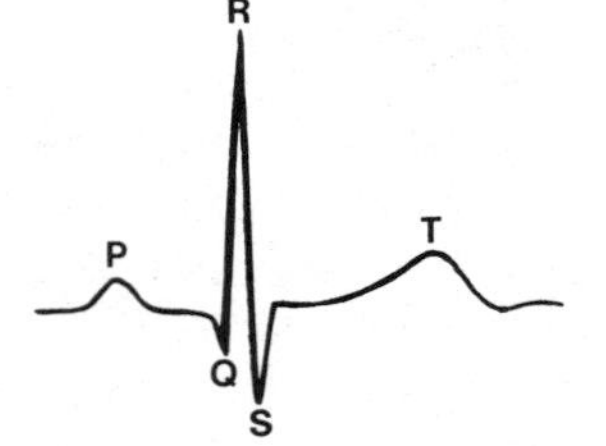

Figure 1-20 Components of the electrocardiogram in a pig after exsanguination (top) and in a diagrammatic form (bottom): (P) atrial contraction; (QRS) ventricular contraction; (T) ventricular repolarization.

The nervous system also has an effect on heart rate. The *thoracic nerve* of the sympathetic nervous system releases *catecholamines*, which increase the heart rate (*tachycardia*), while the *vagus* nerve of the parasympathetic nervous system releases *acetylcholine*, which slows the heart (*bradycardia*). The neural regulation of cardiac activity is a reflex response to inputs from blood pressure receptors or *baroreceptors*, and from *chemoreceptors*, which monitor the concentration of carbon dioxide in the blood. When the heart contracts, it works against the resistance to blood flow created by the peripheral blood vessels in the body tissues. Thus if the peripheral blood vessels decrease their diameter (*vasoconstriction*), the blood pressure tends to rise. Conversely, if peripheral blood vessels are dilated (*vasodilation*), the blood pressure tends to drop.

Exsanguination

The exsanguination or sticking of meat animals in an abattoir is usually performed by severing the carotid arteries and the jugular vein at the base of the neck. The sticking knife must be kept clean or bacteria might be introduced into the venous system and spread through the otherwise relatively sterile muscles of the carcass (Gill, 1979). Once exsanguination has started, the pulse and mean blood pressures rapidly decline because of the reduced stroke volume of the heart. Blood pressure changes are monitored physiologically by *baroreceptors* in the *carotid sinuses* (Booth et al., 1966). During exsanguination, respiratory movements of the thorax may be stimulated, and neurogenic and hormonal mechanisms attempt to restore the blood pressure by increasing the peripheral resistance by *vasoconstriction*. The heart keeps beating for some time after the major blood vessels are emptied, but rapidly stops if exposed and cooled (Thurston et al., 1978). Electrical stunning of pigs may terminate cardiac activity, so that, at the start of exsanguination, the blood escapes by gravity rather than being pumped out. In pigs, cardiac arrest does not affect the rate and extent of exsanguination (Warriss and Wotton, 1981). After exsanguination has started, the heart usually restarts and attempts to pump, until it runs out of energy (Swatland, 1982). Thus, in many cases, there is no reason why pigs cannot be killed by electrocution rather than merely being electrically stunned.

Blood loss as a percentage of body weight differs between species (Ostertag, 1907),

Cows:	4.2% to 5.7%
Calves:	4.4% to 6.7%
Sheep:	4.4% to 7.6%
Pigs:	1.5% to 5.8%

Blood content as a percentage of live weight may decrease in heavier animals since the growth of blood volume does not keep pace with the growth of live weight (Hansard et al., 1953). Approximately 60 per cent of blood is lost at sticking, 20 to 25 per cent remains in the viscera, while a maximum of 10 per cent may remain in carcass muscles (Warriss, 1977). Different stunning methods may modify the physiological conditions at the start of exsanguination (Leach and Warrington, 1976) and the neural responses to exsanguination (Kollai et al., 1973). Electrically stunned sheep lose more blood than those stunned with a captive bolt (Warriss and Leach, 1978), but they also have more blood splashes in their carcasses (Kirton et al., 1981).

Reduction of blood flow to the kidneys causes the release of a proteolytic enzyme, *renin*, which acts on a plasma protein to produce a polypeptide, *angiotensin I.* This polypeptide is enzymatically converted to *angiotensin II*, which then causes widespread vasoconstriction. Vasoconstriction is important because it decreases the retention of blood in meat (Warriss, 1978). Angiotensin II vasoconstriction is operative in both conscious and anesthetized animals (Miller et al., 1979). Catecholamines and ADH may also enhance vasoconstriction during exsanguination. Speed of exsanguination may modify the balance between neural and hormonal vasoconstrictive mechanisms, with hormonal vasoconstriction predominating in rapid exsanguination (Hall et al., 1976). However, asphyxia prior to exsanguination may result in vasoconstriction due to the activity of the sympathetic nervous system (Weissman et al., 1978).

It is traditionally maintained that poor bleeding leads to dark meat with poor keeping qualities due to microbial spoilage and rancidity. However, there is little scientific evidence in support of this view (Warriss, 1977), and it may be false, even in animals that retain massive amounts of blood in their carcasses (Roberts, 1980).

The factors that regulate the balance between extracellular and intracellular fluid compartments in meat are poorly understood. Fluid is delivered to living muscles by arteries, but it may return to the heart by either of two routes, in the venous system or in the lymphatic system. The route taken by intercellular fluid depends primarily on the extent to which fluid is taken up by capillaries and then passsed to the venous system. In living animals, the venous return is far greater than the lymphatic return. The lymphatic capillaries which drain skeletal muscles are mostly located in the connective tissue around bundles of muscle fibers (Korneliussen, 1975). The small amount of lymph that drains from muscles is increased after neural stimulation, and its LDH content (LDH is an enzyme from within the muscle fiber) increases dramatically following muscle damage (Bach and Lewis, 1973). In sheep, the flow of lymph from lymph nodes increases within 15 minutes of stress due to pain (Shannon et al., 1976). Hemorrhage may (Lundvall and Hillman, 1978) or may not (Johnson, 1972) cause absorption of intercellular fluid into the bloodstream, depending on the degree of vasoconstriction and consequent hydrostatic pressure in the vasculature.

Blood to the Brain

The effects of exsanguination on conscious sheep and cattle are reviewed by Baldwin (1971). Arterial blood to the brain is evenly distributed by a circular pattern of arteries called the *circle of Willis*. The circle of Willis receives blood from the *intracranial carotid rete* (a rete is a meshwork of blood vessels). In sheep, the *external carotid arteries* supply the intracranial carotid rete, via the *internal maxillary arteries* since the *internal carotid arteries* are absent in adults. However, blood can also reach the intracranial carotid rete from *vertebral arteries* via the *occipitovertebral anastomosis* (an anastomosis is a communicating link between two vessels). The situation in cattle is similar, but with an additional supply to the intracranial carotid rete from vertebral and occipital arteries. The extent to which intact vertebral arteries might prolong a supply of oxygenated blood to the brain once an animal's throat has been cut is difficult to assess. In sheep, consciousness may persist for 65 to 85 seconds (Newhook and Blackmore, 1982). In pigs, the delay between exsanguination and termination of EEG activity is approximately 20 seconds following proper stunning (Hoenderken, 1978). However, anoxia causes the dilation of cerebral blood vessels (Zeuthen et al., 1979) so that their storage capacity may be increased.

Utilization of Blood

The recovery of animal blood for utilization in food products for human consumption should be attempted. The main problems are to prevent the contamination of collected blood by bacteria from the skin, and to keep the blood of different animals separate until their carcasses have passed veterinary inspection for human consumption. Blood may be collected hygienically with a hollow knife. Coagulation of the blood can be prevented by the addition of *anticoagulants*, such as citric acid or sodium citrate. Alternatively, the *fibrin* which binds blood clots together can be removed by stirring with a paddle (Pals, 1970). When utilized for human food or pet food, blood contains easily assimilated iron. Blood proteins have a high nutritional value and a high water-binding capacity in processed products. The red blood cells burst if water is added to blood. If they are kept intact, red blood cells can be removed by centrifugation in order to prepare *plasma*. Plasma is a yellow liquid, rather like egg white, and it can be dried to a powder for use in human food. If blood is discharged into the abattoir effluent instead of being utilized, it increases the *biochemical oxygen demand* (BOD) of the effluent. *Chemical oxygen demand* is another index of the pollution load of the abattoir effluent; it can be measured in several hours rather than in the 5 days required for BOD determinations.

ENDOCRINE SYSTEM

Communication between cells and organs within the body is essential for the efficient control of body metabolism. There are probably many modes of cellular communication yet to be discovered: nerve impulses and hormones are the two most conspicuous and best known types of communication. Although nerve cells can communicate rapidly by the transmission of action potentials, they rely on chemical transmitters for the final step of the journey to their destination. Endocrine glands have made this last step their whole journey, and they release chemical transmitters or *hormones* directly into the bloodstream to act on cells at remote destinations. Unlike the *exocrine* glands, which release their secretions onto the skin or into the alimentary canal, the endocrine glands do not need a duct for the removal of their secretions. The differences between nerves and endocrine glands are not as well defined as elementary physiology textbooks might suggest. Examples can be found where neural communication is sometimes slow, diffuse, and imprecise. In a few cases, endocrine substances act rapidly, or are sharply localized within restricted circulatory pathways. During the course of evolution, endocrine glands may have become specialized for the production of substances that normal tissues produce in only small amounts.

The systems that regulate animal growth are not yet fully known. When reading Chapter 6, keep in mind the nervelike properties of some endocrine systems and the endocrinelike properties of some nerve cells. It might also be useful to note that the name "hormone" was adapted from the Greek *hormaein*, to excite. This makes an appropriate contrast with the term *chalone*, an internal secretion which depresses or inhibits activity, from the Greek *chalinos*, to curb (Henderson et al., 1966). Jenkin (1970) gives some wise advice concerning the definition of a hormone: "Accept the simpler statement that many members of the animal kingdom have been able in the course of evolution to turn to good physiological use a very wide range of chemical messengers, which cannot at all easily be contained within the limits of any one man made category." Abbreviations for the more readily identifiable hormones are given in Table 1-1.

TABLE 1-1. ABBREVIATIONS FOR HORMONES

ACTH	adrenocorticotropic hormone
ADH	antidiuretic hormone (vasopressin)
CRH	corticotropin releasing hormone
FSH	follicle-stimulating hormone
GnRH	gonadotropin releasing hormone
ICSH	interstitial cell-stimulating hormone
LH	luteinizing hormone
LTH	luteotropic hormone
MSH	melanocyte-stimulating hormone
PTH	parathyroid hormone
STH	somatotropic hormone
TRH	thyrotropin releasing hormone
TSH	thyroid-stimulating hormone

Pituitary Gland

The pituitary gland or *hypophysis* is a small round gland located ventrally to the brain. Embryologically, it is formed from the conjunction of an outgrowth from the floor of the brain (*neurohypophysis* or *posterior pituitary*) and a detached upgrowth from the

TABLE 1-2. PITUITARY (HYPOTHALAMO-HYPOPHYSIAL SYSTEM) COMPONENTS

Component	Product released	Result
Hypothalamus	CRH	Release of ACTH
	GnRH	Release of LTH, FSH, and LH
	TRH	Release of TSH
Posterior pituitary, or neurohypophysis	ADH	Water retention in kidney
	Oxytocin	Uterine contraction and milk release
Anterior pituitary, or adenohypophysis	STH	Stimulates growth
	LTH	Stimulates mammary glands
	ACTH	Stimulates adrenal cortex
	TSH	Activates thyroids and adipose tissue lipase
	FSH	Activates testes or prepares ovarian follicles
	LH	Completes spermatogenesis and stimulates androgen secretion or follicle growth, estrogen secretion, ovulation, formation of corpus luteum, and progesterone secretion
Pars intermedia	MSH	May stimulate pigment cells

roof of the mouth (*adenohypophysis* or *anterior pituitary*). A simplified list of the parts of the hypophysis, their products, and their functions is given in Table 1-2.

Pineal Gland

The pineal gland is a neurosecretory gland whose evolutionary origin can be traced back to the third eye found in the skull roof of certain fossil fishes. It is innervated by sympathetic nerves, and it is located deep in the brain, anterior to the cerebellum. It releases the hormone *melatonin*, which acts on the ovaries to inhibit the estrus cycle. Melatonin also has wider-ranging effects on other neuroendocrine control systems. Melatonin synthesis is inhibited by nerve impulses to the pineal gland; the frequency of impulses is inversely related to the amount of visible light reaching the retinas of the eyes. In poultry, it has been shown that the pineal gland is probably responsible for *circadian rhythms* (24-hour cycles) in physiological activity (Binkley et al., 1977).

Thyroid Gland

The thyroid gland is located around the trachea, near the larynx in mammals. Left and right thyroid glands are joined ventrally in pigs; in cattle and sheep the junction is restricted to a narrow connecting *isthmus*. In poultry, left and right thyroid glands are deep red in color instead of pale brown, and they are completely separated at the base of the neck. The thyroid glands receive an abundant supply of blood, from which they are able to capture iodine. Iodine is used for the synthesis of hormones which contain three or four iodine atoms, *triiodothyronine* and *thyroxine*, usually abbreviated T_3 and T_4, respectively. Thyroid hormones regulate oxidative metabolism and heat production in the body. Some cells in the thyroid also produce the hormone *calcitonin*.

Parathyroid Glands

In mammals, two pairs of very small parathyroid glands are located in or near the thyroid glands. Their position is somewhat variable and may be difficult to identify in the abattoir. In poultry, there is a small parathyroid gland at the posterior end of each thyroid. The *parathyroid hormone* produced by the parathyroid glands forms one circuit of a double feedback system which regulates calcium levels in blood and bone (Figure 2-30). The other circuit is mediated by the hormone calcitonin from the thyroid gland. Parathyroid hormone causes the mobilization of calcium from bone. Calcitonin causes the inhibition of calcium release from bone.

Thymus Gland

The thymus is a large gland, particularly in young animals, and it is located anteriorly to the heart and has lateral extensions into the neck. Thymus glands are sold for human consumption as *sweetbreads*. The thymus gland is composed of lymphoidal tissue and has a vital immunological function in young animals. It produces hormones, not yet completely characterized, which act on other cellular elements of the immune system.

Adrenal Glands

Left and right adrenal glands are located anteriorly to the kidneys; in cattle they are roughly triangular, in sheep and poultry they are oval, and in pigs they are elongated. Each adrenal gland is composed of two distinct endocrine glands. In mammals, the *adrenal cortex* is wrapped around the *adrenal medulla*, although the two glands are mingled in poultry. Table 1-3 gives a simplified list of their products and functions.

TABLE 1-3. ADRENAL COMPONENTS AND SOME OF THEIR FUNCTIONS

Component	Control	Product	Function
Cortex		Steroids	
multiformis	Na and K ions	The following mineralocorticoids: deoxycorticosterone and aldosterone	Homeostasis of extracellular electrolytes
fasciculata and reticularis	ACTH and CRH	The following glucocorticoids: cortisone, hydrocortisone, and corticosterone	To facilitate gluconeogenesis, proteolysis, and release of fatty acids from adipose
		Catecholamines	
Medulla	Neural	Epinephrine and norepinephrine	Rapid response to stress

Pancreas

The *islets of Langerhans* are microscopic areas of the pancreas which have an important endocrine function. The islets contain *alpha cells*, which produce *glucagon*, and *beta cells*, which produce *insulin*. Insulin facilitates the uptake and utilization of blood glucose by body cells. Thus insulin deficiency causes the elevated blood sugar levels that occur in diabetes. Pancreas glands are collected in abattoirs for the commercial isolation of insulin. Insulin concentration is highest in the tail end of the pancreas, and the tissue must be kept dry before being frozen since insulin is water soluble. The action of glucagon is the opposite to that of insulin.

There are also several other organs of the body which produce hormones in addition to their other activities. The testes produce *testosterone*. The ovaries produce *estrogen, progesterone*, and *relaxin*. During gestation, the uterus and placenta secrete *chorionic gonadotropin*. The stomach wall secretes *gastrin*. The kidney produces the hormone *renin*. The liver produces *somatomedins* (Chapter 8).

INTEGUMENT

Animal skin is composed of three basic layers. From outside to inside these layers may be called the (1) *epidermis;* (2) the *dermis;* and (3) the *hypodermis*. The epidermis is formed by layers of flat cells which form a *stratified squamous epithelium*. New cells originate in the lowest layer and become keratinized as they are pushed to the surface. *Keratin* is a fibrous protein that also forms the substance of hair, horns, and hoofs. At the ultrastructural level it is deposited in a fibrillar form which may then be incorporated into a granular form.

Each *hair follicle* develops from an inpushing of the epidermis down into the dermis (Figure 1-21). The formation of hair results from the activity of epithelial cells, which form a papilla at the base of the follicle. There is considerable variation in the rate of hair growth in meat animals. For example, the average length of bovine hair may reach a maximum between 6 and 24 months, and may then decrease (Camek, 1920). The underlying sequence of events in hair growth is due to the periodic shedding of hairs from their follicles. The bulb at the base of the hair eventually becomes hard and clublike. This holds the hair in its follicle for some time, but no further growth is possible. The hair is eventually released when a new hair starts to form in the base of the follicle. This cycle, which determines the average external hair length, is influenced by factors such as climate, age, nutrition, and breed.

Most mammalian hairs and bristles have three layers which appear as concentric rings in a cross section through the hair shaft (Figure 1-21B). From outside to inside these are: (1) a thin *cuticle;* (2) the *cortex;* and (3) the large cells of the *medulla*. Many of the wavy wool fibers of a sheep's fleece lack a medulla but, like strong straight pig bristles, they are still composed of keratin. The high tensile strength and low solubility of keratin in hair and wool fibers is due to the cross-linking of protein chains by disulfide bonds; hence dietary sulfur is important for wool production in sheep (Fraser, 1969). In sheep, the *sebaceous glands* which open into the wool follicles produce an oily secretion called *lanolin*.

In meat animals, most of the *sweat glands* open near the entrance of hair follicles.

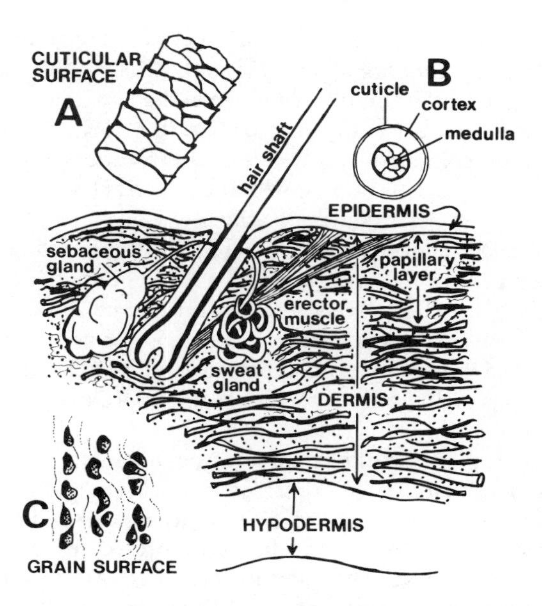

Figure 1-21 Microstructure of the hide, with inset diagrams of (A) the cuticular surface of a hair; (B) a cross section through a hair; and (C) the grain surface of the papillary layer of tanned leather.

Although less conspicuous than the sweat glands of human skin, they still make an important contribution to thermoregulation in meat animals (McDowell et al., 1961).

Feathers are also formed in follicles. The follicles are grouped in *feather tracts* which are easily visible on the eviscerated carcass. In the spaces between the tracts, the follicles produce only *filoplumes*, with a rudimentary feather vane at the end of a hairlike shaft. The large feathers of the wings are called *remiges* while those of the tail are called *retrices*. The *contour feathers* provide the main covering of the body and are interspersed with filoplumes. Young birds have large numbers of *down feathers*. The structure of the *vane* of a typical feather resembles a hollow quill that has been obliquely sliced and unrolled. Thus when it is formed within the follicle it is like a hollow cylinder. The lateral branches or *barbs* of the vane are held together by hooked anterior *barbules* which catch on the sawlike edges of adjacent posterior barbules. The skin of poultry is dry and does not produce its own oil. In poultry, there is an oil gland located dorsally to the stumpy tail of the bird. The oil is distributed when the feathers are kept in order as the bird preens itself.

Pigment cells or *melanocytes* are located in the deepest layers of the epidermis or in the underlying dermis. *Melanin* is a pigment which is formed in organelles called *melanosomes*. Melanin can be passed from melanocytes to skin cells by *cytocrine secretion*. The enzyme *tyrosine hydroxylase* is involved in the formation of melanin. Absence of this enzyme results in an *albino* animal. Variation in the color of farm animals is caused by variations in the amount and distribution of melanin. The distribution of melanin over animals' skin is determined prenatally by an interaction between the migration patterns of *melanocytes* and the diffusion patterns of the messenger substances that either activate or suppress the synthesis of melanin (Bard, 1981). A single dominant gene determines the belt pattern marking that runs over the shoulders and forelimbs of some breeds of pigs (Donald, 1951).

The epidermis is supported on the ridged surface of the underlying dermis. The upper region of the dermis, often called the *papillary layer* of the dermis, is a tightly woven network of collagen fibers with some elastin fibers (collagen and elastin are described in Chapter 2). After the *tanning* of a hide to make leather, the papillary layer becomes the top surface of the leather. With a hand lens, the openings where the hair follicles once penetrated the dermis are easily visible (Figure 1-21C). When the leather is turned over, the much looser coarse fibrous weave of the lower dermis is evident. In pigskin, the follicles of the strong bristles are rooted at the lowest level of the dermis so that many of the follicles almost perforate the leather.

When the hide is removed from the carcass, the separation is made through the deepest layer of the integument, the *hypodermis*. Fat is often deposited in the hypodermis and, particularly in sheep, may even infiltrate the dermis. Numerous blood vessels run through the hypodermis to reach the extensive vascular bed (for heat dissipation) in the dermis. The hide weight of a typical lean steer is about 7 per cent of the live weight, but there is considerable seasonal variation (Dowling, 1964), with colder climates inducing heavier hides.

Beef hides are graded on their cleanliness and degree of damage due to branding or warble fly larvae. If beef hides have been processed with a high standard of hygiene, the collagen of the inner layer of the hide can be used for processed food products such as sausage casings. Green hides (those from recently slaughtered animals) are treated with sodium chloride prior to tanning. The hides are trimmed, are split into left and right sides, and are soaked for several days in water. The hides are then dehaired in a calcium hydroxide (lime) solution which contains sodium or calcium hydrosulfide. The conversion of a hide to leather occurs when it is *tanned*, either with a tree bark extract or with sodium dichromate. Hair remnants are physically forced from the hair follicles (scudding) prior to deliming in sulfuric acid. Elastin fibers are removed enzymatically before the hides are pickled in sodium chloride acidified with sulfuric acid.

RENDERING AND WASTE DISPOSAL

The overall objective of rendering is to produce clarified homogeneous substances such as lard and tallow from a heterogeneous mixture of animal trimmings and scraps. Because of the vast volume of such items produced by a large abattoir, rendering and waste product utilization are major factors in the economics of meat processing. Visceral adipose tissue which can be maintained in a wholesome condition is rendered to produce lard and tallow for cooking or for further processing in the food industry. Intrinsically dirty parts of the carcass such as the head and feet are subjected to inedible rendering as the first step in the manufacture of soap and grease.

In the now-outdated method of *wet rendering*, steam was injected into a pressurized tank full of trimmings with a high fat content. Eventually, the molten fat floated freely on top of an aqueous solution with a high protein content. At the bottom was a slurry of solids. The partial recovery of proteins from the aqueous layer was achieved by a subsequent evaporation process. In *dry rendering*, the steam is confined to a jacket around the tank, and the contents of the tank are held at a negative pressure. This enables a far greater recovery of proteins, which would otherwise greatly elevate the BOD of the abattoir effluent. Thus a general principle of modern abattoir operations is to minimize the amount of water that is added to animal wastes as they are cleaned from the premises. Not only is water expensive, but much of it has to be removed later by evaporation, and this uses a considerable amount of energy. Many inedible waste products, such as clotted blood, bone dust, and manure from the rumen and from stockpens, are best maintained for recovery operations in as dry a condition as is possible. The engineering techniques that may be used to achieve this goal are described by Jones (1974).

REFERENCES

ANDREWS, A. H. 1981. *J. Agric. Sci., Camb.* 97:97.

ANON. 1981a. Salmonellosis in animals and man. *Vet. Rec.* 109:438.

ANON. 1981b. Instant hog chiller. *Meat Ind.* 27(1):52.

ANON. 1982. Litvak tests new liquid shroud that reduces shrink when applied in two edible layers. *Meat Ind.* 28(3):40.

BACH, C., and LEWIS, G. P. 1973. *J. Physiol.* 235:477.

BALDWIN, B. A. 1971. Anatomical and physiological factors involved in slaughter by section of the carotid arteries. In *Human Killing and Slaughterhouse Techniques*, pp. 34–42. Universities Federation for Animal Welfare, Potters Bar, Hertfordshire, England.

BARD, J. B. L. 1981. *J. Theor. Biol.* 93:363.

BARNES, E. M., and IMPEY, C. S. 1975. *Br. Poult. Sci.* 16:319.

BINKLEY, S., RIEBMAN, J. B., and REILLY, K. B. 1977. *Science* 197:1181.

BOOTH, N. H., BREDECK, H. E., and HERIN, R. A. 1966. Baroreceptors and chemoreceptor reflex mechanisms in swine. In L. K. Bustad and R. O. McClellan (eds.), *Swine in Biomedical Research*, pp. 331–346. Batelle Memorial Institute, Pacific Northwest Laboratory, Wash.

BORCHERT, L. 1972. The control of pork quality by rapid chilling pre-rigor. In R. Cassens, F. Giesler, and Q. Kolb (eds.), *The Proceedings of the Pork Quality Symposium*, pp. 169–170. University of Wisconsin Press, Madison.

CAGLE, E. D., and HENRICKSON, R. L. 1970. *J. Food Sci.* 35:270.

CAMEK, J. 1920. *J. Agric. Sci., Camb.* 10:12.

CUTHBERTSON, A. 1980. Hot processing of meat: a review of the rationale and economic implications. In R. A. Lawrie (ed.), *Developments in Meat Science 1*, pp. 61–88. Applied Science Publishers, London.

CUTTING, C. L. 1974. *Inst. Meat Bull., Lond.* 84:8.

DONALD, H. P. 1951. *J. Agric. Sci., Camb.* 41:214.

DOWLING, D. F. 1964. *J. Agric. Sci., Camb.* 62:307.

FORLAND, D. M., LUNDSTROM, K., and ANDRESEN, O. 1980. *Nord. Veterinaer. med.* 32:201.

FRASER, R. D. B. 1969. *Sci. Am.* 221:86.

GILL, C. O. 1979. *J. Appl. Bacteriol.* 47:367.

GODFREY, N. W. 1961. *J. Agric. Sci., Camb.* 57:173.

GOODCHILD, W. M. 1970. *Br. Poult. Sci.* 11:209.

HALL, J. E., SCHWINGHAMER, J. M., and LALONE, B. 1976. *Am. J. Physiol.* 230:569.

HANSARD, S. L., BUTLER, W. O., COMAR, C. L., and HOBBS, C. S. 1953. *J. Anim. Sci.* 12:402.

HEAGERTY, J. J 1928. *Four Centuries of Medical History in Canada*, Vol. 2, p. 50. Macmillan of Canada, Toronto.

HENDERSON, I. F., HENDERSON, W. D., and KENNETH, J. H. 1966. *A Dictionary of Biological Terms*. Oliver & Boyd, Edinburgh.

HOENDERKEN, R. 1978. Electrical Stunning of Pigs for Slaughter. Doctoral thesis, Utrecht.

JENKIN, P. M. 1970. *Control of Growth and Metamorphosis*, Part II of *Animal Hormones*. Pergamon Press, Oxford.

JOHNSON, G. 1972. *J. Surg. Res.* 13:7.

JONES, H. R. 1974. *Pollution Control in Meat, Poultry and Seafood Processing*. Noyes Data Corporation, Park Ridge, N.J.

JUDGE, M. D., SALM, C. P., and OKOS, M. R. 1978. *Proc. Meat Ind. Res. Conf.*, pp. 155–164. American Meat Industry Foundation, Arlington, Va.

JURANEK, D. D., FORBES, L. S., and KELLER, U. 1976. *Am. J. Vet. Res.* 37:785.

KIRTON, A. H., FRAZERHURST, L. F., BISHOP, W. H., and WINN, G. W. 1981. *Meat Sci.* 5:407.

KOLLAI, M., FEDINA, L., and KOVACH, A. G. B. 1973. *Acta Physiol. Acad. Sci. Hung.* 44:145.

KOLLAR, E. J., and FISHER, C. 1980. *Science* 207:993.

KORNELIUSSEN, H. 1975. *Cell Tissue Res.* 163:169.

LAMBOOY, E., and SPANJAARD, W. 1981. *Vet. Rec.* 109:359.

LEACH, T. M., and WARRINGTON, R. 1976. *Med. Biol. Eng.* 14:79.

LEVIN, E. 1970. *Proc. Meat Ind. Res. Conf.*, Chicago, pp. 29–38.

LILLIARD, H. S. 1982. *Food Technol.* 36:58.

LUNDVALL, J., and HILLMAN, J. 1978. *Acta Physiol. Scand.* 102:450.

MAYERSON, H. S. 1963. *Sci. Am.* 208:80.

MCDOWELL, R. E. 1977. *Ruminant Products: More than Meat and Milk*. Winrock International Livestock Research and Training Center, Ark.

MCDOWELL, R. E., MCDANIEL, B. T., BARRADA, M. S., and LEE, D. H. K. 1961. *J. Anim. Sci.* 20:380.

MILLER, E. D., LONGNECKER, D. E., and PEACH, M. J. 1979. *Circ. Shock* 6:271.

MOWAFY, M., and CASSENS, R. G. 1975. *J. Anim. Sci.* 41:1291.

NEWHOOK, J. C., and BLACKMORE, D. K. 1982. *Meat Sci.* 6:295.

OSBORN, J. W. 1978. Morphogenetic gradients: fields versus clones. In P. M. Butler and K. A. Joysey (eds.), *Development, Function and Evolution of Teeth*, pp. 171–199. Academic Press, New York.

OSTERTAG, R. 1907. *Handbook of Meat Inspection*. William R. Jenkins Co., New York.

PALS, C. H. 1970. *Proc. Meat Ind. Res. Conf.*, Chicago, pp. 17–22.

PEEL, B., and SIMMONS, G. C. 1978. *Aust. Vet. J.* 54:106.

ROBERTS, T. A. 1980. *R. Soc. Health J.* 100:3.

RUST, R. E. 1974. *Meat Process.* 13(12):34.

SHANNON, A. D., QUIN, J. W., and JONES, M. A. S. 1976. *Q. J. Exp. Physiol.* 61:169.

SMELTZER, T. I., PEEL, B., and COLLINS, G. 1979. *Aust. Vet. J.* 55:275.

SWATLAND, H. J. 1982. *Can. Inst. Food Sci. Technol. J.* 15:161.

SWINGLER, G. R., and LAWRIE, R. A. 1978. *Meat Sci.* 2:105.

SWINGLER, G. R., and LAWRIE, R. A. 1979. *Meat Sci.* 3:63.

TEN HORN, L. J. 1973. *R. Soc. Health J.* 93:241.

THOMPSON, R. C. A. 1979. *Aust. Vet. J.* 55:93.

THURSTON, J. T., BURLINGTON, R. F., and MEININGER, G. A. 1978. *Cryobiology* 15:312.

WARRISS, P. D. 1977. *J. Sci. Food Agric.* 28:457.

WARRISS, P. D. 1978. *Meat Sci.* 2:155.

WARRISS, P. D., and LEACH, T. M. 1978. *J. Sci. Food Agric.* 29:608.

WARRISS, P. D., and WOTTON, S. B. 1981. *Res. Vet. Sci.* 31:82.

WEISSMAN, M. L., SONNENSCHEIN, R. R., and RUBINSTEIN, E. H. 1978. *Am. J. Physiol.* 235: H72.

ZEUTHEN, T., DORA, E., SILVER, I. A., CHANCE, B., and KOVACH, A. G. B. 1979. *Acta Physiol. Acad. Sci. Hung.* 54:305.

2

The Connective Tissues of the Carcass

INTRODUCTION

The animal body is supported by bone, is held together by fibrous connective tissue, and is protected against starvation and cold by adipose tissue. These three types of tissue, although they differ radically in their appearance and properties, are all classified as types of *connective tissue*. All three types are composed of cells. The cells are located within a *matrix* which contains *fibers*.

In bones, both the matrix and the fibers make an important contribution to mechanical strength. The hardness of a bone is due to its calcified matrix, but its strength is due to embedded collagen fibers. The cells of bone, *osteocytes*, are trapped in small caves called *lacunae*. The gristle of the carcass is formed from *tendons* (by which muscles pull on bones), from *ligaments* (which hold bones together at the joints of the skeleton), from *aponeuroses* (which cover some muscles) and from *fasciae* (which form strong sheets between muscles). The dominant protein in gristle is *collagen*. Since connective tissues permeate nearly all parts of the body at the microscopic level, collagen is the most abundant protein in the animal body. The collagen fibers in meat are converted from strong fibers to jelly (*gelatin*) by the action of moist heat during cooking. The collagen present in bones may be removed by mild hydrolysis to produce gelatin, which can then be used in other food products or in photographic emulsions. In fat or adipose tissue there is little trace of the matrix and fibers, but the cells are bloated with stored *triglyceride*, the chemical component of lard.

This chapter begins with a general description of the bones that occur in the carcasses of meat animals and poultry. Later in the book, this information will be used to describe the anatomical locations of muscles and cuts of meat. After surveying the macroscopic structure of the skeleton, the level of magnification is increased so that we can consider the microstructure and development of the dominant types of connective tissue in the carcass: gristle, bone, and fat.

STRUCTURE OF THE SKELETON

Many parts of the skeleton can be seen or handled through the skin of the live animal (Figures 2-1, 2-2, and 2-3). These points of conformation are important in showing and judging animals. When comparing the conformation of a live animal with the structure

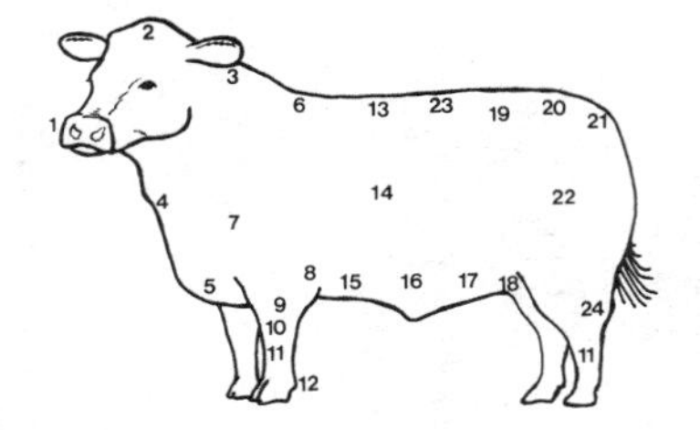

Figure 2-1 Some points of a beef steer: (1) muzzle; (2) poll; (3) crest; (4) dewlap; (5) brisket; (6) top of shoulder; (7) point of shoulder; (8) elbow; (9) arm; (10) knee; (11) shank; (12) dewclaw; (13) back; (14) forerib; (15) foreflank; (16) belly or paunch; (17) hind flank; (18) cod; (19) hip or hook; (20) rump; (21) tail-head between the pins of the pelvis; (22) thigh; (23) loin; (24) hock.

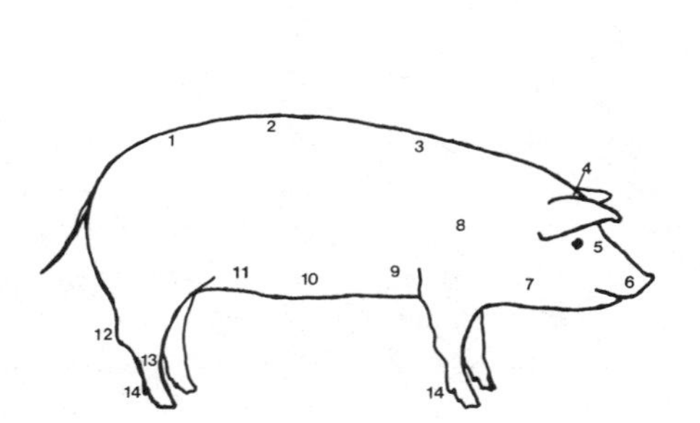

Figure 2-2 Some points of a pig: (1) rump; (2) loin; (3) back; (4) poll; (5) face; (6) snout; (7) jowl; (8) shoulder; (9) foreflank; (10) belly; (11) rear flank; (12) hock; (13) shank; (14) dewclaw.

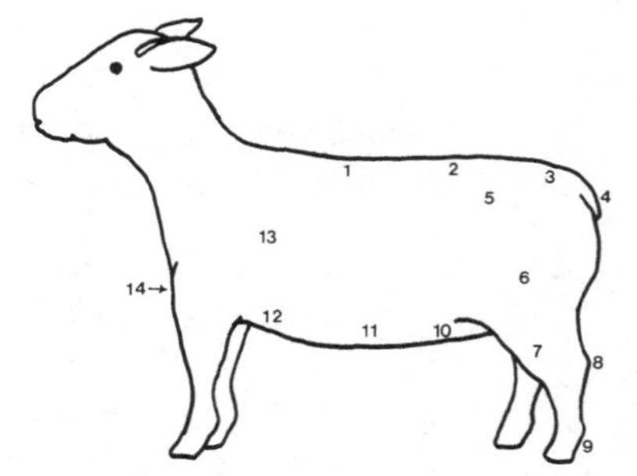

Figure 2-3 Some points of a lamb: (1) back; (2) loin; (3) rump; (4) dock; (5) hip; (6) thigh; (7) leg; (8) hock; (9) dewclaw; (10) hind flank; (11) belly; (12) foreflank; (13) shoulder; (14) breast.

of its carcass, it is important to remember that the hind limb is normally rotated through an angle of nearly 90 degrees when the carcass is suspended in the abattoir. Thus the muscles on the posterior face of the hind limb bulge outward, while the muscles on the anterior face of the hind limb and in the belly are stretched.

Skull

The brain is situated within a boxlike posterior part of the skull called the *cranium*. The brain is connected to the spinal cord through a large hole called the *foramen magnum*. The foramen magnum is flanked by two large knobs or *occipital condyles* which form a joint with the first *cervical* vertebra of the neck. The skulls of meat animals are usually damaged in the region of the *frontal bones* if the animals have been stunned by concussion in an abattoir. Sinuses or spaces are present between the inner and outer cranial walls.

The major muscles involved in chewing are attached to the *coronoid process*, which is a large expansion of the lower jaw or *mandible*. The coronoid process is located medially to the *zygomatic* arch, between the eye and the ear (Figure 2-4). The coronoid process allows muscle leverage to be exerted onto the mandible. The joint between the skull and the lower jaw is formed by a *mandibular condyle*. In cattle and sheep, the mandibular condyle is relatively flat and it allows considerable movement in a horizontal plane. Lateral movement is important in animals whose teeth work with a grinding action.

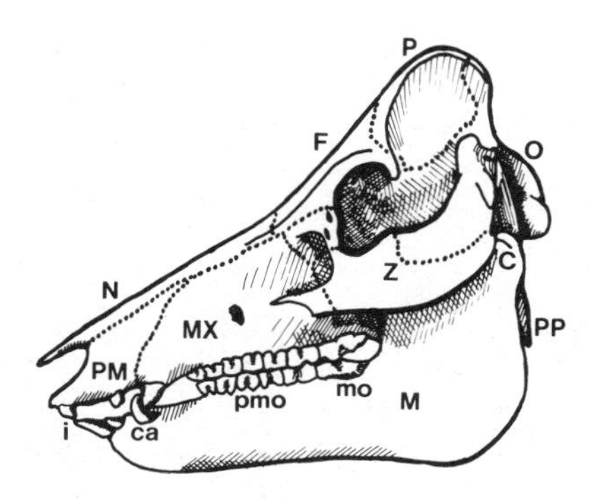

Figure 2-4 Skull of a pig: (C) condyle of mandible; (ca) canine teeth; (F) frontal bone; (i) incisor teeth; (M) mandible; (mo) molar teeth; (MX) maxilla; (N) nasal bone; (O) occipital bone; (P) parietal bone; (PM) premaxilla; (pmo) premolar teeth; (PP) paramastoid process; (Z) zygomatic arch.

The jigsaw pattern of suture joints on the skull surface indicates that the whole skull is formed by the fusion of a number of individual bones. A saw cut made transversely through the face region of the skull reveals the delicate rolls of the *turbinate* bones in the nasal cavity. The turbinate bones support a large area of nasal epithelium which warms and moistens air traveling to the lungs, and which provides a large surface for the sense of smell.

Skeleton of the Neck

The vertebral column or backbone is the main axis of the skeleton and it protects the spinal cord. The spinal cord is located in the *neural canal*, which is formed by a long series of *neural arches*, each contributed by a different vertebra (Figure 2-5). The neural arch of each vertebra is supported on the body or *centrum* of the vertebra. In

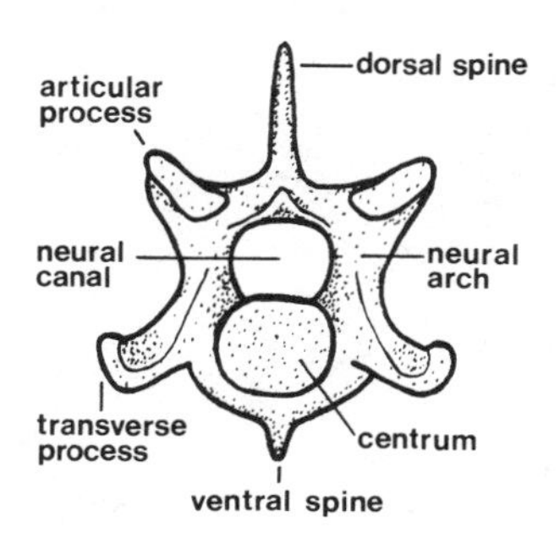

Figure 2-5 Anterior view of a typical cervical vertebra.

some types of vertebrae, the neural arch extends dorsally as a prominent spine which may be called a *dorsal spine*, a *neural spine*, or a *spinous process*. Where movement between vertebrae is possible, the centra are separated by cartilagenous *intervertebral* disks. In mammals, the anterior and posterior faces of the centra have a basically flat shape. The names and numbers of the different types of vertebrae in meat animals are shown in Table 2-1.

TABLE 2-1. NUMBERS OF VERTEBRAE IN MEAT ANIMALS

Name	Region	Beef	Pork	Lamb
Cervical	Neck	7	7	7
Thoracic	Rib cage	13	13–17	13 or 14
Lumbar	Loin	6	5–7	6 or 7
Sacral	Sirloin	5	4	4
Caudal	Tail	18–20	20–23	16–18

Source: Sisson and Grossman (1953), Shaw (1929), and Palsson (1940).

Meat animals, like almost all other living mammals, usually have seven cervical vertebrae in the neck region. However, sheep sometimes have only six cervical vertebrae (Palsson, 1940), and as few as five cervical vertebrae have been reported in pigs (Berge, 1948). In cattle and sheep, but to a lesser extent in pigs, the neck is very mobile, and the cervical vertebrae have a series of interlocking *articular* and *transverse processes* which limit excessive bending of the neck so as to protect the spinal cord (Figure 2-6). The first cervical vertebra is called the *atlas* and it articulates with the skull. The atlas is greatly modified in shape to form a joint which enables the animal to nod its head up and down. Rotation or twisting of the head occurs from the joint between the atlas and the next cervical vertebra, the *axis*. The *ligamentum nuchae* is a very strong elastic ligament in the dorsal midline of the neck, and it relieves the animal of the weight of its head. Were it not for the ligamentum nuchae, the head of the standing animal would droop between its forelimbs.

The ligamentum nuchae is pale yellow in color with a thick cordlike or *funicular* part and a flat sheetlike or *lamellar* part (Figure 2-6). Once the head is removed at slaughter, the elasticity of the ligamentum nuchae causes the neck of the carcass to curve dorsally. Beef and pork carcasses are usually split into right and left halves soon after slaughter. In the meat trade, the series of vertebral centra that is exposed on a split carcass is called the *chine bone*.

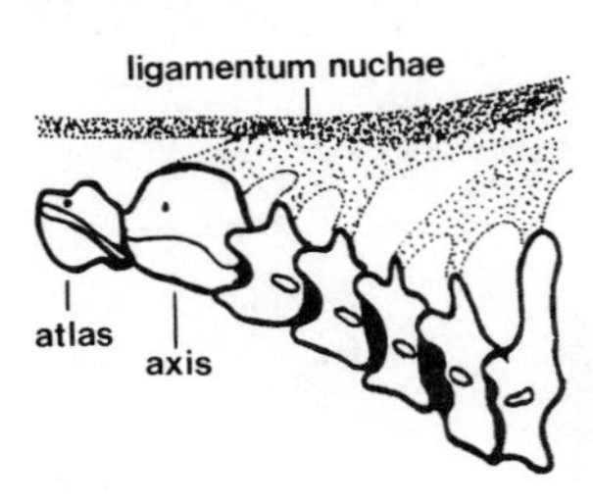

Figure 2-6 Lateral view of cervical vertebrae and the ligamentum nuchae in the neck of a beef carcass.

Rib Cage

The number of vertebrae in pork carcasses is rather variable, particularly in certain breeds (Shaw, 1929, 1930). The heritability of the number of vertebrae is about 0.74 (Berge, 1948). Each extra vertebra adds about 15 mm to the length of the carcasses at slaughter weight (King and Roberts, 1960). The number of thoracic vertebrae, each bearing left and right ribs, ranges from 13 to 17. Breeds with a large size when mature and with heavy bone development tend to have more thoracic vertebrae than do lighter breeds. Sometimes the ribs on extra thoracic vertebrae are only partially formed, but usually they are complete. The minimum number of lumbar vertebrae is generally found in carcasses with the maximum number of thoracic vertebrae. However, the variability of numbers of vertebrae frequently leads to an increase in the total number of vertebrae, so that the phenomenon is not simply due to the substitution of one type of vertebra for another. Lamb carcasses usually have either 13 or 14 thoracic vertebrae and a corresponding number of pairs of ribs (Palsson, 1940).

Experimental studies on embryonic amphibians (Detwiler, 1934) suggest that the number of vertebrae in an animal is determined by the number of *somites* (see Chapter 6) that develop along the length of the spinal cord. By definition, in mammals, the vertebrae that bear ribs are identified as thoracic vertebrae. In embryonic vertebrae, there are ossification centers on each side of the developing vertebrae. In vertebrae that do not normally develop ribs, these lateral ossification centers contribute their bone tissue to the centra of adjacent vertebrae. In the thoracic vertebrae, however, this laterally derived bone tissue remains separate from the centra, and it forms the ribs. Thus the numbers of pairs of ribs and the numbers of thoracic vertebrae are determined by the developmental mechanism that controls the fate of the tissue which is derived from the lateral ossification centers.

The cage which is formed by thoracic vertebrae, the ribs, and the sternum is an essential component of the respiratory system. The thoracic vertebrae in the rib region are distinguished by their tall dorsal spines, many of which point toward the hindquarter. In the meat trade, these spines are known as the *feather bones*. Dorsally, the ribs are joined to the vertebral column so that the head of each rib articulates with the bodies of two adjacent vertebrae. Each rib has a *tubercle* which articulates with the transverse process of the more posterior of its two vertebrae (Figure 2-7). Ventrally, the anterior ribs articulate with the *sternum* and are termed *sternal ribs* (Table 2-2). The more posterior ribs are called *asternal ribs* and they only connect to the sternum indirectly via *costal cartilages*. The most posterior ribs have only small costal cartilages which do not reach all the way to the sternum. Some of the costal cartilages are

TABLE 2-2. STRUCTURE OF THE RIB CAGE IN MEAT ANIMALS

	Beef	Pork	Lamb
Total pairs of ribs	13	13–17	13 or 14
Pairs of sternal ribs	8	7	8
Pairs of asternal ribs	5	6 or 8	5 or 6
Number of sternebrae	7	6	6 or 7

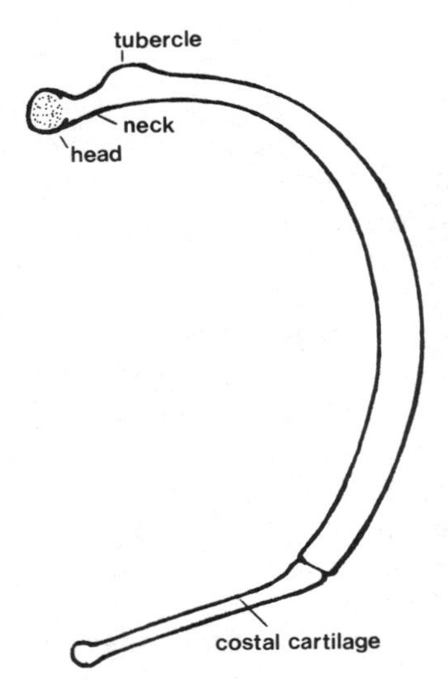

Figure 2-7 Features of a typical rib.

very hard and they may appear to be more like bones than like typical cartilage. The sternum is formed by a number of closely joined bones, the *sternebrae*. When split through the midline, the interior structure of the sternebrae resembles that found in the centra of the vertebrae. In an isolated cut of meat, the distinction between sternebrae and vertebral centra can usually be made by the presence or absence of a neural canal. A neural canal is usually evident in the vertebrae of carcasses that have been symmetrically separated into right and left sides.

The structure of the rib cage is rather variable in lamb carcasses. Carcasses have been found with as few as 12 ribs on one side, and left and right sides of the rib cage may differ in their number of ribs (Palsson, 1940). In lambs, rib length is determined mainly by age, while the plane of nutrition determines rib thickness (Palsson and Verges, 1965).

Skeleton of the Loin, Sirloin, and Rump

In a live animal, the lumbar vertebrae act like a suspension bridge to support the weight of the abdomen. The lumbar vertebrae have flat, winglike transverse processes which broaden the abdominal cavity dorsally (Figure 2-8) so as to provide a strong attachment for the muscles of the abdominal wall, which carry the weight of the viscera. The propulsive thrust which is generated by the hind limb during locomotion is transmitted to the *sacral vertebrae* by the pelvis. To strengthen the sacral vertebrae, they are fused together to form the *sacrum* (Figure 2-9). Fusion is incomplete in young animals, and this provides an important clue to animal age in the dressed carcass.

The pelvis is formed by three bones on each side (Figure 2-10). The most anterior bone on each side is the *ilium*. The shaft of the ilium expands anteriorly to form a flat wing which is attached to the sacrum. In the meat trade, this joint is called the *slip joint*. When seen in a sirloin steak, the ilium may either appear as a small round bone or as a large flat bone. The anterior edges of the ilia form the hooks of the live animal (Figure

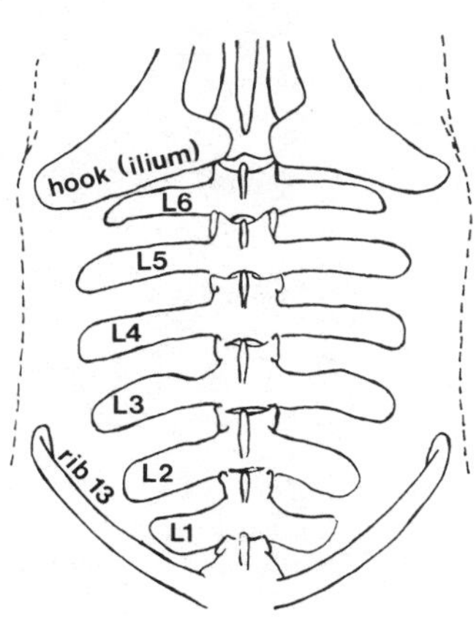

Figure 2-8 X-ray diagram of the bones in the loin of a hanging beef carcass before it is split into left and right sides. The winglike transverse processes of the lumbar vertebrae (L1 to L6) provide a broad support for the abdominal muscles which support the viscera.

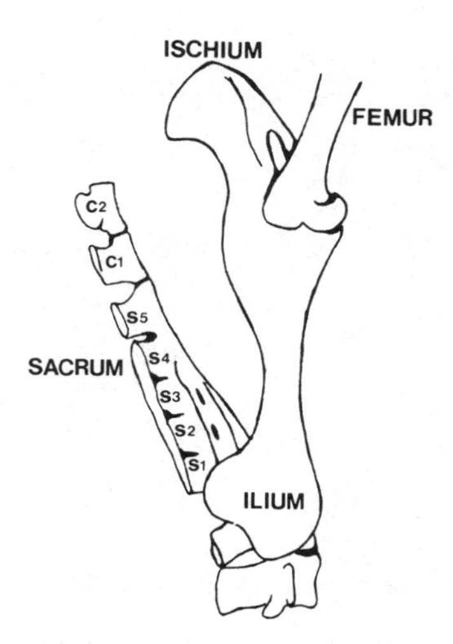

Figure 2-9 Fused sacral vertebrae (S1 to S5) forming the sacrum in a hanging side of beef. Note the extreme angle of the femur to the pelvis and the two caudal vertebrae (C1 and C2) which remain after the tail is removed.

2-1). The most posterior bone of the pelvis on each side is the *ischium*. The pelvis and the sacrum form a ring of bone which is completed ventrally by the *pubes* (Figure 2-11). The *left pubis* is separated from the *right pubis* by fibrocartilage. At parturition this may become soft so as to allow some movement between the bones of the pelvis. The pubes are separated when carcasses are split into left and right sides in the abattoir. The pubic bone which is exposed on a carcass is often called the *aitch bone*. The aitch bone is curved in steer and bull carcasses. In heifers it is curved, but it is straight in cow

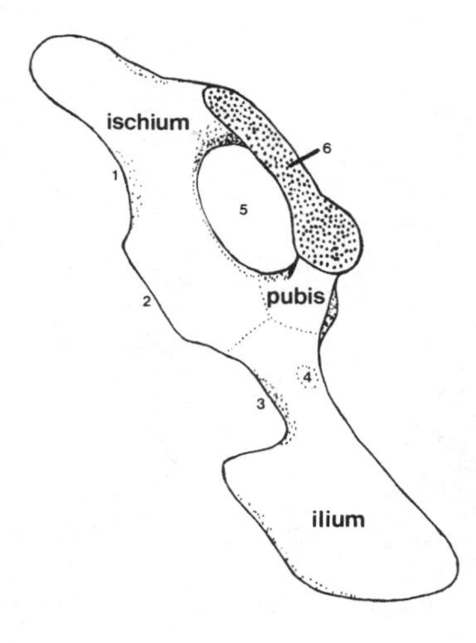

Figure 2-10 Pelvis, seen in its orientation in a hanging side of beef (right side): (1) lesser sciatic notch; (2) ischiatic spine; (3) greater sciatic notch; (4) psoas tubercle; (5) obturator foramen; (6) symphysis pubis.

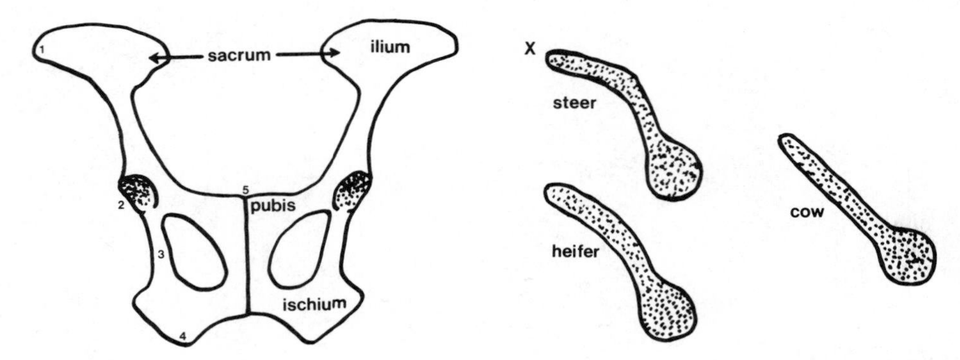

Figure 2-11 Ventral view of the pelvis to show how the sacrum completes a ring of bone in the intact carcass: (1) tuber coxae; (2) acetabulum; (3) acetabular ramus of ischium; (4) tuber ischii; (5) symphysis pubis.

Figure 2-12 Shape of the symphysis pubis in hanging beef carcasses (right sides). The position of the pizzle is marked by an X.

carcasses (Figure 2-12). Only two *caudal* or tail vertebrae are left on a commercial beef carcass. Caudal vertebrae may also be called *coccygeal* vertebrae.

Forelimb Skeleton

The most proximal bone of the forelimb is the blade bone or *scapula*. It is not fused to the vertebral column (like the pelvis in the hind limb), and this allows the muscles that hold the scapula to the rib cage to function as shock absorbers during locomotion. The scapula has a distal socket joint for the next bone in the forelimb, the *humerus*. This socket joint is called the *glenoid cavity* (Figure 2-13). The glenoid cavity is wide and

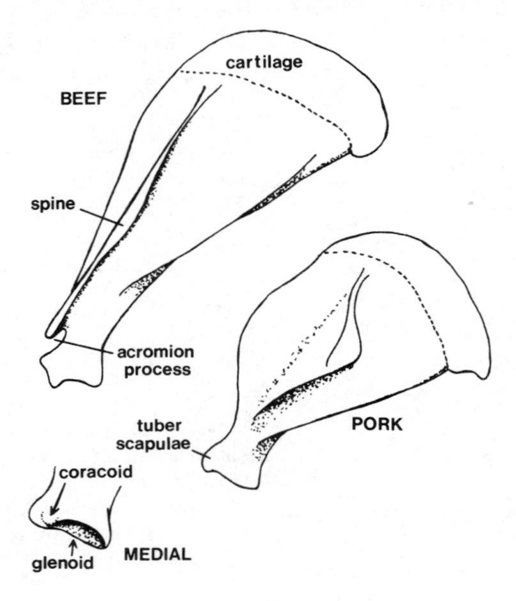

Figure 2-13 Lateral views of the scapula in beef and pork. The medial view of the glenoid cavity shows the coracoid process.

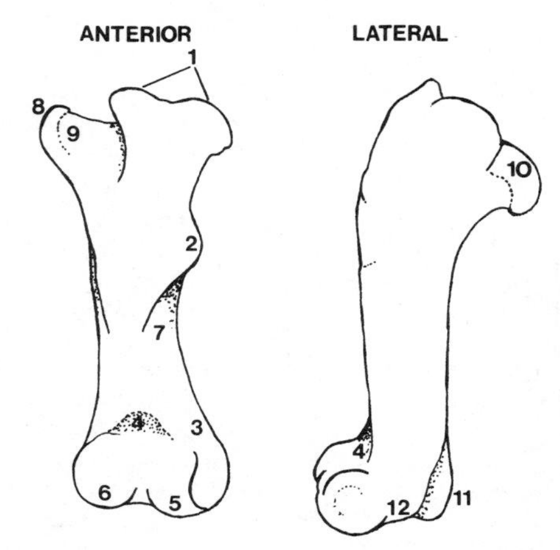

Figure 2-14 Anterior and lateral views of a bovine humerus: (1) lateral tuberosity; (2) deltoid tuberosity; (3) lateral condyloid crest; (4) coronoid fossa; (5) lateral condyle; (6) medial condyle; (7) musculo-spiral groove; (8) medial tuberosity; (9) intertuberal groove; (10) articular head; (11) medial epicondyle; (12) lateral epicondyle.

shallow, unlike the ball-and-socket joint in the hind limb, which is narrow and deep. Along the dorsal edge of the scapula, the bone gives way to flexible *hyaline* cartilage. On the lateral face of the scapula is a prominent ridge of bone called the *spine* of the scapula. In beef carcasses, the scapular spine is extended distally as a prominent *acromion process.*

Proceeding distally down the forelimb, the bone that articulates with the scapula is the humerus (Figure 2-14). Proximally, the humerus has a relatively flat knob or head to fit into the glenoid cavity of the scapula. Two well-defined *condyles* on the distal end of the humerus contribute to the hinge joint at the elbow (Figure 2-15). The elbow is formed by the *olecranon process*, an extension of the *ulna.* The bones of the elbow may be sold by the butcher as "shank knuckle bones," a rather misleading name if compared to the anatomy of the human knuckle bones. The *radius* is joined to the ulna and is the shorter and more anterior bone of the pair.

Beef and lamb carcasses have a set of six compact *carpal* bones which remain on the carcass after slaughter. Before slaughter, the forefeet of cattle and sheep have a large *cannon* bone located distally to the carpal bones. Beef cannon bones are usually removed with the feet at slaughter since there is virtually no meat on them. Cannon bones are sometimes left on lamb carcasses in the abattoir to prevent the meat contracting proximally up the limb. Each forelimb cannon bone in ruminants is derived by the enlargement and fusion of the third and fourth *metacarpal* bones.

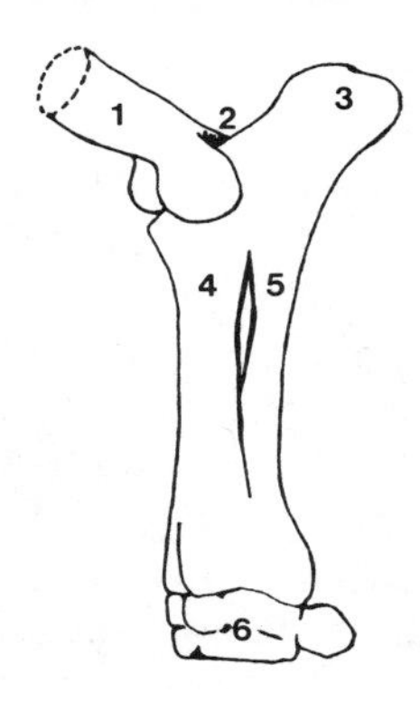

Figure 2-15 Shank bones in a forequarter of beef: (1) distal end of humerus; (2) olecranon fossa; (3) olecranon process; (4) radius; (5) ulna; (6) carpal bones.

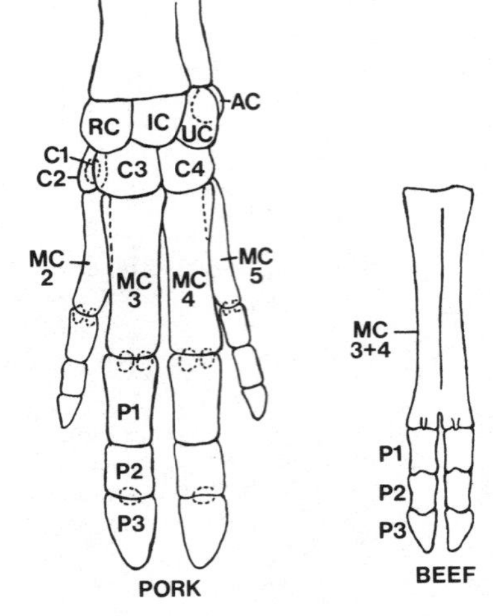

Figure 2-16 Plan view of bones in the left forefoot of the pig compared to a small diagram of the cannon bone and forefoot removed from a beef carcass at slaughter: (AC) accessory carpal; (C) carpal; (IC) intermediate carpal; (MC) metacarpal; (P) phalanges; (RC) radial carpal; (UC) ulnar carpal.

In the human hand, the metacarpal bones lie in the flat part of the hand, between the wrist and the knuckles. If the human hand is placed flat on a desk and is slowly lifted from the wrist, the thumb (digit 1) is the first digit to leave the desk, followed by the index finger and the little finger (digits 2 and 5, respectively). The third and fourth fingers remain on the desk. This demonstrates how the feet of meat animals may have evolved from an original basic plan with five digits. In pigs, digits 3 and 4 on each foot bear most of the body weight, and these digits are larger than the lightly loaded digits 2 and 5 (Figure 2-16). The first digit is absent. The evolutionary trend toward the lifting of the foot and the reduction of digits is even more extensive in cattle and sheep. Cattle and sheep have *cursorial limbs*, long limbs adapted for running. Digits 2 and 5 are reduced to dew claws behind the fetlock. The weight-bearing digits 3 and 4 are enlarged, and their metacarpals are fused to form a long *cannon bone* (Figure 2-16). The small bones in the toes of both fore and hind feet are called *phalanges*.

The feet are left on pork carcasses when they are shipped from the abattoir in shipper's style (unsplit carcasses with head plus perirenal or leaf fat) or in packer's style (split sides with leaf fat and head but not jowls removed). However, in a Wiltshire side for bacon production, the feet are removed, together with the head, pelvis, vertebral column, and psoas muscles.

Hindlimb Skeleton

The proximal bone of the hind limb is the *femur* or round bone (Figure 2-17). Its articular head is deeply rounded and it bears a round ligament which holds it into the acetabulum. Another distinctive feature of the femur is the broad groove between the two *trochlear ridges*, which are located distally. The *patella* or kneecap slides in this groove. The tension generated by the muscles above the knee is transmitted over the knee or stifle joint by the patella in order to avoid having an important tendon in a vulnerable position over the anterior edge of a joint.

In beef and lamb carcasses there is a single major bone, the *tibia* or shank bone, located distally to the femur. In the corresponding position in a pork carcass there are

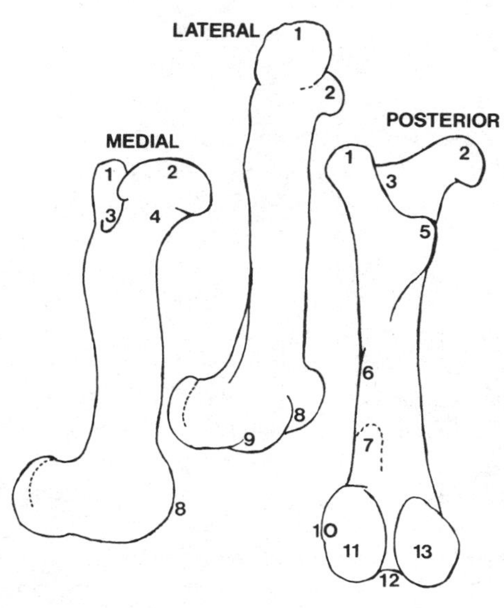

Figure 2-17 Various views of bovine femurs—a medial view of a left femur, a lateral view of a right femur, and a posterior view of a left femur: (1) trochanter major; (2) head of femur; (3) trochanteric fossa; (4) neck of femur; (5) trochanter minor; (6) lateral supracondyloid crest; (7) supracondyloid fossa; (8) trochlea; (9) extensor fossa; (10) lateral epicondyle; (11) lateral condyle; (12) intercondyloid fossa; (13) medial condyle.

two parallel bones, a large tibia and a more slender *fibula* (Figure 2-18). The presence of parallel bones suggests that, at some point in an animal's evolutionary past, rotation of the limb about its axis was possible. For example, rotation of the human wrist involves a partial crossing of the widely spaced ulna and radius. Limb rotation is reduced as animals develop cursorial limbs. In cattle and sheep, one of the parallel bones, the fibula, has lost its shaft. Only a remnant of the head of the fibula can usually be found. In pigs, the fibula retains its shaft and the bone is mobile at birth. After a few years, however, the fibula becomes fused to the tibia.

Distal to the tibia are the tarsal bones of the hock. The structure of the *tarsals, metatarsals*, and *phalanges* of the hind limb is similar to that of the carpals, metacarpals, and phalanges in the forelimb. Pork carcasses are normally suspended by a gambrel or hooked bar which is placed under the tendons of the hind feet. Beef carcasses are normally suspended by a hook under the *fibular tarsal bone*. This bone projects posteriorly and has a rough knob, the *tuber calcis*, for the insertion of the Achilles tendon at the hock.

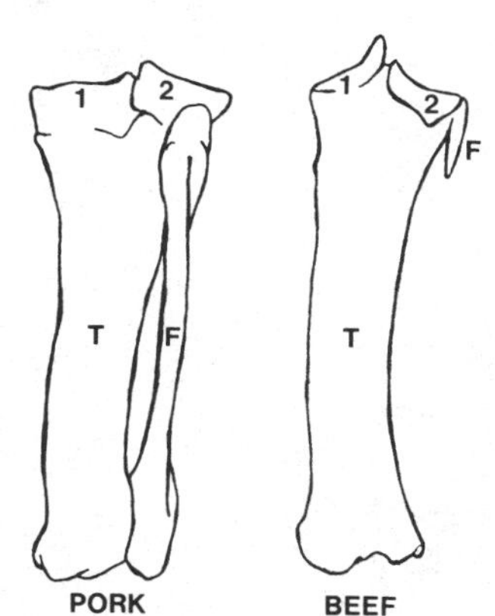

Figure 2-18 Tibia and fibula in pork and beef carcasses—posterior views of the right side: (T) tibia; (F) fibula; (1) medial condyle; (2) lateral condyle.

Poultry Skeleton

The skeletons of poultry are radically different from those of the farm mammals. Not only is the avian skeleton adapted for flight, but birds and mammals are only distantly related zoologically. The skull has very large eye orbits and a small cranial cavity. The long double-curved neck contains 14 cervical vertebrae, and the ringlike atlas articulates to the skull with only a single occipital condyle. The axis has a large odontoid process which projects anteriorly. There are seven thoracic vertebrae, but numbers 2 to 5 are fused. The sixth thoracic vertebrae can move freely, but the last thoracic vertebra is fused to the *synsacrum*. The synsacrum is a fused length of the vertebral column which contains the seventh thoracic vertebra, 14 *lumbosacral* vertebrae, and the first coccygeal or caudal vertebra. Skeletal fusion in the vertebral column does not occur for many weeks after hatching (Hogg, 1982). There are six caudal vertebrae. Apart from the first one, they are free and mobile. However, only numbers 2 to 5 are normal vertebrae, since the last one is formed into a three-sided pyramidal bone called the *pygostyle*.

There are seven ribs: the first two are free while the last five are attached to the sternum. There are no costal cartilages. Ribs 2 to 6 each have an *uncinate process* which overlaps the next posterior rib. The sternum is extremely large. It has a conspicuous ventral ridge in the midline, the *carina*, which increases the area that is available for the attachment of the flight muscles. The dorsal surface of the expanded sternum is concave, and it forms the floor of a continuous thoracic and abdominal cavity.

The bones of the forelimb are greatly modified to form the wing (Figure 2-19). Distal to the humerus are the widely spaced radius and ulna. The carpals, metacarpals, and digits are reduced to form a stiff skeletal unit for the anchorage of the primary flight feathers. The three digits of the wing are equivalent to digits 2, 3, and 4 in other animals (Montagna, 1945). The wing articulates with the body at the glenoid cavity, which is strengthened by the convergence of three bones: the *scapula*, the *coracoid*, and the *clavicle*. In birds the coracoid is a separate bone, whereas in mammals it has been reduced to a small integral part of the scapula. The clavicles of right and left sides are

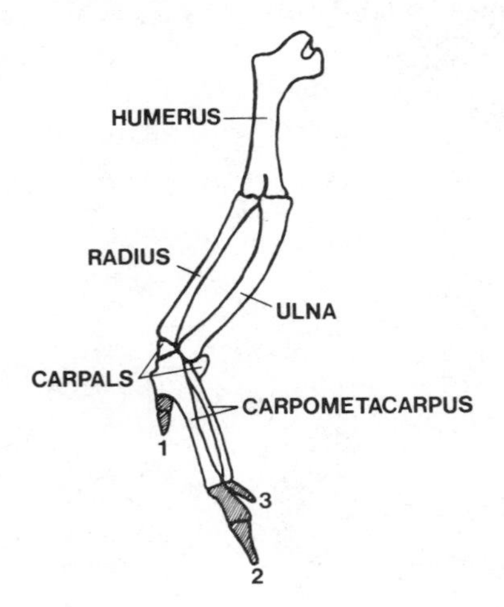

Figure 2-19 Skeleton of chicken wing. Digits are numbered and phalanges are shaded.

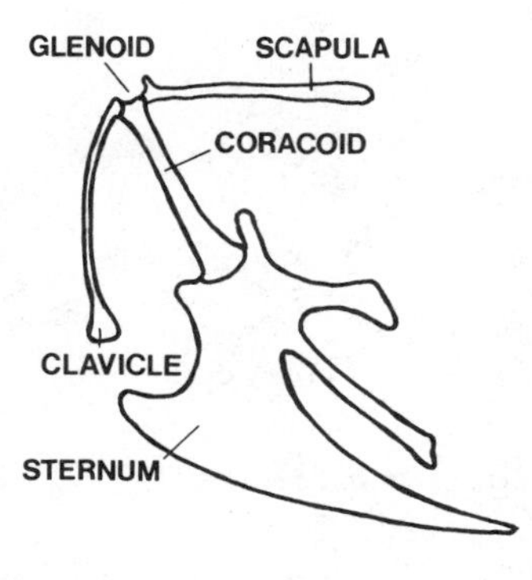

Figure 2-20 Skeletal support for the wing.

fused ventrally to form the *furcula* or wishbone. Although many mammals have a pair of clavicles, they are absent in cattle, sheep, and pigs. The clavicle functions as a strut to support the shoulder joint in animals that have complete mobility of the shoulder joint. Since cattle, sheep, and pigs have cursorial limbs with a restricted fore and aft movement, they do not need clavicles. In poultry, the distal end of the coracoid is braced against the sternum (Figure 2-20). In flight, the body of a bird hangs from its wings at the shoulder joint: hence the more elaborate support for the glenoid cavity.

In poultry, the legs show many cursorial adaptations. Distal to the femur, the fibula is reduced to leave the tibia as the major bone (Figure 2-21). The proximal tarsal bones are fused to the distal end of the tibia to increase its length, and the whole skeletal unit may be called the *tibiotarsus*. The distal tarsal bones are incorporated into the proximal end of a single bone, the *tarsometatarsus*, which also includes the fused metatarsals 2, 3, and 4. Of the four digits that form the bird's claw, digit 1 is directed posteriorly, while digits 2, 3, and 4 are anterior. This adaptation enables the bird to perch. The ilium is fused to the synsacrum. Instead of being fused in the midline, the pubic bones are separate, and they project backward as thin rods (Figure 2-22). The open structure of the pelvis in the ventral region facilitates the passage of eggs from the body cavity. The ilium, ischium, and pubis all contribute to the acetabulum, but the ilium forms more than half of the socket, and the floor is membranous (Harrison, 1975).

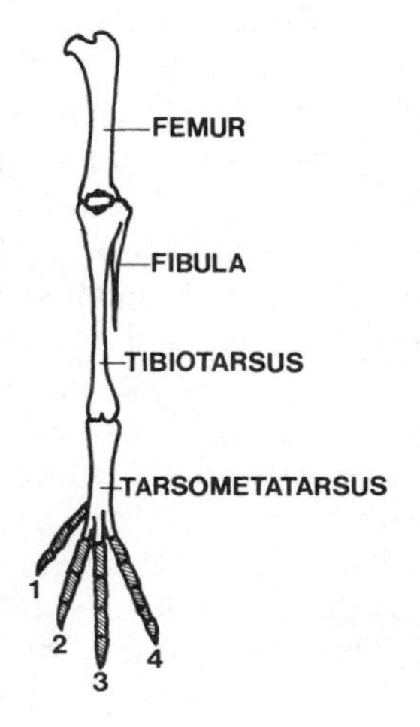

Figure 2-21 Skeleton of chicken leg. Digits are numbered, phalanges are shaded.

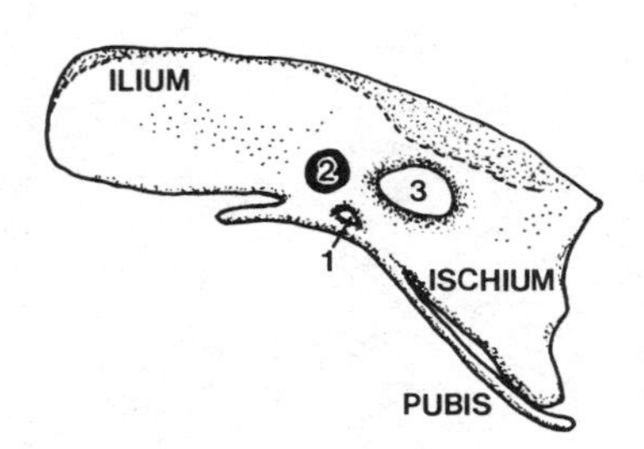

Figure 2-22 Chicken pelvis; (1) obturator foramen; (2) acetabulum; (3) sciatic foramen.

MICROSTRUCTURE AND DEVELOPMENT OF CONNECTIVE FIBERS

Collagen Fibers

Collagen is an elongated protein which forms extremely strong microscopic fibrils. Collagen fibrils may be bound together to form microscopic fibers which, in large numbers, may appear as gristle in raw meat. Collagen is the most abundant protein in the animal body, and the collagen that occurs in meat is an important source of meat toughness. Beef carcasses have to be graded according to age (Chapter 3) mainly because the amount and the strength of their collagen increases with age. Large amounts of collagen are found in animal skin. In pig skin, for example, collagen fibers are tightly woven from two directions so as to form a tight meshwork (Meyer et al., 1982). Collagen is a raw material for major industries in leather, glue, and cosmetics. There are a number of review articles which may be consulted for more detailed information on the chemistry of collagen than is presented here: Veis (1970), Dellman and Brown (1976), Dutson (1976), Gay and Miller (1978), Mecham (1981) and Miller (1982).

Under a light microscope, collagen fibers in the connective tissue framework of meat range in diameter from 1 to 12 μm. They do not often branch and, when branches are found, they usually diverge at an acute angle. Collagen fibers from fresh meat are white, but they are usually stained in histological sections for the light microscope. Eosin gives them a pink color. Unstained collagen fibers can be seen by polarized light since they are *birefringent* (Pimental, 1981). By rotating the plane of polarized light, collagen fibers appear bright against an otherwise dark background. Collagen fibers have a wavy or crimped appearance which disappears when they are placed under tension.

The greater magnification and resolution of electron microscopy reveal that collagen fibers are composed of parallel bundles of small fibrils with diameters ranging from 20 to 100 nm. Collagen fibrils typically have diameters which are multiples of 8 nm, and this may reflect the manner in which they grow radially (Craig and Parry, 1981). Collagen fibrils are formed from long molecules of *tropocollagen* which are staggered in arrangement but tightly bound laterally by *covalent bonds* (Figure 2-23). When negatively stained with heavy metals that spread into the spaces between the ends of molecules, collagen fibrils appear to be transversely striated. The periodicity of these striations is 67 nm, but this often shrinks to 64 nm as samples are processed for microscopic examination. The initial stages of collagen fibril assembly may be

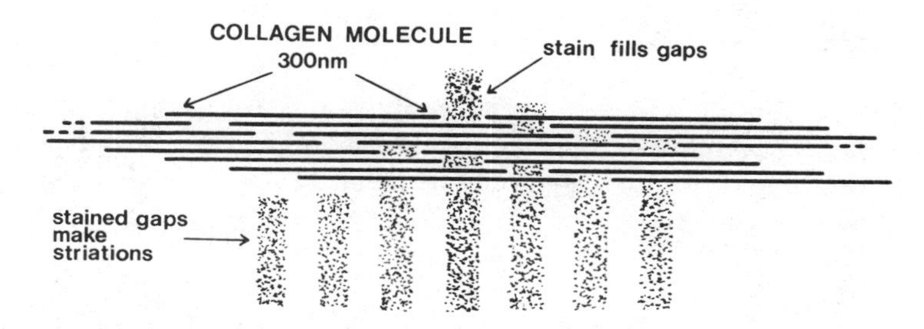

Figure 2-23 Ultrastructural striations in stained collagen fibrils caused by the arrangement of collagen molecules.

intracellular, and fibril morphology may be regulated by a special site on the fibroblast membrane (Trelstad and Hayashi, 1979).

Tropocollagen Molecules

Tropocollagen is a high-molecular-weight protein (300,000) which is formed from three polypeptide strands twisted into a triple helix. Each strand is a left-handed helix twisted on itself, but this is not shown in Figure 2-24, where only the larger right-handed triple helix which involves all three polypeptide strands is shown. The triple helix is responsible for the stability of the molecule and for the property of self-assembly of molecules into microfibrils. The flexible parts of each strand which project beyond the triple helix are responsible for the bonding between adjacent molecules (Kuhn and Glanville, 1980). In other words, the cross-links that bind tropocollagen molecules together laterally are made between the helical shaft of one molecule and the nonhelical extension of an adjacent molecule.

In the polypeptide strands, *glycine* occurs at every third position, and *proline* and *hydroxyproline* account for 23 per cent of the total residues. The regular distribution of glycine is required for the packing of tropocollagen molecules, and this has been claimed as evidence that all animals are derived by evolution from a single ancestral stock, since the chance development of this unique regularity in unrelated animals is extremely unlikely (Finerty, 1981). Since hydroxyproline is quite rare in other proteins of the body, an assay for this imino acid provides a measure of the collagen or connective tissue content in a meat sample (O'Neill et al., 1979). Tropocollagen also contains a fairly high proportion of glutamic acid and alanine as well as some *hydroxylysine*. Methods for the detection and estimation of the collagen content in meat products are reviewed by Etherington and Sims (1981).

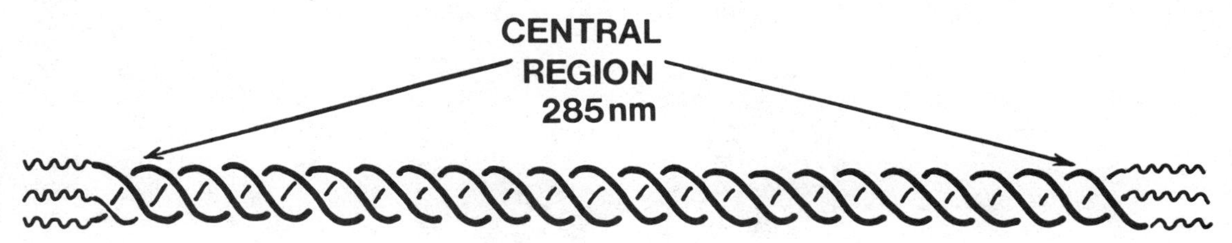

Figure 2-24 Collagen molecule.

Types of Collagen

Several slightly different types of polypeptide strands (also known as alpha chains) have been identified. Since each tropocollagen molecule is composed of three alpha chains, five types of collagen can be formed by the combinations shown in Table 2-3. Different types of collagen can be identified histochemically as well as biochemically (Bock, 1977, 1978).

TABLE 2-3. DERIVATION OF DIFFERENT TYPES OF COLLAGEN FROM DIFFERENT COMBINATIONS OF ALPHA CHAINS

Type of collagen	Type of alpha chains						
	α1-I	α1-II	α1-III	α1-IV	α2	α-A	α-B
I	2				1		
II		3					
III			3				
IV				3			
V						1	2

Sources: Gay and Miller (1978), Bailey and Sims (1977), Burgeson et al. (1976), Bentz et al. (1978), and Bailey et al. (1979).

Type I collagen forms striated fibers which are between 80 and 160 nm in diameter and which occur in blood vessel walls, tendons, bones, skin, and meat. It may be synthesized by fibroblasts, smooth muscle cells, and osteoblasts.

Type II collagen fibers are less than 80 nm in diameter and they occur in hyaline cartilage and in intervertebral disks. It is synthesized by chondrocytes.

Type III collagen forms *reticular fibers* in tissues which have some degree of elasticity, such as spleen, aorta, and muscle. It is synthesized by fibroblasts and smooth muscle cells.

Type IV collagen occurs in the amorphous basement membranes which surround many types of cells, and it may be produced by the cells themselves rather than by fibroblasts (Laurie et al., 1980).

Type V collagen is found prenatally in basement membranes and in cultures of embryonic cells. It is synthesized by myoblasts, smooth muscle cells, and possibly, by fibroblasts (Kuhn and Glanville, 1980). Type V collagen has also been found in the basement membranes of muscle fibers, except at the point where muscle fibers are innervated (Sanes and Cheney, 1982).

Tendons often extend into the belly of a muscle or along its surface before they merge with its connective tissue framework, and types I and III collagen can both be extracted from meat. In fibers composed of collagen types I and II, fibrils have a straight arrangement. In fibers of type III collagen, the fibrils have a helicoidal arrangement (Reale et al., 1981). Small-diameter type III collagen fibers are called *reticular fibers* since, when stained with silver for light microscopy, they often appear as a network or reticulum of fine fibers. The larger-diameter collagen fibers formed from type I collagen are not blackened by silver. The identification of reticular fibers by silver staining has a long and complex history, which has been reviewed by Puchtler and Waldrop (1978).

Collagen fibers shrink when they are placed in hot water, and they are ultimately converted to gelatin. Around 65° C, the triple helix is disrupted and the alpha chains fall into a random arrangement. The importance of this change is that it tenderizes meat with a high connective tissue content. Tropocollagen molecules from older animals are more resistant to heat disruption than are those from younger animals. In early studies, it was suggested that reticular fibers, unlike collagen fibers, did not yield gelatin when treated with moist heat. The original suggestion that reticular fibers survive unchanged after cooking is wrong, but a modification of the idea is plausible. Since a piece of meat may contain different types of collagen, and since these types may differ in the thermal stability of their cross-links, it is possible that, at an intermediate level of cooking around 65° C, endomysial collagen and perimysial collagen may differ in the extent to which they are affected by the cooking treatment (Bailey and Sims, 1977).

Collagen Biosynthesis

The synthesis of the different types of polypeptide strands that are combined to make different types of collagen is genetically regulated by the production of messenger RNA. The synthesis of polypeptide strands occurs on membrane-bounded polysomes, but the hydroxylation of lysine and proline occurs after the strands are assembled. *Ascorbic acid* is required for the hydroxylation of lysine and proline. Polypeptide strands enter the cisternae of the endoplasmic reticulum. The terminal extensions of the strands are aligned, and then the strands spiral around each other. *Procollagen* or immature collagen has long terminal extensions which protrude from each end of the newly formed triple helix. Procollagen moves to the Golgi apparatus and is packaged into vesicles which are moved to the cell surface, probably by microtubules. Except for some type III procollagen molecules, the long terminal extensions are then enzymatically reduced in length.

Outside the cell, collagen molecules become aligned in parallel formations, and then they link up laterally to form fibrils. It is probable that tropocollagen monomers are partially assembled together in groups before they are added to an existing collagen fibril (Trelstad, 1982). The characteristic parallel staggered arrangement of tropocollagen molecules in a collagen fibril is due to the 67-nm repeating pattern of oppositely charged amino acids along the length of the tropocollagen molecule (Miller, 1982). The degree of overlapping of adjacent molecules and the gaps left between the ends of molecules cause the striated appearance of collagen fibers seen by electron microscopy (Figure 2-23). The fibroblasts of young animals are metabolically very active. In older animals, the metabolism of fibroblasts is greatly reduced, particularly their aerobic metabolism (Floridi et al., 1981).

Accumulation of Collagen in Meat

The amount of collagen in muscle increases as animals become older, and this has an important effect on meat toughness. Within a carcass, there are considerable differences in collagen content between different carcass muscles. This is largely responsible for differences in the economic value of cuts of meat (Chapter 3). Collagen content may differ between sexes. For example, the hydroxyproline content is greater in pork from females than from castrated males (Sellier and Boccard, 1971). Many studies on the accumulation of collagen in carcass muscles have used hydroxyproline content as an index of collagen content. The amount of collagen in a meat sample, when expressed as a proportion of wet sample weight, is affected by the fat content. In steaks from a veal carcass, for example, collagen would probably exceed 0.5 per cent, but would be much less in the same region from a market-weight steer carcass in which fat had accumulated to "dilute" the collagen content.

Another method of studying collagen in meat is to measure collagen fibril diameters in electron micrographs (Rowe, 1978). In equine tendons, fibril diameters in the fetus are unimodal but they become bimodal in the adult (Parry et al., 1978b). Large-diameter fibrils may have more intrafibrillar covalent cross-links, while small-diameter fibrils may have more interfibrillar noncovalent cross-links. Thus fibril diameter may be related to fibril strength and elasticity (Parry et al., 1978a). Little is known about the mechanisms by which collagen fibers become arranged in a muscle, or about the interactions that occur between fibroblasts and the fibers which they produce, although it is possible that glycosaminoglycans play some part in this interaction (Parry et al., 1978a).

Collagen has an important effect on muscle development. Myoblasts, the cells that form muscle fibers, develop a parallel alignment when cultured on a substrate of type I collagen, but they do not become elongated or aligned on type V basement membrane collagen (John and Lawson, 1980). Cultured myoblasts can form collagen types I, III, and V (Sasse et al., 1981). Type V collagen (Bailey et al., 1979) is composed of α-A and α-B chains (Burgeson et al., 1976) in a 1:2 ratio (Bentz et al., 1978). Myotubes (immature muscle fibers of the fetus) are able to synthesize collagen (Mayne and Strahs, 1974), but probably in association with fibroblasts (Lipton, 1977). The identification of collagen in developing muscle is complicated by the fact that the tail unit of the *acetylcholinesterase* molecule has a collagenlike sequence which contains hydroxyproline and hydroxylysine (Lwebuga-Mukasa et al., 1976). Acetylcholinesterase is a functional component of the neuromuscular junction (Chapter 6).

Cross-Linking of Collagen Molecules

Within an individual collagen molecule, the three polypeptide strands are linked together by stable intramolecular bonds which originate in the nonhelical ends of the molecule. The great strength of collagen fibers, however, originates mainly from the stable intermolecular covalent bonds between adjacent tropocollagen molecules. Stable disulfide bonds between cystine molecules in the triple helix also occur. During the growth and development of meat animals, covalent cross-links increase in number, and collagen fibers become progressively stronger. Meat from older animals, therefore, tends to be tougher than meat from the same region of carcasses from younger animals. This relationship is complicated in young animals by the rapid synthesis of large amounts of new collagen. New collagen has fewer cross-links, so that if there is a high proportion of new collagen, the mean degree of cross-linking may be low, even though all existing molecules are developing new cross-links. As the formation of new collagen slows down, the mean degree of cross-linking increases.

Another complication is that many of the intermolecular cross-links in young animals are reducible: the collagen is strong but is fairly soluble. In older animals, reducible cross-links are probably converted to nonreducible cross-links; the collagen is strong but is far less soluble and more resistant to moist heat. The chemistry of these changes is still a subject for debate.

Differences in the degree of cross-linking may occur between different muscles of the same carcass, and between the same muscle in different species. For example, collagen from the longissimus dorsi is less cross-linked than is collagen from the semimembranosus, and collagen from the longissimus dorsi of a pork carcass is less cross-linked than is collagen from the bovine longissimus dorsi. Nutritional factors such as high-carbohydrate diet, fructose instead of glucose in the diet, low protein, and preslaughter feed restriction may reduce the proportion of stable cross-links. In general, the turnover rate of collagen is accelerated in cattle that are fed a high-energy diet (Wu et al., 1981). Recent measurements suggest that the synthesis rate of collagen turnover in skeletal muscle is about 10 per cent/day (Laurent, 1982).

Elastin and Elastic Fibers

When placed under tension, collagen fibers stretch to only about 5 per cent of their starting length and, where collagen is formed into cablelike tendons, little elasticity is possible. However, much of the collagen that is present in meat forms a meshwork. Although individual collagen fibers do not stretch, stretching can occur in the whole muscle if the configuration of the meshwork changes. Fibers with truly elastic properties, however, are necessary in structures such as the ligamentum nuchae. Also, large sheets of *elastic fibers* contribute to the abdominal wall, and all arteries, from the aorta to the finest microscopic arterioles, rely on elastic fibers to accommodate the surge of blood that results from contraction of the heart. Elastic fibers can be stretched to several times their original length, and they rapidly resume their original length once they are released. Elastic fibers are made of the protein *elastin*. Elastin resists severe chemical conditions, such as the extremes of alkalinity, acidity, and heat, which destroy collagen. Fortunately, there are relatively few elastic fibers in muscle; otherwise, cooking would do little to reduce meat toughness. The elastin fibers in muscles which are used frequently for locomotion are larger and more numerous than those of muscles used less frequently (Hiner et al., 1955). Some elastic fibers in muscle are involved in the attachment sensory organs called neuromuscular spindles (Cooper and Gladden, 1974). During the digestion of meat in the human gut, elastic fibers are broken down by *elastase*, an enzyme from the pancreas.

Elastic fibers have a pale yellow color as seen in the ligamentum nuchae. Elastic fibers can be stained for light microscopy by special techniques (Munz and Meves, 1974; Brissie et al., 1975; Puchtler et al., 1973, 1976; Gotta-Pereira et al., 1976). When elastic fibers are stretched, they are visible in polarized light without staining. Elastic fibers in meat have a small diameter (approximately 0.2 to 5 μm), although they are larger in the ligamentum nuchae. Elastic fibers in the connective tissue framework of meat are usually branched. Electron microscopy reveals that elastic fibers are composed of bundles of small fibrils (approximately 11 nm thick) which are embedded in an amorphous material. In the bovine ligamentum nuchae, the construction of fibrils from even smaller units or filaments (approximately 2.5 nm thick) has been demonstrated (Serafini-Fracassini et al., 1978). Elastin filaments are bound by noncovalent interactions to form a three-dimensional network (Rucker and Lefevre, 1980). Elastic fibers are probably assembled in grooves on the fibroblast surface. The initial ropelike aggregations of fibrils become infiltrated with an amorphous material, now identified chemically as elastin (Ross and Bornstein, 1971). Unlike the situation in elastic ligaments, where elastin forms fibers, the elastin in the arterial system is formed into sheets. In systems other than ligamentous elastic fibers, the extracellular condensation of elastin may occur in the absence of fibrils.

Although elastin resembles tropocollagen in having a large amount of glycine, it is distinguished by the presence of two unusual amino acids, *desmosine* and *isodesmosine* (Partridge, 1966). Like collagen, elastin contains hydroxyproline, although it does not appear to have the same function of stabilizing the molecule. *Tropoelastin*, the soluble precursor molecule of elastin (molecular weight 70,000 to 75,000), is secreted by fibroblasts after it has been synthesized by ribosomes on the rough endoplasmic reticulum. Elastin is processed and secreted by the Golgi apparatus. In the presence of copper, *lysyl oxidase* links together four lysine molecules to form a desmosine molecule. Isodesmosine is the isomer of desmosine. In animals deprived of copper in their diet, the aorta is often fatally weakened because of a lack of mature elastin. Elastin in the arterial system is produced by smooth muscle cells instead of fibroblasts. The functional properties of elastin in different tissues, such as the lung and the aorta, may be related to differences in the ratio of tropoelastin A to tropoelastin B (Barrineau et al., 1981). However, the diversity of the different genetic types of elastin that are known at present is conspicuously less than the diversity found with collagen. The elastin in elastic cartilage may be of a different genetic type than that found in the vascular system.

CELLS OF FIBROUS CONNECTIVE TISSUE

The dominant type of cell in the fibrous connective tissue of meat is the fibroblast, but other types of cells may also be found. *Macrophages* or *histiocytes* are sometimes quite numerous and, when inactive, may resemble fibroblasts in appearance. However, the motility of macrophages is soon revealed by inflammation of the tissue or by the injection of colloidal dyes. Macrophages migrate through the tissue and act as scavengers by engulfing invasive microorganisms or foreign particles by *phagocytosis*.

Cells from the vascular system may wander through connective tissues. Even compact structures such as tendons have their own lymphatic and vascular supply (Edwards, 1946). These vascular cells include a variety of *lymphocytes* and the *plasma cells* which are responsible for antibody production. *Eosinophils* are cells which have bilobed nuclei and numerous cytoplasmic granules with an affinity for eosin. The skeletal muscles of cattle, and sometimes of sheep (Harcourt and Bradley, 1973), may become inundated with eosinophils (*eosinophilic myositis*). The affected areas appear as irregular pale lesions and are often detected by meat inspectors who are looking for muscle parasites. Eosinophils may be attracted to areas of antibody activity, and eosinophilic myositis might be an allergic response (Oghiso et al., 1977).

Mast cells (Selye, 1965) also occur within the skeletal muscles of meat animals (Swatland, 1978). The cytoplasm of mast cells contains large numbers of *metachromatic granules*. These have the property of changing the color of dyes such as methylene blue (a property called "metachromasia") so that the granules appear purple while the surrounding tissue is stained blue. Mast cells contain *heparin* and *histamine*. Heparin prevents the coagulation of blood and histamine increases the permeability of small blood vessels. Heparin also activates the enzyme *lipoprotein lipase*, which is involved in the accumulation of triglyceride by adipose cells. Mast cells may also release a substance that activates mitosis in nearby cells (Franzen and Norrby, 1980). Thus the development of intramuscular fat in meat might have some relationship to the distribution of mast cells. Mast cells sometimes come into close contact with skeletal muscle fibers (Heine and Forster, 1974), but most mast cells are located along fine branches of the lymphatic system in the perimysium and endomysium (Stingl and Stembera, 1974). Mast cells have also been implicated in the regulation of collagenase activity (Simpson and Taylor, 1974).

CARTILAGE

Cartilage cells or *chondrocytes* occupy lacunae in a stiff flexible matrix which is formed from collagen fibers in a *proteoglycan* ground substance. *Hyaline* cartilage has a white translucent appearance and it occurs on the smooth surfaces of joints. In the larynx, trachea, and bronchi, hyaline cartilage forms the rings that hold these air ducts open during respiration. Flexible units of the skeleton, such as the dorsal part of the scapula and the linkages between the sternum and the posterior ribs, are also formed from hyaline cartilage. Most of the bones of the carcass are initiated prenatally as cartilagenous models which subsequently become ossified. Complete ossification is a slow process, and the bones of young meat animals are more flexible than those of adults. The state of ossification is a useful clue to animal age in carcass grading.

Chondrocytes are derived from mesenchymal cells and they are initially capable of both mitosis and matrix formation. Clusters of related cells are pushed apart by their new matrix in a process called *interstitial growth* (Figure 2-25). Cartilagenous models of prenatal bones are covered by a membrane known as the *perichondrium*. Inner perichondrial cells differentiate into chondrocytes so that, in addition to interstitial growth, new cells and matrix can be added superficially in a process known as *appositional growth* (Figure 2-26). Cartilage may acquire numerous elastic or collagen fibers so that it may become *elastic cartilage* or *fibrocartilage*, respectively.

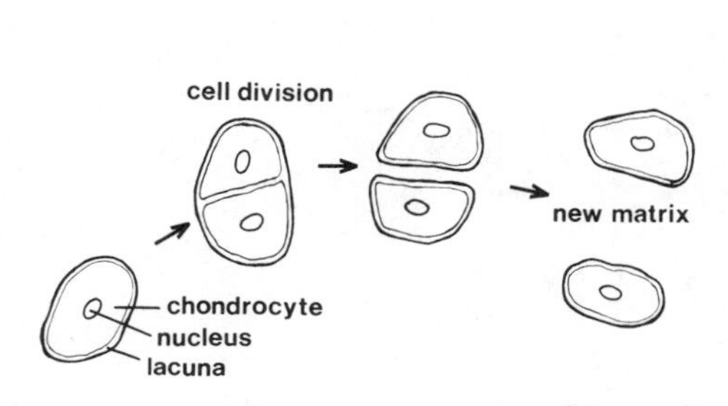

Figure 2-25 Interstitial growth in cartilage by cell division and formation of new matrix.

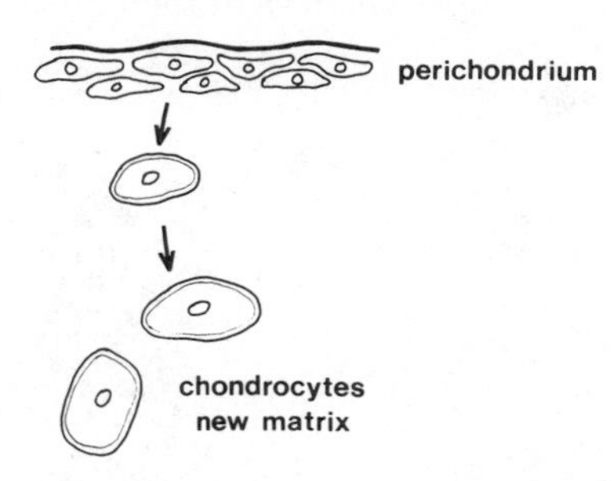

Figure 2-26 Appositional growth in cartilage by formation of new chondrocytes from cells of the perichondrium.

Oxygen, nutrients, and waste products can travel to and from the chondrocytes in cartilage by diffusion through the surrounding matrix. However, when the matrix becomes ossified by the deposition of submicroscopic *hydroxyapatite* crystals, diffusion is reduced. In bone, *osteocytes* can only survive if they develop long cytoplasmic extensions which radiate from the lacunae to regions where exchange by diffusion can take place. These cytoplasmic extensions run through fine tubes or *canaliculi* in the ossified matrix, but these are limited in length. Consequently, large numbers of blood vessels permeate the matrix of bone. Most of these blood vessels run longitudinally through the bone in large *Haversian canals* which are surrounded by concentric rings of osteocytes and bone lamellae (Figure 2-27). Bones are covered by a connective tissue membrane called the *periosteum*.

The prenatal formation of bone is initiated by either of two basic mechanisms: (1) *intramembranous* ossification; or (2) *endochondral* ossification. Intramembramous ossification is typical of the bones that form the vault of the skull, and it occurs when sheets of connective tissue produce osteoblasts which then initiate centers of ossification. Endochondral ossification is more common, and is the process by which cartilagenous models become ossified to form the bones of a commercial meat carcass.

The internal structure of carcass bones becomes visible when they are split longitudinally on a band saw (Figure 2-28). The shaft of a bone is called the *diaphysis*. The knob at each end of a bone is called the *epiphysis*. Between the diaphysis and each epiphysis is a cartilagenous growth plate called the *epiphyseal plate*. In a young animal, the chondrocytes of the epiphyseal plate are constantly dividing to form new matrix. However, on each face of the plate, cartilage is continuously resorbed and is replaced by bone (Howlett, 1980), so that the thickness of the epiphyseal plate tends to remain constant in growing animals. This process allows a bone to grow longitudinally without disrupting the articular surface on the epiphysis. The rate of the longitudinal growth of bones is the product of two factors: (1) the rate of production of new cells; and (2) the size that cells reach before they degenerate at the point of ossification (Thorngren and Hansson, 1981). The strength and thickness of epiphyseal plates is modified by sex hormones (Oka et al., 1979). At puberty, chondrocyte growth slows down and fails to keep pace with ossification on the surface of the epiphyseal plate. Thus epiphyseal plates are lost in mature animals, and the epiphyses become firmly

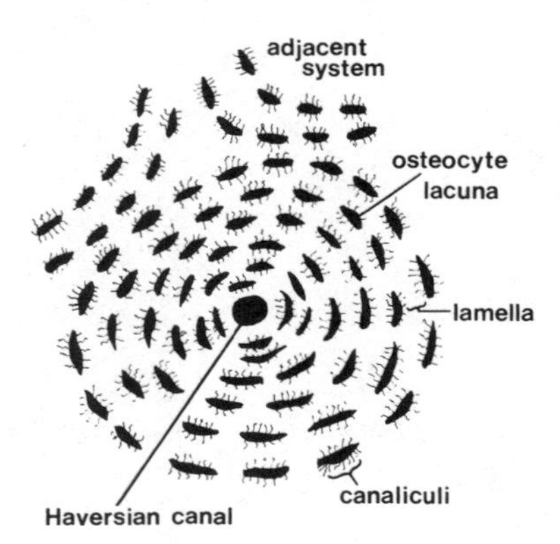

Figure 2-27 Transverse section of a small piece of dried bone showing a Haversian system of concentric lamellae and lacunae around a Haversian canal.

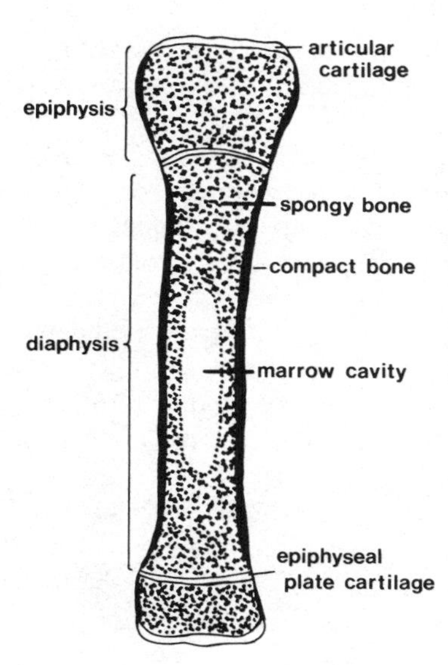

Figure 2-28 Longitudinal section of a long bone.

ossified to their diaphyses. Bone growth in mature animals is restricted to the girth or thickness of the bone, and it occurs by the recruitment of periosteal cells to become osteoblasts.

Calcium, Calcification, and Bone Resorption

Radioactive calcium (^{45}Ca) may be used to study the uptake of calcium by the skeleton. There is a continuous exchange of calcium between the body fluids (mostly plasma) and approximately 1 per cent of the total bone matrix, which has direct access to the circulating fluid. Growth by accretion results from small net gains to the matrix (Figure 2-29). When radioactive calcium is injected into a growing animal, the isotope

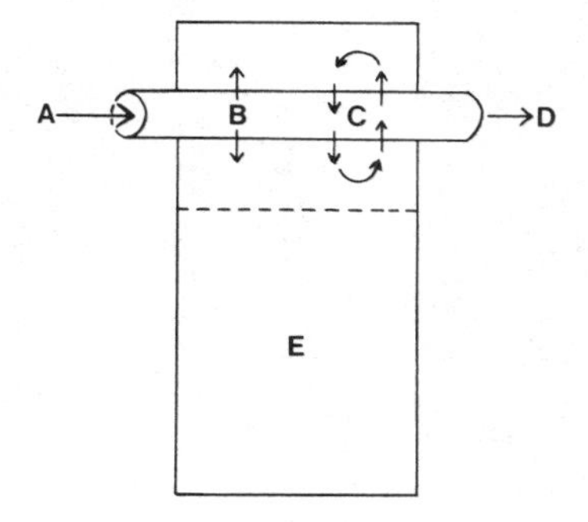

Figure 2-29 Calcium ion movements in bone. Calcium ions bound to serum proteins or as anions (bicarbonate, citrate, phosphate, and so on) enter the bone (A). New bone is added by accretion (B), but there is also continuous calcium ion exchange with the small amount of bone in direct contact with the circulatory system (C). Remaining calcium ions leave the bone (D). Deep in the bone there is little accretion or exchange (E).

is incorporated into new bone, and the concentration of isotope in the plasma declines (Wasserman, 1977). Both calcium exchange and calcium accretion are more rapid in the epiphysis than in the diaphysis.

Calcium, phosphate, and hydroxyl ions are obtained from the extracellular fluid during bone formation. The first stage in ossification is the deposition of a crystal of calcium phosphate. Calcium phosphate is then converted to *hydroxyapatite*. Two histological events associated with ossification have been observed. First, extracellular membrane-bounded vesicles which contain *phosphatase* have been shown to accumulate calcium phosphate and hydroxyapatite. Second, in vivo crystal nucleation has been observed between the ends of tropocollagen molecules in collagen fibers which have been placed in a metastable calcium phosphate solution. The supply of calcium and phosphate from the blood is affected by *vitamin D*. Vitamin D is obtained from the diet or by exposure to ultraviolet light. It is hydroxylated in the liver and then converted to the hormonal form (1,25-dihydroxycholecalciferol) in the kidney. The hormonal form causes the intestine to increase its absorption of calcium and phosphate. The mechanical strength of bone may be used to assess the nutritional availability of minerals (Crenshaw et al., 1981).

The resorption of bone enables bone remodeling in response to local stresses. Bone resorption is also coupled with the maintenance of blood calcium levels (Figure 2-30). The organic components of bone are degraded by the lysosomal enzymes of *osteoclasts*, a process that requires *vitamin A*. The solubilization of hydroxyapatite in response to *parathyroid* hormone is probably achieved by a combination of low pH (due to anaerobic glycolysis) and chelation. Bone resorption is inhibited by *calcitonin* (thyrocalcitonin) and is linked with the maintenance of calcium and phosphorus levels.

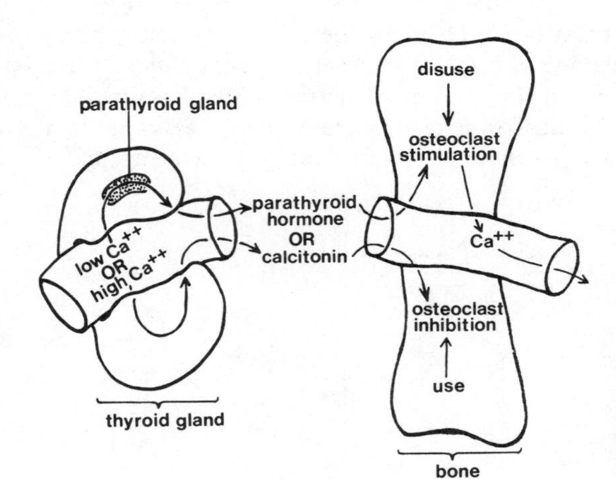

Figure 2-30 Simplified diagram of the hormonal and functional control of bone metabolism. The large tubes in the figure represent blood vessels.

Genetic defects in calcification and bone formation may occur in cattle. Typical signs are lack of calcification of the teeth and hypermobility of joints (Johnston and Young, 1958). In *milk fever* of cows (parturient paresis), paralysis and unconsciousness may occur during parturition or early lactation. The condition is a result of drastically lowered plasma calcium and inorganic phosphorus levels, and it may be treated by the injection of *calcium borogluconate*.

Regulation of Skeletal Growth

Carcasses from young animals have a relatively high bone content because the skeleton is well developed at birth. As an animal grows to market weight, its proportion of bone decreases on a relative basis, due to the growth of muscle and fat. The long-term control of bone growth is superimposed on the short-term regulation of bone metabolism, which occurs in response to changes in blood calcium levels (Figure 2-30) or to remodeling in response to local functional demands.

A number of *hormones* exert secondary effects on skeletal development. Thyroxine, insulin, growth hormone, and gonadal hormones tend to be anabolic. Estrogens may inhibit resorption of bone. Adrenal corticosteroids stimulate resorption of bone and inhibit the formation of new bone. In cattle, castration delays the completion of growth in epiphyseal plates (Brannang, 1971a). This is most noticeable in the distal bones of the limbs, and it enables the continued longitudinal growth of the legs. In the vertebral column, however, castration reduces bone growth. The reduction is centered on the first thoracic vertebra. Removal of the ovaries from heifers also causes an increase in the longitudinal growth of distal bones (Brannang, 1971b).

At localized sites, intrinsic activity in bone (Behari and Andrabi, 1978) probably plays an important role in regulating bone growth, but its current status is uncertain (Moss, 1972). One hypothesis is that loads that are frequently placed on a region of bone cause the transduction of mechanical energy to electrical energy by a *piezoelectric effect*. In a frequently loaded and negatively charged region, growth is stimulated. In an unloaded and positively charged region, resorption is stimulated. In bovine bone, it has been suggested that adaptation to changing load patterns occurs by a viscoelastic "creep" at the cement lines which surround individual Haversian systems (Lakes and Saha, 1979). Differences in the arrangement of hydroxyapatite crystallites, in lacunar structure, and in the transition from spongy to compact bone have been observed between the bones of wild and domesticated sheep. These differences may have accompanied a reduction in exercise (Drew et al., 1971). Severe malnutrition of pigs reduces the formation of Haversian bone (Luke et al., 1980).

The gross anatomy of muscles and skeletal units are closely matched, and mutual or interacting control systems probably exist. Because most farm animals are slaughtered in a fairly immature condition, the relationship between muscles and the bony processes that they pull on may not be immediately obvious. The knobs and wrinkles on bone surfaces become more conspicuous with age, and they are readily seen in the carcasses of old bulls. One possible relationship between muscle activity and bone growth may be that isometric contraction, by stopping or slowing the venous blood flow, may stimulate bone growth. Alternatively, by pulling on the periosteum, the effect of muscle activity may be mediated by connective tissue. The importance of local factors is seen in bone transplants, where growth of the transplant almost immediately becomes regulated by the new local conditions.

Joints

When the movement of two adjacent bones is normally restricted or impossible, the joint between them is called a *synarthrosis*. Different types of connective tissue act as a cement between the two adjacent bones. Skull *suture* joints are cemented by collagen, sternebrae are separated by cartilage, and fibrocartilage occurs in the symphysis pubis and in intervertebral disks. Completely movable joints are called *diarthroses*. Their articular surfaces are covered by smooth hyaline cartilage which is lubricated by *synovial fluid*. Under suitable conditions, diarthroses can be made to develop in organ cultures (Fell, 1956), and development can proceed in the absence of muscular and neural connections, provided that elements of the two diaphyses are present and are separated by undifferentiated tissue. The joint capsule, however, does not develop properly under these conditions, and the proper shaping of articular surfaces develops only with normal use.

Neonatal farm animals are sometimes afflicted by genetic or environmentally induced joint malformations. Rigid joints are most frequently observed in the distal parts of a limb, so that locomotion is difficult or impossible. The name *arthrogryposis* is frequently used to describe this condition, but it is a rather misleading term since arthrogryposis means "crooked joints," whereas, in reality, joints are often fixed in an extended or rigidly straight position. James (1951) preferred the term "multiple congenital articular rigidity" in human medicine. Since farm animals often have only a single affected joint, the term "multiple" is often redundant. Four degrees of immobilization may occur: (1) when the joint can be freed by an externally applied nondestructive force; (2) when muscles and tendons acting around the joint must be severed; (3) when capsular ligaments as well as muscles and tendons must be severed; and (4) when mobilization is impossible due to deformation of the articular surface (Swatland, 1974). The term *ankylosis* is inappropriate for congenital articular rigidity since it implies that bones are fused across the joint; with one possible exception (Schmalstieg and Meyer, 1960) this is not a general feature of the condition.

Hereditary congenital articular rigidity in cattle is often associated with a *cleft palate* in the skull (Leipold et al., 1969; Greeley et al., 1968; Greeley and Jolly, 1968). However, there are a number of reports of hereditary congenital articular rigidity in cattle without cleft palate but with defects such as abnormal limb shape (Hutt, 1934) or hair defects (Tuff, 1948; Nes, 1953). Hereditary bovine congenital articular rigidity is usually caused by a recessive gene without sex linkage. In Australia, congenital articular rigidity sometimes occurs in conjunction with *hydranencephaly* (absence or gross reduction of cerebral hemispheres) in cattle (Blood, 1956; Whittem, 1957). The condition is not heritable and may be induced by ingestion of a plant teratogen (Hindmarsh, 1937). In the western states of the United States, a number of species of the plant *Lupinus* may cause congenital articular rigidity in cattle (Wagnon, 1960; Shupe et al., 1967; Keeler et al., 1969).

In pigs, congenital articular rigidity may occur as a result of either genetic or environmental causes. Hereditary congenital articular rigidity in pigs is probably due to an autosomal recessive gene (Hallqvist, 1933). Hereditary congenital articular rigidity in pigs may be associated with an unusual condition known as the *thick-leg syndrome* (Morrill, 1947; Koch et al., 1957; Kaye, 1962). In this syndrome, certain bones are thickened by the activity of extra osteoblasts, and fibrous connective tissue may invade muscles and joint capsules. Environmentally induced congenital articular rigidity in pigs may result from nutritional deficiencies of vitamin A (Palludan, 1961) or manganese (Miller et al., 1940), or from ingestion of teratogens from plants such as tobacco (Crowe, 1969; Crowe and Pike, 1973), *Conium maculatum* (Edmonds et al., 1972), *Prunus serotina* (Selby et al., 1971), and *Datura stramonium* (Leipold et al., 1973b).

Both genetic (Roberts, 1929; Middleton, 1932, 1934; Zophoniasson, 1929; Morley, 1954) and environmental (Keeler et al., 1967; James et al., 1967; Stamp, 1960; Nisbet and Renwick, 1961) causes of congenital articular rigidity are known in sheep. Judging from the variety of different factors that can give rise to the same end result, congenital articular rigidity, it is probable that different factors exert their effect on the same developmental mechanism—a mechanism that involves the interaction of nerves, muscles, and joints (Chapter 6).

Animal Size and Bone Development

Breeds of cattle with a large size when mature usually produce lean meat at a faster rate than do early-maturing traditional beef breeds which have a relatively small adult size. Differences in adult size are produced by differences in skeletal growth, and relationships between the quantitative anatomy of individual bones and meat production traits in beef cattle have been identified (Wilson et al., 1977). Relationships between skeletal and muscular development may involve meat quality, since large-framed animals produce leaner meat during their production life span. Large-framed breeds mature late and have a later cessation of linear skeletal growth at their epiphyseal plates. The time of maturation is related to the distribution and amount of adipose tissue in the carcass, particularly marbling fat. Differential bone growth between large and small breeds of cattle is usually established prior to a slaughter weight of 500 kg in males (Jones et al., 1978).

Pelvic dimensions in cows of different breeds are related to the incidence of difficult calving or *dystocia* (Laster, 1974; Neville et al., 1978). Dystocia may be particularly serious when a homozygous double-muscled calf is born. Double muscling in the calf is due to an increase in the number of muscle fibers, so that the shape of the calf is very bulky. The dam, which may be either a heterozygous carrier or completely double muscled, may also exhibit some reduced bone development (Hendricks et al., 1973).

Although one might expect the proportion of bone in a carcass to affect specific gravity measurements, this relationship may not show up in practice (Preston et al., 1974). Growth promotants apparently have little or no effect on skeletal development (Ralston et al., 1975), but environmental factors may affect bone development, since certain confinement conditions may cause lameness involving skeletal joints (Murphy et al., 1975).

The early research on bone development in beef carcasses is reviewed by Preston and Willis (1974). In the early 1950s, attempts were made to use measurements of isolated carcass bones such as the cannon bone to predict the *muscle-to-bone ratios* of carcasses. Although the method worked satisfactorily when applied to a wide range of dissimilar carcasses, it was of no practical use when applied to more uniform commercial carcasses. Muscle-to-bone ratios improve as animals grow older or fatter, since longitudinal bone growth slows down in older animals and muscles start to accumulate appreciable amounts of intramuscular fat. Animal age is the dominant factor determining muscle-to-bone ratios (Preston and Willis, 1974). However, when adjustments are made for animal age and carcass weight, considerable unexplained variation is found in muscle-to-bone ratios (Dolezal et al., 1982).

The recent emphasis on larger, leaner breeds of cattle has obscured the fact that, some years ago, the emphasis was in the other direction. The desire to produce small compact animals with bulging muscles began to favor the survival of dwarf animals with impaired longitudinal bone growth. Although mildly affected animals looked very muscular (Marlowe, 1964), severely affected animals became increasingly common and were poorly suited for beef production. Dwarfism due to the impaired longitudinal growth of bones is a recessive trait which affects males more strongly than females (Bovard and Hazel, 1963).

From research on the allometric growth of the skeleton in pigs (Doornenbal, 1975; Davies, 1975), it appears that the relationship between mature body size and the potential for lean meat production may also exist in pigs. In other words, selection for leaner pork may favor greater skeletal dimensions in the mature animal, due to later maturation.

Lameness and locomotor disorders are a particularly serious problem in pigs. Optimum dietary levels of calcium and phosphorus, and the required levels of vitamin D, vitamin A, copper, and manganese, have been investigated extensively (Pond et al., 1975). Connective tissue metabolism is affected by the levels of iron (Prockop, 1971), manganese (Leach, 1971), copper (Carnes, 1971), and zinc (Westmorland, 1971). Since bones contain collagen fibers, factors that affect collagen synthesis may also affect bone growth (Bengtsson and Hakkarainen, 1975). Although exercise levels do not produce much noticeable effect on bone growth (Murray et al., 1974), there is evidence that certain confinement conditions do affect the breaking strength of bone (Elliot and Doige, 1973). The pathogenesis of malfunctioning joints in the absence of any readily apparent infection remains a difficult problem (Hogg et al., 1975). A predisposition to such conditions as a result of selective breeding for meat production has been proposed, and pathological changes have been found in the skeletal cartilage of pigs prior to, or without evidence of lameness (Thurley, 1969).

The avian epiphyseal plate has blood vessels that penetrate deeply into the zones of cellular proliferation, whereas in mammals, the epiphyseal plate is almost devoid of such vessels. There is considerable sexual dimorphism in the longitudinal growth of bones in chickens and turkeys, and males have longer bones with a later completion of growth (Latimer, 1927; Sullivan and Al-Ubaidi, 1963). Wise (1977) found that the bones of broiler-type chickens tend to be shorter and thicker than those of layer-type birds. A relationship between mature body size and the potential for lean meat production has also been observed in poultry. Thus at equal body weights, the skeletal system may be less mature in broilers, which have been selected on the basis of meat yield, than in layers. As well as skeletal defects due to nutritional deficiencies (vitamin D, vitamin A, phosphorus, manganese, choline, biotin, nicotinic acid, folic acid, zinc, and pyridoxine), poultry are afflicted by a number of genetic skeletal defects. Asymmetrical or abnormal development of the sternum may occur in birds at market weight and is a cause for downgrading.

MICROSTRUCTURE AND DEVELOPMENT OF ADIPOSE TISSUE

Introduction

In terms of the overall balance sheet for energy in agriculture, relatively large amounts of feed energy are used when animals deposit fat in their bodies. Much of this fat is then removed and wasted after the animal is slaughtered. About 6 per cent of the live weight of a steer may be removed as fat in the abattoir, and an equal amount may be trimmed from the dressed carcass by the butcher. Adipose tissue serves its proper function only when an animal uses the energy and insulation provided by its adipose tissue to survive a period of inadequate feed intake or cold weather. Adipose tissue in meat, however, is not altogether undesirable and wasteful. It is desirable in moderation to give a "finished" appearance to a carcass and, without at least some subcutaneous fat, a carcass is judged to be unattractive by traditional standards.

Fat that is deposited within muscles (*intramuscular* adipose tissue) appears as a delicate pattern of wavy lines in the meat—hence its common name, *marbling fat*. It is traditionally maintained that marbling fat contributes to the juiciness of cooked meat because it melts away from between bundles of muscle fibers to make the meat more tender and more succulent. However, it is difficult to find much scientific evidence in support of this traditional view (Breidenstein et al., 1968), except in poultry, where subcutaneous and intermuscular fat baste the meat as it is being roasted. Some of the characteristic flavor associated with different types of meat originates from carbonyl compounds which are concentrated in the adipose tissue (Sink, 1979; Wasserman, 1979). The flavor of meat is sometimes modified by an animal's diet (Park et al., 1972). Not all the fat in the carcass is macroscopically visible. Many muscle fibers, particularly those in postural muscles, contain large numbers of microscopic fat droplets (Bullard, 1912).

The ancestors of present-day farm animals were once used to feed populations of people, many of whom were manual workers who expended large amounts of energy in tasks which we now perform by machine. Fat contains a lot of stored energy relative to lean, and the high fat content of meat once supplied much of the energy that people expended in their daily work. Now this extra energy is undesirable, and a reduction of the fat content of meat is a major goal in the continued improvement of meat animals. Most of the meat that was eaten in the nineteenth century was derived from animals older than those which are marketed now, and most of these animals had already given long service in the production of milk or wool, or had been used to pull plows or wagons. The presence of marbling fat would have greatly improved the palatability of the tough and strong-tasting meat derived from these mature animals. Also, large families were more common a hundred years ago, and large joints of meat cut from large carcasses were well suited to domestic requirements.

This following section reviews the origin, metabolism, and proliferation of adipose cells. There are a number of review articles which can be used to gain access to more detailed information on these topics (Allen et al., 1976; Leat, 1976; Garton, 1976; Evans, 1977).

White Adipose Cells

A mature adipose cell or *adipocyte* has a diameter of about 100 μm and is filled with *triglyceride* (Plate 1). Thus its nucleus and cytoplasm are restricted to a thin layer under the cell membrane. When isolated from their surroundings, adipose cells are rounded in shape, but when crowded together as they normally are in adipose depots, they are compressed with flattened sides. Pockets of very small adipose cells sometimes appear between normal-sized cells. Thus the measured mean values for cell size may be biased by the extent to which these small adipose cells are recognized or measured (Kirtland et al., 1975; Ashwell et al., 1975). When making histological measurements on frozen sections of adipose tissue, it is important to make sure that each cell has a nucleus. Large adipose cells are easily fragmented, and when a section is thawed, the cell fragments may become rounded like oil droplets. Some of these fragments may appear to be bounded by part of the original cell membrane, and they can only be distinguished from real, small adipose cells by their lack of a nucleus.

Mature adipose cells have very little cytoplasm and contain few organelles. The Golgi complex is small, there are only a few ribosomes and mitochondria, and the endoplasmic reticulum is sparse. The large triglyceride droplet which fills the bulk of each cell is not directly bounded by a membrane, although it may be restrained in position by a delicate meshwork of very thin filaments approximately 10 nm in diameter. These filaments are most conspicuous in the adipose cells of poultry (Wood, 1967).

In many locations in the body, large numbers of adipose cells are grouped together to form adipose depots. Adipose cells are kept in place by a meshwork of fine *reticular* fibers (Motta, 1975; Swatland, 1975). Large adipose depots are usually subdivided into layers or lobules by partitions or *septa* of fibrous connective tissue. In the layered subcutaneous fat of a pork carcass, the septa may follow the body contours and may create a weak boundary layer echo in the ultrasonic estimation of fat depth. Adipose depots are well supplied by blood capillaries (Gersh and Still, 1945).

Brown Adipose Cells

Brown adipose cells may occur in cold-adapted, hibernating, or newborn mammals. Apart from their brown color, which is due to greater vascularity and a high cytochrome concentration, brown adipose cells have a different appearance microscopically. Brown adipose cells are small with abundant cytoplasm and *mitochondria*, and their triglyceride is subdivided into a number of small droplets (Figure 2-31). At the ultrastructural level, the mitochondria of brown adipose cells may have a distinct appearance resulting from parallel arrays of long *cristae* (internal partitions), most of which extend across the full width of the mitochondrion. Brown adipose cells generate heat by a process called *nonshivering thermogenesis*. Instead of releasing fatty acids into the bloodstream as do white adipose cells, brown adipose cells oxidize their own fatty acids to release heat instead of synthesizing adenosine triphosphate (ATP). Cold animals may also generate heat by shivering, using their muscles to convert chemical energy to heat. When animals with brown fat are placed in a thermoneutral environment, nonshivering thermogenesis is increased by the intravenous infusion of *norepinephrine* (noradrenaline).

Unfortunately, the study of brown adipose tissue in farm mammals is complicated by the fact that, relative to some other mammalian species, the distinctions

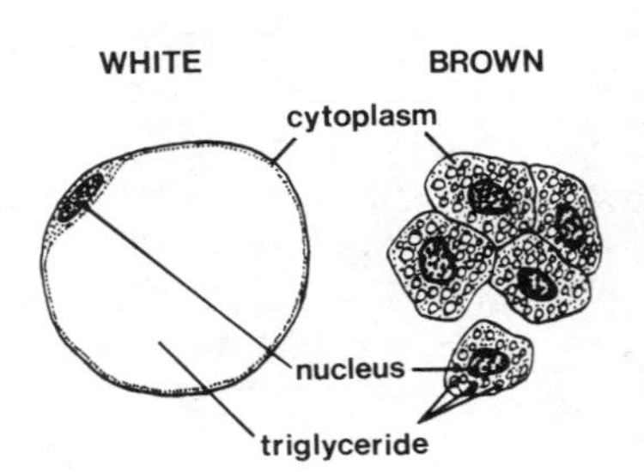

Figure 2-31 Comparison of an isolated white adipose cell with some isolated brown adipose cells.

between brown adipose cells and immature cells of ordinary white adipose tissue are seldom distinct. During their development, white adipose cells pass through a *multilocular* stage with many small triglyceride droplets before they reach their final *unilocular* stage of development with one large triglyceride droplet. Thus it is often difficult to decide on morphological grounds alone whether a multilocular adipose cell in a neonatal animal is a brown adipose cell or an immature white adipose cell. In newborn farm animals, it is possible that immature white adipose cells exhibit some of the heat-generating properties of brown adipose cells. In hamsters, multilocular brown adipose cells may develop either from unilocular adipose cells or by proliferation from endothelial cells (Nechad and Barnard, 1979).

In lambs (Thompson and Jenkinson, 1969; Gemmel et al., 1972), calves (Alexander et al., 1975), and kids (Thompson and Jenkinson, 1970), brown adipose cells have been identified on the basis of their mitochondrial morphology and norepinephrine response, although other expected features, such as brown color and a multilocular condition, were indistinct. In calves, Alexander et al. (1975) found that brown fat formed approximately 2 per cent of the body weight, but none of the adipose tissue located subcutaneously exhibited the properties of brown adipose tissue. In farm mammals, any brown adipose tissue that is present at birth to warm the newborn animal is seldom maintained for very long. Dauncey et al. (1981) identified brown adipose tissue in neonatal pigs and found that small amounts persisted until at least 3 weeks after birth. Where brown fat does occur in newborn farm animals, it is probably converted to ordinary white adipose tissue later in development. The transition from brown to white adipose cells has been observed among cells in culture (Dyer and Pirie, 1978). However, when white adipose cells lose their triglyceride, they do not revert back to being brown adipose cells (Cox et al., 1978).

Origin of Adipose Cells

Speculation concerning the origin of adipose cells has continued since the end of the nineteenth century, by which time most of the feasible possibilities had already been proposed (Clark and Clark, 1940). In certain locations in the body of a fetus, where adipose tissue will develop later in life, *mesenchyme* cells congregate in lobules which resemble glandular tissue. In an embryo, the mesenchyme is a diffuse tissue composed of cells which may differentiate to form the connective tissues of the body, as well as blood and lymphatic vessels. The mesenchyme cells in glandlike lobules begin to accumulate small droplets of triglyceride. The droplets coalesce as they are crowded together and, finally, they form a single large mass in the center of each cell. In other locations where adipose tissue is about to develop, the adipose cell precursors may resemble fibroblasts. In this case, cells are spindle shaped with an oval nucleus, and the cytoplasm contains rough endoplasmic reticulum, microtubules, microfilaments, and spherical mitochondria (Slavin, 1979). In pigs, presumptive adipose cells contain one to three prominent nucleoli in their otherwise pale nuclei (Hausman and Martin, 1981).

It is difficult to prove that the glandlike cells and fibroblastlike cells that give rise to adipose cells are distinct types of cells with a rigid pattern of development. However, the possibility that cells are rigidly programmed to develop exclusively into adipose cells is supported by the behavior of transplanted cells. Cells that have been transplanted from precursor adipose lobules to future low-fat regions will store triglyceride and become adipose cells, even in an inappropriate location (Le Gros Clark, 1945). Similarly, precursor adipose cells from certain sites will continue their programmed development into adipose cells, even if they are removed from the body and cultured in vitro (Van and Roncari, 1977, 1978; Poznanski et al., 1973; Van et al., 1976; Dardick et al., 1976). In neonatal pigs, there are four types of cells, any or all of which could be adipose cell precursors (Mersmann et al., 1975).

An alternative origin of adipose cells is thought to be by the recruitment of fibroblasts. Fibroblasts can increase in number by mitosis. When recruited to become adipose cells, they give up the shape and activities of a fibroblast, and pass through a multilocular stage of triglyceride accumulation, finally to become adipose cells resembling those derived from precursor cells (Clark and Clark, 1940). Adipose cells sometimes originate from the endothelial cells of the vascular system (Tavassoli, 1976). Although extra adipose cells may be recruited when animals become obese, the extra cells are retained when an animal returns to its normal level of fatness (Faust et al., 1978). Macrophages have also been suggested as a source of new adipose cells (McCullough, 1944; Smith et al., 1952).

Adipose Tissue Distribution

Adipose depots range in size from small groups of adipose cells located between muscle fiber bundles, to the vast numbers of adipose cells that are located subcutaneously and viscerally. It is important to distinguish between *anatomical sites* and *systemic locations*. Specific muscles or regions of the carcass are anatomical sites. Intermuscular, intramuscular, visceral, and subcutaneous are systemic locations. For example, fat from a specified anatomical site such as the shoulder can be separated into different systemic depots (subcutaneous, intermuscular, and intramuscular). The distinction between anatomical sites and systemic locations is important commercially. For example, bovine intramuscular marbling fat sometimes first becomes noticeable in rump and loin muscles, where it adds to their value. In the same systemic location, but at a different anatomical site, such as the brisket, marbling fat confers no economic advantage and is wasteful. In cattle, the relative growth of subcutaneous fat is similar in both the forequarter and the hindquarter. However, the relative growth of intermuscular fat is higher in the forequarter than in the hindquarter (Berg et al., 1979). There is no guarantee that systemic fat depots are homogeneous, even at a single anatomical site. For example, the backfat seen on pork carcasses is subdivided into three layers which differ in their composition and pattern of growth (Moody and Zobrisky, 1966).

The systemic deposition of fat in a carcass influences the commercial indices of carcass composition, such as the dressing percentage.

$$\text{Dressing percentage} = \frac{\text{carcass weight}}{\text{live weight}} \times 100$$

Thus most of the fat that is deposited around the viscera is removed with the viscera at slaughter, and this reduces the dressing percentage. Fat that is deposited between or within carcass muscles increases the dressing percentage.

In cattle, it was traditionally maintained that fat deposition followed three systemic phases (Hammond et al., 1971). In the first phase, fat was thought to be deposited around the viscera and kidneys, and within the caul and mesenteries. *Caul* fat is a thin sheet of adipose tissue that is contained in a large fold of connective tissue over the stomach and adjacent organs. In the second phase, fat was thought to be deposited subcutaneously and intermuscularly. The third phase was thought to be the deposition of marbling fat within muscles. Sometimes, however, no simple chronological separation of the three phases of fat deposition can be detected, and the relative amounts of fat in the main systemic locations may remain constant (Berg and Butterfield, 1976). On a high-energy ration, cattle deposit subcutaneous fat at a greater rate than they deposit intermuscular fat. However, this difference does not appear when animals are on a low-energy ration (Fortin et al., 1981). Breeds of cattle differ in the way they develop their systemic fat depots. For example, Herefords produce more subcutaneous fat and less perirenal and pelvic fat than Angus, Friesian, and Charolais crossbred cattle (Charles and Johnson, 1976). However, when adipose growth at different anatomical sites is examined, relative to total fat, only minor differences may be found between breeds (Berg et al., 1978).

In pigs, subcutaneous fat grows at the same rate as total body fat, intermuscular fat grows more slowly, and visceral fat grows faster (Kempster and Evans, 1979). The separation of phases in adipose deposition is complicated by the fact that, whereas the experimenter usually works in terms of calendar days and weeks, the experimental animal is following its own physiological calendar. The animal's physiological calendar is based on events that mark the progress through its life cycle (Chapter 8). For example, skeletal and reproductive development follow an orderly sequence of events, and different breeds may progress through this sequence at different rates. Measured in calendar weeks and months, early-maturing breeds deposit noticeable amounts of marbling fat before late-maturing breeds. Thus the introduction of late-maturing breeds with a large adult size may be used to delay fat deposition and to enhance lean growth in cattle populations.

Sex hormones can produce large and economically important differences in the overall fatness of beef, mutton, and pork carcasses. Provided that comparisons are made at equal fatness, however, the distribution of fat within beef carcasses may be similar for both sexes (Berg et al., 1979). Bulls, rams, and boars generally produce leaner carcasses than do steers, wethers, and barrows, respectively. Steers and wethers generally produce leaner carcasses than do cows and ewes, respectively. In pigs the situation is reversed, and gilts generally produce leaner carcasses than do barrows. The deposition of fat tends to occur at a lighter weight in heifers than in steers, and at a lighter weight in steers than in bulls.

Adipose Tissue Metabolism

Although adipose tissue depots are primarily for energy storage or for mechanical and thermal insulation, they are far from passive. Anabolic processes are very active when an animal is depositing fat, while catabolic processes are very active when an animal is living off its energy reserves, either during prolonged starvation or between periods of feeding. The metabolism of adipose cells is altered in cows after calving, due to the mobilization of adipose reserves for milk production: lipogenesis is reduced and lipolysis is increased (Pike and Roberts, 1980). *Metabolic water* can be produced by

TABLE 2-4. STRUCTURE OF TRIGLYCERIDE (TRIACYLGLYCEROL)

```
G—Fatty acid
L
Y
C—Fatty acid
E
R
O
L—Fatty acid

      H     O
      |     ||
  H—C—O—C—R1
      |     O
      |     ||
  H—C—O—C—R2
      |     O
      |     ||
  H—C—O—C—R3
      |
      H
```

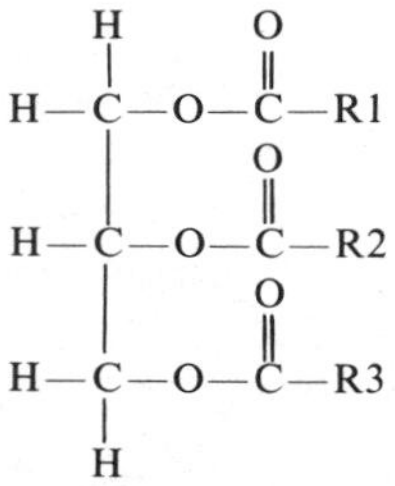

oxidation of fat when animals are deprived of water. Thus the depth of backfat and the total weight of subcutaneous fat may be reduced in pigs on a restricted water intake (Skipitaris, 1981).

Energy is stored in molecules of *triglyceride* (Tables 2-4 and 2-5). The long-chain *fatty acids* found in animal triglycerides (R1, R2, and R3 in Table 2-4) vary in length, as shown by the values of n in the general formula for fatty acids. The value for n in animal fat usually ranges from 5 to 20. These fatty acids are insoluble in water, like the triglycerides that are formed from them. Adjacent carbon atoms along a chain may be linked together with an unsaturated double bond. Unsaturated bonds produce a bend in the chain and this lowers the melting point. Unless complicated by the presence of unsaturated bonds, melting points are proportional to chain length.

The degree of saturation of carcass fat is affected by a number of factors. Within a breed, steers may have more saturated fat than heifers (Terrell et al., 1969). About

TABLE 2-5. STRUCTURE OF FATTY ACIDS

General structure:

$$CH_3—(CH_2)_n—COOH$$

Names for values of n:

n	Name
2	butyric
4	caproic
6	caprylic
8	capric
10	lauric
12	myristic
14	palmitic
16	stearic
18	arachidic
20	behenic

12 months after birth, cattle normally start to fatten rapidly, and progressively more unsaturated fatty acids are deposited (Leat, 1975, 1977). Genetically obese pigs have a greater degree of saturation of their fatty acids than do normal pigs (Scott et al., 1981).

Most plant oils contain a high proportion of unsaturated fatty acids with low melting points, but animals have a high proportion of saturated fatty acids and their fat is solid or semisolid at room temperature. Thus when a beef carcass is cooled from body temperature down to nearly 0° C after slaughter, subcutaneous fat passes from a liquid to a solid state. In beef carcasses, a greater proportion of fatty acids may be saturated in the summer than in the winter (Link et al., 1970). Seasonal changes of this type are quite marked in wild sheep, where they may help the animal to adapt to climatic changes (Turner, 1979). Fat tends to be soft and yellow in ram carcasses relative to wethers, and fat also tends to be soft and yellow in lambs fed a high-energy diet relative to those on a low-energy diet (Busboom et al., 1981). In general, triglycerides, which are located subcutaneously where they may be relatively cool even in the live animal, have a lower melting point. Conversely, perirenal or suet fat in the beef carcass is brittle at room temperature since it comes from a warm place in the body and thus has a high melting point. In the abattoir, shrouds are often pinned over beef carcasses so that the molten subcutaneous fat solidifies with a smooth surface. In animal fat, most fatty acids contain an even number of carbon atoms, and unsaturated bonds are either absent or few in number. In meat, the most common unsaturated fatty acids are *oleic*, *linoleic*, and *linolenic acids* with 18 carbons and with one, two, and three double bonds, respectively.

Systems Involved in Triglyceride Deposition

The integration of adipose metabolism with that of the rest of the body is rather complicated. At a greatly simplified level, the major systems involved are shown in Figure 2-32. At the top of the diagram is an adipose cell, complete with a diagrammatic

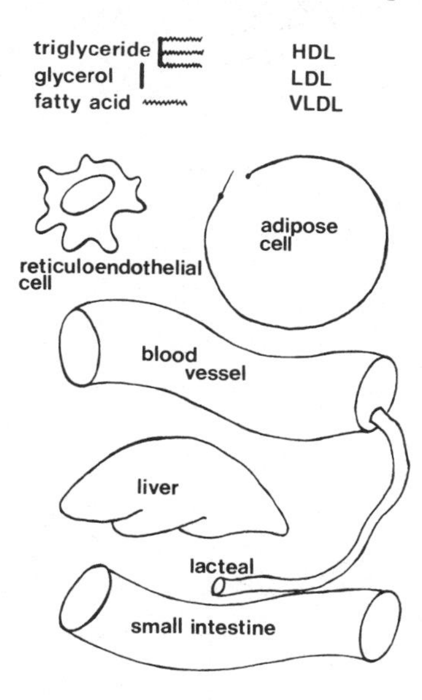

Figure 2-32 Main components involved in triglyceride deposition.

trapdoor to represent the countless membrane sites by which glucose gains access to the cell. In the diagram, two short lengths of tube represent the blood vascular system and the small intestine. The liver is also shown diagrammatically. Figure 2-32 also contains a diagrammatic representation of a *reticuloendothelial* cell. The reticuloendothelial system is formed from a heterogeneous collection of individual cells which are widely dispersed through the body. Although the cells of the reticuloendothelial system may differ in appearance and location, they share certain common properties, many of which are related to defense against infectious microorganisms. Reticuloendothelial cells are involved in adipose tissue metabolism. Figure 2-32 contains some letter "E" symbols that represent glycerol, fatty acids, and triglycerides.

Dietary Intake

In the case of a pig which has some triglyceride in its diet, the triglyceride reaches the small intestine and is hydrolyzed in the presence of *pancreatic lipase* (Figure 2-33A). *Colipase*, a protein secreted by the pancreas, binds to the surface of fat droplets in the presence of *bile salts* from the liver and facilitates the attachment of lipase (Patton and Carey, 1979). The partially or completely separated fatty acids may pursue a number of different pathways, one of which involves absorption by an intestinal cell and reassembly into a new triglyceride inside the cell. Large numbers of reassembled triglyceride molecules may be wrapped in protein and phospholipid to form a small globule about 1 μm in diameter called a *chylomicron*. The triglycerides contained in the chylomicra then pass to the *lacteals* (Figure 2-33B). Lacteals are blind-ending lymphatic capillaries in the axes of the intestinal villi. The lymphatic system conducts chylomicra to the venous system where, after passing into the general circulation, they are made available to adipose cells (Figure 2-33C). However, chylomicra do not enter directly into adipose cells. Their triglyceride is first hydrolyzed to fatty acids and glycerol by *lipoprotein lipase* (clearing factor lipase), which is located on the surfaces of adipose cells or capillary cells (Figure 2-33D). The transfer of the fatty acids to the adipose cell is facilitated by the continuity of intercellular connections with intracellular channels (Blanchette-Mackie and Scow, 1981). Yet again, triglycerides

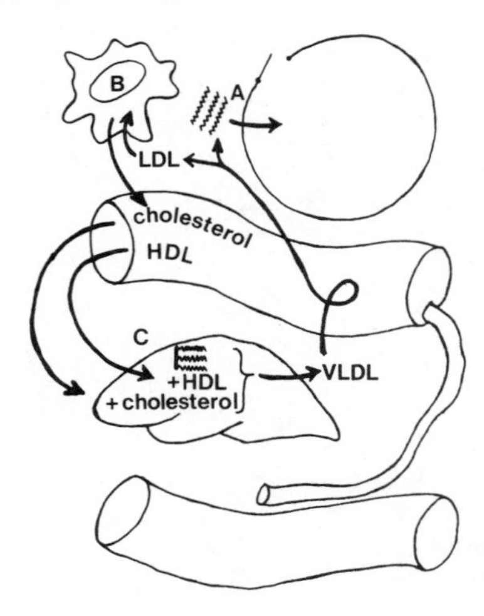

Figure 2-33 Triglyceride deposition from chylomicrons. Events at points A to G are described in the text.

are now reassembled, this time within the adipose cell (Figure 2-33E). Lipoprotein lipase activity in adipose tissue is linearly related to cell size. In young obese rats, lipoprotein lipase activity is deficient in red muscle fibers (Hartman, 1980).

Most of the glycerol that is released by lipoprotein lipase outside the adipose cell returns to the bloodstream (Figure 2-33F). Inside the adipose cell, a new carbon backbone for the triglyceride is formed from *glycerophosphate* which has been derived from *glucose* (Figure 2-33G). The entry of glucose into a cell is facilitated by *insulin.* By a simple route, such as that just outlined, ingested fatty acids may be deposited in an unaltered form in their new "host." Thus, in pigs, the degree of saturation of carcass fat is influenced by the nature of the fatty acids that are ingested. This may affect meat quality since fat with a low melting point appears greasy. Fat with a relatively large number of unsaturated bonds is also more likely to undergo autooxidation, with the formation of unpleasant odors (Lawrie, 1974). Commercial problems often arise when highly unsaturated fats such as linseed oil or fish oil are fed to monogastric animals, since these oils may cause unpleasant odors in the meat. Unpleasant odors may not become apparent until the meat has been cooked and its unsaturated fatty acids have been oxidized (Reineccius, 1979). Such problems are rare in ruminants since unsaturated fats are reduced in the rumen. However, prior treatment of the feed with formaldehyde inhibits this reduction and enables unsaturated fatty acids to pass through the digestive system relatively unchanged (Cook et al., 1970). In cattle and sheep, dietary lipids are extensively modified in the rumen. Esterified fatty acids are released by hydrolysis, the residual glycerol and galactose are fermented, and unsaturated fatty acids are hydrogenated (Garton, 1967). Large amounts of stearic acid are formed from oleic, linoleic, and linolenic acids. These, together with other fatty acids derived from the microflora, are absorbed in the intestine.

Alternative Inputs and Lipogenesis

An alternative carbon input is available to adipose cells from *very low-density lipoproteins* (VLDL) from the liver (Figure 2-34). In VLDL, a central core of triglyceride is packaged by a layer of protein, phospholipid, and cholesterol. The

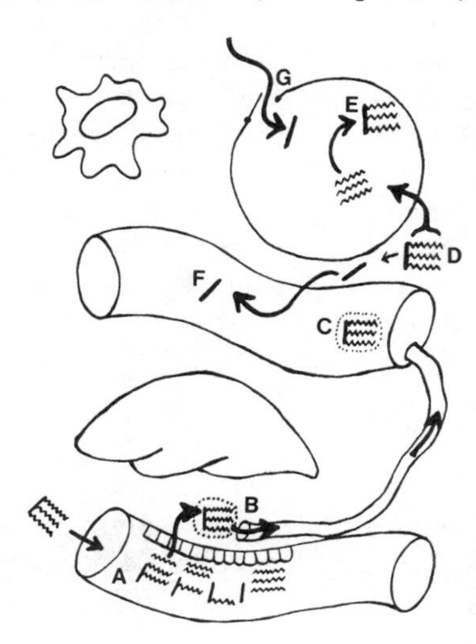

Figure 2-34 Triglyceride deposition from very low-density lipoproteins (VLDL). Events at points A to C are described in the text.

leaching of triglyceride by lipoprotein lipase near adipose cells (Figure 2-34A) leads to a reorganization of VLDL components, and a *low-density lipoprotein* (LDL) is produced. The cholesterol that remains in LDL is released into the bloodstream by cells of the reticuloendothelial system (Figure 2-34B). In the liver, cholesterol, triglyceride, and circulating *high-density lipoproteins* (HDL) are assembled into new VLDL (Figure 2-34C).

Meat animals may form new fat by *lipogenesis*, as shown in the classical experiments of Lawes and Gilbert (Hall, 1905). In pigs, most of the new triglyceride is synthesized from glucose, and the fatty acids that are produced are dominated by stearic acid and its desaturated form, oleic acid. In ruminants, *acetate* is the main substrate, and stearic acid is the dominant fatty acid. In poultry, lipogenesis occurs in the liver rather than in adipose cells, and triglyceride reaches the adipose cells as VLDL in the blood (Figure 2-34). Oleic and palmitic acids are dominant when the dietary intake of triglyceride is low (Evans, 1977).

Adipose Cell Hyperplasia and Hypertrophy

The genetic and nutritional factors that regulate the amount and distribution of adipose tissue can be described in terms of cellular *hyperplasia* (an increase in cell numbers) and cellular *hypertrophy* (an increase in cell size). An early stimulus to research on these topics was provided by the hopeful hypothesis that adipose cell numbers might be genetically regulated, whereas adipose cell size might be nutritionally regulated. Had this been the case in meat animals, selection for animals with low numbers of adipose cells might have provided a method for enhancing the production of lean meat. It was also suggested that a low plane of nutrition early in life might reduce the numbers of fat cells, so that animals would be less likely to deposit fat in the final stages of growth to market weight. Once adipose cells start to store triglyceride, they are no longer capable of mitosis (Desnoyers et al., 1980), and *insulin* causes increased adiposity by increasing the size rather than the number of adipose cells (Salans et al., 1972). It has also been proposed that adipose cell size is involved in the regulation of feed intake and triglyceride deposition (Faust et al., 1977).

The basic cellular problem that has daunted these early hopes is that populations of morphologically identical adipose cells in adult animals appear to be formed from an initial population of specific precursor cells which is variably supplemented by the recruitment of fibroblastlike cells. Thus the apparently limitless possibilities for the recruitment of extra adipose cells may allow compensatory growth to offset any reduction in the initial population of specific precursor cells. In double-muscled cattle, adipose cells are greatly reduced in number (cellular *hypoplasia*). Since double muscling is a genetic condition, there could be a simple genetic mechanism that regulates adipose cell numbers in these animals. Adipose cell number and size can be modified by the selective breeding of mice for either high or low postweaning growth rate (Martin et al., 1979).

Adipose Cellularity in Pigs

Newborn pigs have very little fat relative to other mammals. Between birth and 4 weeks of age, the percentage of fat increases from about 1 per cent to about 18 per cent of empty live weight (Manners and McCrea, 1963). Adipose tissue lobules in regions that will later deposit large amounts of fat contain few cells, relative to neonatal ruminants, and they are separated by areas of loose mesenchyme or undifferentiated tissue. Triglyceride deposition in adipose cells first becomes microscopically detectable at about the third month of gestation. After this time and for the remainder of gestation, the numbers of cells in each lobule show no great increase, although the numbers of lobules and the total numbers of adipose cells do increase (Hermans, 1973a). In pigs, fibroblastlike cells and glandlike cells are not the only source of adipose cells, since perirenal adipose cells may originate from endothelial cells of the vascular system (Desnoyers and Vodovar, 1977). Fat is deposited very rapidly around the kidneys, and perirenal fat has high enzyme activity, large adipose cells, and a low connective fiber content (Anderson et al., 1972). The activities of lipoprotein lipase and hormone-sensitive lipase increase markedly after birth (Steffen et al., 1978), and the pig then starts its lifelong process of triglyceride accumulation. If a pregnant sow is diabetic, its offspring may have an increased amount of body fat (Ezekwe and Martin, 1980).

The heritability of adipose deposition is quite high in pigs, typically about 50 per cent for backfat thickness. Pigs readily respond to selection for increased adipose deposition in medical experiments on obesity, and for decreased adipose deposition in commercial agriculture. Lean breeds of pigs might have a greater ability to mobilize their fat. This might, perhaps, be due to a greater release of norepinephrine from their sympathetic nerve endings (Gregory and Lister, 1981). Metz and Dekker (1981), however, have shown that genetic differences in the rate of fat deposition may occur without marked differences in fat mobilization.

Anderson and Kauffman (1973) found that, up to 5 months after birth, subcutaneous adipose tissue grew by a combination of hyperplasia and hypertrophy. Growth after this time was almost completely due to adipose cell hypertrophy. This was confirmed by Hood and Allen (1977) for both perirenal and extramuscular (basically subcutaneous) adipose tissue. Measurements of adipose cell numbers and diameters in this type of research are usually made on cells that have been released from their connective tissue framework. In histological sections, it is difficult to measure adipose cell diameters since most cells are cut off-center and have reduced cross-sectional areas. Adipose cells are easily fragmented, and the extent of fragmentation may vary with experimental conditions. Fragments may round up and form small globules, and without cytological examination these are difficult to distinguish from small adipose cells. Similarly, cells have to be captured once they have been released from their connective tissue framework, and different techniques vary in their efficiency.

Bimodality in frequency histograms of adipose cell diameters may suggest that there are two populations of cells: large ones and small ones. The large ones might be mature adipose cells filled to capacity with triglyceride, and the small ones might be young adipose cells in the process of filling up. On the basis of these possibilities, bimodality is often regarded as evidence for the formation of new adipose cells. Although this may be a correct assumption in many cases, other possibilities cannot be excluded automatically. A unimodal population might be continuously supplemented by new cells which are too few in number at any time to form their own peak in the frequency distribution of cell size. Over the animal's lifetime, however, this might be a major source of cells. Similarly, a bimodal population might be static, and the populations of large and small adipose cells might be related to an unrecognized factor.

For example, adipose cells near blood vessels might be larger than distant cells, or vice versa.

When Anderson and Kauffman (1973) measured the diameters of adipose cells from the back fat of Chester White barrows, they found unimodal populations in the frequency distribution of different-sized cells. When Mersmann et al. (1973) attempted the same measurements with cells isolated from the neck region of crossbred pigs, they found that cell diameters formed bimodal populations. Wood et al. (1975) found that the rate of adipose cell hypertrophy in subcutaneous shoulder fat depended on the depth at which cells were taken from the fat. Carpenter et al. (1961) found considerable variation in adipose distribution along the length of the longissimus dorsi muscle. Enser et al. (1976) found differences in growth patterns between shoulder and midback adipose depots; hyperplasia persisted for longer in the shoulder depot, and new cells appeared to originate from near the muscle surface. Subcutaneous fat from the shoulder region of Large White pigs exhibits both hypertrophy and hyperplasia to at least 188 days of age (Wood et al., 1978). Etherton (1980) reported bimodal distributions for adipose cell diameters as late as 1 year of age in both lean and obese breeds. Histological differences exist between breeds, as in the case of Large Whites and Pietrains (Moody et al., 1978).

Kirtland and Gurr (1980) used a thymidine label to assess the extent of DNA synthesis associated with the formation of new cells in fat depots. Formation of new cells proceeded rapidly between 2 and 40 days of age, but beyond this time the growth of backfat layers was due primarily to the filling of preexisting empty adipose cells.

Despite this confusion over the relative contributions of hyperplasia and hypertrophy, the case for the genetic regulation of adipose cell numbers and diameters is quite strong. Steele et al. (1974) measured adipose cell numbers and diameters in strains of the Duroc breed which had either a high or a low backfat thickness, and they found considerable differences, particularly in the contribution from hyperplasia. Breeds and strains of pigs differing in backfat thickness may also exhibit hormonal differences in fat mobilization (Standal et al., 1973; Steele and Frobish, 1976; Wood et al., 1977), but whether these differences can be meaningfully related to differences in cell numbers and diameters is a difficult question. What does it mean if the concentration of lipolytic enzymes is high when expressed on the basis of tissue weight? In rats, for example, the volume of cytoplasm per adipose cell decreases from 8% to 0.8% during cellular hypertrophy (Goldrick, 1967). Adipose cell cytoplasm is restricted to a thin layer beneath the plasma membrane and, as an adipose cell grows in size, the plasma membrane increases in area at a slower rate than the volume of the cell. Thus, if geometrical proportionality is maintained, cytoplasmic components will decline in their overall concentration in the tissue as cells grow in size.

The development of intramuscular adipose tissue in pork may depend on muscle structure and fascicular arrangement. Kauffman and Safanie (1967) proposed that muscles with long parallel fasciculi may deposit intramuscular fat more readily than pennate muscles with short fasciculi inserted at an angle to their tendon. The deposition of intramuscular fat is later than in subcutaneous, visceral, and intermuscular locations, and it does not occur to any marked extent until 16 weeks in trapezius and semitendinosus muscles of Duroc and Hampshire pigs (Lee and Kauffman, 1974a). Thus growth from this time onward involves the formation of new adipose cells in these locations. The small mean diameters typically found among intramuscular adipose cells (Lee and Kauffman, 1974b) may be due to recruitment of new small-diameter cells while older cells continue to grow by hypertrophy. Lee et al. (1973a, 1973b) examined the effect of an early low plane of nutrition on adipose tissue development, and they compared animals at a standard age and at a standard weight. At 24 weeks, the reduced amount of intramuscular fat in pigs on a low plane of nutrition was because of a decrease in cell size and cell number. Reduction in subcutaneous fat, however, was due to a decrease in cell size but not in cell number. On a weight-constant basis (80 kg), an early low plane of nutrition had little effect on subcutaneous adipose tissue growth, but there were reductions in adipose cell numbers and diameters intramuscularly.

Adipose Cellularity in Cattle

An extensive study of adipose tissue in bovine fetuses was undertaken by Bell in 1909; although his work is dated technically, many of his observations and conclusions are still valid. The morphological distinction between mesenchyme cells and adipose precursor cells was then, and still is, a rather subjective separation. Bell (1909) identified the first adipose precursor cells around the kidneys of 4.7-cm fetuses (approximately 50 days gestation). By 30-cm length (approximately 125 days gestation), masses of adipose cells were found in close association with blood vessels. By this time, adipose cells had also developed in sternal intermuscular and subcutaneous depots. An interesting point observed by Bell (1909) was that, in different anatomical sites, there was a considerable variation in the delay between the time of appearance of precursor cells and the time of triglyceride deposition. Perirenal adipose precursor cells showed a long delay, whereas sternal cells showed a short delay. Bell (1909) sketchily followed adipose tissue development into the postnatal period, and he found that thin cattle had smaller subcutaneous adipose cells than did fat cattle. He concluded that an increase in adipose cell numbers was partly responsible for the postnatal accumulation of carcass fat. Much the same view still prevails (Allen, 1976).

In cattle, as in pigs, visceral adipose tissue develops before intramuscular adipose tissue. Adipose precursor cells make their greatest contribution to cell numbers viscerally, while the recruitment of fibroblastlike cells is more important intramuscularly. Hood and Allen (1973) found that hyperplasia was nearly complete in subcutaneous and perirenal depots by approximately 8 months. Intramuscularly, progressive cellular recruitment, as revealed by bimodality of cell diameters, still occurred at 14 months. Hood and Allen (1973) also found smaller-diameter subcutaneous adipose cells in lean Holstein cattle relative to well-finished Hereford × Angus crossbreds. The subjective assessment of marbling fat is very important in the commercial grading of beef carcasses. However, it is difficult to relate the histological frequency and distribution of intramuscular adipose cells to marbling scores (Cooper et al., 1968; Moody and Cassens, 1968). Lipoprotein lipase distribution in bovine adipose cells is described by Plaas et al. (1978).

Adipose Cellularity in Sheep

In fetal lambs, Wensvoort (1967) found adipose cell precursors in 5- to 8-cm fetuses (approximately 44 to 52 days gestation). The initial storage of triglyceride occurred at 9 cm (approximately 55 days gestation). According to Wensvoort (1968), adipose tissue formation in fetal lambs differs from that in fetal calves. In fetal lambs, two extra types of cells are involved, pleoprotoplasmic and plurivacuolar cells. Pleoprotoplasmic cells are large polygonal cells which have central nuclei and abundant granulated cytoplasm without lipid droplets. Possibly, a morphological overlap between the developmental stages of white and brown adipose cells may account for this. Lambs have little prenatal development of subcutaneous fat, although it develops rapidly immediately after birth (Vezinhet and Prud'hon, 1975).

Burton et al. (1974) found that the amount of perirenal fat increased between 50 and 70 kg live weight in Suffolk ewes, and that this was due to adipose cell hypertrophy without evidence of new cells being formed. Nougues and Vezinhet (1977) found the same thing in the perirenal fat of Merino sheep. Haugebak et al. (1974) examined the effect of maintenance and ad libitum early diets combined with either of two protein levels in a finishing diet. In general, adipose cell hypertrophy was found in all conditions that allowed fat deposition. An interesting exception was that sternal intermuscular adipose cells reached their final diameters at a much earlier time than in the rest of the carcass. In well-fed animals, hyperplasia was complete by the time that the finishing diet was fed, whereas an early low plane of nutrition prolonged hyperplasia into the finishing period. Thus, in the growth of adipose tissue, it appears that the prolongation of hyperplasia may contribute to the mechanism of compensatory growth or catch-up growth. However, comparable studies on compensatory growth mechanisms by Burton et al. (1974) yielded no statistically valid evidence of changes in cell sizes and numbers. Nougues and Vezinhet (1977) found that hypertrophy coupled with hyperplasia persisted to 100 days in intermuscular and subcutaneous fat of Merino sheep. After 100 days, adipose tissue growth was solely due to hypertrophy. *Somatotropic* hormone has a strong effect on adipose accumulation and distribution in sheep. Hypophysectomy causes an increase in fat deposition, and this can be reversed by administering STH (Vezinhet et al., 1974).

Adipose Cellularity in Poultry

Avian adipose cells differ from those of farm mammals since they have only a limited capacity for lipogenesis. Thus they rely mainly on the capture of circulating lipids which have been synthesized in the liver or released by digestion in the gut. Brown fat is absent or at least difficult to locate in birds (Johnston, 1971). By day 11 or 12 of incubation, precise anatomical sites of future adipose tissue can be identified in the chick embryo (Liebelt and Eastlick, 1952). The distribution of subcutaneous fat depots in the chick embryo is described by Liebelt and Eastlick (1954). From the few available studies of adipose cellularity in poultry, Evans (1977) concluded that hyperplasia continued after hatching until the onset of sexual maturity. Beyond this time, adipose growth was due to hypertrophy. In ducks, Evans (1977) found a ninefold increase in adipose cell volume between hatching and 8 weeks of age.

There are marked differences in adipose cell numbers and diameters between layer-type and broiler-type fowl, and the lean broiler-type birds have fewer and smaller cells (March and Hansen, 1977). These authors also found that an early restriction of nutrient intake, although it inhibited hypertrophy, had only a slight effect on adipose cell hyperplasia. Thus adequate cell numbers for fat deposition were present in birds once they were returned to an ad libitum diet.

Boar Taint

Pork from boar carcasses sometimes yields an unpleasant odor called *boar taint*. However, this is detectable by only a small percentage of consumers (Walstra, 1974). The odor is most often detected in meat from old boars which have been used for breeding, and it need not be a problem in boars that have been slaughtered at a relatively light weight. This enables pork producers in some countries to take advantage of the rapid growth of young males that have not been castrated. Boar taint becomes more noticeable when pork is cooked or when carcass fat is tested with a hot iron, and it is due to the concentration of sex steroids in the fat (Patterson, 1968; Beery et al., 1969). Patterson (1968) identified the major factor as 5α-androst-16-ene-3-one. This *androstenone* is produced in the testes. It passes into the bloodstream and is accumulated by adipose tissue and by parotid and submaxillary salivary glands (Gower, 1972). Androstenone is transmitted in the boar's breath or saliva to the sow during mating. It acts as a chemical messenger or *pheromone* (Sink, 1967). The enthusiasm with which sows root out truffles from deep in the ground may be due to the fact that these underground mushrooms also produce androstenone (Claus et al., 1981). Jonsson and Wismer-Pedersen (1974) found that the intensity of boar taint was a heritable trait ($h^2 = 0.54$). Claus (1975) showed that boar taint could be suppressed immunologically. *Skatole* and *indole* (which are produced from the amino acid tryptophan in the gut) may also make a small contribution to boar taint (Hansson et al., 1980).

Abnormal Development of Adipose Tissue

Replacement of muscle fibers by adipose cells is a common end result of a number of pathological conditions that affect skeletal muscle. In most of these conditions there is usually some evidence of muscle fiber regeneration. In meat animals, however, it is not uncommon to find that muscle fibers have been replaced by adipose cells without any evidence of muscle fiber regeneration and with no decrease in overall muscle volume. The most appropriate name for this condition is *muscular steatosis* (Hadlow, 1962).

Muscular steatosis is most frequent in cattle and pigs, but it can also occur in sheep (Hartman and Shorland, 1957). Sometimes the occurrence of muscular steatosis is indicated before slaughter by an abnormal gait (Leipold et al., 1973a), but usually the condition is not found until a carcass is butchered. The dividing line between excessive marbling fat and muscular steatosis is sometimes difficult to establish, and it is often only the restriction of muscular steatosis to a single muscle or muscle group in an otherwise poorly marbled carcass that makes it conspicuous. Muscular steatosis often occurs in conditions which suggest that it has been caused by muscle damage or denervation. Strenuous muscle exertion can cause extensive muscle damage, particularly in those muscles that are used when an animal rears up on its hind legs. Link et al. (1967) found that muscular steatosis sometimes occurred in muscles from which biopsy samples had been taken. Naturally occurring muscular steatosis has also been linked to vascular abnormalities in muscle (MacKenzie, 1912; Nowicki and Zajac, 1964). Another possibility is that muscular steatosis is the end result of a lipid accumulation myopathy, similar to those which occur in humans (Harriman and Reed, 1972; Johnson et al., 1973). In pigs, there is evidence that the onset of muscular steatosis is accompanied by lipid accumulation in muscle fibers (Allen et al., 1967; Montroni and Testi, 1965).

Pigs sometimes develop nodular lesions in their subcutaneous and visceral adipose tissue. Nodular lesions of fat necrosis may also be found in the pancreas and in other visceral organs. Ito (1973) suggested that this condition is caused by the release of *esterase* from degenerating pancreatic cells, and elevated levels of esterase can be found in the adipose tissue of pigs with "yellow fat disease" (Danse and Steenbergen-Botterweg, 1974). Adipose degeneration in pigs is usually accompanied by a marked increase in yellow pigmentation, probably due to accumulation of ceroid pigment (Hermans, 1973b). In cattle, yellow *carotenoid* pigment normally accumulates in the healthy adipose tissues of grass-fed animals. The depth of coloration increases as animals become older.

REFERENCES

ALEXANDER, G., BENNETT, J. W., and GEMMELL, R. T. 1975. *J. Physiol.* 244:223.

ALLEN, C. E. 1976. *Fed. Proc.* 35:2302.

ALLEN, C. E., BEITZ, D. C., CRAMER, D. A., and KAUFFMAN, R. G. 1976. *Biology of Fat in Meat Animals*. North Central Regional Research Publication 234. University of Wisconsin, Madison.

ALLEN, E., CASSENS, R. G., and BRAY, R. W. 1967. *J. Food Sci.* 32:146.

ANDERSON, D. B., and KAUFFMAN, R. G. 1973. *J. Lipid Res.* 14:160.

ANDERSON, D. B., KAUFFMAN, R. G., and KASTENSCHMIDT, L. L. 1972. *J. Lipid Res.* 13:593.

ASHWELL, M. A., PRIEST, P., and SOWTER, C. 1975. *Nature* (*Lond.*) 256:724.

BAILEY, A. J. and SIMS, T. J. 1977. *J. Sci. Food Agric.* 28:565.

BAILEY, A. J., SHELLSWELL, G. B., and DUANCE, V. C. 1979. *Nature* (*Lond.*) 278:67.

BARRINEAU, L. L., RICH, C. B., and FOSTER, J. A. 1981. *Connect. Tissue Res.* 8:189.

BEERY, K. E., SINK, J. D., PATTON, S., and ZIEGLER, J. H. 1969. *J. Am. Oil Chem. Soc.* 46:439A.

BEHARI, J., and ANDRABI, W. H. 1978. *Connect. Tissue Res.* 6:181.

BELL, E. T. 1909. *Am. J. Anat.* 9:412.

BENGTSSON, S. G. and HAKKARAINEN, R. V. J. 1975. *J. Anim. Sci.* 41:106.

BENTZ, H., BACHINGER, H. P., GLANVILLE, R., and KUHN, K. 1978. *Eur. J. Biochem.* 92:563.

BERG, R. T. and BUTTERFIELD, R. M. 1976. *New Concepts of Cattle Growth*. John Wiley & Sons, New York.

BERG, R. T., ANDERSEN, B. B., and LIBORIUSSEN, T. 1978. *Anim. Prod.* 27:63.

BERG, R. T., JONES, S. D. M., PRICE, M. A., FUKUHARA, R., BUTTERFIELD, R. M., and HARDIN, R. T. 1979. *Can. J. Anim. Sci.* 59:359.

BERGE, S. 1948. *J. Anim. Sci.* 7:233.

BLANCHETTE-MACKIE, E. J., and SCOW, R. O. 1981. *J. Ultrastruct. Res.* 77:277.

BLOOD, D. C. 1956. *Aust. Vet. J.* 32:125.

BOCK, P. 1977. *Mikroskopie* 33:332.

BOCK, P. 1978. *Histochemistry* 55:269.

BOVARD, K. P., and HAZEL, L. N. 1963. *J. Anim. Sci.* 22:188.

BRANNANG, E. 1971a. *Swed. J. Agric. Res.* 1:69.

BRANNANG, E. 1971b. *Swed. J. Agric. Res.* 1:79.

BREIDENSTEIN, B. B., COOPER, C. C., CASSENS, R. G., EVANS, G., and BRAY, R. W. 1968. *J. Anim. Sci.* 27:1532.

BRISSIE, R. M., SPICER, S. S., and THOMPSON, N. T. 1975. *Anat. Rec.* 181:83.

BULLARD, H. H. 1912. *Am. J. Anat.* 14:1.

BURGESON, R. E., EL ADLI, F. A., KAITILA, I. I., and HOLLISTER, D. W. 1976. *Proc. Natl. Acad. Sci., USA* 73:2579.

BURTON, J. H., ANDERSON, M., and REID, J. T. 1974. *Br. J. Nutr.* 32:515.

BUSBOOM, J. R., MILLER, G. J., FIELD, R. A., CROUSE, J. D., RILEY, M. L., NELMS, G. E., and FERRELL, C. L. 1981. *J. Anim. Sci.* 52:83.

CARNES, W. H. 1971. *Fed. Proc.* 30:995.

CARPENTER, Z. L., BRAY, R. W., BRISKEY, E. J., and TRAEDER, D. H. 1961. *J. Anim. Sci.* 20:603.

CHARLES, D. D., and JOHNSON, E. R. 1976. *J. Anim. Sci.* 42:332.

CLARK, E. R., and CLARK, E. L. 1940. *Am. J. Anat.* 67:255.

CLAUS, R. 1975. Neutralization of pheromones by antisera in pigs. In E. Nieschlag (ed.), *Immunization with Hormones in Reproductive Research*, pp. 189–197. North-Holland Publishing Company, Amsterdam.

CLAUS, R., HOPPEN, H. O., and KARG, H. 1981. *Experientia* 37:1178.

COOK, L. J., SCOTT, T. W., FERGUSON, K. A., and MCDONALD, I. W. 1970. *Nature (Lond.)* 228:178.

COOPER, C. C., BREIDENSTEIN, B. B., CASSENS, R. G., EVANS, G., and BRAY, R. W. 1968. *J. Anim. Sci.* 27:1542.

COOPER, S., and GLADDEN, M. H. 1974. *Q. J. Exp. Biol. Physiol.* 59:367.

COX, R. W., LEAT, W. M. F., CHAUCA, D., PEACOCK, M. A., and BLIGH, J. 1978. *Res. Vet. Sci.* 25:58.

CRAIG, A. S., and PARRY, D. A. D. 1981. *Proc. R. Soc. Lond. B* 212:85.

CRENSHAW, T. D., PEO, E. R., LEWIS, A. J., and MOSER, B. D. 1981. *J. Anim. Sci.* 53:827.

CROWE, M. W. 1969. *Med. Vet. Pract.* 50(13):54.

CROWE, M. W., and PIKE, H. T. 1973. *J. Am. Vet. Med. Assoc.* 162:453.

DANSE, L. H. J. C., and STEENBERGEN-BOTTERWEG, W. A. 1974. *Vet. Pathol.* 11:465.

DARDICK, I., POZNANSKI, W. J., WAHEED, I., and SETTERFIELD, G. 1976. *Tissue Cell* 8:561.

DAUNCEY, M. J., WOODING, F. B. P., and INGRAM, D. L. 1981. *Res. Vet. Sci.* 31:76.

DAVIES, A. S. 1975. *Anim. Prod.* 20:45.

DELLMAN, H. D., and BROWN, E. M. 1976. *Textbook of Veterinary Histology*. Lea & Febiger, Philadelphia.

DESNOYERS, F., and VODOVAR, N. 1977. *Biol. Cell.* 29:177.

DESNOYERS, F., DURAND, G., and VODOVAR, N. 1980. *Biol. Cell.* 38:195.

DETWILER, S. R. 1934. *J. Exp. Zool.* 67:395.

DOLEZAL, H. G., MURPHEY, C. E., SMITH, G. C., and CARPENTER, Z. L. 1982. *Meat Sci.* 6:55.

DOORNENBAL, H. 1975. *Growth* 39:427.

DREW, I. M., PERKINS, D., and DALY, P. 1971. *Science* 171:280.

DUTSON, T. R. 1976. *Proc. Reciprocal Meat Conf.* 29:336.

DYER, H. MCM., and PIRIE, B. J. S. 1978. *J. Anat.* 125:519.

EDMONDS, L. D., SELBY, L. A., and CASE, A. A. 1972. *J. Am. Vet. Med. Assoc.* 160:1319.

EDWARDS, D. A. 1946. *J. Anat.* 80:147.

ELLIOT, J. I., and DOIGE, C. E. 1973. *Can. J. Anim. Sci.* 53:211.

ENSER, M. B., WOOD, J. D., RESTALL, D. J., and MACFIE, H. J. H. 1976. *J. Agric. Sci., Camb.* 86:633.

ETHERINGTON, D. J., and SIMS, T. J. 1981. *J. Sci. Food Agric.* 32:539.

ETHERTON, T. D. 1980. *Growth* 44:182.

EVANS, A. J. 1972. *Br. Poult. Sci.* 13:615.

EVANS, A. J. 1977. The growth of fat. In K. N. Boorman and B. J. Wilson (eds.), *Growth and Poultry Meat Production*, pp. 29–64. British Poultry Science, Edinburgh.

EZEKWE, M. O., and MARTIN, R. J. 1980. *Horm. Metab. Res.* 12:136.

FAUST, I. M., JOHNSON, P. R., and HIRSCH, J. 1977. *Science* 197:393.

FAUST, I. M., JOHNSON, P. R., STERN, J. S., and HIRSCH, J. 1978. *Am. J. Physiol.* 235:E279.

FELL, H. B. 1956. Skeletal development in tissue culture. In G. H. Bourne (ed.), *The Biochemistry and Physiology of Bone*, pp. 401–441. Academic Press, New York.
FINERTY, M. 1981. *J. Theor. Biol.* 93:279.
FLORIDI, A., IPPOLITO, E., and POSTACCHINI, F. 1981. *Connect. Tissue Res.* 9:95.
FORTIN, A., REID, J. T., MAIGA, A. M., SIM, D. W., and WELLINGTON, G. H. 1981. *J. Anim. Sci.* 53:982.
FRANZEN, L., and NORRBY, K. 1980. *Cell Tissue Kinet.* 13:635.
GARTON, G. A. 1967. *World Rev. Nutr. Diet.* 7:225.
GARTON, G. A. 1976. Physiological significance of lipids. In D. Lister, D. N. Rhodes, V. R. Fowler, and M. F. Fuller (eds.), *Meat Animals: Growth and Productivity*, pp. 159–176. Plenum Press, New York.
GAY, S., and MILLER, E. J. 1978. *Collagen in the Physiology and Pathology of Connective Tissue*. Gustav Fischer Verlag, Stuttgart.
GEMMELL, R. T., BELL, A. W., and ALEXANDER, G. 1972. *Am. J. Anat.* 133:143.
GERSH, I., and STILL, M. A. 1945. *J. Exp. Med.* 81:219.
GOLDRICK, R. B. 1967. *Am. J. Physiol.* 212:777.
GOTTA-PEREIRA, G., RODRIGO, F. G., and DAVID-FERREIRA, J. F. 1976. *Stain Technol.* 51:7.
GOWER, D. B. 1972. *J. Steroid Biochem.* 3:45.
GREELEY, R. G., and JOLLY, D. G. 1968. *Southwest Vet.* 21:189.
GREELEY, R. G., BOYD, C. L., and JOLLY, D. G. 1968. *Southwest Vet.* 21:277.
GREGORY, N. G., and LISTER, D. 1981. *Proc. Nutr. Soc.* 40:11A.
HADLOW, W. J. 1962. Diseases of skeletal muscle. In J. R. M. Innes and L. Z. Saunders (eds.), *Comparative Neuropathology*, pp. 147–243. Academic Press, New York.
HALL, A. D. 1905. *The Book of the Rothamsted Experiments*. John Murray (Publishers), London.
HALLQVIST, C. 1933. *Hereditas* 18:215.
HAMMOND, J., MASON, I. L., and ROBINSON, T. J. 1971. *Hammond's Farm Animals*, 4th ed., p. 89. Edward Arnold (Publishers), London.
HANSSON, K.-E., LUNDSTROM, K., FJELKNER-MODIG, S., and PERSSON, J. 1980. *Swed. J. Agric. Res.* 10:167.
HARCOURT, R. A., and BRADLEY, R. 1973. *Vet. Rec.* 92:233.
HARRIMAN, D. G., and REED, R. 1972. *J. Pathol.* 106:1.
HARRISON, T. J. 1975. *J. Anat.* 120:625.
HARTMAN, A. D. 1980. *Am. J. Physiol.* 241:E108.
HARTMAN, L., and SHORLAND, F. B. 1957. *J. Sci. Food Agric.* 8:428.
HAUGEBAK, C. D., HEDRICK, H. B., and ASPLUND, J. M. 1974. *J. Anim. Sci.* 39:1016.
HAUSMAN, G. J., and MARTIN, R. J. 1981. *J. Anim. Sci.* 52:1442.
HEINE, H., and FORSTER, F. J. 1974. *Acta Anat.* 89:387.
HENDRICKS, H. B., ABERLE, E. D., JONES, D. J., and MARTIN, T. G. 1973. *J. Anim. Sci.* 37:1305.
HERMANS, P. G. C. 1973a. *Tijdschr. Diergeneeskd.* 98:662.
HERMANS, P. G. C. 1973b. *Tijdschr. Diergeneeskd.* 98:668.
HINDMARSH, W. L. 1937. *Vet. Res. Rep., NSW* 7:58.
HINER, R. L., ANDERSON, E. E., and FELLERS, C. R. 1955. *Food Technol.* 9:80.
HOGG, D. A. 1982. *J. Anat.* 135:501.
HOGG, A., ROSS, R. F., and COX, D. F. 1975. *Am. J. Vet. Res.* 36:965.
HOOD, R. L., and ALLEN, C. E. 1973. *J. Lipid Res.* 14:605.
HOOD, R. L., and ALLEN, C. E. 1977. *J. Lipid Res.* 18:275.
HOWLETT, C. R. 1980. *J. Anat.* 130:745.
HUTT, F. B. 1934. *J. Hered.* 25:41.
ITO, T. 1973. *Jap. J. Vet. Sci.* 35:299.
JAMES, L. F., SHUPE, J. L., BINNS, W., and KEELER, R. F. 1967. *Am. J. Vet. Res.* 28:1379.
JAMES, T. 1951. *Edinb. Med. J.* 58:565.

JOHN, H. A., and LAWSON, H. 1980. *Cell Biol. Int. Rep.* 4:841.
JOHNSON, M. A., FULTHROPE, J. J., and HUDGSON, P. 1973. *Acta Neuropathol.* 24:97.
JOHNSTON, D. W. 1971. *Comp. Biochem. Physiol.* 40A:1107.
JOHNSTON, W. G., and YOUNG, G. B. 1958. *Vet. Rec.* 70:1219.
JONES, S. D. M., PRICE, M. A., and BERG, R. T. 1978. *Can. J. Anim. Sci.* 58:151.
JONSSON, P., and WISMER-PEDERSEN, J. 1974. *Livestock Prod. Sci.* 1:53.
KAUFFMAN, R. G., and SAFANIE, A. H. 1967. *J. Food Sci.* 32:283.
KAYE, M. M. 1962. *Can. J. Comp. Med.* 26:218.
KEELER, R. F., JAMES, L. F., BINNS, W., and SHUPE, J. L. 1967. *Can. J. Comp. Med.* 31:334.
KEELER, R. F., BINNS, W., JAMES, L. F., and SHUPE, J. L. 1969. *Can. J. Comp. Med.* 33:89.
KEMPSTER, A. J., and EVANS, D. G. 1979. *J. Agric. Sci., Camb.* 93:349.
KING, J. W. B., and ROBERTS, R. C. 1960. *Anim. Prod.* 2:59.
KIRTLAND, J., and GURR, M. I. 1980. *J. Agric. Sci., Camb.* 95:325.
KIRTLAND, J., GURR, M. I., SAVILLE, G., and WIDDOWSON, E. M. 1975. *Nature (Lond.)* 256:723.
KOCH, P., FISCHER, H., and SCHUMANN, H. 1957. *Erbpathologie der landwirtschaftlichen Haustiere*, pp. 189–190. Verlag Paul Parey, Berlin.
KUHN, K., and GLANVILLE, R. W. 1980. Molecular structure and higher organization of different collagen types. In A. Viidik and J. Vuust. (eds.), *Biology of Collagen*, pp. 1–14. Academic Press, New York.
LAKES, R., and SAHA, S. 1979. *Science* 204:501.
LASTER, D. B. 1974. *J. Anim. Sci.* 38:496.
LATIMER, H. B. 1927. *Am. J. Anat.* 40:1.
LAURENT, G. J. 1982. *Biochem. J.* 206:535.
LAURIE, G. W., LEBLOND, C. P., COURNIL, I., and MARTIN, G. R. 1980. *J. Histochem. Cytochem.* 28:1267.
LAWRIE, R. A. 1974. *Meat Science*. Pergamon Press, Oxford.
LEACH, R. M. 1971. *Fed. Proc.* 30:991.
LEAT, W. M. F. 1975. *J. Agric. Sci., Camb.* 85:551.
LEAT, W. M. F. 1976. The control of fat absorption, deposition and mobilization in farm animals. In D. Lister, D. N. Rhodes, V. R. Fowler, and M. F. Fuller (eds.), *Meat Animals: Growth and Productivity*, pp. 177–193. Plenum Press, New York.
LEAT, W. M. F. 1977. *J. Agric. Sci., Camb.* 89:575.
LEE, Y. B., and KAUFFMAN, R. G. 1974a. *J. Anim. Sci.* 38:532.
LEE, Y. B., and KAUFFMAN, R. G. 1974b. *J. Anim. Sci.* 38:538.
LEE, Y. B., KAUFFMAN, R. G., and GRUMMER, R. H. 1973a. *J. Anim. Sci.* 37:1312.
LEE, Y. B., KAUFFMAN, R. G., and GRUMMER, R. H. 1973b. *J. Anim. Sci.* 37:1319.
LE GROS CLARK, W. E. 1945. *The Tissues of the Body*. Clarendon Press, Oxford.
LEIPOLD, H. W., CATES, W. F., RADOSTITS, O. M., and HOWELL, W. E. 1969. *Can. Vet. J.* 10:268.
LEIPOLD, H. W., BLAUGH, B., HUSTON, K., EDGERLY, C. G. M., and HIBBS, C. M. 1973a. *Vet. Med. Small Anim. Clin.* 68:645.
LEIPOLD, H. W., OEHME, F. W., and COOK, J. E. 1973b. *J. Am. Vet. Med. Assoc.* 162:1059.
LIEBELT, R. A., and EASTLICK, H. L. 1952. *Anat. Rec.* 112:422.
LIEBELT, R. A., and EASTLICK, H. L. 1954. *Poultry Sci.* 33:169.
LINK, B. A., CASSENS, R. G., BRAY, R. W., and KOWALCZYK, T. 1967. *J. Anim. Sci.* 26:694.
LINK, B. A., BRAY, R. W., CASSENS, R. G., and KAUFFMAN, R. G. 1970. *J. Anim. Sci.* 30:726.
LIPTON, B. H. 1977. *Dev. Biol.* 61:153.
LUKE, D. A., TONGE, C. H., and REID, D. J. 1980. *J. Anat.* 130:859.
LWEBUGA-MUKASA, J. S., LAPPI, S., and TAYLOR, P. 1976. *Biochemistry* 15:1425.
MCCULLOUGH, A. W. 1944. *J. Morphol.* 75:193.
MACKENZIE, L. E. 1912. *Virchows Arch. Pathol. Anat. Physiol.* 210:57.
MANNERS, M. J., and MCCREA, M. R. 1963. *Br. J. Nutr.* 17:495.

MARCH, B. E., and HANSEN, G. 1977. *Poult. Sci.* 56:886.
MARLOWE, T. J. 1964. *J. Anim. Sci.* 23:454.
MARTIN, R., WHITE, J., HERBEIN, J., and EZEKWE, M. O. 1979. *Growth* 43:167.
MAYNE, R., and STRAHS, K. R. 1974. *J. Cell Biol.* 63:212a.
MECHAM, R. P. 1981. *Connect. Tissue Res.* 8:155.
MERSMANN, H. J., UNDERWOOD, M. C., BROWN, L. J., and HOUK, J. M. 1973. *Am. J. Physiol.* 224:1130.
MERSMANN, H. J., GOODMAN, J. R., and BROWN, L. J. 1975. *J. Lipid Res.* 16:269.
METZ, S. H. M., and DEKKER, R. A. 1981. *Anim. Prod.* 33:149.
MEYER, W., NEURAND, K., and RADKE, B. 1982. *J. Anat.* 134:139.
MIDDLETON, D. S. 1932. *Edinb. Med. J.* 39:389.
MIDDLETON, D. S. 1934. *Edinb. Med. J.* 41:401.
MILLER, A. 1982. *Trends Biochem. Sci.* 7(1):13.
MILLER, R. C., KEITH, T. B., MCCARTY, M. A., and THORP, W. T. S. 1940. *Proc. Soc. Exp. Biol. Med.* 45:50.
MONTAGNA, W. 1945. *J. Morphol.* 76:87.
MONTRONI, L., and TESTI, F. 1965. *Acta Med. Vet., Napoli* 11:505.
MOODY, W. G., and CASSENS, R. G. 1968. *J. Food Sci.* 33:47.
MOODY, W. G., and ZOBRISKY, S. E. 1966. *J. Anim. Sci.* 25:809.
MOODY, W. G., ENSER, M. B., WOOD, J. D., RESTALL, D. J., and LISTER, D. 1978. *J. Anim. Sci.* 46:618.
MORLEY, F. H. W. 1954. *Aust. Vet. J.* 30:237.
MORRILL, C. C. 1947. *N. Am. Vet.* 28:738.
MOSS, M. L. 1972. The regulation of skeletal growth. In R. J. Goss (ed.), *Regulation of Organ and Tissue Growth*, pp. 127–142. Academic Press, New York.
MOTTA, P. 1975. *J. Microsc.* 22:15.
MUNZ, K., and MEVES, C. 1974. *Histochemistry* 40:181.
MURPHY, P. A., WEAVERS, E. D., and BARRETT, J. N. 1975. *Vet. Rec.* 97:445.
MURRAY, D. M., BOWLAND, J. P., BERG, R. T., and YOUNG, B. A. 1974. *Can. J. Anim. Sci.* 54:91.
NECHAD, M., and BARNARD, T. 1979. *Biol. Cell.* 36:43.
NES, N. 1953. *Nord. Veterinaermed.* 5:869.
NEVILLE, W. E., MULLINIX, B. G., SMITH, J. B., and MCCORMICK, W. C. 1978. *J. Anim. Sci.* 47:1080.
NISBET, D. I., and RENWICK, C. C. 1961. *J. Comp. Pathol.* 71:177.
NOUGUES, J., and VEZINHET, A. 1977. *Ann. Biol. Anim. Biochem. Biophys.* 17:799.
NOWICKI, L., and ZAJAC, H. 1964. *Med. Wet.* 20:279.
OGHISO, Y., LEE, Y.-S., TAKAHASHI, R., and FUJIWARA, K. 1977. *Jap. J. Vet. Sci.* 39:101.
OKA, M., MIKI, T., HAMA, H., and YAMAMURO, T. 1979. *Clin. Orthop.* 145:264.
O'NEILL, I. K., TRIMBLE, M. L., and CASEY, J. C. 1979. *Meat Sci.* 3:223.
PALLUDAN, B. 1961. *Acta Vet. Scand.* 2:32.
PALSSON, H. 1940. *J. Agric. Sci., Camb.* 30:1.
PALSSON, H., and VERGES, J. B. 1965. *J. Agric. Sci., Camb.* 64:247.
PARK, R. J., SPURWAY, R. A., and WHEELER, J. L. 1972. *J. Agric. Sci., Camb.* 78:53.
PARRY, D. A. D., BARNES, G. R. G., and CRAIG, A. S. 1978a. *Proc. R. Soc. Lond. B* 203:305.
PARRY, D. A. D., CRAIG, A. S., and BARNES, G. R. G. 1978b., *Proc. R. Soc. Lond. B* 203:293.
PARTRIDGE, S. M. 1966. Elastin. In E. J. Briskey, R. G. Cassens, and J. C. Trautman (eds.), *The Physiology and Biochemistry of Muscle as a Food*, pp. 327–339. University of Wisconsin Press, Madison.
PATTEN, B. M. 1931. *The Embryology of the Pig*, 2nd ed. P. Blakiston's Son, Philadelphia.
PATTERSON, R. L. S. 1968. *J. Sci. Food Agric.* 19:31.
PATTON, J. S., and CAREY, M. C. 1979. *Science* 204:145.

PIKE, B. V., and ROBERTS, C. J. 1980. *Res. Vet. Sci.* 29:108.
PIMENTAL, E. R. 1981. *Acta Histochem. Cytochem.* 14:35.
PLAAS, H. A. K., HARWOOD, R., and CRYER, A. 1978. *Biochem. Soc. Trans.* 6:596.
POND, W. G., WALKER, E. F., and KIRTLAND, D. 1975. *J. Anim. Sci.* 41:1053.
POZNANSKI, W. J., WAHEED, I., and VAN, R. 1973. *Lab. Invest.* 29:570.
PRESTON, R. L., VANCE, R. D., CAHILL, V. R., and KOCK, S. W. 1974. *J. Anim. Sci.* 38:47.
PRESTON, T. R., and WILLIS, M. B. 1974. *Intensive Beef Production*. Pergamon Press, Oxford.
PROCKOP, D. J. 1971. *Fed. Proc.* 30:984.
PUCHTLER, H., and WALDROP, F. S. 1978. *Histochemistry* 57:177.
PUCHTLER, H., WALDROP, F. S., and VALENTINE, L. S. 1973. *Histochemie* 35:17.
PUCHTLER, H., MELOAN, S. N., and POLLARD, G. R. 1976. *Histochemistry* 49:1.
RALSTON, A. T., KENNICK, W. H., and DAVIDSON, T. P. 1975. *J. Anim. Sci.* 40:1211.
REALE, E., BENAZZO, F., and RUGGERI, A. 1981. *J. Submicrosc. Cytol.* 13:135.
REINECCIUS, G. A. 1979. *J. Food Sci.* 44:12.
ROBERTS, J. A. F. 1929. *J. Genet.* 21:57.
ROSS, R., and BORNSTEIN, P. 1971. *Sci. Am.* 224:44.
ROWE, R. W. D. 1978. *Meat Sci.* 2:275.
RUCKER, R. B., and LEFEVRE, M. 1980. Chemical changes in elastin as a function of maturation. In J. R. Whitaker and M. Fujmaki (eds.), *Chemical Deterioration of Proteins*. Symp. Ser. 123, pp. 63–82. American Chemical Society, Washington, D.C.
SALANS, L. B., ZARNOWSKI, M. J., and SEGAL, R. 1972. *J. Lipid Res.* 13:616.
SANES, J. R. and CHENEY, J. M. 1982. *J. Cell Biol.* 93:442.
SASSE, J., MARK, H. VON DER, KUHL, U., DESSAU, W., and MARK, K. VON DER. 1981. *Dev. Biol.* 83:79.
SCHMALSTIEG, R. VON, and MEYER, H. 1960. *Dtsch. Tierarztl. Wochenschr.* 67:41.
SCOTT, R. A., CORNELIUS, S. G., and MERSMANN, H. J. 1981. *J. Anim. Sci.* 53:977.
SELBY, L. A., MENGES, R. W., HOUSER, E. C., FLATT, R. E., and CASE, A. A. 1971. *Arch. Environ. Health* 22:496.
SELLIER, P., and BOCCARD, R. 1971. *Ann. Genet. Sel. Anim.* 3:433.
SELYE, H. 1965. *The Mast Cells*. Butterworth & Company (Publishers), London.
SERAFINI-FRACASSINI, A., FIELD, J. M., and HINNIE, J. 1978. *J. Ultrastruct. Res.* 65:190.
SHUPE, J. L., BINNS, W., JAMES, L. F., and KEELER, R. F. 1967. *J. Am. Vet. Med. Assoc.* 151:198.
SHAW, A. M. 1929. *Sci. Agric.* 10:29.
SHAW, A. M. 1930. *Sci. Agric.* 10:690.
SIDMAN, R. L. 1956. *Anat. Rec.* 124:581.
SIMPSON, J. W., and TAYLOR, A. C. 1974. *Proc. Soc. Exp. Biol. Med.* 145:42.
SINK, J. D. 1967. *J. Theor. Biol.* 17:174.
SINK, J. D. 1979. *J. Food Sci.* 44:1.
SISSON, S., and GROSSMAN. J. D. 1953. *The Anatomy of the Domestic Animals*, 4th ed. W. B. Saunders Company, Philadelphia.
SKIPITARIS, C. N. 1981. *J. Agric. Sci., Camb.* 97:83.
SLAVIN, B. G. 1979. *Anat. Rec.* 195:63.
SMITH, C., SAMSON, E. H., PADYKULA, H. A., LASKEY, M. G., and LOEWENTHAL, L. A. 1952. *J. Morphol.* 90:103.
STAMP, J. T. 1960. *J. Comp. Pathol.* 70:296.
STANDAL, N., VOLD, E., TRYGSTAD, O., and FOSS, I. 1973. *Anim. Prod.* 16:37.
STEELE, N. C., and FROBISH, L. T. 1976. *Growth* 40:369.
STEELE, N. C., FROBISH, L. T., and KEENEY, M. 1974. *J. Anim. Sci.* 39:712.
STEFFEN, D. G., BROWN, L. J., and MERSMANN, H. J. 1978. *Comp. Biochem. Physiol.* 59B:195.
STINGL, J., and STEMBERA, O. 1974. *Lymphology* 7:160.
SULLIVAN, T. W., and AL-UBAIDI, Y. Y. 1963. *Poult. Sci.* 42:46.
SWATLAND, H. J. 1974. *Vet. Bull.* 44:279.

SWATLAND, H. J. 1975. *J. Anim. Sci.* 41:78.
SWATLAND, H. J. 1978. *Zentralbl. Veternaermed.* 25:556.
TAVASSOLI, M. 1976. *Acta Anat.* 94:65.
TERRELL, R. N., SUESS, G. G., and BRAY, R. W. 1969. *J. Anim. Sci.* 28:449.
THOMPSON, G. E., and JENKINSON, D. M. 1969. *Can. J. Physiol. Pharmacol.* 47:249.
THOMPSON, G. E., and JENKINSON, D. M. 1970. *Res. Vet. Sci.* 11:102.
THORNGREN, K.-G., and HANSSON, L. I. 1981. *Acta Anat.* 110:121.
THURLEY, D. C. 1969. *Pathol. Vet.* 6:217.
TRELSTAD, R. L. 1982. *Cell* 28:197.
TRELSTAD, R. L. and HAYASHI, K. 1979. *Dev. Biol.* 71:228.
TUFF, P. 1948. *Skand. Vet. Tidskr.* 38:379.
TURNER, J. C. 1979. *Comp. Biochem. Physiol.* 62A:599.
VAN, R. L. R., and RONCARI, D. A. K. 1977. *Cell Tissue Res.* 181:197.
VAN, R. L. R., and RONCARI, D. A. K. 1978. *Cell Tissue Res.* 195:317.
VAN, R. L. R., BAYLISS, C. E., and RONCARI, D. A. K. 1976. *J. Clin. Invest.* 58:699.
VEIS, A. 1970. Collagen. In E. J. Briskey, R. G. Cassens, and B. B. Marsh (eds.), *The Biochemistry and Physiology of Muscle as a Food*, Vol. 2, pp. 455–470. University of Wisconsin Press, Madison.
VEZINHET, A., and PRUD'HON, M. 1975. *Anim. Prod.* 20:363.
VEZINHET, A., PRUD'HON, M., and BENEVENT, M. 1974. *Ann. Biol. Anim. Biochem. Biophys.* 14:117.
WAGNON, K. A. 1960. *J. Range Mgmt.* 13:89.
WALSTRA, P. 1974. *Livestock Prod. Sci.* 1:187.
WASSERMAN, A. E. 1979. *J. Food Sci.* 44:6.
WASSERMAN, R. H. 1977. Bones. In M. J. Svenson (ed.), *Duke's Physiology of Domestic Animals*, 9th ed., pp. 413–432. Comstock Publishing Associates, Ithaca, N.Y.
WENSVOORT, P. 1967. *Pathol. Vet.* 4:69.
WENSVOORT, P. 1968. *Pathol. Vet.* 5:270.
WESTMORLAND, N. 1971. *Fed. Proc.* 30:1001.
WHITTEM, J. H. 1957. *J. Pathol. Bacteriol.* 73:375.
WILSON, L. L., ROTH, H. B., ZIEGLER, J. H., and SINK, J. D. 1977. *J. Anim. Sci.* 44: 932.
WISE, D. R. 1977. The growth of the skeleton. In K. N. Boorman and B. J. Wilson (eds.), *Growth and Poultry Meat Production*, pp. 65–78. British Poultry Science, Edinburgh.
WOOD, E. M. 1967. *Anat. Rec.* 157:437.
WOOD, J. D., ENSER, M. B., and RESTALL, D. J. 1975. *J. Agric. Sci., Camb.* 84:221.
WOOD, J. D., GREGORY, N. G., HALL, G. M., and LISTER, D. 1977. *Br. J. Nutr.* 37:167.
WOOD, J. D., ENSER, M. B., and RESTALL, D. J. 1978. *Anim. Prod.* 27:1.
WU, J. J., KASTNER, C. L., HUNT, M. C., KROPF, D. H., and ALLEN, D. M. 1981. *J. Anim. Sci.* 53:1256.
ZOPHONIASSON, P. 1929. *Nord. Jordbrugsforsk.* 11:327.

3

The Commercial Structure of the Carcass

INTRODUCTION

Carcasses are separated into cuts of meat following a pattern that reflects differences in tenderness, taste, shape, and size between carcass muscles. A few years ago, the major subdivisions of carcass cutting were dictated by the need to produce primal or wholesale cuts which could be easily transported and which could be conveniently subdivided at retail outlets. More recently, there has been a trend toward centralized meat cutting, with carcasses being subdivided into retail cuts in a location near the abattoir. The retail cuts are then packed into boxes for refrigerated distribution. In this way, the bones and the fat that are removed during the preparation of retail cuts can be handled more efficiently. Thus the collection of fat and bones for further processing is simplified, and the distribution of edible meat is cheaper and more hygienic.

Meat cutting is more standardized in North America than on any other continent, but subtle differences that reflect consumer preferences still exist. The cuts of meat that are described later in this chapter follow a generalized North American pattern. In other parts of the World, the patterns of meat cutting vary enormously, even within individual countries. Some European patterns of meat cutting are shown in Figures 3-1 and 3-2. Further information on the geographical variation of meat-cutting patterns may be obtained from OEEC (1961) and ADSRI (1981) publications. The yields of various meat cuts on an international basis are given by Strother (1975).

This chapter begins by describing some large and easily recognized muscles that appear in cuts of meat. A knowledge of these muscles, and of the bones described in Chapter 2, greatly facilitates the recognition of different cuts of meat. The anatomical

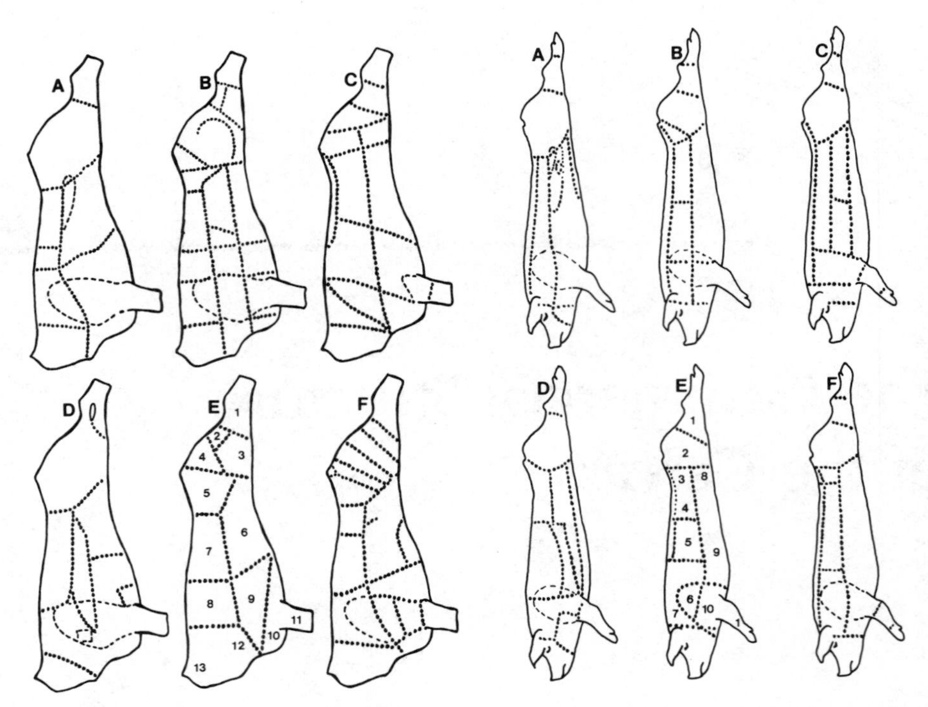

Figure 3-1 Some European methods for beef cutting: (A) Germany; (B) France; (C) Spain; (D) Italy; (E) Great Britain; (F) Norway; (1) leg; (2) topside; (3) silverside; (4) rump; (5) rump-steak; (6) thick flank; (7) sirloin; (8) best rib; (9) rolled rib; (10) brisket; (11) shin; (12) clod; (13) sticking. (Data from OEEC, 1961.)

Figure 3-2 Some European methods for pork cutting: (A) Germany; (B) France; (C) Spain; (D) Italy; (E) Great Britain; (F) Norway; (1) knuckle; (2) leg; (3, 4, and 5) chump, middle and foreloin; (6) blade; (7) sparerib; (8) flank; (9) belly; (10) hand and spring. (Data from OEEC, 1961.)

location and structure of the muscles of beef, lamb, and pork carcasses are considered in more detail in Chapter 4. The latter part of the present chapter describes the federal grading systems that are used for meat and poultry carcasses in the United States and Canada, and it concludes with a review of the principles of electronic grading.

MAJOR MUSCLES OF THE CARCASS

The most tender cuts of beef and lamb do not require complete cooking; in fact, extended cooking may even reduce their tenderness. Moist heat converts the strong collagen fibers of connective tissue to gelatin, but heat may make the muscle fiber proteins more tough. Thus extensive cooking may make tough muscles more tender, but it may also make tender muscles more tough. The same effect occurs in poultry meat, where prolonged cooking decreases the toughness of leg muscles but increases the toughness of breast muscles (Marini and Goodwin, 1973).

Neck Muscles

The neck muscles produce tough meat because of their high connective tissue content. Like other tough muscles in the carcass, they are usually removed from their bones (cervical vertebrae) and are used as stewing meat or ground beef. Since the anatomy of the neck muscles is rather complex, and since their individual identity is of no consequence once the carcass is subdivided, they need not be considered in detail here. The only muscle worthy of passing note is a straplike muscle called the *sternomandibularis*. This muscle forms the superficial part of a compound muscle which runs in the ventral throat region from the head to the sternum. Because this muscle is severed when the head of a beef carcass is removed at slaughter, samples can be obtained immediately after slaughter without damaging a commercial carcass. Many of the published studies in meat science have been based on this muscle, and a great deal is now known about its metabolism.

Shoulder Muscles

Shoulder muscles (Figures 3-3, 3-4, and 3-5), are intermediate in their level of toughness and they are usually completely cooked in order to make them tender. An easily recognized group of muscles and bones enables meat from the shoulder or chuck region to be readily identified. The *supraspinatus* is dorsal to the spine or ridge on the

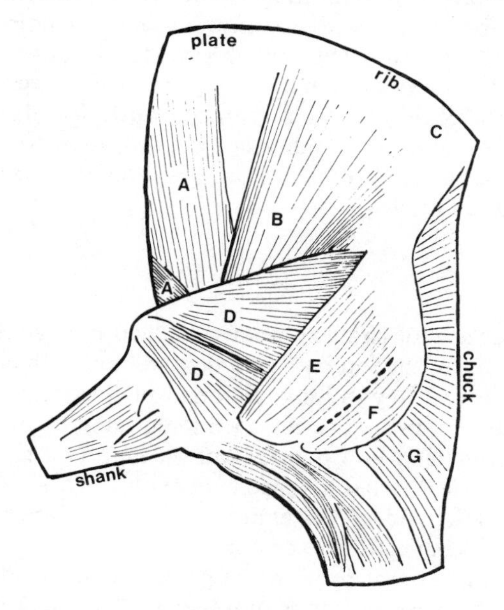

Figure 3-3 Location of some major muscles in a lateral view of a trimmed forequarter of beef: (A) pectoralis; (B) latissimus dorsi; (C) position of longissimus dorsi; (D) triceps brachii; (E) infraspinatus; (F) supraspinatus; (G) trapezius.

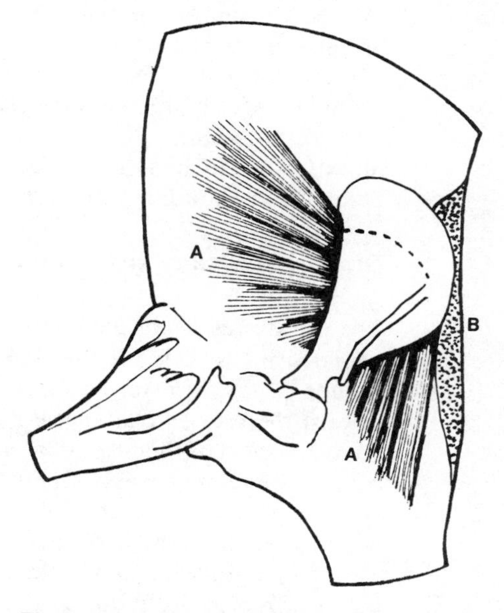

Figure 3-4 Location of the serratus ventralis (A) and rhomboideus (B) deep in the beef forequarter.

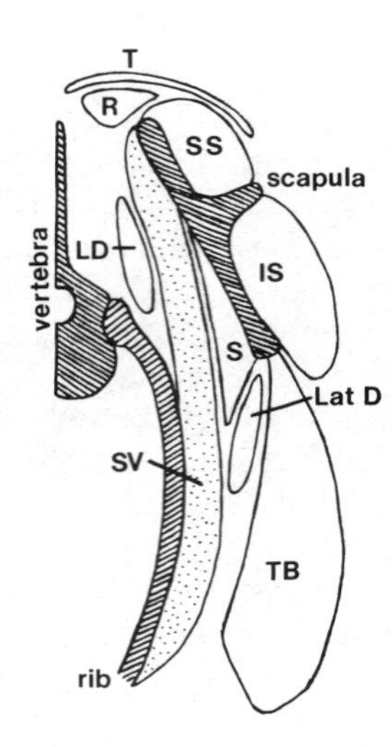

Figure 3-5 Some muscles of the shoulder region seen in a section through the blade bone or scapula. Supraspinatus (SS) is dorsal to the spine of the scapula, while infraspinatus (IS) is ventral. Subscapularis (S) is medial to the blade of the scapula. Triceps brachii (TB) is ventral to the scapula. Trapezius (T) and rhomboideus (R) are dorsal to the scapula. Serratus ventralis (SV) forms a sling to suspend the body from the forelimbs. Longissimus dorsi (LD) is relatively small in this section. Latissimus dorsi (Lat D) is sectioned at the narrow end of its fanlike shape.

scapula, while the *infraspinatus* is ventral to the scapular spine. The *trapezius* is located superficially between the left and right scapular blades. The *rhomboideus* is ventral to the trapezius. Figuratively speaking, if one were to stab a standing animal between its shoulder blades, the knife would pass through the trapezius first and then through the rhomboideus. The *subscapularis* is located on the flat medial face of the scapula, toward the ribs. The *biceps brachii* is anterior to the humerus in an equivalent position to the biceps muscle in the human arm. The adjective brachii is needed to indicate the biceps muscle of the arm, since there is another biceps muscle, the biceps femoris, which is located in the hind limb. The *triceps brachii* is a large muscle located in the triangular area bounded by the ventral edge of the scapula and the posterior edge of the humerus. Its name, "tri-ceps" or "three-heads," indicates that this large triangular muscle is subdivided. Thus when seen in a cut of meat from the shoulder, the triceps brachii may look like more than one muscle.

Distal Muscles of the Limbs

The distal muscles of the limbs produce tough meat because of their high content of connective tissue. In the distal regions of the forelimb (shank) and hind limb (leg) are groups of *fusiform* (cigar-shaped) muscles with long tendons which extend distally toward the toes or phalanges. If these muscles are cut open longitudinally, it can be seen that most of them contain internal tendons which fan outward for the attachment of short bundles of muscle fibers. This featherlike arrangement of muscle fiber bundles is called a *pennate* structure. In meat animals, the meat derived from pennate muscles is quite tough, because of their high connective tissue content.

Most of the distal muscles located anteriorly in the limb are *extensors* and, when they contract in the living animal, they cause the toes to move forward, as in the start of a new stride. Most of the muscles located posteriorly in the distal part of the limb are *flexors* which bend the limb during locomotion. In beef and lamb carcasses, tendons from distal muscles pass down the length of the cannon bone and are kept in place by ligamentous rings. Thus the extremities of a limb are moved by remote control. Beef cannon bones are discarded in the abattoir because they have virtually no meat on them. Flexors and extensors from the distal parts of the limbs are difficult to identify individually once they have been removed from the skeleton.

Muscles of the Rib Cage

Located between the scapula and the rib cage are several muscles that hold the forelimb onto the body. Like other shoulder muscles around the scapula, most of these muscles are intermediate in their level of toughness. The *serratus ventralis* is a large fanlike muscle which radiates from the medial face of the scapula and which attaches to the lateral surfaces of the ribs (Figure 3-5). The serratus ventralis is the major component of a muscular sling that suspends the thorax of an animal from between its forelimbs. This muscular suspension system has no counterpart in the hind limb since the pelvis is fused to the vertebral column. The muscular sling that holds the forelimb onto the body serves as a shock absorber during locomotion.

The *longissimus dorsi* is an extremely important muscle. It forms the eye of meat that is seen when chops and steaks are cut from the posterior rib region and from the loin (Figures 3-6 and 3-7). The naming of this muscle is something of a problem since it is really a compound muscle composed of many subunits. Each subunit acts over the length of several vertebrae and helps to flex the vertebral column. The longissimus dorsi is also involved in respiratory movements, as well as helping to move the neck. Because of its compound structure, the longissimus dorsi may be called by an alternative name, *longissimi thoracis et lumborum*. In agricultural journals the longissimus dorsi is often simply called the eye muscle or longissimus muscle. The muscle fiber bundles of the longissimus dorsi are arranged at an acute angle to the vertebral column. The cross-sectional area of the longissimus dorsi increases toward the posterior part of the rib cage, but it has an approximately constant cross-sectional area through the loin. Beef carcasses are usually split into forequarters and hindquarters between ribs 12 and 13. The area of the longissimus dorsi seen at this point is often measured or examined to assess the amount of meat in a carcass. This may be a useful guide to

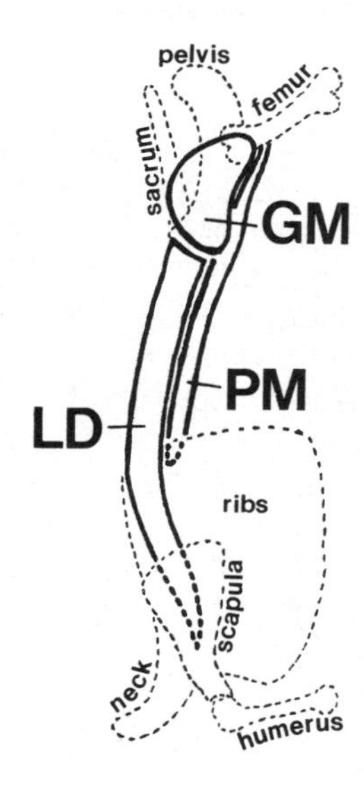

Figure 3-6 Distribution of the major axial muscles of the carcass: (GM) gluteus medius; (LD) longissimus dorsi; (PM) psoas major.

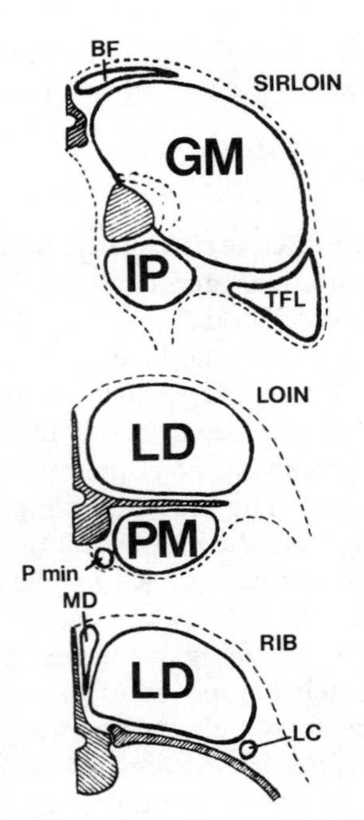

Figure 3-7 Muscles of the rib, loin, and sirloin. The major muscles are longissimus dorsi (LD), gluteus medius (GM), and psoas major (PM) and its continuation as iliopsoas (IP). Other muscles shown are: (BF) biceps femoris; (TFL) tensor fascia lata; (P min) psoas minor; (MD) multifidus dorsi; (LC) longissimus costarum.

muscularity when comparing animals with a similar carcass length. However, in the comparison of long carcasses with short carcasses, a smaller cross-sectional area does not necessarily indicate a smaller muscle mass, since the mass is spread over a greater length.

The *pectoralis* muscle is located over the sternum in the brisket, and it extends posteriorly into the plate. The pectoralis is composed of deep and superficial layers. The *intercostal* muscles are located between adjacent ribs in the wall of the chest, and there are two layers in the depth of the muscle. Intercostal muscles make an important contribution to the meat content of North American pork spareribs.

Loin Muscles

The loin muscles (Figures 3-6 and 3-7) give rise to tender meat with a desirable taste, and they command a high price when presented for sale as steaks or chops. The longissimus dorsi extends posteriorly from the rib region, it runs through the loin, and most of the muscle terminates on the anterior face of the ilium. Thus the longissimus dorsi is seen in cuts of meat taken through the ribs and loin, but not in cuts of meat such as sirloin steaks that are posterior to the anterior face of the ilium. The longissimus dorsi is dorsal to the transverse processes of the lumbar vertebrae, and it is dorsal to the ribs in the thoracic region. For most of the length of the rib cage, there are no major muscles immediately ventral to the heads of the ribs. Thus in a rib steak, there is only one main eye of meat, and that is the longissimus dorsi, which is dorsal to the ribs. However, in the loin, there are muscles both above and below the level of the transverse processes of the lumbar vertebrae. The dorsal muscle above the transverse processes is the longissimus dorsi. The ventral muscle below the transverse processes is the *psoas major* or filet muscle. The psoas major originates ventrally to the last couple of ribs in the rib cage. The cross-sectional area of the psoas major increases toward the sirloin. Medial to the psoas major, almost under the centra of the vertebrae, is a smaller psoas muscle called the *psoas minor*.*

Immediately lateral to the dorsal spines of the vertebrae (medial to the longissimus dorsi) are some small *multifidus dorsi* muscles. The *longissimus costarum* is a relatively small ropelike muscle, dorsal to the ribs. It appears as a small eye of meat at the separation between the forequarter and the hindquarter. The multifidus dorsi and the longissimus costarum have little commercial significance, since they are such small muscles, but they can create a problem in the measurement of rib-eye areas since they might be accidentally included with the longissimus dorsi.

Hindlimb Muscles

The muscles of the hind limb (Figures 3-8 and 3-9) are relatively large, and they produce a large volume of moderately tender meat. The *gracilis* is a thin sheet of muscle which is spread over the medial face of the hind limb. The gracilis, together with the small *sartorius* muscle which is anterior to the gracilis, may be used to orient hindlimb cuts of meat such as a beef round steak or a slice of ham. Orientation is necessary for the identification of the remaining muscles of the hind limb, which are grouped around the femur. The *quadriceps femoris* muscles form a group of four large muscles that

*The letter P in the word "psoas" is silent when the word is spoken.

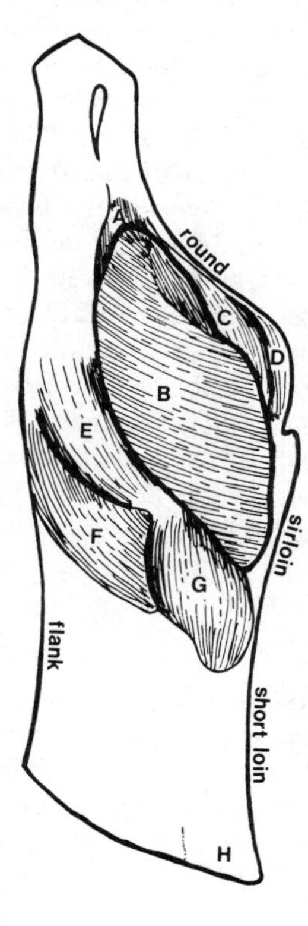

Figure 3-8 Location of some major muscles in a lateral view of a trimmed hindquarter of beef: (A) gastrocnemius; (B) biceps femoris; (C) semitendinosus; (D) semimembranosus; (E) vastus lateralis; (F) tensor fascia lata; (G) gluteus medius; (H) longissimus dorsi.

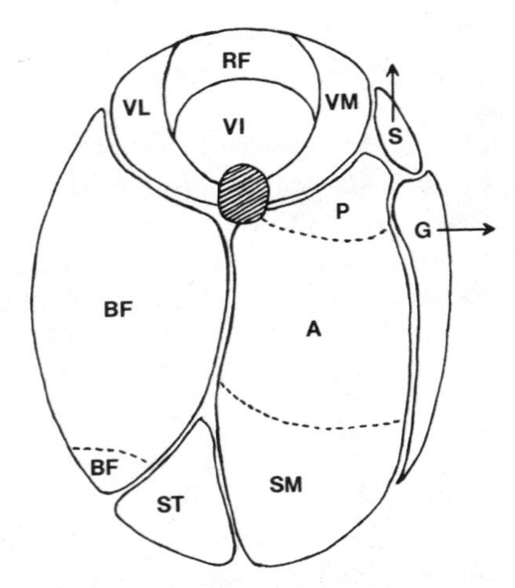

Figure 3-9 Arrangement of hindlimb muscles around the femur (shaded). The sartorius (S) indicates the anterior direction, while the gracilis (G) indicates the medial direction. Other muscles are: (VL) vastus lateralis; (RF) rectus femoris; (VI) vastus intermedius; (VM) vastus medialis; (P) pectineus; (A) adductor; (SM) semimembranosus; (ST) semitendinosus; (BF) biceps femoris, which sometimes appears as two parts when sectioned. Separations between P, A, and SM may be difficult to see.

pull on the patella when the leg is extended. The *vastus medialis* is medial, the *vastus lateralis* is lateral, the *vastus intermedius* covers the anterior face of the femur, and the *rectus femoris* covers the vastus intermedius.

The *biceps femoris* is a single large muscle on the lateral face of the hind limb. In cross section, it often appears to be divided into two parts because it has a very deep cleft along part of its length. The biceps femoris appears as a single muscle in cuts of meat that miss the cleft, but sections through the cleft make the muscle appear double. To add to the possibility of confusion, the small segment of muscle cut off by the cleft is often paler than the main part of the muscle.

The *semitendinosus* and *semimembranosus* are two large muscles that are located on the posterior face of the hind limb. The semimembranosus is medial to the semitendinosus. The *adductor* and *pectineus* are located in the medial part of the hind limb, near to the femur. The pectineus is anterior to the adductor. In lean carcasses, it may be difficult to separate the adductor from the semimembranosus, and these two muscles may appear as a single muscle. Given the possibility that the biceps femoris muscle on the other side of the limb may look like two muscles, caution is needed in the identification of these muscles for the first time. The *gastrocnemius* is a large muscle that is located deep in the hind limb. It is covered by distal extensions of some of the proximal muscles of the limb. The gastrocnemius pulls on the Achilles tendon at the hock, and is the equivalent of the calf muscle in the human leg. A major anatomical difference between the human leg and the hind limbs of meat animals is that the human gastrocnemius is not covered by the posterior thigh muscles.

Sirloin Muscles

Between the hind limb and the loin, and located laterally to the pelvis, are several large muscles that form the rump and sirloin of the carcass (Figure 3-7). The positioning and naming of the first of these muscles is rather misleading. Despite its name, the *gluteus medius* is located laterally in meat animals. It covers the lateral face of the ilium and it appears as the large muscle area in sirloin steaks and chops. From lateral to medial, there are three layers of muscles: gluteus medius, *gluteus accessorius* and *gluteus profundus*. The gluteus medius is the largest of the three.

The psoas major continues posteriorly from the loin into the sirloin. It is joined by another muscle, the *iliacus*, and the two together may be given a compound name, the *iliopsoas*. Little, if any, of the longissimus dorsi appears in the sirloin since most of the muscle terminates on the anterior face of the ilium.

Flank Muscles

The flank and belly of the animal are formed by sheets of muscle and connective tissue. The muscles are relatively tough and need not be identified individually. The layers of abdominal muscles appear as the parallel layers of lean that may be seen in a slice of side bacon. The *tensor fascia lata* is a triangular muscle located in the angle between the animal's flank and its hind limb. The fascia lata is a sheet of connective tissue that covers the anterior surface of the hind limb, and it is stretched by the tensor fascia lata muscle during locomotion. It is important to remember that, although an animal walks into the abattoir on four legs, it is suspended by its hind limbs when it leaves the abattoir. As a carcass is hoisted onto the overhead rail, there is an extreme rotation of the hind limb relative to the vertebral column. Consequently, the tensor fascia lata is spread through the stretched muscle mass of the sirloin.

Poultry Muscles

The major power stroke of the wing during flight originates from the large *pectoralis* muscle located superficially in the breast region (Figure 3-10). The wing is elevated by the *supracoracoideus* muscle located between the pectoralis and the sternum. The supracoracoideus, although adjacent and parallel to the pectoralis, is able to cause an opposite movement (raising instead of lowering the wing) because its tendon is inserted onto the opposite side of the humerus to the pectoralis tendon. In poultry, the pectoralis muscles are the largest muscles of the body and they comprise approximately 8 per cent of the body weight. The small muscles that are located within the wing are concerned with controlling the shape and degree of rotation of the wing during flight.

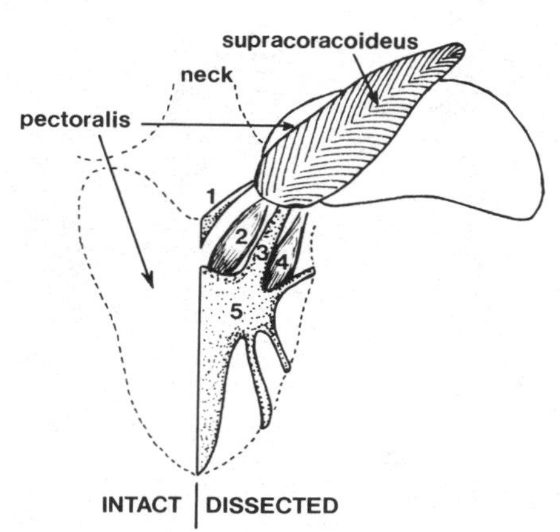

Figure 3-10 Ventral dissection of the chicken breast. The pectoralis is lifted back to show the major head of the supracoracoideus. (1) Clavicle; (2) minor head of supracoracoideus; (3) coracoid; (4) posterior coracobrachialis; (5) sternum.

The insertion of the flight muscles in the breast region should be examined critically. At first sight, it might appear that most of the expanded sternum in poultry is used for the attachment of the pectoralis. In fact, the only part available for the pectoralis is the zone around the supracoracoideus muscle. The bulk of the pectoralis originates from the furcula and from a membrane that is stretched between the furcula and the coracoid.

The muscles in the leg of the fowl are adapted for bipedal locomotion (Figure 3-11). The sartorius (iliotibialis cranialis), instead of being a medial straplike muscle, is

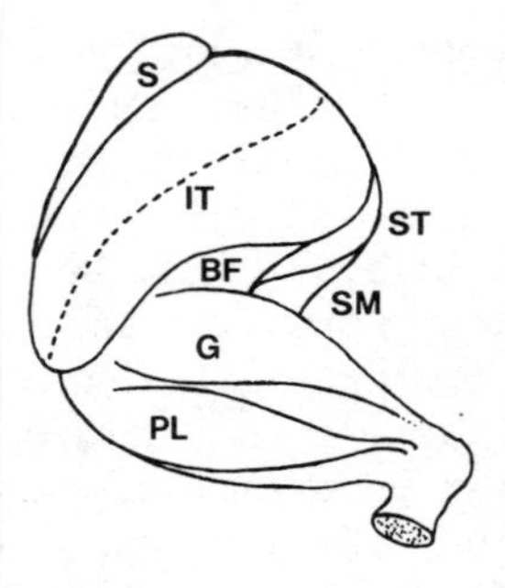

Figure 3-11 Lateral view of chicken leg muscles: (BF) biceps femoris; (G) gastrocnemius; (IT) two main parts of iliotibialis; (PL) peroneus longus; (S) sartorius; (SM) semimembranosus; (ST) semitendinosus.

triangular in cross section and is the most anterior muscle in the leg. Posterior to the sartorius are a series of muscles (tensor fascia lata, biceps femoris, semitendinosus, and semimembranosus) that occupy similar general positions to their mammalian equivalents. They are not, however, exactly equivalent muscles, and mammalian muscle names are used here solely for convenience. The anatomically correct names for avian muscles are given by Vanden Berge (1979).

Carcass development in poultry is reviewed by Moran (1977). In broiler chickens and turkeys, the dressing percentage increases as birds grow older and heavier. In ducks and in small game birds, such as partridge and quail, the dressing percentage remains fairly constant during growth. In broiler chickens and turkeys, males have a higher dressing percentage than females of the same age (Table 3-1). Pigs often have a dressing percentage of 75 per cent, but a value near 70 per cent would be exceptionally high for cattle.

TABLE 3-1. APPROXIMATE DRESSING PERCENTAGES OF POULTRY AT TYPICAL SLAUGHTER WEIGHTS

Type	Age (weeks)	Dressing percentage	
		Males	Females
Broiler chicken	8	73	71
Small turkey	14	79	76
Large turkey	21	79	78
Pekin ducks	8	66	
Geese	11	75	

Source: Data from Moran (1977).

CUTS OF MEAT

Cuts of Beef

Except for the detachment of the forequarter from the hindquarter, motor-driven band saws are normally used in North America to detach many of the primal and retail cuts of meat. Heat is generated by friction when bones are cut, and the resulting paste of bone dust, fat, and meat juice must be scraped off the meat to reduce bacterial spoilage and to preserve the appearance of the meat. Cleavers, choppers, hand saws, and knives are used more extensively in the traditional methods of meat cutting in other countries. Although traditional methods are slower and more difficult to learn, they allow greater scope for craftsmanship and curvilinear cutting. In traditional methods, the soft tissues of the carcass should be cut with a knife or chopper, since hand sawing leaves a ragged edge.

The first step in breaking up the carcass is to separate it into primal cuts that can be handled more easily. The primal cuts correspond fairly closely to the units which a retail butcher might order from a wholesaler or abattoir. The primal cuts of beef are shown in Figure 3-12. It is important to note that the separation of the forequarter and the hindquarter shown in Figure 3-12 leaves only the last rib on the hindquarter. This is the common practice in the United States. In Canada, the separation has routinely been made between ribs 11 and 12, instead of between ribs 12 and 13. However, this point of separation may soon be changed to follow the pattern used in the United States.

On the hanging side of beef, count seven vertebral centra down from the sacral-lumbar junction, add on just less than the length of a half a centrum, and cut perpendicularly through the vertebral column at this point with a saw. Separate the *forequarter* from the *hindquarter* by cutting through the intercostal and abdominal muscles, following the curvature of the twelfth rib. The forequarter can be dropped onto a table or held suspended by its own hook from a hoist. Separate the *chuck* from

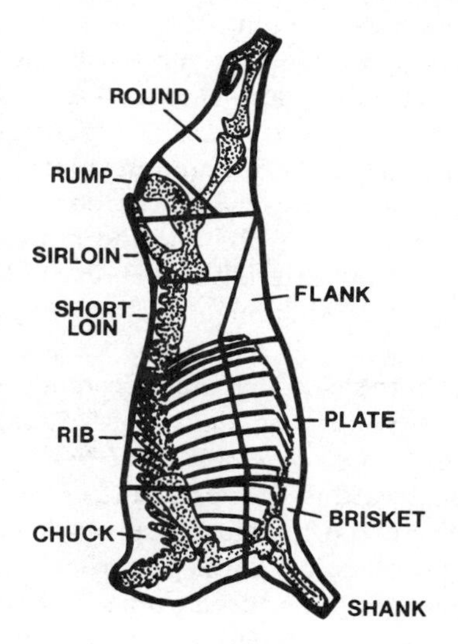

Figure 3-12 Primal cuts of beef.

the *rib* with a perpendicular cut through the vertebral column, level with the intercostal muscles between the dorsal parts of ribs 4 and 5. Separate the rib from the *plate* by an anterior to posterior cut. This separation may be made much nearer to the vertebral column than the separation shown in Figure 3-12. Separate the chuck from the *brisket* by a cut that is perpendicular to the fourth rib at a point about 1 cm proximal to the olecranon process of the elbow. The *shank* may be cut into thick slices; the *shank knuckle* slices are proximal.

Before breaking up the hindquarter, trim off the excess fat near the pubis and over the posterior part of the abdominal muscles. Anterior to the rectus femoris, at a point where the tensor fascia lata reaches its most distal extent, start a separation that ends on rib 13, about 20 cm from the vertebral column. This detaches the *flank*. Separate the *round* from the *rump* with a cut that passes about 1 cm distal to the ischium and which terminates just after passing through the head of the femur. Separate the rump from the *sirloin* with a cut that passes between sacral vertebrae 4 and 5, and which terminates just ventral to the acetabulum of the pelvis. Separate the *sirloin* from the *short loin* with a cut that is perpendicular to the vertebral column and passes between lumbar vertebrae 5 and 6.

The primal cuts are next separated into retail cuts. In the following paragraphs they are given an approximate rating according to tenderness:

*	Less tender cuts to braise, stew, or pot roast
**	Medium tender cuts, good for cooking by moist heat
***	Tender meat for roasting, broiling, or frying

The major basis for this rating system is the connective tissue content of the major muscles in each cut (Strandine et al., 1949). Other important muscle properties that affect meat tenderness are described in Chapter 5.

The *rib* cut is separated into *rib steaks**** or *standing rib roasts**** by cuts made perpendicularly to the vertebral column. *Rib-eye**** or *delmonico**** steaks are composed of sections of the spinalis dorsi together with the longissimus dorsi muscle. The *chuck* is sliced in planes that are parallel to rib 4 to make *blade steaks*** or *blade pot roasts***. *Arm steaks**, *arm pot roasts**, or *cross-cut ribs** are sliced off perpendicularly to the humerus. *Brisket** is sold in chunks to be braised or cooked in liquid. The *shank** is cut into thick slices that are perpendicular to the radius and ulna. The plate may be divided into cubes of rib bone and muscle, and sold as *short ribs**. The flat mass of meat located ventrolaterally to the rib cage is usually rolled, tied, and cut into cylindrical cuts of *plate**. Abdominal muscles may be isolated from the flank to make *flank steaks**.

The short loin is sliced into steaks that are perpendicular to the vertebral column. The most anterior steaks are the *wing* or *club steaks****, and nearly all their muscle is derived from the longissimus dorsi. Next are the *T-bone steaks****; these gain extra meat from the psoas major toward the posterior end of the loin. Last are two or three *porterhouse steaks****. These have large areas of meat derived from both the longissimus dorsi and the psoas major. In the porterhouse region at the posterior end of the short loin, the vertebrae can be removed from the steaks to create *New York strip steaks**** (longissimus dorsi) and *tenderloin* or *filet steaks**** (psoas major and minor). In a restaurant with a French menu, the longissimus dorsi may appear as "biftek de contre filet" and the psoas muscles as "filet mignon."

The steaks that are cut perpendicularly to the shaft of the ilium in the sirloin are named by the shape of the sectioned ilium. These steaks are, from anterior to posterior, (1) *pin bone sirloin steaks****, named from the oval section of the anterior projection of the ilium; (2) *flat bone* or *double bone sirloin steaks****, named from the flat sections of the wing of the ilium where it joins with the wing of the sacrum; (3) *round bone sirloin steaks****, named from the round sections of the slender shaft of the ilium; and (4) *wedge bone sirloin steaks****, named from the triangular cross section of the ilium near the acetabulum. The triangular shape of the rump and the complex shape of the pubis, ischium, and the head of the femur make this cut difficult to handle. If the bones are carefully removed, slices of *rump steak**** can be cut quite easily, or the cut can be left in large chunks as *standing rump**** or *boneless rump****. The round may be cut into full cut *round steaks*** that are perpendicular to the femur, or it can be cut into large pieces of meat parallel to the femur so as to create the *inside* or *top round*** (mostly semimembranosus and adductor) and the *outside* or *bottom round*** (mostly semitendinosus and biceps femoris). The semitendinosus is sometimes detached and slices may be sold as the *eye of the round***. The *sirloin tip*** is a cut from the round which includes the muscles that pull on the patella.

Cuts of Veal

Veal carcasses are smaller than beef carcasses and there is less need to subdivide the carcass into primal cuts. Typical primal cuts are the *forequarter, loin* (from scapula to ilium), *flank* (from midsternum to tensor fascia lata), and *leg* (including sirloin). The cuts of veal are quite small, and many of the beef names are used since the overall pattern for beef is followed. The brisket is usually called the *breast* in the veal carcass. The region equivalent to the T bone may be called a *kidney chop* if the kidney has been left in place and sectioned with the chop. Differences in tenderness between cuts of meat from various parts of the veal carcass are far less pronounced than for the beef carcass (Hiner and Hankins, 1950).

Cuts of Pork

The primal cuts of the pork carcass are shown in Figure 3-13. Remove the *hind foot* with a cut through the tuber calcis. Remove the *front foot* with a cut that is just distal to the ulna and radius. Remove the *leg* with a cut that starts between sacral vertebrae 2 and 3 and which is then directed toward the tensor fascia lata. The cutting line is then changed so that most of the tensor fascia lata is incorporated into the leg (Figure 3-13). The *butt* and *picnic* are removed together as a *shoulder*, by a cut that is perpendicular to the vertebral column and which starts between thoracic vertebrae 2 and 3. The butt is separated from the picnic by a cut that skims past the ventral region of the cervical vertebrae at a tangent. This keeps the top of the picnic relatively square. The *jowl* is removed from the picnic with a cut that follows the crease lines in the skin. The remainder of the side of pork is split into the *loin* and *belly* by a curved cut that follows the curvature of the vertebral column. One end of the curve is just ventral to the ilium, the other end is just ventral to the blade of the scapula.

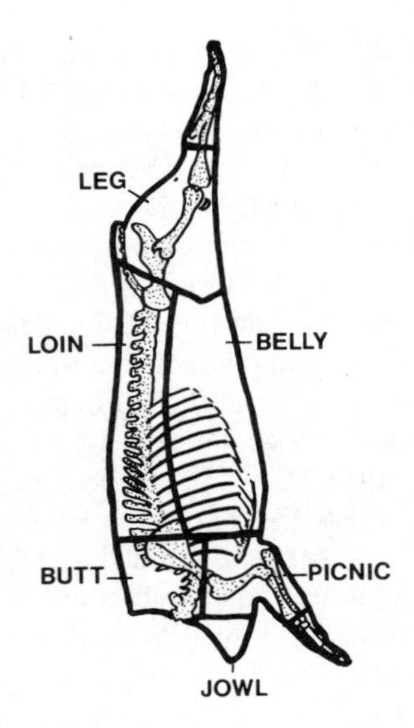

Figure 3-13 Primal cuts of pork.

The patterns of consumer use for retail cuts of beef are determined primarily by the considerable differences in tenderness between the various muscles of the bovine carcass. Tough meat that has been removed from a beef carcass can be ground into hamburger and rendered tender enough for rapid dry cooking. A similar pattern of consumer use applies to the pork carcass, but problems with meat toughness are far less severe. All pork cuts must be thoroughly cooked to eliminate the risk of trichinosis, as described in Chapter 1.

The loin can be divided into a continuous sequence of chops. From anterior to posterior these are the *rib chops, center loin chops*, and *tenderloin chops*. They can all be satisfactorily cooked by dry heat. Alternatively, the thoracic, lumbar, and iliac regions can be left intact as large roasts, the *rib end roast, center loin roast*, and *tenderloin end roast*. The psoas muscles can be removed from the lumbar region to make *tenderloin*, and the longissimus dorsi and adjacent small muscles can be removed from the vertebral column, and rolled and tied to make *boned and rolled loin roast*. A *crown roast* can be made by twisting the thoracic vertebral column into a circle so that the stumps of the ribs radiate outward like the points of a crown. This facilitates the rapid carving and distribution of portions at a banquet.

The longissimus dorsi can be cured and smoked to make *Canadian style bacon* or (in Canada) *peameal bacon* and *back bacon*. The rib cage plus its immediately adjacent muscles are removed from the belly to make the *spareribs*. The remaining muscles of the abdomen, together with those that overlap the rib cage for their insertion, constitute the *side of pork*. Side of pork may be cured and smoked to make *slab bacon*. The picnic can be sliced through the humerus to make *picnic shoulder chops*, or it can be partly subdivided to make *picnic shoulder roasts*. Picnic shoulder roasts can be boned and rolled, or smoked and cured in a variety of ways. The butt, or *Boston butt*, is usually divided into a number of *blade steaks* which are cut from dorsal to ventral through the scapula. The more anterior part then forms a *butt roast*. The leg can be subdivided to create, from proximal to distal, the *butt end roast* and the *shank end roast*. Alternatively, the leg may be cured and smoked to make *ham*. The feet, the hocks, the knuckles, and the tail can be baked or cooked in liquid.

Cuts of Lamb

The primal cuts of lamb are shown in Figure 3-14. The *sirloin* plus *leg*, or *pin bone leg*, is removed by cutting perpendicularly through the vertebral column at a point level with the anterior face of the ilium. In the lamb carcass, the *loin* includes part of the abdominal wall. The loin is removed by a cut that passes between ribs 12 and 13 and which then continues perpendicularly through the vertebral column. Sometimes the whole *breast* and the *shank* are removed with a single cut from the anterior of the sternum to the ventral part of rib 11. Alternatively, the dominant cut may be made between ribs 5 and 6, to separate the *rib* from the shoulder, and to divide the breast into anterior and posterior sections. In Figure 3-14, note how the metacarpal cannon bone is fixed back so that the carcass can be more easily transported.

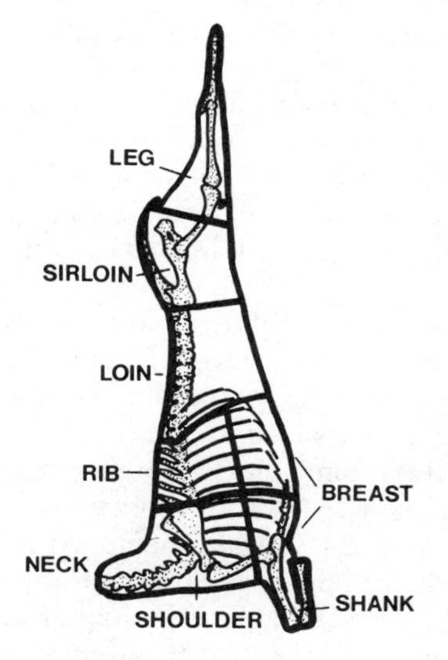

Figure 3-14 Primal cuts of lamb.

Differences in the tenderness of lamb muscles may become apparent in carcasses from older animals, and the pattern of consumer use reflects the method of cooking required. The notation of asterisks (*) that was used for beef is used again in this paragraph. The leg can be divided a number of ways, either into *leg chops**** or *steaks**** that are cut perpendicularly to the femur, or into large or small roasting cuts. Like many other decisions made by the butcher, seasonal preferences are taken into account. Steaks and chops are popular in the summer, while large roasts are more popular in the winter. Similarly, the sirloin can either be cut into *sirloin chops****, or left as a *sirloin roast****. The flap of abdominal muscle on the loin is removed, and is added to the breast meat. The loin is sliced into *loin chops**** or left whole as a *loin roast****. The rib or *rack* of lamb can be subdivided into *rib chops****, or left whole as a *rib roast*. The rack makes an excellent *crown roast* when the vertebral column is trimmed and bent back on itself. There are a number of ways in which to divide the shoulder. It can be made into *blade chops****, or left largely intact as a *square shoulder roast****. Parts of the shoulder can be boned and rolled to make *Saratoga chops****. The *neck** is usually sliced perpendicularly to the vertebral column. The *fore shank** is removed intact, and the remaining *breast** is subdivided in an arbitrary manner. Much of the fat on the breast may be removed, and the remaining lean can be rolled or cut into riblets to conform to local preferences.

Recognition of Cuts of Meat

The recognition of the species of meat when cuts of beef, pork, and lamb are displayed for sale as top-quality fresh meat is based on the color of the lean and on the size of whole muscles and bones. Beef lean has the deepest color, and pork has the lightest color. Lamb and veal are intermediate, depending on the age of the animal. Veal from entirely milk-fed calves is extremely pale. If marbling fat is present as wavy lines and dots of white fat in the lean, it is very conspicuous against the dark color of the lean in beef, but is sometimes less visible in pork. Pork exhibits the greatest variation in depth of color between different muscles. Pork often has the whitest fat, and beneath the subcutaneous fat may be seen the thick cutaneous muscles of the pork carcass. Some pork cuts retain their skin.

To identify a cut of meat, first decide whether an unidentified cut is from the left side or from the right side of the carcass. Then ascertain its position and orientation in the carcass. Do not forget that left and right sides of the carcass form mirror images, and that the two flat surfaces of a chop or steak from one side of the carcass may also form mirror images. This is particularly important when identifying muscles from diagrams.

Examine the surfaces of the cut of meat, and look for a surface that might have been medial, as indicated by vertebrae, sternum, pubis, ribs, adductor muscle, gracilis, and so on. Surfaces that were once part of the lateral surface of the carcass usually bear traces of trimmed or untrimmed subcutaneous fat, often with a grade stamp. The orientation of a cut of meat is often indicated by the extent to which the cut of meat is tapered. The abdomen is narrower than the thorax in an eviscerated carcass, and the limbs are tapered from proximal to distal. The dorsal spines of most of the thoracic vertebrae project posteriorly. The anterior ribs are shorter than the posterior ribs. The muscle groups that were surveyed earlier in this chapter were selected because they are useful in placing a cut of meat. Some further information relating to the skeleton is listed below.

1. Look for a series of exposed blocks of porous bone. If a deep groove (neural canal) runs through the series, the bones are vertebrae from along the animal's backbone. If no groove is present, the bones may be part of the sternum. If a carcass has been poorly split into sides, however, the midline cut may miss the neural canal.
2. Look for rounded cross sections of bone that might be from a limb, but remember that part of the shaft of the ilium is also round in cross section. The whole hind limb is rounded in cross section, but the forelimb is flattened because it is located against the rib cage. When the ilium has a rounded cross section in a whole sirloin, the muscle mass is lopsided, and there is some trace of the sacrum on the edge of the cut of meat. The more posterior part of the shaft of the ilium is triangular in cross section (wedge bone of sirloin). When the femur has a rounded cross section in the round, ham, or hind leg, it is almost in the center of a circle of meat (Figure 3-15A). When the humerus or the shaft of the scapula have a

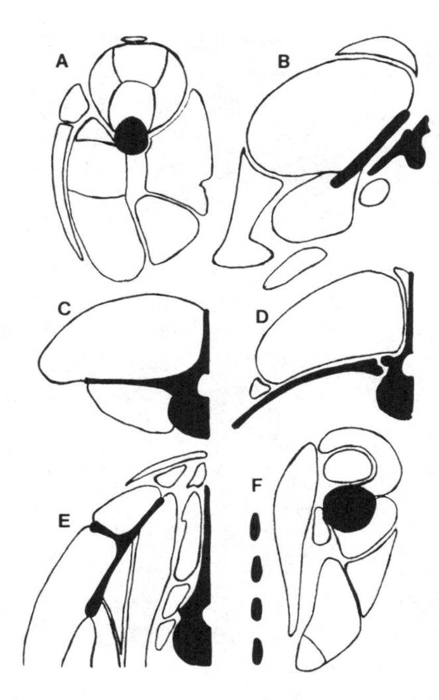

Figure 3-15 Pattern recognition of cuts of meat. In beef: (A) round; (B) sirloin; (C) T-bone or porterhouse; (D) rib; (E) chuck or blade; (F) arm steak.

rounded cross section in the chuck or arm region, it is alongside a series of transected ribs, and the muscle mass of the limb is oval in cross section (Figure 3-15F).

3. Look for a section that has been cut through a flat bone. If it is rigidly part of the body of a vertebra, and if it is narrow, it may be a winglike transverse process of a lumbar vertebra from the loin (Figure 3-15C). If it is rigidly part of a vertebra and is dorsal to the neural canal, and if it is one of a series of wide porous sections of bone, it may be a dorsal spine of a thoracic vertebra from the blade or rib region of the carcass. If it is curved and if it is movably jointed to a vertebra, it is probably the dorsal part of a rib (Figure 3-15D). If it is parallel to a vertebral process, or if it is joined by cartilage to a vertebra, it may be the flat part of the ilium from the sirloin (Figure 3-15B). If it is isolated by itself in the meat, or if it has a ⊦ shape, it is probably the scapula (Figure 3-15E).
4. If there are no bones in the cut of meat, and if it is a flat slab of meat composed of several layers of flat muscles, it is probably part of the flank or abdominal wall.
5. If the cut of meat has large vertebrae with a complex shape, and if the outer surface of the meat is dark and ragged, the meat is probably from the neck.
6. If the outer surface of the cut of meat contains a flat rounded area of bone with a dimpled surface and traces of dried cartilage, the bone is the pubis from the rump region.
7. Look for a hole in the meat where the carcass might have been suspended from a large hook or gambrel. This indicates a hind leg, or the heel of the round in beef. In beef, the achilles tendon is hard, dry, pale yellow in color, and extremely strong.

8. Look for a series of parallel ribs. The anterior ribs are shorter than the posterior ribs, and anterior ribs connect directly to the sternum.
9. Look for a long flap of muscle that runs diagonally over the medial surfaces of the ribs. This flap of muscle is the diaphragm. The ventral part of the diaphragm is anterior to the posterior part. In the beef carcass, the anterior part of the diaphragm appears in the plate, and the posterior part of the diaphragm appears at the start of the short loin, in the wing or club steak region.
10. Look for a ball-and-socket joint. The socket of the scapula in the chuck region of the carcass is wide and shallow. The socket that forms the acetabulum of the pelvis is narrow and deep, and there may be a trace of the ligament that holds the head of the femur into the socket. The acetabulum occurs at the junction of the rump, the round, and the sirloin. In pork and lamb, the acetabulum may be contained in the top of the ham or leg.
11. Look for a small loose bone that would fill a cupped hand. This is the patella of the hind limb.
12. Look for the stump of the tail, with its small, simple caudal vertebrae.
13. Look for a series of small round sections of white cartilages. These are the costal cartilages from the plate, flank, belly, or breast.
14. Look for groups of several small muscles, each surrounded by white fibrous tissue. These are the extensor and flexor muscles from the distal part of a limb. The Achilles tendon indicates the hind limb.

Parts of the Poultry Carcass

Poultry carcasses are often cut into pieces so that consumers may purchase half carcasses (right or left sides), white or red meat, and premium or economy parts of the carcass. Right and left sides may be *quartered* with a cut that follows the posterior edge of the last rib and which is continued through the vertebral column. *Wings* may be removed with a cut through the shoulder joint (glenoid). Sometimes the wing is subdivided by a cut that is immediately distal to the radius and ulna. The *breast* is composed of the sternum and its associated muscles. The sternum may be intact (*whole breast*) or split into left and right halves. There is some variation with regard to the inclusion of the clavicles and ventral parts of the ribs. Sometimes the whole breast is split into three parts with the *wishbone* (pulley bones) as the central unit. Neck skin is excluded from all breast cuts.

The *leg* may be removed at the acetabulum so that the pelvic muscles, but not the pelvic bones, accompany the leg. The proximal part of the leg, the *thigh*, may be separated from the distal part, the *drumstick*, at the joint between the femur and the tibia. There is considerable variation in the exact structure of the *back* of the poultry carcass, depending on how the other parts have been removed. Whole back includes the pelvic bones, scapulae, dorsal parts of the ribs, and the vertebrae from the posterior part of the neck to the tail.

CARCASS GRADING

The primary objective of carcass grading is to describe the value of a carcass in clearly defined in terms that are useful to the meat industry. It is advantageous to both the buyer and the seller if the task of grading the carcass is left to an impartial third party—

the federal government. If the buyer and the seller have worked out their own system of payment for high- and low-value carcasses, they can save time or money by not having the carcass federally graded. The federal grading of carcasses facilitates long-distance transactions and contracts for future shipments in which one or both parties have not yet examined the carcasses.

Quantity and Quality

Three major factors determine the value of a carcass relative to market conditions: (1) *carcass weight;* (2) the *cutability* or *yield* of salable meat; and (3) the *quality* of the lean meat. All three factors are continuous variables which can be accurately measured in either absolute terms, such as weight, or in relative terms, such as those used by a taste panel. In scientific experiments, accurate carcass evaluation is necessary to search for minor differences between carcasses. A less accurate system is adequate for commercial transactions, and the continuous spectrum of carcass properties is subdivided into a relatively small number of *grades* that progress in a stepwise sequence. Thus carcasses that are placed in the same grade may exhibit small differences, but carcasses that are placed into different grades should exhibit much larger, and commercially significant differences.

Beef Grades in the United States

In the United States, the eating quality of beef is thought to be affected by (1) the sex or type of animal from which the carcass originated; (2) the age or maturity of the animal; and (3) the amount of intramuscular or *marbling fat* within the muscles. After deciding the sex of the carcass from the shape of the pubis (Figure 2-12) and from any trace of a penis, the carcass is placed into one of five possible *maturity groups*. This is done by examining a number of the features of the skeleton and lean meat. In young animals around 1 year of age, the interiors of the vertebral centra are soft, red, and porous in appearance. The medial surfaces of the ribs are rounded and streaked with red. As animals grow older, the interiors of bones become harder, more white, and less porous. Carcasses from young animals exhibit a lot of relatively soft cartilage, particularly on the tips of the dorsal spines of the thoracic vertebrae. As animals grow older, cartilage in such locations becomes hard and ossified. In young animals, the sacral vertebrae are only loosely fused together. As animals grow older, their sacral vertebrae become solidly fused together. The concentration of myoglobin increases as animals grow older, and old animals tend to have dark muscles. However, young cattle may also produce carcasses with dark meat if they have been severely stressed or exhausted prior to slaughter. These carcasses are usually called *dark-cutters*. In older animals, the fasciculi or bundles of muscle fibers that form the grain of the meat are grouped into large and easily detected units, and the lean is said to have a *coarse texture*. The smaller and more tightly packed fasciculi of younger animals give the lean a *firm texture*. Cattle may acquire carotenoid pigments from their feed. These pigments accumulate in the carcass fat so that the fat from older animals tends to have a yellow or amber color. However, high pigment levels in feed such as fresh grass may cause the fat to become yellow at an earlier age.

The US Department of Agriculture (USDA) maturity groups range from A to E in order of increasing maturity. Young steers and heifers are included in groups A and

TABLE 3-2. SUBJECTIVE TERMS USED TO DESCRIBE DEGREES OF MARBLING, AND THEIR APPROXIMATE RELATIONSHIP TO THE PERCENTAGE AREA OF THE MEAT FILLED BY MARBLING FAT

Subjective term	Area of fat (per cent)
Very abundant	25
Abundant	21
Moderately abundant	18
Slightly abundant	15
Moderate	11
Modest	7.5
Small	2.5
Slight	1.5
Trace	0.5
Practically devoid	0

B. Mature dairy cows, old breeding stock, animals with retarded growth, and those sent to market long after their prime spread down into groups C, D, and E.

The longissimus dorsi muscle is examined at the point where the forequarter and hindquarter are separated. Provided that the texture and color of the lean are acceptable, the main factor that determines the quality grade of the carcass is the amount of marbling fat. The degrees of marbling that occur in a beef carcass are normally described by a series of subjective terms. These terms are related in an approximate manner to the percentage area of meat that contains marbling fat (Table 3-2).

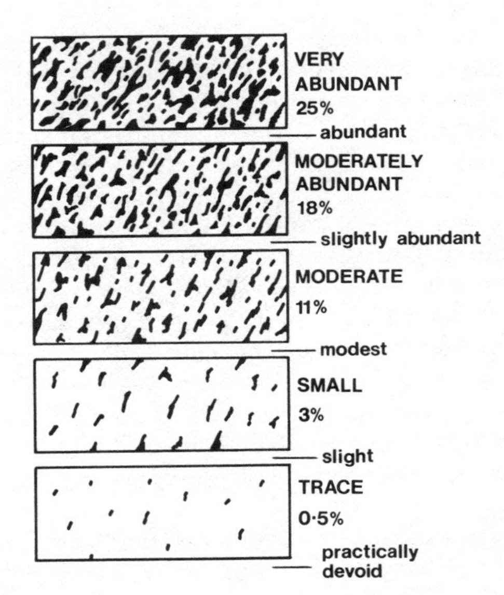

Figure 3-16 Degrees of marbling. Intramuscular fat is shown in black in this diagram, and the subjective terms are those used by the USDA in beef carcass grading. The percentage of the area filled by fat is indicated.

The visual appearance of some of these degrees of marbling is shown in Figure 3-16. The subjective terms have a precise meaning only to the experienced meat graders and experts who use the USDA system. For the rest of us, the salient points of the USDA grading system are more easily grasped by using the equivalent percentages of marbling. Table 3-3 shows how the degree of marbling is used to grade carcasses in each maturity group. By scanning across the vertical columns, it may be seen that more marbling is required to qualify for any particular grade as animal age is increased. It must be emphasized that, in practice, the degree of marbling is subjectively estimated by the grader, and that the equivalent percentage areas of marbling used here are approximations. Although the percentages were measured with a computer from photographs of the USDA standards, different computer methods give slightly different results, depending on how the muscle-fat boundaries are recognized.

TABLE 3-3. DISTRIBUTION OF PERCENTAGE MARBLING AREA IN USDA BEEF GRADES

	Maturity group				
Grade	A	B	C	D	E
Prime	>16	>17.5			
Choice	16–5	17.5–7.5			
Good	5–1.5	7.5–2.5			
Standard	<1.5	2.5–0.5			
Commercial		<0.5	>7.5	<11	>14.5
Utility			7.5–0.5	11–1	14.5–2.5
Cutter			<0.5	<1	<2.5

In the United States, beef carcasses receive two separate grades, one for quality as just outlined, and one for the predicted yield of edible meat. The yield grade is calculated from:

1. An estimate of the subcutaneous fat thickness
2. An estimate of the visceral fat that remains on the carcass
3. An estimate of the cross-sectional area of the longissimus dorsi muscle at the separation of the forequarter and the hindquarter
4. The hot carcass weight, or an estimate made from the cold carcass weight (cold weight × 1.02)

1. The subcutaneous fat is measured in units of 0.1 inch at a particular point over the sectioned area of the longissimus dorsi (Figure 3-17). If the fat at this point is considered by the grader to be a negatively biased sample of the overall subcutaneous fat, 0.1 or 0.2 inch may be added to this figure.

2. The grader estimates the amount of fat that will have to be removed from regions ventral to the vertebral column and around the pelvis for the preparation of salable retail cuts of meat. The amount of this fat is expressed as a percentage of the carcass weight.

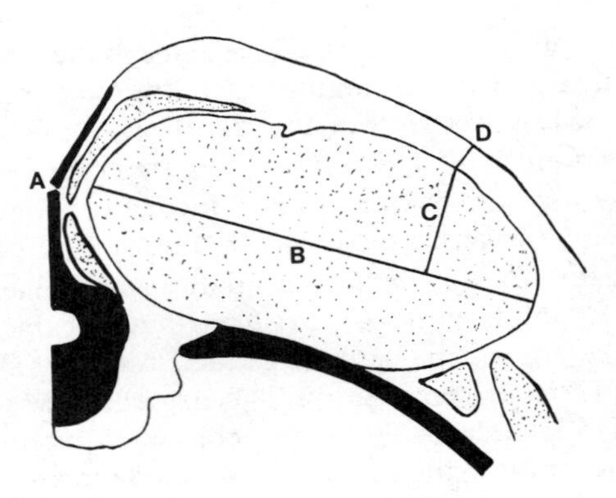

Figure 3-17 Measurement of subcutaneous fat depth for the USDA yield grade. The dorsal spines of the vertebrae are often cut or scribed (A) in the commercial carcass to give the longissimus dorsi area a rounder shape. (B) Long axis of the longissimus dorsi area; (C) perpendicular at three-fourth length; (D) position of fat depth measurement.

3. The cross-sectional area of the longissimus dorsi, expressed in square inches, is measured or estimated subjectively.

These data are combined as follows:

$$\begin{aligned} & 2.5 \\ & + (2.5 \times \text{fat thickness}) \\ & + (0.2 \times \text{per cent trim fat}) \\ & + (0.0038 \times \text{hot carcass weight}) \\ & - (0.32 \times \text{longissimus dorsi area}) \end{aligned}$$

The result of this calculation is adjusted downward to the next whole number (integers from 1 to 5). Carasses with the lowest number yield grade have the highest yield of edible meat.

Beef Grades in Canada

The Canadian federal beef grading system differs from the USDA system in a number of ways. Instead of two separate grades, one for quality and one for yield, there is a single system. In all the Canadian grades, meat quality is indicated by a letter (A to E). In the top two grades, A and B, the letter is followed by a number from 1 to 4 which indicates the predicted yield. Two of the lower grades, C and D, are subdivided into classes. In the Canadian system, there is no measurement of marbling or longissimus dorsi area, but minimum standards for both factors must be satisfied for a carcass to be placed in the A grade. The carcass is placed into one of three possible *maturity divisions* (Table 3-4).

The thickness of the subcutaneous fat is now subjectively estimated or measured between ribs 12 and 13 at the minimum point of thickness in the fourth quarter from the vertebrae along the longitudinal axis of the longissimus dorsi and perpendicular to the outside surface of the fat (Figure 3-18). Provided that there is some evidence of marbling, subcutaneous fat depth becomes the primary determinant of yield grade. This greatly facilitates the prediction of carcass grade from live cattle, since subcutaneous fat can be assessed quite accurately. Table 3-5 gives a brief description of Canada grade A carcasses. The B1, B2, B3, and B4 grades are used for a few quite good

TABLE 3-4. DESCRIPTION OF THE MATURITY DIVISIONS FOR BEEF CARCASSES IN THE CANADIAN BEEF GRADING SYSTEM

Maturity Division I
- (a) Bones are soft, red, and porous when split.
- (b) Cartilagenous tips occur on the lumbar vertebrae.
- (c) Dorsal spines of thoracic vertebrae have no more than slight ossification.
- (d) Distinct divisions exist between sternebrae.

Maturity Division II
- (a) Sacral vertebrae are completely fused.
- (b) Lumbar vertebrae are capped with cartilage or they have a red line present on the tips.
- (c) Cartilagenous caps on the dorsal processes of the thoracic vertebrae are partially ossified.
- (d) Dorsal spines of vertebrae with some degree of redness as blood cells recede from the periphery.

Maturity Division III
- (a) Dorsal spines of vertebrae are hard, white, and flinty when split.
- (b) Sacral vertebrae are completely fused.
- (c) Cartilagenous caps on the lumbar vertebrae are ossified and have lost their red color.
- (d) Ribs are wide, flat, and white.
- (e) The sternum is in an advanced state of ossification.

Source: data from Canada Gazette (1983).

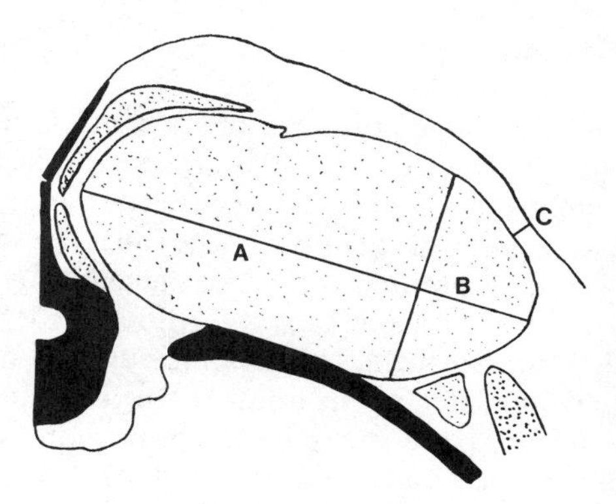

Figure 3-18 Measurement of subcutaneous fat depth for the Canadian beef grade: (A) long axis of the longissimus dorsi area; (B) fourth quarter of the area; (C) minimum fat depth over the fourth quarter measured perpendicularly to the outer surface.

TABLE 3-5. DESCRIPTION OF CANADA GRADE A BEEF CARCASSES

Maturity: division I.

Lean (rib-eye): firm, bright red in color, and with at least some marbling.

Fat cover: extensive, but sparse on hips or chuck.

Fat type: firm, white in color, or with a slight red or amber tinge.

Muscling: without angularity, depressed hip or chuck, or narrowness along the loin or rib.

Fat depth:	Canada A1	5 to 10 mm
	Canada A2	>10 to 15 mm
	Canada A3	>15 to 20 mm
	Canada A4	>20 mm

Source: data from Canada Gazette (1983).

carcasses which have just missed the A grades because of a lack of fat covering, because they have absolutely no marbling, because their lean is too dark, or because their fat is tinged with yellow. Like those in grade A, carcasses in the B grade must satisfy the requirements of maturity division I. The fat depth limits are the same as the A grades, except that the lower limit of fat depth for B1 is 2mm instead of 5mm.

C1 and C2 are two classes of Canada grade C carcasses. The numbers describe subdivisions on the basis of meat quality. The C grades are carcasses from either maturity divisions I or II that have failed to meet the minimum fat depth for the B grades. Those with grade B quality meat are called C1, and those with poorer meat quality are called C2. Thus a C2 carcass may have fat that is very soft or yellow, or the whole carcass may be poorly muscled. Like the B grade, only relatively small numbers of carcasses are placed in the C grade.

The D grades are for cows and steers in maturity division III. Cows with good muscle development, only pale-colored fat, and not too much fat are graded D1. There are two grades below this, D2 and D3, which are for cows and steers with yellower fat and poorer muscling. Cows or steers with excessive fat are graded D4, while carcasses with some degree of emaciation are graded D5.

Grade E is for mature stags or bulls that have dark, sticky meat. Stags are male cattle which have been castrated after the development of secondary sexual characteristics.

Grade Stamps on the Carcass

In both the USDA and the Canadian grading systems, a carcass requires a veterinary meat inspection stamp before it can be graded. In both systems, carcasses are marked in a preliminary manner immediately after being graded. Next, the length of the carcass is marked with the appropriate roller brand. The roller brand resembles a paint roller, and as it rotates it leaves a continuous track of the marking. The USDA roller brand uses abbreviations for two of the grades (STNDRD and CMRCL), and the grade appears inside the US federal shield. Canners and cutters are not stamped. The number of the yield grade is indicated inside another shield-shaped stamp.

Canadian carcasses have a square stamp on the primal cuts, and a roller brand down the length of the carcass and along the limbs. Red ink is used for grade A, blue ink for B, brown ink for C, and black ink for D and E. Thus, only a fraction of the roller brand need be visible to alert consumers to the quality grade of the carcass. In both the USDA and Canadian systems, codes are included in some of the grade stamps to identify the individual grader.

Pork Grades in the United States

Only two classes of pork carcasses are federally graded: (1) barrows and gilts; and (2) sows. Barrows and gilts are graded on the basis of meat quality and yield. However, unless an obvious defect such as oily fat is readily apparent, it is difficult to grade intact sides of pork for meat quality. Carcasses with poor-quality meat are graded as *utility*. The yield is predicted from a combination of average backfat thickness and carcass length or weight. Length takes precedence over weight when length and weight do not agree in the determination of the grade. The degree of muscle development is estimated subjectively on a six-point scale (very thick, thick, moderately thick, slighty thin, thin, and very thin). The degree of muscling may be used to adjust the backfat thickness for the calculation of the grade. For example, one degree of muscling greater than that expected on the basis of backfat thickness and carcass length allows 0.1 inch to be subtracted from the average backfat measurement. The depth of the backfat plus skin is measured at the level of the first rib, the last rib, and the last lumbar vertebra. Carcass length is measured from the anterior edge of the pubis to the anterior edge of rib 1. Carcass weight is measured as the actual, or estimated, hot carcass weight of a packer-style dressed carcass. The relationship between average backfat and carcass length in the determination of the grade is summarized in Table 3-6. Sows are graded on the basis of backfat thickness; *cull* (<1.1 inches), *medium* (1.1 to 1.5), U.S. No. 1 (1.5 to 1.9), U.S. No. 2 (1.9 to 2.3), and U.S. No. 3 (>2.3). The federal grading of pork carcasses is used far less often than the federal grading of beef carcasses.

TABLE 3-6. RELATIONSHIP OF AVERAGE BACKFAT THICKNESS AND CARCASS DIMENSIONS TO THE USDA YIELD GRADES FOR BARROWS AND GILTS WITH NORMAL MEAT QUALITY AND A CARCASS LENGTH OF 27 INCHES[a]

Grade	Backfat thickness (inches)
US No. 1	<1.3
US No. 2	1.3–1.6
US No. 3	1.6–1.9
US No. 4	>1.9

[a]For each additional inch of length, add 0.033 inch to the allowable backfat.

Pork Grades in Canada

Pork grades are used extensively in Canada to pay a producer for the amount of salable meat that has been produced. The system is based on the inverse linear relationship that exists between total backfat and the percentage yield of the ham and loin (Fredeen and Bowman, 1968). The dorsal spines of the thoracic vertebrae remain on the left side of the carcass when it is split into sides. Two maximum backfat depth measurements used to be taken and added together. Now only a single maximum measurement is made with no exact skeletal reference point. A grading table is used to calculate the grade or index from the combination of the backfat measurement and the warm carcass weight. The table is designed so that any carcass with an expected yield of 65.5 per cent trimmed cuts will receive a grade of 100. A few years ago, this used to be the national average. Carcasses with a lower expected yield receive a grade lower than 100, and carcasses with a higher expected yield receive a grade greater than 100. For example, if the combination of backfat and carcass weight indicates that a carcass will yield 70.5 per cent trimmed cuts, it might receive a grade of 105. This direct equivalence cannot be guaranteed, however, because the relationship between predicted yield and grade also contains a weighting factor for carcasses within a preferred weight range. With some exceptions, the relationship between predicted yield and grade is followed for other combinations of backfat and carcass weight. The exceptions are:

1. Ridgelings (cryptorchids) all grade at 67.
2. Carcasses with excessive backfat all grade at 80.
3. Carcasses of 210 pounds or more all grade at 80.
4. Three index points may be deducted for a badly shaped belly.
5. Ten index points may be deducted for abnormal fat color or texture.
6. Tissue trimmed off by a meat inspector because of defects of farm origin reduces the carcass weight.

The farm of origin is identified by a shoulder tattoo on the pork carcass, and the producer is paid the numerical product of the reported market price, the grade, and the carcass weight.

Veal and Lamb Grades

The USDA system has five grades for veal carcasses: *prime, choice, good, standard,* and *utility*. Carcasses in the higher grades have good muscle development and some evidence of the deposition of fat within the intercostal muscles. The Canadian system for veal grades is a greatly abbreviated version of the Canadian beef grading system, and is also based largely on the degree of muscle development and the degree of fat deposition.

In the USDA system, lamb carcasses are graded as *prime, choice, good, utility*, or *cull*. Similar grades are used for yearling and mutton carcasses, but there is no prime grade for mutton. A yield grade is calculated from the degree of muscle development in the leg, the amount of fat to be trimmed off the carcass, and the subcutaneous fat depth over the rib. The Canadian system for lamb and mutton is an abbreviated version of the beef grading system (A1 to A4, B, C1 and C2, D1 to D4, and E). Neither veal nor lamb grades are used very often.

Carcass Grading of Poultry

The basic principles of poultry carcass grading are similar in the United States (USDA, 1977) and Canada, although nearly all the details are different. The different species of poultry are separated into classes; the main ones are shown in Table 3-7.

Poultry with tender meat are identified by their soft, pliable, smooth-textured skin, and by their flexible sternal cartilage. The classes of poultry that have meat of an intermediate level of tenderness are identified by their greater age, coarse skin, and less flexible sternal cartilage. These include (1) *stag*, male chicken under 10 months; and (2) *yearling hen* and *tom turkeys* under 15 months. The mature classes of poultry are the

TABLE 3-7. COMMON CLASSES OF POULTRY CARCASSES WITH TENDER MEAT IN NORTH AMERICA

Species	Class	Age (weeks)	Carcass weight (kg)	Sex
Chicken	Rock Cornish	4–5	<0.8	M or F
Chicken	Broiler/fryer	5–8	0.8–1.8	M or F
Chicken	Roaster	>9	>1.8	M or F
Chicken	Capon	>9	>1.8	was M
Turkey	Fryer/roaster	12–16	<4.5	M or F
Turkey	Medium/young hen	18–20	4.5–7.5	F
Turkey	Heavy/young tom	20–24	>7.5	M
Duck	Broiler/fryer	<8	1.8–2.8	M or F
Duck	Roaster	<16		M or F
Goose	Young	15–20	2.5–6.5	M or F

rooster, mature hen chickens, mature hen or *tom turkeys, mature duck*, and *mature goose*. These classes have tough meat, coarse skin, and stiff sternal cartilages.

There are a number of additional features seen during and after slaughter that indicate the age of poultry.

1. In chickens, young birds have unwrinkled combs with sharp points. In older birds, the comb becomes wrinkled with blunt points.
2. Young ducks have soft bills, while older ducks have hard bills.
3. Plumage becomes worn and faded in older birds (unless they have just molted).
4. With age, subcutaneous fat becomes darker and becomes lumpy under the main feather tracts.
5. With age, the pelvic bones become less pliable.
6. With age, scales become larger, rough, and slightly raised.
7. In old birds, the oil sac becomes enlarged and sometimes hardened.
8. With age, male chickens and turkeys may develop long spurs.
9. With age, the cartilagenous rings of the trachea become stiff in ducks and geese.

After poultry have been placed in a class, they are graded for *condition* and *quality* after evisceration. Birds are rejected from grading or are trimmed if they exhibit bruises or pathological conditions of the carcass, particularly of the skin and musculature. By this time, the carcasses of sick and emaciated birds have already been eliminated by veterinary inspection. *Breast blisters* are caused by pressure or irritation as the bird rests on its sternum. Subcutaneous connective tissues are thickened and, since there is rarely a fluid-filled cavity, the condition should really be termed a *breast cyst* (McCune and Dellmann, 1968). The incidence of breast cysts increases with body weight and with the severity of environmental conditions that may cause skin irritation (Mayes, 1980).

Chickens may suffer from genetic *muscular dystrophy*, which leads to muscle fiber atrophy (Cardinet et al., 1972; McMurtry et al., 1972; Ashmore et al., 1973), and ducks may suffer from nutritional *myodegeneration* (George et al., 1973). Affected birds that survive to a market weight are easily detected. A more serious problem with regard to carcass grading is caused by degenerative conditions that occur deep in the pectoral muscle mass, particularly in turkeys (Sutherland, 1974; Jones et al., 1974). The early stages look like deep bruises with a bloody discoloration of the meat over the sternum. In later stages, whole tracts of muscle degenerate and are discolored by the green breakdown products of muscle and blood. The condition can be induced experimentally by a surgical disruption of the blood supply to the supracoracoideus muscle, and it is currently thought that *deep pectoral myopathy* may be vascular in origin (Orr and Riddell, 1977). In heavy birds, the supracoracoideus outgrows the space that is available within its osteofascial compartment. Strenuous muscle activity may then cause a lethal rise in intramuscular pressure (Martindale et al., 1979; Siller et al., 1979). The condition is difficult to detect in an intact carcass and its occurrence soon leads to justifiable complaints from consumers or processors. A bright light placed in the eviscerated body cavity may enable the condition to be detected if the resulting translucency of the flesh can be examined in a darkened room. Other causes for rejection on the basis of condition are incomplete evisceration, incomplete removal of pinfeathers, or surface contamination by blood or feces.

The principles of *quality* grading for poultry carcasses are shown in Table 3-8. In the Canadian grading system there are two grades, *utility* and *canner*, for carcasses with moderate and extensive removal of damaged parts, respectively.

TABLE 3-8. PRINCIPLES OF QUALITY GRADING FOR POULTRY CARCASSES

	Grade		
Quality parameter	A	B	C
Bilateral symmetry of sternum and pectoral muscles	Straight	Can be slightly crooked	Can be seriously curved
Lateral convexity and distal extension of pectoral muscles	Muscular	At least moderate	Can be poor
Fat cover over the clavicle, breast, and leg	Good	Enough to conceal muscles	Can be poor

Principles of Meat Grading

The essential features of the USDA and Canadian grading systems for beef and pork have been described in the preceding paragraphs. However, these features may well have been changed by the time that these pages are in print. In the past, grading systems have been changed to appease one or more segments of the meat industry, and in response to the slow emergence of provable facts from traditional beliefs.

Ideally, the difference in price between a carcass grade with a high yield and one with a low yield should be equal to the actual value of the difference in meat yield. If the price differential between a high-yield and a low-yield grade is less than it should be, progressive farmers who produce animals with high-yielding carcasses will not be fairly rewarded. Thus, as the amount of money at risk becomes greater, the accuracy and repeatability of grading must be increased.

If we work forward from the taxonomic studies of Mumford (1902), who might reasonably be regarded as the Linnaeus of carcass evaluation, we are faced with an overwhelmingly complex history of meat grading legislation. The main problem with the quality grading of beef carcasses relates primarily to animal age. Nearly all young cattle produce meat with an acceptable quality, provided that the meat is properly treated, but many older animals also produce good-quality meat. On average, over a wide range of conditions, only about 5 per cent of the variation in beef tenderness is due to marbling (Blumer, 1963). The traditional viewpoint is that intramuscular marbling fat contributes to the juiciness of meat, as perceived by consumers. There may be a relationship between these two parameters, but it is neither simple nor linear. Gaddis et al. (1950) showed that subjective scores for the quantity of juice in cooked meat increased in proportion to the percent fat in the fluid that could be pressed from the meat under pressure. However, this relationship was curvilinear. The two parameters were correlated up to 2 per cent fat in the juice but, beyond this point, additional fat in the juice produced no further increase in the subjective perception of juiciness. Given (1) that the high heat capacity and mobility of intramuscular fat might increase the thermal conductivity of meat during cooking; (2) that intramuscular fat might reduce the evaporation of water during cooking; and (3) that fatness might be associated with differences in carcass cooling rates after slaughter, we can hardly regard the degree of marbling as a simple variable factor.

At present, all the older animals that are detected are graded downward, just because some of them are deficient in quality. Some years ago, the presence of marbling fat was a reassurance to the consumer. It indicated that an animal had been deliberately reared or fattened to please the consumer. At worst, marbling fat was an indication that meat toughness due to the maturity of an animal would be offset by increased succulence. It is wrong to conclude from this that a young animal must have a lot of marbling fat to produce good-quality meat. Romans et al. (1965) found that members of a taste panel who could distinguish between beef from different USDA maturity groups could not distinguish any differences in tenderness due solely to the degree of marbling. Also, with taste panel testing, Breidenstein et al. (1968) were unable to detect any significant relationship between tenderness and marbling.

If we could produce animals whose total body fat was concentrated in their muscles, there would be no problem. At present, however, high levels of marbling fat can only be obtained by allowing the wasteful accumulation of fat around the viscera, between the muscles, and under the skin. More than any other factor, it is the arbitrary requirement for marbling fat that binds the cost of beef production to the price of fossil fuels (Ward et al., 1977). Breeds of cattle with a large size when they are mature are known to be the most efficient producers of muscle protein when they are young. Even though the meat from these young animals might be quite acceptable to the average consumer, there is little financial incentive to produce such meat if it is downgraded because it lacks marbling.

Marbling fat is quite difficult to measure. If the longissimus dorsi muscle is examined along its length, marked differences in the amount of marbling may be seen, even between nearby parts of the muscle (Blumer et al., 1962). The subjective evaluation of degrees of marbling is poorly understood in terms of visual psychology.

Some objective methods for the measurement of the area of marbling fat, such as the transmittance of light through high-contrast photographic transparencies (Cook and Bray, 1961), are only poorly related to subjective scores for marbling.

There is no guarantee that the range in subjective degrees of marbling is linearly related to the relative area of marbling fat or to the chemically determined triglyceride content. Some aspects of human visual perception are logarithmic in nature. How do we respond to steaks with finely particulate marbling relative to those with coarsely clumped marbling when the actual amount of marbling fat is the same? In histological sections, the per cent area occupied by marbling fat does not conform very well with the subjective assessment of degrees of marbling (Cooper et al., 1968). Transverse and longitudinal histological sections even yield different results. The magnitude of these problems is difficult to assess. Perhaps deviations from linearity are not too important and are restricted to the lower levels of marbling, where the bias that they create is unimportant (as in the data shown in Figure 3-16).

In the USDA system, beef is downgraded if it lacks high levels of marbling fat. In Canada, a similarly unreasonable response is invoked by yellow fat. The yellow carotene pigments in fat originate from the animal's feed (Palmer and Eckles, 1914). Yellow fat once indicated to the wary consumer that the beef was from an old animal and that it might be tough. White fat indicated that the animal had been deliberately reared to please the consumer. At worst, it indicated that the fat pigments of an old animal had been diluted by new fat (Forrest, 1981), which would tend to offset the toughness of mature beef. It is wrong to conclude from this that beef from a young animal will be tough simply because its fat has acquired extra pigment from a feed such as corn or grass.

Carotene is often added to butter to make it appear more yellow. The presence of natural carotene in fat causes no changes in the keeping properties of meat fat (Craig et al., 1959). Thus there is no reason to reject carotene in raw meat fat, especially since cooking methods such as roasting and broiling give much of the fat an amber color, regardless of its whiteness when raw. An overwhelming weight of evidence indicates that fat color by itself has no undesirable relationship to the eating qualities of beef (Bull et al., 1941). Although consumers can detect differences in taste between beef with yellow fat and beef with white fat, both tastes are equally acceptable (Malphrus, 1957).

A cynical person might suspect that the survival of standards for marbling and fat color in the beef grading systems of North America is a tacit admission that there is something lacking in the evaluation of animal age from the degree of skeletal ossification. A pessimist might suspect that the standards for marbling and fat color are allowed to persist in the grading system because it is too difficult to explain to consumers that yellow fat and low levels of marbling are acceptable in beef from young animals.

In cattle, the *dentition* can be used as an alternative to skeletal ossification for the estimation of animal age or maturity (Luckock, 1976). Since the animal's head is removed before the carcass is eviscerated, the information on the state of the dentition must be attached to the carcass in some way. A major advantage of the estimation of age from animal dentition is that it can also be undertaken on live animals before slaughter.

Much of the data on tooth eruption found in textbooks of veterinary anatomy was collected in the last century and relates to what we would now call primitive types

TABLE 3-9. AGES AT WHICH PERMANENT INCISOR TEETH ERUPT IN BEEF COWS

Numbers of pairs of incisor teeth	Average age (days)	Standard deviation
1	732	22.5
2	977	76.5
3	1220	129.9
4	1512	170.4

Source: data from Graham and Price, 1982.

of cattle. Some more recent data are given in Table 3-9. Considerable variation is caused by the season of birth and by animal breed. For data on dairy breeds, consult Wiener and Forster (1982). Graham and Price (1982) conclude that dentition is a more accurate measure of animal age than skeletal ossification. For grading purposes, the most critical age group for estimation would be between 1.5 and 2.5 years of age. The accuracy of estimation can be increased by incorporating some measure of the exposure of the second molar, since the anterior part of the tooth erupts before the posterior part (Andrews, 1981). However, this might be difficult to assess under abattoir conditions.

Yield grades are based on the simple principle that meat yield is directly proportional to carcass weight, but is inversely proportional to carcass fatness (Figure 3-19). Thus once both carcass weight and fatness have been taken into account, the inclusion of further data in the grading formula may be a wasted effort (Thompson and Atkins, 1980). The main problem in the application of this simple principle (Figure 3-19) to yield grading is that it is difficult to find a simple measure of carcass fatness. Although grading systems contain precise instructions on where fat depth measurements are to be made, the grader is also required to adjust any objective measurement that does not agree with his or her subjective evaluation of fat deposition in other areas. This is an attempt to deal with a complex problem—fat is not always spread uniformly through the carcass.

Consider a hypothetical case in which a student of geometry is given some cubes of fat, and is requested to estimate their weight from a linear measurement. A reasonable solution to this problem would be to multiply the third power of the height of each cube by the density of fat. The geometry of subcutaneous fat deposition is more complex, and it resembles the geometry of a puddle of rainwater which collects on the

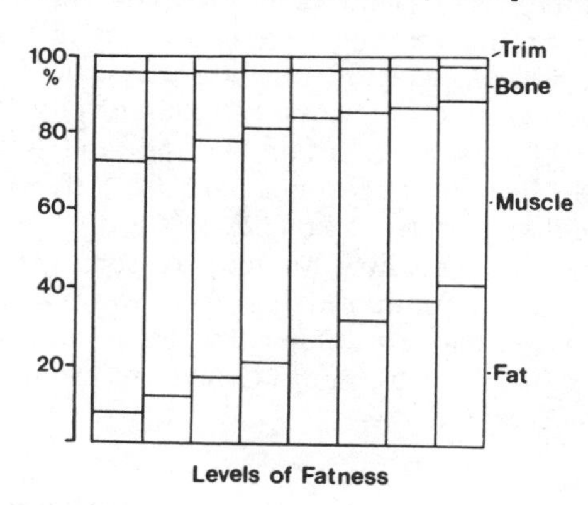

Figure 3-19 Decrease in per cent muscle with increasing fatness in beef carcasses. (Data from Callow, 1948.)

surface of an irregularly concave road. As it rains, the puddle gets deeper and spreads sideways. The depth of the puddle at this stage of flooding is a poor measure of the volume of water. When the puddle reaches a sidewalk or curb and can no longer spread sideways, its depth becomes a better measure of volume. In a similar manner, as subcutaneous fat accumulates in the carcass, it gets deeper and spreads over more of the carcass. When the spreading of fat becomes limited by anatomical constraints, fat depth becomes a better measure of fat volume. Thus both the sensible decisions made by the meat grader on a subjective basis and the geometry of fat deposition suggest that the relationship of subcutaneous fat depth and carcass fatness is very complex. If empirical evidence shows that fat depth measurements are sometimes directly related to carcass fatness by a straight-line relationship, this is a fortuitous departure from the laws of simple geometry and is probably due to the irregular shape of the carcass fat. In some cases, such as in the evaluation of meat yield in beef and lamb carcasses, subjective scoring of the external fat cover can be quite accurate (Chadwick and Kempster, 1979; Kempster et al., 1976).

In meat grading, the position at which a fat depth measurement is made is critical. In pork carcasses, for example, fat depth over the rump may be more strongly correlated with total carcass fat than is fat depth over the shoulder (McMeekan, 1941). Woodman et al. (1936) found that the distribution of subcutaneous backfat in very fat pigs was weighted toward the shoulder region. Differences in the depth of backfat between the shoulder and the loin are more extreme in late-maturing breeds relative to early-maturing breeds of pigs (Hammond and Murray, 1937). Although this research is outdated, it may indicate the biases that are likely with very fat pigs at the present time.

Some degree of irregularity in the distribution of subcutaneous adipose tissue is generated by the manner in which the carcass is handled as it is chilled, since fat is fluid at body temperature. Reference to Figures 3-17 and 3-18 will demonstrate how the exact position of measurement might affect the fat depth measurement. In the rib shown in Figure 3-18, USDA and Canadian fat depth measurements would differ. In other national grading systems, fat depth measurements made over both the medial and lateral edges of the longissimus dorsi muscle have been considered (Gifford, 1973).

Electronic Grading

The future of the carcass grading systems in any particular country is usually a subject for lively debate. Some experts may claim that their national grading system is basically sound and that only minor revisions are needed; at the same time, other experts may be demanding radical changes in the same system. The future of the USDA grading system has been debated by Cross (1980), Smith (1980), and Carpenter (1980). The grading systems of Australia, Britain, Ireland, and New Zealand were considered in an Australian Meat Board Symposium (1976). The future of carcass grading in Canada has been discussed by Price (1982). It is difficult to predict what changes will actually be made in each country, but there are some general features that must be taken into account in any future developments. No modern industry can afford to ignore the advantages that are offered by the application of computer technology to routine tasks. The automation of fat depth measurements is possible, but automation brings its own problems.

At a practical level, the future of electronic grading will depend on the type of probe that is adopted as being most reliable, and on the anatomical position at which the probe is most reliable. In pork carcasses, measurements of the fat depth over the posterior thoracic region of the longissimus dorsi are particularly useful in the prediction of carcass lean content (Kempster and Evans, 1979b; Fahey et al., 1977). Kempster and Evans (1979a) showed that optical probe measurements made over the longissimus dorsi could give a more accurate measurement of carcass fatness than could direct measurement of the exposed fat on the split carcass. Fredeen and Weiss (1981) compared the results from three instruments: ruler measurements, an ultrasonic probe, and a conductivity probe. The ruler and the conductivity probe worked best in the midline over the gluteus medius, while the ultrasonic probe worked best lateral to the midline in the lumbar region. All three measurements gave equally valid data, but the incorporation of further measurements on longissimus muscle depth improved the accuracy of prediction. The depth of the longissimus dorsi can be measured quite easily in an eviscerated carcass if a probe is pushed right through from the dorsal surface to the body cavity. The main importance of fat depth probes is the estimation of the lean meat yield of the carcass. However, it has also been suggested that measurements of subcutaneous fat depth could be used to replace the present USDA quality grading system for the prediction of beef palatability (Dolezal et al., 1982).

Probes for the measurement of subcutaneous fat depth have been available for many years, but only recently have they attracted serious attention for meat grading. Two general factors have contributed to this development: (1) grading systems are becoming less subjective and more reliant on objective measurements; and (2) the miniaturization of electrical components has greatly reduced the power requirements, the size, and the weight of meat probes.

Figure 3-20 summarizes the operating principles of fat depth probes. Information is needed on the depth of the probe tip in the carcass. In early prototypes, this information was obtained simply by looking at ruler markings which were engraved along the probe. In automated probes, the information on probe depth may be obtained electrically by any of the methods normally used to measure the relative position of moving parts in physiological transducers. Three simplified possibilities based on systems used in physiological transducers are shown in Figure 3-20. The

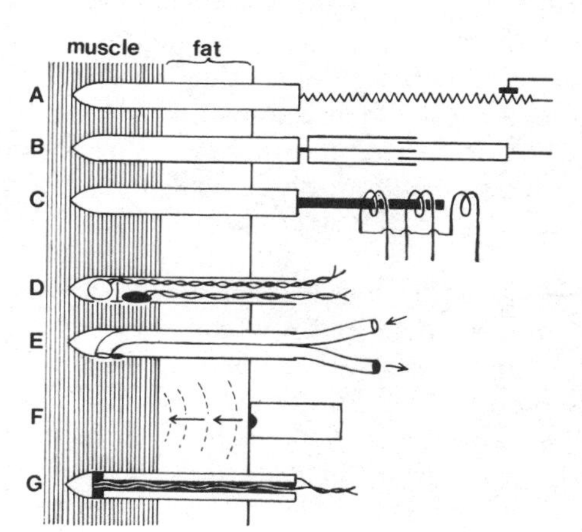

Figure 3-20 Fat depth probes: (A) penetration depth measured with a resistor; (B) penetration depth measured with a capacitor; (C) penetration depth measured with a transformer; (D) boundary between muscle and fat detected optically by an internal photometer; (E) boundary between muscle and fat detected by fiber optics; (F) ultrasonic detection of the boundary between muscle and fat; (G) boundary between muscle and fat detected by electrical conductance.

extent of probe penetration may be monitored by (A) movement of a sliding contact along a *resistor* (potentiometer), (B) displacement of the plates of a *capacitor*, or (C) displacement of the sliding core of a *linear variable differential transformer* (LVDT). With the LVDT, an alternating current in the central energizing coil induces equal voltages in the left and right sensor coils. The sensor coils are wired in reversed series so that the induced current is canceled when the core is situated symmetrically between the sensor coils. When the core is displaced, a voltage is generated in the sensor coil circuit and this is proportional to the depth of the probe in the carcass.

With *optical probes*, two basic configurations are possible (Figure 3-20). The light source and the *photometer* may be located within the barrel of the probe (Figure 3-20 D), or they may be external to the barrel of the probe and connected by *fiber optics* (Figure 3-20E). Optical fibers are composed of long fibers of transparent plastic, glass, or silica which are coated with a substance, the cladding, which has a slightly lower refractive index than the fiber core. Light waves cannot leave the core because the angle at which they strike the cladding causes the cladding to act as a mirror. Thus light is conducted along the fibers, following the contours of the waveguide or bundle of optical fibers. The principle of optical probes is that they detect the transition from fat (white) to lean (dark). More recent fat depth probes are *radiometric* in nature and they use invisible *infrared light*. The peak emission wavelength of commonly available infrared-emitting diodes is in the near-infrared range. Fat reflects more infrared light than muscle does, probably because fat contains less water. The advantage of infrared light is that it allows better resolution of boundaries between subcutaneous fat and abnormally pale muscles. However, even a thin paste of meat juice over the radiometer aperture can change the infrared reflectance of fat.

One of the basic technical problems found in fat depth probes is the *window effect*. The window effect can be explained by a simple analogy. Consider a situation in which the sun is shining through the window of a room onto a large photometer. The window is fitted with an opaque blind that can be drawn vertically downward to block the light coming through the window. By experiment, it would be possible to find a photometer reading that corresponded to the position of the blind when it is halfway down the window. This photometer reading might be used to identify the halfway position so that a digital electronic circuit could then recognize the "blind-up" and the "blind-down" positions. In just the same way, the photometer of a meat probe recognizes the depth within the carcass at which the dark-colored lean is halfway across the photometer aperture. In the analogous situation of the window blind, if a translucent blind is substituted for the opaque blind in the window, the photometer will give biased readings. The translucent blind will have traveled more than halfway down the window before the photometer threshold (half-light intensity) is reached. In just the same way, the photometer in a fat-depth probe may give biased reading with abnormally pale meat.

To minimize the magnitude of the bias created by the window effect, the window size or aperture over the photometer in the probe is made as small as possible relative to the expected fat depth. Although this does minimize the window effect, two further problems are then encountered. First, the small window now reacts to small boundaries such as blood clots in the fat and marbling fat in the lean. Second, the reduction in the size of the photometer aperture reduces the amount of light that reaches the photometer. Although this reduction can be offset by increasing the sensitivity of the photometer, it is difficult to increase the sensitivity to a high level without causing an undesirable change in the signal-to-noise ratio. Thus, as the background noise rises, it becomes progressively more difficult to identify the meaningful information.

To overcome the problem caused by the reduction of the light intensity that reaches the photometer, the intensity of the light source may be increased. The photometer window, however, now reacts to an enlarged cone of diffuse light outside the window, so that the operational window size is increased and the window effect again becomes noticeable. A more effective solution to this problem is to position the light source exit and the photometer aperture input side by side in a plane that is perpendicular to the long axis of the probe. The *y* axis of the window effect is then reduced to the width of the exit or the aperture, rather than being the distance between the exit and the aperture. In other words, both the light source and the photometer aperture now cross the muscle-fat boundary simultaneously. However, with this configuration, we now find that we have a probe with a large diameter which tends to deform carcass tissues on initial penetration. Three of the commercially available fat depth probes were evaluated by Kempster et al. (1981) and were found to be similar in accuracy. Thus the main technical problem is not how to make a fairly accurate probe, but how to make one that will not fall apart after being used a few hundred thousand times.

A general problem with all probes is the extent to which they depress the surfaces and tissue boundaries that they penetrate. Warm carcasses are more easily depressed than cold ones. More accurate measurements can usually be made when a probe is withdrawn from a carcass rather than when the probe first enters the carcass. The future technical evolution of the meat probe may be as a double-purpose instrument that measures meat paleness on entry to the muscle, and muscle and subcutaneous fat depth as it is withdrawn.

Another contender for fat depth measurements is the *ultrasonic* method (Figure 3-20). The time interval between the start of a high-frequency sound pulse and the return of the echo from the fat to lean boundary is measured and is used to calculate the depth of fat. The main technical problem is that weak echoes are returned from layers or septa of connective tissue in the depth of the subcutaneous fat.

The working principle of the *conductivity probe* (Figure 3-20) is that fat has a low electrical conductivity, whereas lean meat has a high conductivity because of its greater content of aqueous electrolytes. Thus as the conductivity probe passes through the subcutaneous fat and enters a muscle, an appreciable current can be carried between the otherwise electrically isolated electrodes of the probe. However, conductivity measurements are readily biased by any water that is carried inward from the wet surface of the carcass. The conductivity probe was originally developed for measuring the backfat of live pigs (Pearson et al., 1957), but was superseded by ultrasonic methods. In actual practice, the electrodes are usually formed by narrow metal rings separated by insulating rings along the shaft of the probe. The window effect is also found with conductivity probes, and its magnitude increases in direct proportion to the insulated distance between the tip and shaft electrodes. The conductivity probe can be constructed in an alternative configuration with separate parallel electrodes. However, penetration of layers of connective tissue with a high conductivity may then simulate the penetration of a muscle. A technical problem in the automation of the conductivity probe is that conductivity is directly proportional to temperature. Thus the conductivity of lean meat at body temperature soon after slaughter is considerably greater than that of the lean in a chilled carcass (Swatland, 1980).

Fat depth probes can be used to predict the lean content of the carcass by a process of elimination. A constant proportion of the carcass weight is assumed to be bone and, from the remainder, the estimated amount of fat is subtracted to leave the amount of lean. A more direct approach is to measure the amount of muscle directly. For many years, this has been a standard technique in agricultural research. A number of methods have been developed based on three possibilities:

1. linear measurements (typically the depth and width of certain muscles);
2. area measurements of cross-sectioned muscles;
3. weight measurements of specific muscles that have been removed from the carcass

In grading, the need for speed and nondestructive assessment rules out the third possibility. The first two possibilities, however, can be automated by *image analysis* using a computer and a television scanner, instead of the ruler (McMeekan, 1941) and grid (Henderson et al., 1966) methods used manually in research. Butterfield et al. (1977) investigated the use of image analysis as a research technique. Frozen lamb carcasses were sectioned on a band saw and the cut surfaces were photographed. From scans of 5 and 11 sections, 89 and 99 per cent (respectively) of the variation in percentage muscle could be accounted for or predicted. However, with just one scan of the cross-sectioned longissimus dorsi of a beef carcass, the accuracy of prediction in carcass grading would be very much lower (Cole et al., 1960).

Despite the fact that area measurements of the longissimus dorsi are or have been widely used in carcass grading, it is difficult to show that such measurements are reliable predictors of meat yield (Goll et al., 1961). Correlations ($r < 0.5$) of longissimus dorsi area with retail yields can be identified (Cole et al., 1960; Brungardt and Bray, 1963) but when adjustments are made for differences in carcass weight and fatness, the residual correlations are of little practical value. As one would expect on geometrical grounds, the relationship of longissimus *area* to carcass *weight* is nonlinear (Field et al., 1963). Thus the use of image analysis in carcass grading is limited unless, somehow, data can be collected from extensive cross-sectional areas of the carcass. A more feasible application of image analysis techniques in the meat industry is the quantification of the areas of muscle and fat that appear in cuts of meat. This might lead to the widespread use of automated techniques for meat cutting, for the recognition of cuts of meat, and for the on-line measurement of the muscle-to-fat ratio.

From an operational viewpoint, meat grading can be resolved into three main activities:

1. the collection of data that are thought to be related to meat quality and to meat yield;
2. the arithmetic calculation of the grade;
3. the recording of the grade, either by marking the carcass or by paying the producer.

The collection of data may change with the improvement of electronic methods for the measurement of carcass properties, or by a shift in the consensus of expert opinion. The marking of carcasses may change as new methods are invented. Our present grading stamps are technologically no further advanced than branding irons. Alphanumeric systems require a large keyboard for data entry and could easily be simplified to annotated bar codes or matrix patterns. The arithmetic calculation of the grade is the most flexible part of the system. It provides the means for modifying the market strategy of producers and, ultimately, the way in which most of the meat is produced. In Canada, for example, slight arithmetic changes in the pork grading table were soon followed by changes in the average weights of pigs sent to market. In this case, neither the collection of carcass measurements nor the method of payment were changed.

At the beginning of this review of meat grading, the federal grading system was cast in the role of an impartial third party whose objective was to facilitate commerce. With hindsight, it can now be seen that the federal grading of meat also directs, as well as facilitates commerce. The existence of a meaningful grading system is an essential prerequisite to the development of computerized trading systems at the wholesale level (Engelman et al., 1979). In the future, we may look back at most of the present grading systems as being *fixed systems*, in which the grade was fixed only by the nature of the carcass. *Flexible systems* may become more common in the future, and the grade may be determined by both the nature of the carcass, and changeable specifications in accordance with planned policies or market conditions. The logical outcome of this would be to control a variable pattern of meat cutting by the use of information on carcass yield and market conditions. The practical feasibility of this has already been demonstrated (Anon., 1975). Electronic systems for the identification of individual live animals are just starting to be used on a large scale (Holm, 1981). Such systems could easily be carried on to the time of carcass grading, thus linking the complete production, grading, and meat trading system. Perhaps in the future, instead of waiting for the meat grader, we may be waiting for the electronics repairperson.

REFERENCES

ADSRI, 1981. *The Cuts of a Beef Carcass.* Anim. Dairy Sci. Res. Inst., S. Afr. Dept. Agric. Fish., Tech. Commun. 170.

ANDREWS, A. H. 1981. *Res. Vet. Sci.* 31:65.

ANON. 1975. *Natl. Provisioner* 173(7):14.

ASHMORE, C. R., ADDIS, P. B., DOERR, L., and STOKES, H. 1973. *J. Histochem. Cytochem.* 21:266.

Australian Meat Board. 1976. *Proceedings of a Symposium on Carcase Classification.* Australian Meat Board, Sydney.

BLUMER, T. N. 1963. *J. Anim. Sci.* 22:771.

BLUMER, T. N., CRAIG, H. B., PIERCE, E. A., SMART, W. W. G., and WISE, M. B. 1962. *J. Anim. Sci.* 21:935.

BREIDENSTEIN, B. B., COOPER, C. C., CASSENS, R. G., EVANS, G., and BRAY, R. W. 1968. *J. Anim. Sci.* 27:1532.

BRUNGARDT, V. H., and BRAY, R. W. 1963. *J. Anim. Sci.* 22:177.

BULL, S., SNAPP, R. R., and RUSK, H. P. 1941. *Univ. Ill. Agric. Exp. Stn. Bull. 475.*

BUTTERFIELD, R. M. PINCHBECK, Y., ZAMORA, J., and GARDNER, I. 1977. *Livestock Prod. Sci.* 4:283.
CALLOW, E. H. 1948. *J. Agric. Sci., Camb.* 38:174.
Canada Gazette. 1983. Regulations respecting the grading of beef carcasses. Canada Gazette, Part I, January 8, pp. 288-298.
CARDINET, G. H., FREEDLAND, R. A., TYLER, W. S., and JULIAN, L. M. 1972. *Am. J. Vet. Res.* 33:1671.
CARPENTER, Z. L. 1980. *Proc. Reciprocal Meat Conf.* 33:100.
CHADWICK, J. P., and KEMPSTER, A. J. 1979. *Anim. Prod.* 28:442.
COLE, J. W., ORME, L. E., and KINCAID, C. M. 1960. *J. Anim. Sci.* 19:89.
COOK, C. F., and BRAY, R. W. 1961. *Food Technol.* 15:540.
COOPER, C. C., BREIDENSTEIN, B. B., CASSENS, R. G., EVANS, G., and BRAY, R. W. 1968. *J. Anim. Sci.* 27:1542.
CRAIG, H. B., BLUMER, T. N., and BARRICK, E. R. 1959. *J. Anim. Sci.* 18:241.
CROSS, H. R. 1980. *Proc. Reciprocal Meat Conf.* 33:87.
DOLEZAL, H. G., SMITH, G. C., SAVELL, J. W., and CARPENTER, Z. L. 1982. *J. Food Sci.* 47:397.
ENGELMAN, G., HOLDER, D. L., and PAUL, A. B. 1979. *The Feasibility of Electronic Marketing for the Wholesale Meat Trade.* USDA AMS-583.
FAHEY, T. J., SCHAEFER, D. M., KAUFFMAN, R. G., EPLEY, R. J., GOULD, P. F., ROMANS, J. R., SMITH, G. C., and TOPEL, D. G. 1977. *J. Anim. Sci.* 44:8.
FIELD, R. A., SCHOONOVER, C. O., NELMS, G. E., and KERCHER, C. J. 1963. *J. Anim. Sci.* 22:826.
FORREST, R. J. 1981. *Can. J. Anim. Sci.* 61:575.
FREDEEN, H. T., and BOWMAN, G. H. 1968. *Can. J. Anim. Sci.* 48:117.
FREDEEN, H. T., and WEISS, G. M. 1981. *Can. J. Anim. Sci.* 61:319.
GADDIS, A. M., HANKINS, O. G., and HINER, R. L. 1950. *Food Technol.* 4:498.
GEORGE, J. C., JOHN, T. M., MORAN, E. T., SWEENY, P. R., BROWN, R. G., and STANLEY, D. W. 1973. *Can. J. Zool.* 51:383.
GIFFORD, D. R. 1973. *Beef Carcass Appraisal.* S. Aust. Dept. Agric. Ext. Bull. 19.
GOLL, D. E., HAZEL, L. N., and KLINE, E. A. 1961. *J. Anim. Sci.* 20:264.
GRAHAM, W. C., and PRICE, M. A. 1982. *Can. J. Anim. Sci.* 62:745.
HAMMOND, J., and MURRAY, G. N. 1937. *J. Agric. Sci., Camb.* 27:394.
HEDRICK, H. B., MILLER, J. C., THOMPSON, G. B., and FREITAG, R. R. 1965. *J. Anim. Sci.* 24:333.
HENDERSON, D. W., GOLL, D. E., STROMER, M. H., WALTER, M. J., KLINE, E. A., and RUST, R. E. 1966. *J. Anim. Sci.* 25:334.
HINER, R. L., and HANKINS, O. G. 1950. *J. Anim. Sci.* 9:347.
HOLM, D. M. 1981. *J. Anim. Sci.* 53:524.
JONES, J. M., KING, N. R., and MULLINER, M. M. 1974. *Br. Poult. Sci.* 15:191.
KEMPSTER, A. J., and EVANS, D. G. 1979a. *Anim. Prod.* 28:87.
KEMPSTER, A. J., and EVANS, D. G. 1979b. *Anim. Prod.* 28:97.
KEMPSTER, A. J., AVIS, P. R. D., CUTHBERTSON, A., and HARRINGTON, G. 1976. *J. Agric. Sci., Camb.* 86:23.
KEMPSTER, A. J., CHADWICK, J. P., JONES, D. W., and CUTHBERTSON, A. 1981. *Anim. Prod.* 33:319.
LUCKOCK, C. R. 1976. The development of an objective beef carcase classification scheme for Australia. In *Proceedings of a Symposium on Carcass Classification.* Australian Meat Board, Sydney.
MALPHRUS, L. D. 1957. *Food Res.* 22:342.
MARINI, P. J., and GOODWIN, T. L. 1973. *Poult. Sci.* 52:2191.
MARTINDALE, L., SILLER, W. G., and WIGHT, P. A. L. 1979. *Avian Pathol.* 8:425.
MAYES, F. J. 1980. *Br. Poult. Sci.* 21:497.
MCCUNE, E. L., and DELLMANN, H. D. 1968. *Poult. Sci.* 47:852.

McMeekan, C. P. 1941. *J. Agric. Sci., Camb.* 31:1.
McMurtry, S. L., Julian, L. M., and Asmundson, V. S. 1972. *Arch. Pathol.* 94:217.
Moran, E. T. 1977. Growth and meat yield in poultry. In K. N. Boorman and B. J. Wilson (eds.), *Growth and Poultry Meat Production*. Proc. 12th Poult. Sci. Symp. Longman, Edinburgh.
Mumford, H. W. 1902. *Univ. Ill. Agric. Exp. Stn. Bull. 78*.
OEEC. 1961. *Meat Cuts in OEEC Member Countries*. Eur. Prod. Agen., Org. Eur. Econ. Co-op., Doc. 32.
Orr, J. P., and Riddell, C. 1977. *Am. J. Vet. Res.* 38:1237.
Palmer, L. S., and Eckles, C. H. 1914. *J. Biol. Chem.* 17:223.
Pearson, A. M., Price, J. F., Hoefer, J. A., Bratzler, L. J., and Magee, W. T. 1957. *J. Anim. Sci.* 16:481.
Price, M. A. 1982. *Can. J. Anim. Sci.* 62:3.
Romans, J. R., Tuma, H. J., and Tucker, W. L. 1965. *J. Anim. Sci.* 24:681.
Siller, W. G., Martindale, L., and Wight, P. A. L. 1979. *Avian Pathol.* 8:301.
Smith, G. C. 1980. *Proc. Reciprocal Meat Conf.* 33:89.
Strandine, E. J., Koonz, C. H., and Ramsbottom, J. M. 1949. *J. Anim. Sci.* 8:483.
Strother, J. W. 1975. The commercial preparation of fresh meat at wholesale and retail levels. In D. J. A. Cole and R. A. Lawrie (eds.), *Meat*. pp. 183–204. AVI Publishing Company, Westport, Conn.
Sutherland, I. R. 1974. *Can. Vet. J.* 15:77.
Swatland, H. J. 1980. *J. Anim. Sci.* 50:67.
Thompson, J. M., and Atkins, K. D. 1980. *Aust. J. Exp. Agric. Anim. Husb.* 20:144.
USDA 1977. *Poultry Grading Manual*. Agricultural Handbook 31, USDA, Washington, D.C.
Vanden Berge, J. C. 1979. Myologia. In J. J. Baumel (ed.), *Nomina Anatomica Avium*, pp. 175–220. Academic Press, New York.
Ward, G. M., Knox, P. L., and Hobson, B. W. 1977. *Science* 198:265.
Wiener, G., and Forster, J. 1982. *Anim. Prod.* 35:367.
Woodman, H. E., Evans, R. E., and Callow, E. H. 1936. *J. Agric. Sci., Camb.* 26:546.

4

Anatomical Distribution of Carcass Muscles

INTRODUCTION

The amount and distribution of meat on a carcass can be determined accurately only by complete carcass dissection. The distribution of meat is important because it affects the commercial value of a carcass. A carcass with heavy muscling in the chuck and brisket is worth less than a carcass of the same weight that has a well-muscled rump and loin, and a light forequarter. Differences in muscle distribution between breeds are now thought to be relatively minor in magnitude, but commercially important differences are found between sexes and between animals at different physiological ages. At first sight, carcass myology may appear to be a difficult subject to learn. However, if it is tackled in a deductive manner, it becomes relatively simple. This chapter contains, as a reference source, a condensation of the information that is required for the identification of carcass muscles in cattle, sheep, and pigs.

The muscles are considered here in groups that correspond approximately to the different regions of the carcass. Never cut a muscle attachment until a note has been made of the part about to be severed, since many muscles cannot be identified with any degree of certainty once they have been removed from the carcass. In a surface view of a carcass, many muscles are concealed, either by overlying muscles or by fat. Thus subcutaneous fat should be removed before carcasses are dissected. Large thin sheets of cutaneous muscle may be located within the subcutaneous fat. An arbitrary decision must be made as to whether the fat located medially to a cutaneous muscle is to be regarded as subcutaneous fat or as intermuscular fat. In favor of the former identification is the continuity of this fat depot with unquestionable subcutaneous fat beyond the limits of the cutaneous muscle. In favor of the latter identification is that, by definition, fat located between two striated skeletal muscles is intermuscular. Cross sections through the carcass are useful for the identification of hidden muscles. Atlases of cross-sectional views are available for beef (Tucker et al., 1952), pork (Kauffman and St. Clair, 1965; Briskey et al., 1958), and lamb (Kauffman et al., 1963). Generalized plans of carcass cross sections were given in Chapter 3.

The terms *origin* and *insertion* are used in a particular manner for the points at which a muscle is attached to the skeleton or to a sheet of connective tissue (fascia). Most muscles move parts of the body relative to the main trunk of the body. The origin of a muscle is its attachment onto the main trunk of the body, or onto something that is not far removed from the main trunk. The insertion of a muscle is its attachment, directly or indirectly, onto the body part that is being moved. In a simple case, such as a small muscle located in the distal part of a limb, this nomenclature is uncomplicated: muscle origins are proximal (toward the body) while muscle insertions are distal (away from the body). However, with muscles that are located within the trunk of the body and which serve to flex the body, it is not easy to distinguish origins from insertions.

This introduction to carcass myology follows the format of an identification key, and is similar in principle to those that are used to identify animals and plants. The key is of the "either-or" type, where the appropriate answer to each question leads to a further question and, ultimately, to the name of a muscle. The key is constructed so as to enable the identification of any muscle that might be found in a commercial beef, pork, or lamb carcass. The muscle to be identified is first partly dissected to reveal its skeletal attachments. The slight differences between the myology of cattle and sheep, and the greater differences between cattle and pigs, create some problems. This key has been constructed for beef muscles. When it is used for pork or lamb, the brief notes indicated in the text should be consulted since they may modify the choice of an answer. If less than a whole side of a carcass is dissected, reference may be made to the diagrams in Chapter 3. An attempt has been made to avoid identifications based on easily missed attachments or on damaged areas of the commercial carcass.

Order of Dissection

The general order in which muscles can be removed from the carcass is determined by their depth, since superficial muscles must be removed to gain access to deeper ones. Many of the major muscles of the beef carcass were identified in the diagrams of Chapter 3 so that, after removal of the subcutaneous fat from the carcass, many muscles can be visually identified and removed. Be careful not to remove by accident any of the separate small muscles which have been omitted from the figures in Chapter 3. The identification key then provides a logical sequence on which to base the removal of all the remaining muscles around the major units of the skeleton. Keep a checklist of muscles that have been removed. If the key fails at any point, list all the "downstream" possibilities and eliminate them individually by working the key in the reverse sequence.

Unidentified slips of muscle can be collected as a category called trimmings. Remember that muscles lose weight by evaporation. Parts of the carcass that are awaiting dissection should be covered with a damp shroud. Similar precautions should be taken during any delay between the dissection and weighing of muscles. The external tendons that are present on many muscles should be severed where the muscle tissue starts, even though this may leave an extensive intramuscular tendon. A category called connective tissue trimmings can be used for tendons, aponeuroses, and fasciae. Ligaments and joint capsules are best considered as part of the skeleton.

The cutaneous muscles cover the flank or shoulder of the dressed carcass. They are quite thin and, except in the case of the porcine cutaneus muscle that attaches to the sternum, the cutaneous muscles have no direct attachment onto the skeleton (*group 1 muscles*). Many limb muscles are restricted to a distal location and have no direct origin either onto the scapula, in the case of forelimb muscles, or onto the pelvis, in the case of hindlimb muscles. Forelimb and hindlimb muscles of this type are described as *group 2* or *group 3 muscles*, respectively. The remaining muscles of the proximal or upper parts of the limbs mostly have a major attachment to either the scapula (*group 4 muscles*) or to the pelvis (*group 5 muscles*). There are a number of skeletal muscles that move the lips, cheek, nostrils, and tongue, but only the jaw muscles have any commercial importance (*group 6 muscles*). The remaining muscles of the carcass are all associated with part of the axial skeleton. *Group 7 muscles* are those with a direct attachment to the vertebrae of the tail. Those with no direct caudal attachment are either attached directly to the ribs or body of the sternum (*group 8 muscles*) or are associated with the neck (*group 9 muscles*).

There are differences between beef, pork, and lamb carcasses. These differences are indicated by a series of notes to be taken into account when answering certain questions. For example, the note "(pig 32)" in the first question should be consulted if a pork carcass is being dissected, and so on, through the key. The notes relating to pigs and sheep are located toward the end of this chapter. In the notes for the pig, a separate key is given for the distal limb muscles. This key is intended for intact carcasses that retain their feet. If the feet are missing or damaged, proceed with the main key.

KEY TO THE MUSCLE GROUPS

1. either: a large flat muscle covering the flank or shoulder of the carcass and without any direct attachment to any part of the skeleton .... GROUP 1 CUTANEOUS MUSCLES (pig 32)
   or: a muscle excluded from the above 2
2. either: a muscle with an attachment to a limb bone but without any attachment to either scapula or pelvis 3
   or: a muscle excluded from the above 4
3. either: a muscle of the forelimb .... GROUP 2 DISTAL FORELIMB MUSCLES
   or: a muscle of the hind limb .... GROUP 3 DISTAL HINDLIMB MUSCLES
4. either: a muscle with direct attachment to either the scapula or the pelvis 5
   or: with no direct attachment to scapula or pelvis but may have an attachment to part of the axial skeleton (that is, vertebrae, ribs, or sternum) 6
5. either: a muscle associated with the scapula .... GROUP 4 PROXIMAL FORELIMB MUSCLES
   or: associated with the pelvis .... GROUP 5 PROXIMAL HINDLIMB MUSCLES

6. either: a muscle with no direct attachment to vertebrae, ribs, or sternum .... GROUP 6 HEAD MUSCLES
   or: a muscle with direct attachment to vertebrae, ribs, or sternum 7
7. either: a muscle with a direct attachment to vertebrae of the tail .... GROUP 7 TAIL MUSCLES
   or: without direct attachment to caudal vertebrae 8
8. either: a muscle with an attachment to either the ribs or the sternum .... GROUP 8 RIBCAGE AND FLANK MUSCLES
   or: a muscle with no attachment to ribs or sternum .... GROUP 9 NECK MUSCLES

Group 1 Muscles
Cutaneous Muscles

1. either: covering the greater part of the animal's flank ...... CUTANEUS TRUNCI
   or: a small platelike muscle, often appearing as several digitations, which is situated on the shoulder anteriorly to the above (pig 31; sheep 2, 14) .... CUTANEUS OMOBRACHIALIS
   or: a small straplike muscle with an origin on the iliac fascia and a cut surface indicating that it was once inserted onto the testis .... CREMASTER EXTERNUS

Group 2 Muscles
Distal Forelimb Muscles

(Pig—See Separate Key for Distal Muscles)

1. either: with any attachment to the humerus 2
   or: without any attachment to the humerus 14
2. either: a muscle lying completely within the limb 3
   or: extending to an attachment not on the limb skeleton 6
3. either: a muscle whose belly does not lie associated with the radius in the lower limb 4
   or: a muscle whose belly is associated with the radius 7
4. either: a muscle attached to the radius and lying in the musculospiral groove of the humerus .... BRACHIALIS
   or: not attached to the radius but to the olecranon instead 5
5. either: inserted laterally on the olecranon .... TRICEPS BRACHII CAPUT LATERALE
   or: inserted medially on the olecranon .... TRICEPS BRACHII CAPUT MEDIALE
   or: a small muscle, difficult to separate from the above and covering the olecranon fossa .... ANCONEUS (pig 1)
6. either: a compound straplike muscle with an attachment to the skull which has been cut by dressing the carcass .... BRACHIOCEPHALICUS
   or: a broad, flat muscle which extends over the sternum to the midventral line and is divisible into two parts .... PECTORALIS SUPERFICIALIS

7. either: a conspicuous muscle with a tendinous attachment to the foot, severed at the carpometacarpal joint in the dressed carcass 8

or: a feeble muscle on the medial surface of the elbow .... PRONATOR TERES

8. either: a muscle with any direct attachment to the ulna or radius or both 9
or: with no direct attachment to the ulna or radius 10
(pig 34, 42)

9. either: a muscle with an attachment to the coronoid fossa of the humerus .... EXTENSOR DIGITORUM COMMUNIS COMMON HEAD
(pig 6)

or: with an attachment to the lateral epicondyle of the humerus .... EXTENSOR DIGITORUM LATERALIS
(pig 8)

or: with an attachment to the medial epicondyle of the humerus .... FLEXOR CARPI ULNARIS
(pig 43)

10. either: a muscle with any direct attachment to the coronoid fossa of the humerus 11

or: without any direct attachment to the coronoid fossa 12

11. either: a large muscle lying in the anterior part of the leg with an attachment to the lateral condyloid crest of the humerus ...... EXTENSOR CARPI RADIALIS
(pig 4)

or: with no attachment to the lateral condyloid crest .... EXTENSOR DIGITORUM COMMUNIS MEDIAL PART
(pig 7)

12. either: a muscle with an attachment to the medial epicondyle of the humerus 13

or: with an attachment to the lateral epicondyle of the humerus .... EXTENSOR CARPI ULNARIS

13. Since there are four muscles attached to the medial epicondyle of the humerus and one that overlies the others has already been identified (the flexor carpi ulnaris), the remaining muscles are distinguishable in the dressed carcass by their position relative to each other.

either: the most medial .... FLEXOR CARPI RADIALIS
(pig 11)

or: largest and most lateral .... FLEXOR DIGITORUM PROFUNDUS HUMERAL HEAD

or: intermediate in position ...... FLEXOR DIGITORUM SUPERFICIALIS

14. either: a muscle with a belly level with the radius in the lower limb 15
or: a thin superficial muscle lying to the posterior of the limb which extends dorsally toward the scapula .... TENSOR FASCIA ANTIBRACHII
(pig 44)

15. either: a muscle with any attachment to the radius 16
or: without any attachment to the radius .... FLEXOR DIGITORUM PROFUNDUS ULNAR HEAD
(pig 35)

16. either: a muscle attached to the ulna .... ABDUCTOR POLLICIS LONGUS (pig 8)
    or: with no ulnar attachment .... FLEXOR DIGITORUM PROFUNDUS RADIAL HEAD

Group 3 Muscles
Distal Hindlimb Muscles
(Pig—See Separate Key for Distal Muscles)

1. either: a muscle with a direct attachment to the fibula or tibia or both .... 2 (pig 41)
   or: without an attachment to the tibia or fibula 8
2. either: a muscle with an attachment to the lateral epicondyle of the femur, to the fibrous fibula, or to a ligament connecting to the femur 3
   or: a muscle with an attachment to the lateral condyle of the tibia but with no attachment to the lateral epicondyle of the femur, to the fibrous fibula, or to a ligament connecting to the femur 6
   or: a small muscle whose only skeletal attachment is to the lateral surface of the head of the fibula and which blends distally onto the lateral face of a separate larger muscle .... SOLEUS
3. either: a small triangular muscle behind the joint between the femur and tibia and with an attachment to the femur .... POPLITEUS
   or: attached to a ligament connecting to the femur or to the fibrous fibula .. 4
4. either: attached to the fibrous remnant of the fibula (pig 9, 46) 5
   or: no attachment to the fibula (pig 9) .... EXTENSOR DIGITORUM LATERALIS (PEDIS) (pig 40)
5. either: a muscle with two heads located anteriorly in the limb .... TIBIALIS CRANIALIS
   or: a muscle located laterally in the limb, but with a tendon that runs diagonally to the posterior distal part of the limb .... PERONEUS LONGUS
6. either: one of three obvious muscles with tendons to the foot severed at the tarsometatarsal joint .... 7 (pig 38, 39)
   or: a small muscle which blends into another larger muscle .... SOLEUS in the sheep
7. The remaining muscles reached here are all part of the flexor digitorum profundus pedis (pig 12, 13, 14)
   either: most medial in position .... FLEXOR DIGITORUM LONGUS
   or: deepest, and in the axis of the limb ..... FLEXOR HALLUCIS LONGUS
   or: superficial and lateral in position .... TIBIALIS CAUDALIS
8. either: a muscle attached to the patella or to its immediate ligaments 9
   or: not attached to the patella or its immediate ligaments 11
9. either: a small muscle situated immediately proximal to the patella .... ARTICULARIS GENU
   or: a large muscle located either medially, anteriorly, or laterally to the femur 10

10. either: a double-headed muscle wrapped around the anterior face of the femur and lying covered by other muscles in the intact limb .... VASTUS INTERMEDIUS

or: a large muscle located over and medial to the muscle listed above; difficult to separate from its medial surface .... VASTUS MEDIALIS

or: a large muscle lying laterally in the leg with part of its surface located subcutaneously .... VASTUS LATERALIS

11. either: a small muscle lying distally in the limb with its origin from the tibial tarsal bone .... EXTENSOR DIGITORUM BREVIS (pig 5)

or: with no origin from the tibial tarsal bone 12

12. either: a muscle with an origin from a tendon attached in the extensor fossa of the femur 13

or: without such an origin 14

13. There are three quite large extensor muscles originating from the tendon attached in the extensor fossa of the femur; in the dressed carcass they can be distinguished by their relative positions.

either: the most superficial and situated medially .... PERONEUS TERTIUS

or: deeper and medial .... EXTENSOR DIGITORUM LONGUS MEDIAL PART (pig 10)

or: deeper and lateral .... EXTENSOR DIGITORUM LONGUS LATERAL PART

14. either: a muscle with an attachment to the vertebrae and the trochanter of the femur .... PSOAS MAJOR (pig 22)

or: without a vertebral attachment 15

15. either: a large double-headed muscle (although it sometimes appears to have three heads in cross section) situated posteriorly to the femur and originating on the lateral and medial supracondyloid crests and from the medial epicondyle of the femur (pig 33, sheep 8) .... GASTROCNEMIUS

or: a smaller muscle covered by the muscle listed above and originating from the supracondyloid fossa of the femur only .... FLEXOR DIGITORUM SUPERFICIALIS (pig 15)

Group 4 Muscles
Proximal Forelimb Muscles

1. either: with a direct attachment to the humerus 2

or: without a humeral attachment 11

2. either: with any attachment to the lateral surface of the humerus 3

or: without any attachment to the lateral surface of the humerus 8

3. either: with any attachment to the medial surface of the humerus 4

or: without attachment to the medial surface of the humerus 5

4. either: a large muscle filling the supraspinous fossa dorsal to the scapular spine .... SUPRASPINATUS
   or: a large muscle extending over the sternum .... PECTORALIS PROFUNDUS (sheep 7, pig 21)
5. either: with direct attachment to the deltoid tuberosity of the humerus 6
   or: without such an attachment 7
6. either: a muscle attached to the acromion (pig 3) .... DELTOID PARS ACROMIALIS
   or: attached to the posterior edge of the scapula blade .... DELTOID PARS SCAPULARIS
7. either: a large muscle almost filling the infraspinous fossa ventral to the scapular spine .... INFRASPINATUS
   or: a small muscle closely associated with the above muscle but having a small insertion on the humerus, posterior to that of the muscle listed above .... TERES MINOR
8. either: with any attachment to the ribs or to the lumbodorsal fascia (a large connective tissue sheet over the flank and loin) .... LATISSIMUS DORSI
   or: without such an attachment 9
9. either: a muscle with any attachment to the lip of the glenoid cavity or to the coracoid process .... CORACOBRACHIALIS
   or: without such an attachment 10
10. either: a muscle with any direct attachment to the central region of the costal surface of the blade of the scapula (that is, facing the ribs) .... SUBSCAPULARIS
    or: a muscle attached to the posterior corner and adjacent posterior border of the scapula .... TERES MAJOR
11. either: a muscle with any attachment to the radius or to the olecranon process of the ulna 12
    or: without such an attachment 14
12. either: attached to the radius .... BICEPS BRACHII (pig 2, sheep 1)
    or: attached to the olecranon 13
13. either: a large muscle with an origin from the posterior border of the scapula .... TRICEPS BRACHII CAPUT LONGUM
    or: a small muscle attached mostly to the surface of another larger muscle but just reaching the posterior angle of the scapula .... TENSOR FASCIAE ANTIBRACHII
14. either: with any attachment to the ribs 15
    or: without any attachment to the ribs 16
15. The serratus ventralis is a large fanlike muscle that radiates out from the medial surface of the scapula. It is composed of easily visible radiating subunits which end ventrally in a sawtooth pattern, hence the muscle's name. It can be subdivided as follows:
    either: with an insertion to ribs 4 to 9 and no contact with the transverse processes of the cervical vertebrae .... SERRATUS VENTRALIS THORACIS

or: without any insertion to ribs 4 to 9 but having instead insertions on ribs 1 to 3; it originates on the transverse processes of some of the cervical vertebrae .... SERRATUS VENTRALIS CERVICIS

16. either: a large flat muscle, directly subcutaneous in position 17
or: a large muscle lying deeper in the shoulder region and with no subcutaneous exposure 18
or: a straplike muscle, partly subcutaneous and with a cervical vertebral attachment .... OMOTRANSVERSARIUS

17. either: a muscle with an origin from the median raphe of the cervical vertebrae (a connective tissue seam in the dorsal midline of the neck) ..... TRAPEZIUS PARS CERVICIS (sheep 13, pig 30)
or: a muscle with any attachment to the dorsal spines of the thoracic vertebrae (excepting the first thoracic vertebra) or to the lumbodorsal fascia (connective tissue sheet over the loin) .... TRAPEZIUS PARS THORACIS

18. either: a muscle with any contact with the ligamentum nuchae or the median raphe (seam) in the neck .... RHOMBOIDEUS PARS CERVICIS (sheep 10, pig 26)
or: any attachment to the dorsal spines of thoracic vertebrae 2 to 5 .... RHOMBOIDEUS PARS THORACIS

Group 5 Muscles
Proximal Hindlimb Muscles

1. either: with any direct attachment to the femur 2
or: without attachment to the femur 11

2. either: with any attachment in the trochanteric fossa of the femur 3
or: without such an attachment 5

3. either: a small fan-shaped muscle arising from the lateral surface of the acetabular ramus of the ischium below the lesser sciatic notch, but having no direct attachment to the lip of the obturator foramen .... GEMELLUS
or: not resembling the above 4

4. either: a muscle arising from the pelvic face of the ischium and coming through the obturator foramen .... OBTURATORIUS INTERNUS
or: a muscle arising from the ventral surface of the ischium and pubis and from the lip of the obturator foramen .... OBTURATORIUS EXTERNUS

5. either: with an attachment to the wing of the ilium 6
or: not attached to the wing of the ilium 8

6. either: a muscle with an attachment to the medial face of the ilium and which may appear to converge with a long muscle running ventral to the lumbar vertebrae .... ILIACUS
or: with any attachment to the lateral face of the wing of the ilium 7

7. either: a muscle with a long area of attachment to the pelvis, extending anteriorly from the acetabular region .... GLUTEUS PROFUNDUS

or: with its insertion restricted to the wing of the ilium and lying mainly dorsal and anterior; this is a large muscle with much of its outer face being subcutaneous .... GLUTEUS MEDIUS

or: a muscle intermediate to the above two muscles in both its position and insertion onto the ilium; its attachment to the lateral face of the ilium is central .... GLUTEUS ACCESSORIUS

8. either: with any attachment to the pubis 9
or: without any attachment to the pubis 10

9. Two muscles have been defined so far and now they are most easily separated by their size and position (pig 20, sheep 6).
either: a very large posterior muscle .... ADDUCTOR FEMORIS
or: a smaller anterior muscle .... PECTINEUS

10. either: a large muscle with an origin from the vertebral part of the tuber ischii .... SEMIMEMBRANOSUS
or: a small muscle (pig 23) with an origin from the ischium just posterior to the obturator foramen .... QUADRATUS FEMORIS

11. either: a small muscle severed during the dressing of the carcass 12
or: not as above 13

12. either: a small muscle which before slaughter originated near the tuber coxae and inserted on the testis; it may look like a detached part of one of the abdominal muscles .... CREMASTER EXTERNUS
or: a muscle originally attached in the root of the penis removed at slaughter and now only remaining at its origin on the medial surface of the tuber ischii .... ISCHIOCAVERNOSUS
or: a muscle originally attached to the external sphincter muscle of the anus before its removal and now remaining only at its origin from the ischiatic spine and sacrosciatic ligament .... LEVATOR ANI

13. either: with any direct vertebral attachment 14
or: without a vertebral attachment 17

14. either: a sheetlike muscle forming part of the abdominal wall .... OBLIQUUS ABDOMINIS INTERNUS
or: not like the above 15

15. either: a long muscle that forms the large round eye of meat in rib and loin steaks or chops .... LONGISSIMUS THORACIS ET LUMBORUM traditionally called the LONGISSIMUS DORSI
or: a muscle lying ventrally to the transverse processes of the vertebrae 16

16. either: a thin muscle lying directly ventral to the transverse processes and with an insertion to the medial (vertebral) face of the wing of the ilium .... QUADRATUS LUMBORUM
or: a long ropelike muscle attached to the psoas tubercle of the ischium (pig 22) .... PSOAS MINOR

17. either: a muscle with any direct attachment to the tibia or patella 18
or: without such an attachment 22

18. either: a large muscle with no attachment to the tibia but attached directly to the patella (pig 25) .... RECTUS FEMORIS
or: attached to the tibia 19

19. either: attached to the tuber ischii 20
 or: not attached to the tuber ischii 21

20. either: a large flat muscle located laterally on the thigh and which has an attachment to the lateral patellar ligament .... BICEPS FEMORIS
 or: a large muscle located posteriorly in the leg and with a medial attachment to the tibia .... SEMITENDINOSUS

21. either: a narrow straplike muscle with two heads originating near the psoas tubercle of the ilium .... SARTORIUS
 or: a large flat muscle covering a large part of the medial surface of the thigh .... GRACILIS

22. The rectus abdominis (group 8, question 13) might have been categorized into this group because of its attachment to the pubis is via the prepubic tendon. This muscle is easily identified since it is a large abdominal muscle distinguished by its unusual transverse bands of connective tissue (sheep 9).
 either: a muscle contributing largely to the abdominal wall and with attachment to ribs 5 to 13 and the linea alba (white connective tissue in the midline of the belly) .... OBLIQUUS ABDOMINIS EXTERNUS
 or: a triangular muscle situated anteriorly and anterolaterally on the thigh; its pulls on the fascia lata, which is the tight sheet of connective tissue covering the anterior face of the hind limb .... TENSOR FASCIA LATA

Group 6 Muscles
Muscles of the Head

These muscles, as defined so far, include the muscles of the head and throat that have no attachment to the cervical vertebrae. These include the muscles of the lips, cheek, nostrils, eyelids, and tongue. Only the mandibular muscles are considered further.

1. either: a large muscle on the lateral surface of the cranium filling the temporal fossa and attached to the coronoid process of the mandible .... TEMPORALIS
 or: not like the above 2

2. either: a muscle situated laterally to the jaws, composed of three layers and with a wide attachment to the malar and maxillary bones and to the zygomatic arch .... MASSETER
 or: not as above, being situated medially to the jaws 3

3. either: a muscle with two bellies in series which attach to the paramastoid process .... DIGASTRICUS
 or: attached to the palatine bone and located laterally .... PTERYGOIDEUS LATERALIS
 or: attached to the palatine bone and located medially .... PTERYGOIDEUS MEDIALIS

Group 7 Muscles
Muscles of the Tail

1. either: ventral to the transverse processes of the caudal vertebrae 2
 or: dorsal to the transverse processes of the caudal vertebrae 3
 or: located between the transverse processes of the remaining caudal vertebrae .... INTERTRANSVERSARII CAUDAE

2. either: a muscle extending into the tail with an attachment to transverse processes posterior to the third coccygeal vertebra .... SACROCOCCYGEUS VENTRALIS

or: a flat muscle attached to the sacrosciatic ligament which does not extend posterior to the third caudal vertebra .... COCCYGEUS

3. either: the most dorsal muscle of the tail, which has contact with its fellow on the other side .... SACROCOCCYGEUS DORSALIS MEDIALIS

or: a muscle located laterally .... SACROCOCCYGEUS DORSALIS LATERALIS

Group 8 Muscles
Rib-Cage and Flank Muscles

1. either: with any direct attachment to any vertebrae or connected to the lumbar vertebrae via the deep lumbar fascia or the ventral longitudinal ligament ... 2

or: without any vertebral attachment (pig 32) 8

2. either: with any attachment to cervical vertebrae 3

or: without any attachment to cervical vertebrae 6

3. either: with any attachment to the lumbar vertebrae 4

or: located in the neck region 5

4. either: an extensive muscle closely related to the dorsal spines of the vertebrae and included here as it shows irregular attachment to ribs 10 to 15 .... SPINALES CERVICIS ET THORACIS

or: a long muscle lying dorsally along the rib cage from the transverse processes of cervical vertebra 7 (pig 16) to a weak attachment to the lumbodorsal fascia .... ILIOCOSTALIS

5. either: attached to the fourth rib .... SCALENUS DORSALIS (sheep 11, pig 27)

or: attached to the first rib and to the manubrium .... SCALENUS VENTRALIS

6. either: a series of muscles attached to the transverse processes of thoracic vertebrae 1 to 12 .... LEVATORES COSTARUM

or: with lumbar attachments 7

7. either: a large flat muscle contributing to the abdominal wall and extending from an origin on the deep lumbar fascia to the linea alba (connective tissue in the midline of the belly) .... TRANSVERSUS ABDOMINIS

or: a flat triangular muscle situated in the angle between the last rib and the lumbar vertebrae .... RETRACTOR COSTAE

or: the dorsal remains of the lumbar part of the diaphragm situated in the abdominal cavity

8. either: situated between successive ribs 9

or: outside the rib cage and extending into the abdomen or neck 10

or: situated within the rib cage (pig 36, 37) 15

9. either: an intercostal muscle situated laterally and not extending between the costal cartilages .... INTERCOSTALES EXTERNI (pig 17, sheep 3)

or: situated medially and extending between the costal cartilages .... INTERCOSTALES INTERNI

10. either: a muscle originally extending to the head or throat but now severed in the dressed carcass 11
or: not as described above 12

11. either: with an attachment to the first costal cartilage as well as to the manubrium .... STERNOCEPHALICUS
deep part = STERNOMASTOIDEUS
superficial part = STERNOMANDIBULARIS
or: a slender muscle which is attached to the cartilage of the sternum and which was ventral to the trachea .... STERNOTHYROHYOIDEUS

12. either: with an attachment to the sternum 13
or: not attached to sternum 14

13. either: a large muscle contributing to the abdominal wall and having transverse bands of connective tissue along its length .... RECTUS ABDOMINIS
or: a flat muscle confined to the anterior, ventral surface of the ribcage .... RECTUS THORACIS

14. either: a series of muscular digitations on the dorsal parts of ribs 5 to 9 (approximately) .... SERRATUS DORSALIS CRANIALIS (pig 29, sheep 12)
or: a series of muscular digitations on the dorsal parts of ribs 9 to 13 (approximately) .... SERRATUS DORSALIS CAUDALIS
or: a flat muscle forming part of the abdominal wall and having attachment to the posterior ribs .... OBLIQUUS EXTERNUS ABDOMINIS

15. either: situated directly dorsal to the sternum in the midline of the rib cage .... TRANSVERSUS THORACIS
or: situated diagonally on the medial face of the body wall at the junction of the thoracic and abdominal cativities .... DIAPHRAGM

Group 9 Muscles
Neck Muscles

1. either: the muscle, its tendon, or obvious fascia has been severed during dressing of the carcass (pig 45) 2
or: the muscle is intact in the dressed carcass 10

2. either: with its attachment restricted to the atlas 3
or: its attachment not restricted to the atlas 6

3. either: attached to the dorsal or lateral surface of the wing of the atlas 4
or: attached to the ventral surface of the wing of the atlas 5

4. either: attached to the lateral border of the wing of the atlas .... OBLIQUUS CAPITIS CRANIALIS
or: attached anterodorsally to the neural arch of the atlas .... RECTUS CAPITIS DORSALIS MINOR (pig 24)
or: attached dorsally to the spine of the atlas .... RECTUS CAPITIS DORSALIS MAJOR (pig 24)

5. either: a small muscle attached laterally on the ventral surface of the wing of the atlas .... RECTUS CAPITIS LATERALIS
or: a large muscle attached to the ventral surface of the wing of the atlas, medially to the above .... RECTUS CAPITIS VENTRALIS

6. either: without any attachment to thoracic vertebrae 7
or: with some attachment to thoracic vertebrae 8

7. either: originating from a fascia near the shoulder and crossing the trachea in the living animal .... CERVICOHYOIDEUS
or: originating ventrally to the transverse processes of cervical vertebrae 3 to 7 and severed in the dressed carcass as it passes forward to the base of the skull (sheep 5) .... LONGUS CAPITIS

8. either: attached to the articular processes of the cervical vertebrae .... SEMISPINALIS CAPITIS (pig 28)
or: attached to the transverse processes of the cervical vertebrae 9

9. either: a thin, flat muscle sharing a median raphe (seam) with its fellow from the other side .... SPLENIUS
or: not as above .... LONGISSIMUS CAPITIS

10. either: with an attachment to the thoracic vertebrae 11
or: not attached to thoracic vertebrae 15

11. either: a series of muscles that attach to the articular processes and dorsal spines of the lumbar vertebrae .... MULTIFIDI DORSI
or: with no lumbar attachments 12

12. either: with any obvious tendon (shared) attaching to the wing of the atlas; the muscle extends posteriorly to thoracic vertebra 7 .... LONGISSIMUS ATLANTIS (pig 19)
or: not like the above 13

13. either: attached to the lateral and ventral surfaces of the bodies of the thoracic vertebrae .... LONGUS COLLI
or: attached to the transverse processes of the thoracic vertebrae 14

14. either: a series of muscles attached to the posterior articular processes, posterior surfaces of bodies, or the dorsal spines of the cervical vertebrae .... MULTIFIDI CERVICIS
or: not attached as above but attached to the transverse processes of the cervical vertebrae .... LONGISSIMUS CERVICIS (pig 19)

15. either: a muscle restricted in position to between its attachments to the spine and posterior articular process of the axis and to the dorsal side of the wing of the atlas .... OBLIQUUS CAPITIS CAUDALIS
or: a series of muscles restricted in position to between the transverse processes of the lumbar vertebrae .... INTERTRANSVERSARII LUMBORUM (pig 18, sheep 4)
or: not as either of the above 16

16. either: a large neck muscle with dorsal and ventral parts, both of which are attached to cervical vertebra 4 and to the wing of the atlas .... INTERTRANSVERSARIUS LONGUS
or: a series of muscles either between the transverse processes of successive cervical vertebrae or between a cervical articular process and its preceding transverse process .... INTERTRANSVERSARII CERVICIS
or: a superficial muscle of the neck region which has an origin from the shoulder fascia and an insertion onto the wing of the atlas .... OMOTRANSVERSARIUS

NOTES ON THE PIG

1. There are two anconeus muscles.
2. Biceps brachii has a medial insertion to the ulna as well as to the radius.
3. Deltoideus is undivided.
4. Extensor carpi radialis is divided into two parts, longus and brevis.
5. Extensor digitorum brevis is a well-developed muscle inserting to both tibial and fibular tarsal bones.
6. The common head of the extensor digitorum communis has an insertion to the lateral epicondyle of the humerus and is subdivided into medial, middle, and deep parts.
7. The extensor digitorum communis has an insertion to the lateral epicondyle of the humerus. Its tendon ends mainly on the third digit.
8. Extensor digitorum lateralis is subdivided into a large dorsal part (extensor digit quarti proprius) and a small volar part (extensor digiti quinti proprius).
9. Extensor digitorum lateralis is attached to the lateral surface of the fibula.
10. Functionally represented by a branch of the long extensor part of extensor digitorum brevis.
11. Flexor carpi radialis is well developed.
12. Tibialis caudalis has an origin from the plantar surface of the fibula.
13. Flexor hallucis longus attaches to the medial and plantar surfaces of the fibula as well as being lateral in position.
14. Flexor digitorum longus pedis attaches to the fibula.
15. Flexor digitorum superficialis has a large belly, closely fused with the lateral head of the gastrocnemius.
16. Iliocostalis extends to the wing of the atlas.
17. External intercostals may be absent in places.
18. Intertransversarii lumborum are well developed.
19. Longissimus cervicis fuses with longissimus atlantis.
20. Pectineus is quite large.
21. Pectoralis profundus is subdivided into cranial and caudal parts.
22. Psoas major and minor are fused anteriorly.
23. Quadratus femoris is large.
24. Rectus capitis dorsalis major and minor are fused.
25. Rectus femoris shares a tendon with the other muscles of the quadriceps femoris group (rectus femoris + three vastus muscles).
26. Rhomboideus can be divided into three parts which together extend from the anterior part of the neck, as far back as thoracic vertebrae 9 or 10.
27. Scalenus dorsalis attaches to the third rib.
28. Semispinalis capitis is subdivided into a dorsal part (marked with connective tissue) and a ventral part.
29. Serratus dorsalis cranialis is situated on ribs 5 to 8.

30. Trapezius is inseparable.
31. Cutaneus omobrachialis is absent but the cervical cutaneus muscle is very well developed.
32. The inferior part of the cervical cutaneus muscle is attached to the anterior extremity of the sternum. The inferior part crosses obliquely over the superior part to continue to the pig's face.
33. The porcine soleus muscle has been categorized so far. It is a thick muscle that has an origin from the lateral epicondyle of the femur as well as the deep fascia at the stifle joint. It merges with the lateral edge of the lateral head of the gastrocnemius.
34. Extensor digitorum communis is a three-part muscle which originates from the lateral epicondyle of the humerus and the lateral ligament of the elbow.
35. This description also covers extensor digiti secundi proprius. Although this muscle arises on the ulna, its tendon and part of its belly are fused with extensor digitorum communis (see pig 34).
36. Sternocephalicus attaches to the sternum but not costally. Since it runs by way of a round tendon to the mastoid, it is severed during dressing.
37. Sternothyrohyoideus attaches to the manubrium of the sternum.
38. Tibialis cranialis is a small muscle with an origin on the lateral surface of the tibial tuberosity and on the lateral condyle of the tibia.
39. Peroneus longus has been categorized so far since it is only attached to the lateral condyle of the tibia.
40. Extensor digitorum lateralis is divided into a large anterior part (extensor digitorum quarti) and a posterior part (extensor digitorum quinti).
41. Extensor hallucis longus and the three heads of flexor digitorum profundus may be included in this group.
42. Flexor carpi ulnaris originates only from the medial epicondyle of the humerus.
43. Flexor digitorum profundus is described so far and is distinguished by its attachment to the radius, ulna, and olecranon.
44. Supinator is a thin muscle slip across the face of the radius and originates from the lateral border of the radius. It attaches to the radial head of flexor digitorum profundus.
45. Although a pork carcass may retain its head (shipper's style), follow the key as if the head had been removed (packer's style).
46. Although the fibula is well developed, tibialis anterior and peroneus longus originate on the tibia rather than on the fibula.

DISTAL EXTENSORS AND FLEXORS IN THE INTACT PORK CARCASS

1.	either:	situated in the forelimb	2
	or:	situated in the hind limb	17
2.	either:	with any direct attachment to the ulna or radius	3
	or:	without any direct attachment to the ulna or radius	10

3. either: without any direct or shared insertion onto the phalanges of the digits ... 4
or: with a direct or shared insertion onto the phalanges 5

4. either: a well-covered muscle inserting onto the second metacarpal
.... EXTENSOR CARPI OBLIQUUS
or: a small muscle placed medially on the elbow with an insertion to the medial epicondyle of the humerus .... PRONATOR TERES

5. either: with its tendon situated on the volar surface of the digits; a strong muscle with three heads and large tendons to digits 3 and 4 6
or: with its tendon situated on the dorsal surface of the digits 7

6. either: with an origin from the medial epicondyle of the humerus
.... FLEXOR DIGITORUM PROFUNDUS HUMERAL HEAD
or: with an origin from the medial, caudal, and lateral parts of the proximal region of the ulna .... FLEXOR DIGITORUM PROFUNDUS ULNAR HEAD
or: with an origin from the radius
.... FLEXOR DIGITORUM PROFUNDUS RADIAL HEAD

7. either: a small muscle without any direct insertion onto the phalanges but contributing its insertion to another muscle or tendon 8
or: a laterally situated muscle with its own tendon to the phalanges 9

8. either: a small muscle situated medially and partly covered by another small muscle to which its tendon is attached
.... EXTENSOR DIGITI SECUNDI PROPRIUS
or: a muscle slip with a lateral origin from the radius and a medial insertion onto the head of another muscle .... SUPINATOR

9. These muscles are both part of the extensor digitorum lateralis and they can be separated as follows:
either: a large dorsal part inserting onto digit 4
.... EXTENSOR DIGITI QUARTI PROPRIUS
or: a small part with no insertion onto digit 4
.... EXTENSOR DIGITI QUINTI PROPRIUS

10. either: with a lateral origin from the humerus 11
or: with a medial origin from the humerus 14

11. either: with any direct insertion onto a metacarpal bone 12
or: with a direct insertion onto the phalanges of the digits 13

12. either: a large muscle inserted onto the metacarpal of the third digit
.... EXTENSOR CARPI RADIALIS
or: inserted onto the metacarpal of the fifth digit
.... EXTENSOR CARPI ULNARIS

13. These muscles are parts of the extensor digitorum communis, which can be subdivided as follows:
either: with a large flat tendon running centrally down the dorsal surface of the carpal region (back of hand) and inserting onto digit 3
.... EXTENSOR DIGITORUM COMMUNIS MEDIAL PART
or: with an insertion to digit 5
.... EXTENSOR DIGITORUM COMMUNIS DEEP PART

or: with a "Y" branch between digits 3 and 4
.... EXTENSOR DIGITORUM COMMUNIS MIDDLE PART

14. either: with any connection or insertion onto metacarpal or accessory bones ... 15
or: with no insertion onto metacarpal or accessory bones but with an insertion onto the phalanges 16

15. either: a well-developed muscle inserting onto the metacarpal of digit 3 and with a tendon tending to wrap round another tendon
.... FLEXOR CARPI RADIALIS
or: a narrow muscle inserted onto an accessory carpal bone only
.... FLEXOR CARPI ULNARIS
or: the superficial part of another muscle inserting onto an accessory carpal bone but also having a double tendon to digit 4
.... FLEXOR DIGITORUM SUPERFICIALIS SUPERFICIAL PART

16. either: joining to form a large tendon with branches to digits 2, 3, 4, and 5
.... FLEXOR DIGITORUM PROFUNDUS HUMERAL HEAD
or: with a tendon on the volar surface (palm of hand) of the third digit
.... FLEXOR DIGITORUM SUPERFICIALIS DEEP PART

17. either: With an origin from the femur or a ligament connecting with the femur
.... 18
or: without such an origin 23

18. either: with an origin from the extensor fossa of the femur 19
or: without such an origin 20

19. either: of the two muscles originating from the extensor fossa, this is a well-developed muscle inserted onto the third metatarsal .... PERONEUS TERTIUS
or: a deeper muscle, the tendon of which has an obvious "Y" branch between digits 3 and 4 to insert on their third phalanges
.... EXTENSOR DIGITORUM LONGUS

20. either: with a tendon terminating at the fibular tarsal bone 21
or: continuing to the digits 22
or: inserting onto the tibia .... POPLITEUS

21. either: originating posteriorly from the shaft of the femur
.... GASTROCNEMIUS
or: with an origin from the lateral epicondyle of the femur and a fascia at the knee
.... SOLEUS

22. either: with an origin from the lateral femorotibial ligament and also from the lateral part of the fibula (this muscle has a part to digit 4, extensor digiti quarti, and a part to digit 5, extensor digiti quinti)
.... EXTENSOR DIGITORUM LATERALIS
or: with an origin one-third the way up the shaft of the femur, posteriorly surrounded by another large muscle and with a tendon running over a groove on the fibular tarsal bone
.... FLEXOR DIGITORUM SUPERFICIALIS

23. either: with an origin from the fibula 24
or: without an origin from the fibula 27

24. either: the whole or part of a three-headed muscle (flexor digitorum profundus) origi-

nating from the tibia and from a large common tendon on the plantar surface of the foot 25

or: without an origin from the tibia 26

25. either: a medial muscle with an origin from the proximal part of the fibula .... FLEXOR DIGITORUM LONGUS

or: a lateral muscle with its fibular origin medial and plantar .... FLEXOR HALLUCIS

26. either: a small muscle with a tendon to the dorsal surface of digit 2 .... EXTENSOR HALLUCIS LONGUS

or: a muscle contributing to a large compound tendon on the plantar surface .... TIBIALIS CAUDALIS

part of flexor digitorum profundus

27. either: a muscle originating from the tibial and fibular tarsal bones .... EXTENSOR DIGITORUM BREVIS

or: without such an origin 28

28. either: with a tendon to the plantar surface, crossing over to insert on the first tarsal bone .... PERONEUS LONGUS

or: with a tendon passing down the dorsal face of the foot to insert onto the second tarsal and metatarsal bones .... TIBIALIS CRANIALIS

NOTES ON THE SHEEP

1. Biceps brachii has an ulnar insertion as well as to the radius.
2. Omobrachialis is absent.
3. External intercostals may be greatly reduced, particularly ventrally.
4. Intertransversarii lumborum are poorly developed.
5. Longus capitis is attached to the transverse processes of the second to the sixth cervical vertebrae.
6. Pectineus is quite large.
7. Pectoralis profundus is subdivided into cranial and caudal parts.
8. The piriformis muscle (caudal lobe of gluteus accessorius) is also categorized so far. It may be identified by its origin from the lateral surface of the sacrosciatic ligament in the region of the ischiatic spine and by its insertion onto the medial aspect of the greater trochanter, caudal to the insertion of the deep gluteal.
9. Rectus abdominis is particularly likely to be included here rather than in group 8 because of its more direct attachment to the pubis in sheep. The muscle is easily recognized by the transverse bands of connective tissue along its length.
10. Rhomboideus can be subdivided into three parts. Pars cervicalis is inserted onto the second cervical to the sixth thoracic vertebrae. Pars capitis is inserted with splenius on the occipital bone. Pars thoracalis is inserted as far posteriorly as the ninth or tenth thoracic vertebrae.
11. Scalenus dorsalis might be attached to the dorsal part of the first rib.
12. Serratus dorsalis cranialis is situated on ribs 4 to 5.

13. Trapezius may be difficult to subdivide.
14. Cutaneus omobrachialis is absent as a separate muscle.

SYNONYMS FOR SOME MUSCLE NAMES USED IN THE KEY

Abductor pollicis longus = extensor carpi obliquus
Adductor = adductor femoris
Brachiocephalicus = cleidobrachial + cleidoccipital + cleidomastoideus
Cervicohyoideus = omohyoideus
Cutaneus trunci = panniculus carnosus
Extensor carpi ulnaris = ulnaris lateralis
Extensor digitorum brevis = extensor digitalis brevis
Extensor digitorum communis = common digital extensor + medial digital extensor
Extensor digitorum lateralis = extensor digiti quarti proprius
Extensor digitorum longus = extensor digitorum longus + medial digital extensor
Flexor digitorum profundus = flexor digitalis profundus + ulnar head + radial head
Flexor digitorum profundus pedis = tibialis caudalis (posterior) + flexor hallucis longus + flexor digitorum longus pedis
Flexor digitorum superficialis = flexor digitorum sublimis
Flexor digitorum superficialis pedis = plantaris
Iliocostalis = longissimus costarum
Intertransversarii cervicis = intertransversales colli
Intertransversarius longus = intertransversales colli (cranioventral bundle)
Ischiocavernosus = erector penis
Levator ani = retractor ani
Longissimi thoracis et lumborum = longissimus dorsi (when spinalis dorsi and longissimus cervicis are separable)
Longissimus cervicis = anteroventral subdivision of longissimus dorsi
Longus capitis = rectus capitis ventralis major
Obturatorius externus = obturator externus
Obturatorius internus = obturator internus
Piriformis = lobe of gluteus accessorius
Rectus capitis ventralis = rectus capitis ventralis minor
Rectus thoracis = transversus costarum
Sacrococcygeus dorsalis lateralis = sacrococcygeus lateralis
Sacrococcygeus dorsalis medialis = sacrococcygeus dorsalis

Semispinalis capitis = complexus
Serratus ventralis cervicis = serratus ventralis
Serratus ventralis thoracis = serratus thoracis
Spinales cervisi et thoracis = spinalis dorsi
Sternocephalicus = sternomastoideus + sternomandibularis

REFERENCES

BOURDELLE, E. 1920. *Anatomie régionale des animaux domestiques*, Vol. 3. J. B. Baillière et Fils, Paris.

BRISKEY, E. J., KOWALCZYK, T., BLACKMON, W. E., BREIDENSTEIN, B. B., BRAY, R. W., and GRUMMER, R. H. 1958. *Porcine Musculature—Topography*. Wis. Agric. Exp. Stu. Res. Bull. 206.

BUTTERFIELD, R. M., and MAY, N. D. S. 1966. *Muscles of the Ox*. University of Queensland Press, St. Lucia, Australia.

FOURIE, P. D. 1962. *N.Z. J. Agric. Res.* 5:190.

KAUFFMAN, R. G., and ST. CLAIR, L. E. 1965. *Porcine Myology*. Univ. Illinois Coll. Agric., Agric. Exp. Stn. Bull. 715.

KAUFFMAN, R. G., ST. CLAIR, L. E., and REBER, R. J. 1963. *Ovine Myology*. Univ. Illinois Agric. Exp. Stn. Bull. 698.

MAY, N. D. S. 1964. *The Anatomy of the Sheep*. University of Queensland Press, St. Lucia, Australia.

SISSON, S., and GROSSMAN, J. D. 1953. *The Anatomy of the Domestic Animals*. W. B. Saunders Company, Philadelphia.

TUCKER, H. Q., VOEGELI, M. M., and WELLINGTON, G. H. 1952. *A Cross Sectional Muscle Nomenclature of the Beef Carcass*. Michigan State College Press, East Lansing.

5

The Structure and Properties of Meat

INTRODUCTION

A large proportion of the body mass of farm animals is composed of *striated skeletal muscle*. Striated muscle is a contractile tissue which is distinguished by its microscopic transverse striations. Skeletal muscles are responsible for the body movements which produce locomotion, breathing, chewing, and the maintenance of posture. The meat on the carcasses of cattle, sheep, pigs, and poultry is derived from striated skeletal muscles. Cardiac muscle is striated and edible, but it is not directly involved with skeletal movement. The smooth muscle of the viscera is not striated.

The conventional order for teaching the structure of meat is to start with the physiology of living muscle. Most people, however, are more familiar with the everyday properties of meat rather than with muscle physiology. This account of muscle structure, therefore, proceeds from the familiar to the unfamiliar, and starts with a close look at the structure of ordinary meat. It then digresses into a few aspects of muscle physiology that are particularly useful in understanding the properties of meat.

THE FIBROUS NATURE OF MEAT

A characteristic feature of meat is its fibrous structure or texture. In live animals, the arrangement and sliding interaction of microscopic filaments enable a muscle to contract while, in meat, an ordered arrangement of *fasciculi* (bundles of fibers), *fibers*, *fibrils*, and *filaments* creates a characteristic texture which is difficult to imitate with processed plant proteins. It is essential to note that these components of meat (*fasciculi*, *fibers*, *fibrils*, and *filaments*) constitute a *descending* series with respect to size. Fasciculi are the largest units, while filaments are the smallest. The prefix "myo" is used by some authors to create the terms *myofiber, myofibril,* and *myofilament*. These terms are identical to muscle fiber, muscle fibril, and muscle filament, respectively.

In introductory physiology books, descriptions of muscle structure usually begin with a diagram of a simple muscle with an elongated shape. The upper end, usually detached from its skeletal attachment, is clamped into the jaws of a *myograph.* The lower end of the muscle usually narrows down to a tendon which activates the recording apparatus to measure muscle contraction. For students of meat science, this is rather a misleading view of muscle structure since nearly all the muscles of a meat carcass have either a complex shape or are attached to the skeleton by a variety of oddly shaped tendons or sheets of connective tissue. The complex arrangement of muscle fasciculi is often seen when meat is carved.

The muscle fibers within a piece of meat often have a complex arrangement which is related to muscle function in the live animal. It is often stated that muscles which contract to a relatively small fraction of their resting length usually have fibers which are parallel to the long axis of the muscle, and that muscles in which the strength of contraction is more important than the distance over which contraction occurs usually have fibers that are arranged at an angle to the direction in which the muscle contracts. In this way, muscle fibers might gain strength by leverage, but the distance over which the muscle contracts is reduced. Experimental data, however, offer little or no support to this traditional concept and a pennate structure may in fact serve to enhance the overall range of muscle excursion (Muhl, 1982).

Muscle Fibers

If a piece of muscle or meat is placed under a dissecting microscope and is teased apart with a pair of needles, the smallest fasciculi that are visible without magnification are found to be composed of a bundle of muscle fibers. Muscle fibers are the basic cellular units of living muscle and of meat. The muscle fibers of meat are unusual cells because they are *multinucleate* (with many cell nuclei) and because they are extremely long (commonly several centimeters) relative to their microscopic diameter (usually less than one-tenth of a millimeter). The muscle fibers that are found in most commercial cuts of meat seldom run the complete length of the muscle in which they are located. Individual fibers within a fasciculus may terminate at a point along the length of their fasciculus by means of a tapered ending (Plate 3) which is anchored in the connective tissue on the surface of an adjacent fiber (Le Gros Clark, 1971). Apart from such tapered *intrafascicular* endings, the diameter of a fiber is assumed to be approximately constant along its length. Fiber diameters slowly increase during the growth of a muscle, but they also increase temporarily when a fiber contracts. Thus when measuring fiber diameters in a growth study, special care must be taken to avoid or to correct for differences in the degree of muscle contraction.

The simplest way to examine individual muscle fibers is to place some meat fragments together with some water in a kitchen blender. After running the blender for a few seconds, the connective tissue that holds the muscle fibers together is disrupted to leave a pale red suspension of broken muscle fibers in water. The red color is due to *myoglobin*, a soluble red pigment found inside muscle fibers. The essential features of muscle fiber structure can be observed with an ordinary light microscope if a drop of the macerated muscle suspension is mounted on a microscope slide beneath a coverslip. The transverse striations of fibers become visible if the iris diaphragm of the substage condenser is almost closed (the gain in contrast is offset by a loss of resolution, which is why the diaphragm is normally open wider). With a high-magnification microscope objective, fibrils can be seen if they protrude from the broken end of a fiber, or if they have escaped from a broken fiber (Figure 5-1). Under the surface membranes of muscle fibers may be seen some flattened bubblelike inclusions. These are the nuclei of the muscle fiber, and their DNA can be stained by treating the macerated suspension with dyes such as *hematoxylin.* On the surfaces of any fibers that have retained some of their surrounding connective tissue may be seen branching *capillaries* which were once part of the vascular bed of the muscle. Red blood cells (*erythrocytes*) may or may not remain in the capillaries, depending on the efficiency with which the animal was exsanguinated. If they do remain in the meat, they appear pale yellow in unstained preparations, and they are often distorted or are piled together like a stack of coins.

When meat animals are slaughtered, they are normally shackled and suspended from their hind limbs. Some muscles, such as the filet or psoas muscles ventral to the vertebral column, are stretched. Other muscles, such as those in the posterior part of the hind limb, are free from skeletal restraint and may contract weakly as the carcass

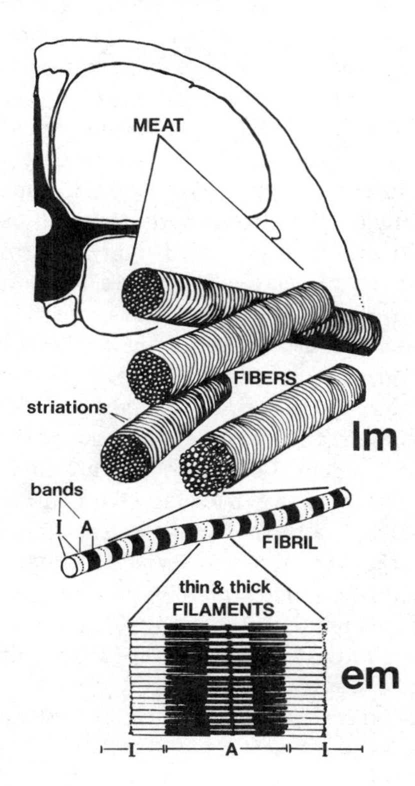

Figure 5-1 Microstructure of meat seen by light microscopy (lm) and by electron microscopy (em).

becomes stiff after death (*rigor mortis*). If samples from stretched and contracted muscles are compared, their transverse striations will appear relatively far apart in the stretched sample and closer together in the contracted sample. The distance between the transverse striations is called the *sarcomere length*.

If a drop of saturated sodium chloride solution is mixed with a drop of macerated muscle suspension, the muscle fiber fragments undergo some marked changes. Fiber fragments may slowly swell and disappear, or they may expand so violently that their interiors are extruded from their broken ends. Sarcomere length and the solubility of meat proteins in salt solutions are two commercially important properties of meat: meat with short sarcomeres tends to be tough, and salt-solubilized proteins are used to bind together the meat fragments in many types of processed meat products.

Transverse Striations

After prolonged maceration, muscle fiber fragments disintegrate and their fibrils are released into suspension. With an ordinary light microscope at its highest magnification, it is often possible to see the *transverse striations* on individual fragments of fibrils (Figure 5-1). The lenses from an old pair of Polaroid sunglasses can be used to make a simple *polarizing microscope*. One spectacle lens is placed in front of the light source and the other is placed on top of the microscope eyepiece. By rotating one spectacle lens relative to the other, the amount of transmitted light can be greatly reduced. The first lens only transmits light waves that vibrate in a certain plane (*polarized light*), but these are unable to get through the second lens, whose transmitting plane is now at 90 degrees to the first lens. Thus the field of view is dark, except for alternate transverse striations on certain fibrils in the field. Alternate striations are able to rotate the waves of polarized light strongly enough for the light waves to get through the second spectacle lens: These striations are strongly *birefringent* and appear bright. Rotation of the microscope slide generally allows this property to be observed in any particular fibril. Striations that appear bright in polarized light are termed *anisotropic* or A bands. Those that appear dark or relatively dim are termed *isotropic* or I bands (Plate 2).

The transverse striations on muscle fibers (Figure 5-1) are due to the remarkably precise alignment of the A and I bands on the parallel fibrils which run longitudinally within the fiber. In most stained preparations for light and electron microscopes, A bands appear darker than I bands (the reverse appearance to that seen with polarized light). This is because the birefringent A bands contain a greater density of protein. A bands also appear darker than I bands when unstained preparations are observed with a *phase-contrast* light microscope. Further details on the staining of transverse striations for light microscopy are given by Dempsey et al. (1946).

Other features of the fibril are detectable by light microscopy under optimum conditions, but these details are more clearly seen by electron microscopy. A thin *Z line* or *Z disc* occurs in the middle of the I band. The repeating unit of a regular series of transverse striations is termed the *sarcomere*, and it is usually considered to be the structural unit from Z line to Z line (Figure 5-2). The Z line resembles a woven disk, like the bottom of a wicker basket, and it extends as a partition across the fibril.

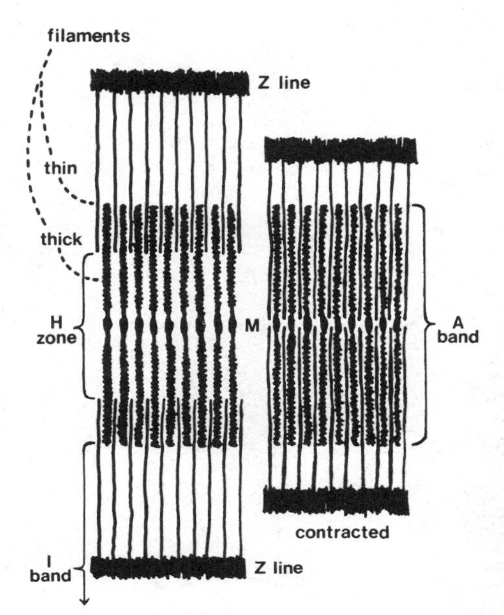

Figure 5-2 Structure of rest length and contracted sarcomeres (Z line to Z line) seen by electron microscopy. (From Swatland, 1982a.)

Filaments (Myofilaments)

At the ultrastructural level (Figure 5-2), the transverse striations of fibrils are caused by the regular longitudinal arrangement of sets of *thick filaments* (10 to 12 nm or more in diameter) and *thin filaments* (about 5 to 7 nm in diameter). In a transverse section that has been cut through a region of overlapping thick and thin filaments, each thick filament is surrounded by six thin filaments (Figure 5-3). When a muscle fiber contracts, the thick filaments slide between the thin filaments so that the I band gets shorter. The length of the A band remains constant. These events are explained by the *sliding filament theory* of muscle contraction (Huxley, 1972). If a muscle is at its resting length, the gap between opposing thin filaments at the midlength of the sarcomere gives rise to a pale *H zone* in the A band. Although the sliding filament theory is now almost universally accepted, there remain many unsolved problems in the mechanism of the system.

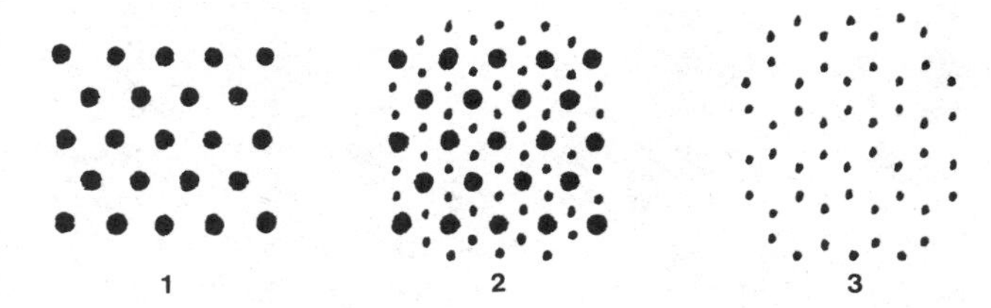

Figure 5-3 Thick and thin filaments seen by electron microscopy in transverse sections along a sarcomere: (1) section through the H zone; (2) section through the overlap of thick and thin filaments at the edge of the A band; (3) section through the I band.

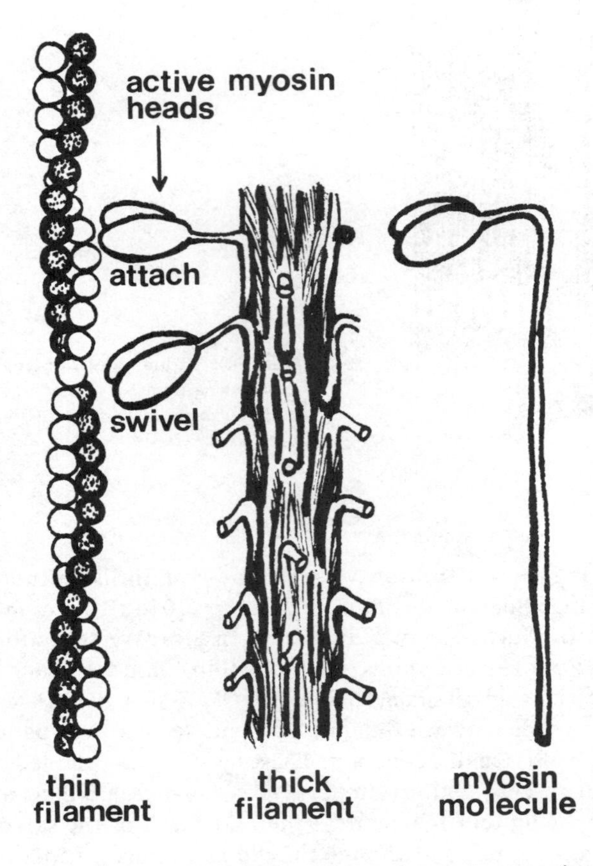

Figure 5-4 Diagram to show (from right to left of the figure) how myosin molecules might be bound together to form thick filaments, and how the protruding heads of myosin molecules might interact with thin filaments. (From Swatland, 1982a.)

Contraction is an active process that requires energy. This energy is provided by the release of phosphate from *adenosine triphosphate* (ATP). Contraction by filament sliding is achieved by the rowing action of numerous *cross bridges.* These cross bridges protrude from the thick filaments and are formed by the *active heads of myosin molecules* whose backbones are bound into the thick filament (Figure 5-4). Thin filaments are formed from a *double helix of actin molecules.* Thin filaments are held in place by the Z line. A myosin molecule head which is "charged" with ATP can bind to an actin molecule. Energy is released by the hydrolysis of ATP, and this powers the rowing stroke of the myosin head and contributes to filament sliding.

Filament sliding and muscle contraction are due to the rowing action of very large numbers of active myosin molecule heads. Each individual stroke by a myosin molecule head takes about 1 millisecond and produces a 12-nm movement (Barden and Mason, 1978). Although this is only a microscopic distance, many thousands of sarcomeres are arranged in a series, and in a very short time the sum of all these

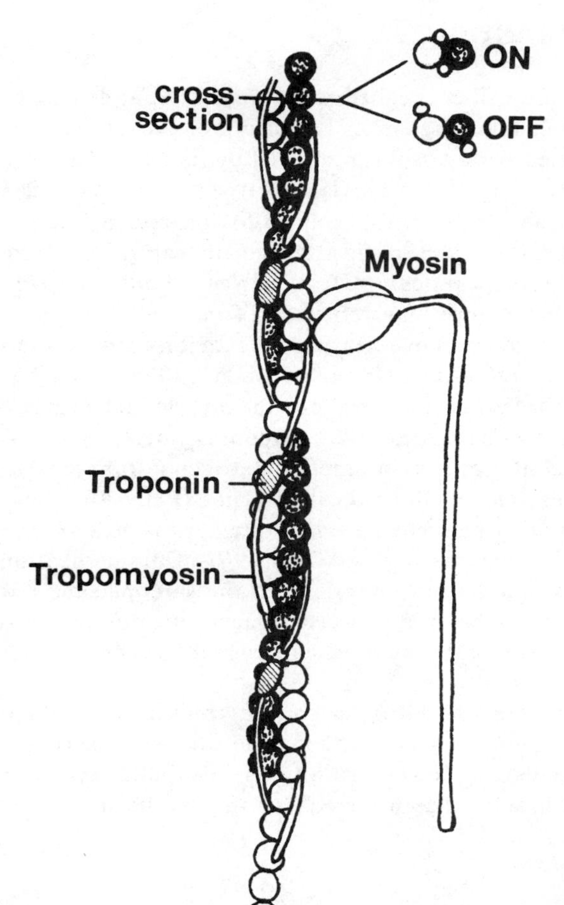

Figure 5-5 Thin filament showing, in the transverse section at the top of the diagram, how tropomyosin molecules in the two grooves of the thin filament might allow (ON) or prevent (OFF) access of myosin heads to actin molecules. (From Swatland, 1982a.)

microscopic distances may be measured in centimeters. The myosin head only releases its grip on the actin, and swings back for another power stroke with another actin, if it is recharged with another ATP molecule (Murray and Weber, 1974). Thus when muscles are converted to meat and no more ATP is available, thick and thin filaments become locked together wherever they overlap. This prevents any further filament sliding and the muscle becomes almost inextensible; this condition is called *rigor mortis.* Further details are given in Chapter 9.

The arrangement of actin molecules in a thin filament is analogous to two strings of pearls that are twisted around each other; a cross section would reveal two pearls, one from each string, separated by grooves (Figure 5-5, top). The two grooves are important because, located in them, are other strings of proteins (Figure 5-5). One of these proteins, *troponin*, responds to the presence of *calcium ions* and causes the other protein, *tropomyosin*, to change its depth in the groove. This is the switching mechanism which allows myosin heads to row against actin molecules; it is the calcium-activated trigger mechanism for muscle contraction (Cohen, 1975).

CONTROL OF MUSCLE CONTRACTION

The cytoplasm within a muscle fiber is called *sarcoplasm.* The aqueous phase of the cytoplasm that surrounds cellular organelles is called the *cytosol.* The calcium ion concentration of the muscle fiber cytosol is regulated by the *sarcoplasmic reticulum* (Porter and Franzini-Armstrong, 1965; Hoyle, 1970). The sarcoplasmic reticulum surrounds the fibrils within a fiber (Plate 29; Figure 5-6). In a resting fiber, the cytosol is kept almost free of calcium ions by the rapid and vigorous capturing or sequestering of calcium ions by the sarcoplasmic reticulum. A protein called *calsequestrin* binds and stores calcium ions within the sarcoplasmic reticulum (Tume, 1979). The sarcoplasmic reticulum triggers muscle contraction by releasing calcium ions when prompted to do so by the *transverse tubular* or *T system* (Inesi and Malan, 1976). Each T tubule is a fingerlike inpushing from the surface membrane of the muscle fiber (Plate 28), and it comes into very close contact with elements of the sarcoplasmic reticulum (Walker and Schrodt, 1966). In longitudinal sections, a junction between a transverse tubule and two elements of the sarcoplasmic reticulum is called a triad (Plate 30). Thus T tubules may conduct *action potentials* which have been initiated at the *neuromuscular junction* deep into the interior of the fiber (Sugi and Ochi, 1967). Communication between a transverse tubule bearing an action potential and the sarcoplasmic reticulum is mediated by protein-containing bridges between adjacent membranes of the sarcoplasmic reticulum and the transverse tubule (Eisenberg and Eisenberg, 1982; Somlyo, 1979; Cadwell and Caswell, 1982).

An extracellular marker, *horseradish peroxidase*, can move into the transverse tubules, thus proving that they open onto the muscle fiber surface (Eisenberg and Eisenberg, 1968). In the striated muscles of farm animals, T tubules occur at the level of each A-I junction. After contraction has occurred for a short while, it is turned off if the

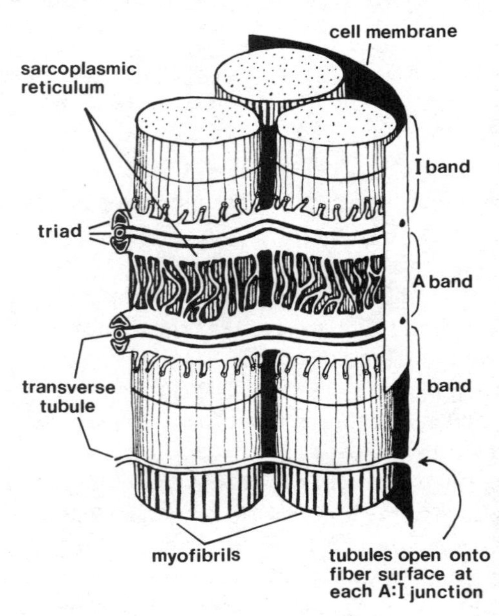

Figure 5-6 A small part of the sarcoplasmic reticulum is shown inside part of a muscle fiber. The transverse tubules that run inward from the surface of the fiber are sandwiched between parts of the sarcoplasmic reticulum to form triads. The sarcoplasmic reticulum does not open onto the surface of the muscle fiber.

sarcoplasmic reticulum resequesters all the calcium ions it initially released. Sustained muscle contraction or *tetanus* is the result of the fusion of individual muscle twitches. The peak tension generated by a single twitch occurs a few milliseconds after the action potential on the muscle fiber membrane, and when about 60 per cent of the maximum amount of calcium ion release has occurred (Bendall, 1969).

The main steps involved in a single muscle contraction can be summarized as follows (refer back to Figure 1-18).

1. The sequence of events leading to contraction is initiated somewhere in the central nervous system, either as voluntary activity from the brain or as reflex activity from the spinal cord.
2. A motor neuron in the ventral horn of the spinal cord is activated, and an action potential passes outward in a ventral root of the spinal cord.
3. The axon branches to supply a number of muscle fibers called a motor unit, and the action potential is conveyed to a motor end plate on each muscle fiber.
4. At the motor end plate, the action potential causes the release of packets or quanta of *acetylcholine* into the *synaptic clefts* on the surface of the muscle fiber.
5. Acetylcholine causes the electrical *resting potential* under the motor end plate to change, and this then initiates an action potential which passes in both directions along the surface of the muscle fiber.
6. At the opening of each transverse tubule onto the muscle fiber surface, the action potential spreads inside the muscle fiber.
7. At each point where a transverse tubule touches part of the sarcoplasmic reticulum, it causes the sarcoplasmic reticulum to release calcium ions.
8. The calcium ions result in the movement of troponin and tropomyosin on their thin filaments, and this enables the myosin molecule heads to "grab and swivel" their way along the thin filament. This is the driving force of muscle contraction.

Contraction is turned off by the following sequence of events:

9. Acetylcholine at the neuromuscular junction is broken down by *acetylcholinesterase*, and this terminates the stream of action potentials along the muscle fiber surface.
10. The sarcoplasmic reticulum ceases to release calcium ions, and immediately starts to resequester all the calcium ions that have been released.
11. In the absence of calcium ions, a change in the configuration of troponin and tropomyosin then blocks the action of the myosin molecule heads, and contraction ceases.
12. In the living animal, an external stretching force, such as gravity or an antagonistic muscle, pulls the muscle back to its original length.

Muscle contraction may take either of two forms, or a combination of both. In *isotonic* contraction, the muscle shortens against a constant load which is moved. In *isometric* contraction, the load is too great to be moved, and the muscle generates tension as it attempts to shorten.

MUSCLE PROTEINS

Myofibrillar Proteins

The proteins of the myofibril account for somewhat more than half of the protein content of most muscle fibers. Within the myofibril, myosin accounts for 55% to 60% of the protein content and actin accounts for about 20% (Ebashi and Nonomura, 1973). The structure of the myosin molecule is shown in Figure 5-7. Experimentally isolated myosin molecules are just large enough to be seen by electron microscopy after they have been sprayed with atoms of metal. The two active heads of each molecule are difficult to resolve separately, so the molecule usually has the appearance of a miniature matchstick.

If the myosin molecule is briefly treated with the digestive enzyme *trypsin*, myosin splits into two large fragments: (1) *light meromyosin* with a molecular weight of about 140,000, and (2) *heavy meromyosin* with a molecular weight of about 340,000. When light meromyosins are placed together under conditions similar to those inside a muscle fiber, they *self-assemble* into a bundle that resembles the backbone of a thick filament. Under precise experimental conditions, *papain* can be used to split heavy meromyosin into two subfragments. Papain is a plant enzyme which is obtained from tropical melons and it is sometimes used in a crude manner to make meat more tender. Subfragment 1 of heavy meromyosin is composed of the two active heads of the myosin molecule, and these retain their ability to bind to actin and to hydrolyze ATP.

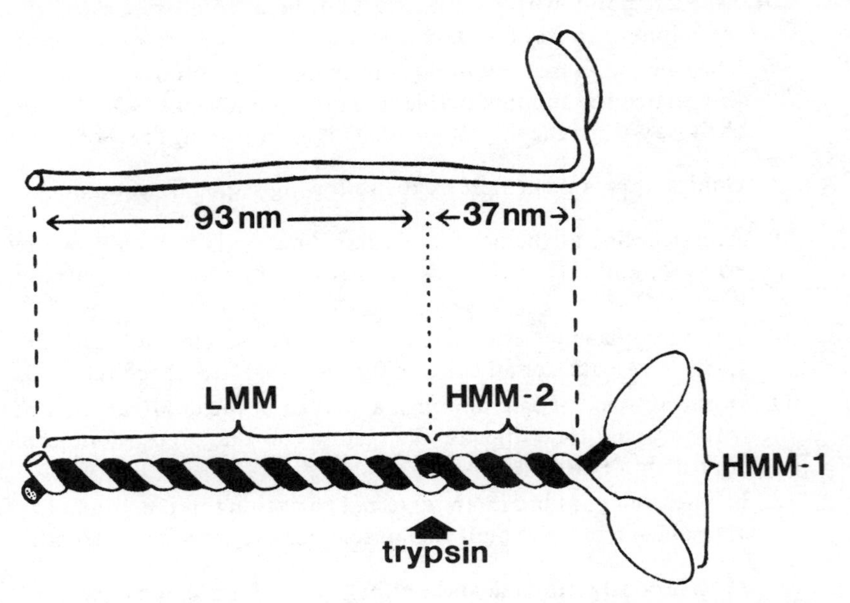

Figure 5-7 Structure of a myosin molecule: (LMM) light meromyosin; (HMM-1) heavy meromyosin subfragment 1; (HMM-2) heavy meromyosin subfragment 2. Approximate dimensions are indicated, as well as the point at which trypsin attacks the myosin molecule.

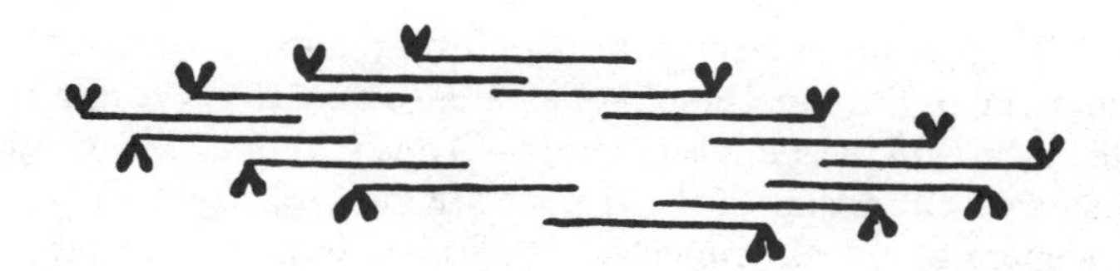

Figure 5-8 Greatly simplified plan of how myosin molecules might be packed to form a thick filament, as proposed by Huxley (1972). Only a few myosin molecules are shown and the diameter of the filament is expanded to show the molecular arrangement.

The exact way in which myosin molecules are arranged in a thick filament is difficult to determine, but the general principle is shown in Figure 5-8. Half of the molecules in a filament face toward one Z line of their sarcomere, while the remainder face toward the opposite Z line. Thus, although the heads protrude at regular intervals along the thick filament, there is a short headless zone at the midlength of the filament where the light meromyosins of opposite sides overlap by their tails. There are several hundred myosin molecules in a thick filament, and the length of a whole filament is about 1.5 μm.

Myosin is the dominant protein of thick filaments, but at the midlength of each thick filament there is an enlargement due to *M line proteins* (Mani et al., 1980). Collectively, these enlargements give rise to the appearance of *M lines* which run parallel to Z lines (Figure 5-2). At a high magnification, cross bridges that link adjacent thick filaments may be seen. However, when the preservation of myofibrils for electron microscopy is enhanced by the use of low-temperature techniques, the M line may appear as several lines, all perpendicular to the long axis of the sarcomere and located near its midlength (Squire et al., 1982). The appearance of these M lines differs between the different physiological types of fibers in an individual mucle. M-protein disappears after meat is placed under great pressure (Macfarlane and Morton, 1978). M-protein is composed of two (Porzio et al., 1979) or more proteins, such as glycogen debranching enzyme (Trinick and Lowey, 1977), creatine kinase (Wallimann et al., 1978), and *myomesin* (Eppenberger et al., 1981).

About 40 molecules of *C-protein* occur in each thick filament. C-protein has no ATPase activity, although it may inhibit actomyosin independently of calcium ion concentration (Offer et al., 1973). Small amounts of *I-protein* (Ohashi et al., 1977a) are distributed along A bands, except at their midlength (Ohashi et al., 1977b). I-protein binds to myosin and inhibits its ATPase activity in the absence of calcium ions; its function may be to prevent unnecessary hydrolysis of ATP in resting muscles (Maruyama et al., 1977b).

The helical arrangement of actin molecules in the thin filament was shown in Figure 5-4. Individual molecules or monomers of actin have a globular shape and are called *G-actin.* The polymerization of G-actin to form filamentous *F-actin* requires energy from ATP, and the resulting ADP is bound into the filament. G-actin molecules have a diameter of about 5.5 nm. Two strands of F-actin are twisted around each other to form the thin filament. The crossover points are about 35.5 nm apart (Bendall, 1969). Each complete thin filament has a length of about 1 μm and contains 300 to 400 G-actins. Although G-actin is globular, it has an exact site on its surface where it can react with myosin molecule heads during muscle contraction. When G-actin is polymerized to F-actin, these sites all face toward the H zone and away from the Z line in which the thin filament is anchored. Thus within an individual sarcomere, the thin filaments on either side of the H zone are mirror images and are compatible with the arrangement of myosin molecles in the thick filament. Thus if a thick filament should penetrate the Z line, the immediate thin filaments in the next sarcomere are facing the wrong way for the myosin molecule heads of the intruding thick filament.

The Z lines of red and white pork muscles contain similar amounts of *alpha-actinin*, although Z lines are thicker in red muscle and so probably contain substantial amounts of some other protein as well (Suzuki et al., 1973). Alpha-actinin from red muscles is slightly different to that from white muscles. In chicken muscle, the filamentous component of the Z line has been identified as alpha-actinin and the amorphous residue has been identified as a protein called *amorphin* (Chowrashi and Pepe, 1982). *Beta-actinin* is probably located at the ends of thin filaments, and it may participate in their assembly. It accelerates polymerization of F-actin from G-actin, but inhibits the irregular assembly of F-actin (Maruyama et al., 1977a). *Gamma-actinin* has the opposite action, and it inhibits polymerization of G-actin (Kuroda and Maruyama, 1976). Beta-actinin appears in cultured muscle cells that are starting to assemble their first myofibrils (Heizmann and Hauptle, 1977).

Many types of cells maintain their shape and internal structure by means of intermediate-diameter (10 nm) filaments which are often woven into a *cytoskeleton*. In skeletal muscle fibers, very small amounts of the protein *desmin* occur near the Z line. Desmin may form 10-nm-diameter filaments in vitro (Robson et al., 1980). The proteins desmin and *vimentin* probably link together the Z lines of adjacent fibrils (Granger and Lazarides, 1979).

Connectin or titin is an elastic protein that occurs in myofibrils (Maruyama, 1976; Maruyama et al., 1976, 1981) and which is thought to cause longitudinal elasticity. In parallel with thick and thin filaments it holds Z lines together (Maruyama et al., 1978). In A bands that have been enzymatically liberated from their fibrils, connectin is concentrated at the edges of the isolated A bands where the A-I junction used to be (Ohashi et al., 1981). In series with the thin filaments, connectin binds to the muscle fiber membrane at muscle-tendon junctions (Maruyama and Shimada, 1978).

From time to time, electron microscopists have observed that myofibrils may contain some thin filaments with elastic properties. Hoyle (1967) thought that these filaments might run parallel to the readily visible thick and thin filaments, and that they might link adjacent Z lines. In highly stretched beef muscle, Locker and Leet (1975, 1976) found *gap filaments* which linked thick filaments to the Z line. Other researchers have been unable to find gap filaments in vertebrate muscle, but have found them in insect muscle (Ullrick et al., 1977). At the present time, it seems likely that these filaments within the myofibril correspond to some of the minor filamentous proteins within the myofibril that have been found by biochemists (possibly connectin). At present, gap filaments are thought to be both elastic in nature and heat stable (Locker and Wild, 1982). When meat has been cooked, actomyosin depolymerizes and forms a coagulum at the midlength of the sarcomere. The coagulum is formed around the relatively unchanged gap filaments. Thus, gap filaments may be a major factor in the determination of the toughness of cooked meat.

Sarcoplasmic Proteins

In muscles such as the longissimus dorsi of the pig, the proteins that are dissolved in the sarcoplasm amount to almost half the weight of the myofibrillar proteins (Scopes, 1970). The sarcoplasmic proteins are dominated by the *glycolytic enzymes* (73 per cent of sarcoplasmic protein), *creatine kinase* (9 per cent), and myoglobin (which increases with age). The function of creatine kinase and the glycolytic enzymes is considered in more detail in Chapter 9. Their main function is to provide the immediate energy for muscle contraction, and they can work anaerobically (without oxygen). The major enzymes in the sarcoplasm are too large to fit between the thick and thin filaments of working myofibrils (Scopes, 1970). The sarcoplasmic proteins do not simply drift around in the sarcoplasm that surrounds the fibrils. The glycolytic enzymes are concentrated at the level of the I bands of fibrils (Pette, 1975), and the three proteins that form the bulk of the thin filament (F-actin, tropomyosin, and troponin) can all bind to a number of glycolytic enzymes (Stewart et al., 1980; Walsh et al., 1980). The enzymes involved in breaking down glycogen as the substrate for glycolysis are bound to parts of the sarcoplasmic reticulum (Wanson and Drochmans, 1972). The localized distribution of these various enzymes and substrates is quite important. When they are mixed in dilute solutions for biochemical analysis, the results obtained may be quite different from events that actually occur in meat. In intact muscle fibers, biochemical reactions may occur much more rapidly since all the appropriate components are tightly fitted together at a high concentration.

The water-soluble components of meat flavor are concentrated in the sarcoplasm, although extra components may be added from the myofibrils as the meat is aged. Water-soluble components that have been purified by *dialysis* can be used to recreate the general taste of meat when they are heated. Some of the most active compounds when they are heated with glucose and phosphate are *inosinic acid*, *glycoprotein*, and *amino acids*. There is a progressive degradation of nucleotides postmortem, from ADP to AMP to inosinic acid to *ribose* + *hypoxanthine* (Tsai et al., 1972b). This may account for the increase in flavor that occurs as beef is aged or conditioned. *Pentose* sugars such as ribose may be more important in the generation of a meaty flavor than *hexose* sugars such as glucose.

PHYSICAL PROPERTIES OF MEAT

The most important physical property of meat is its degree of tenderness when eaten, usually after some degree of cooking. "Tenderness" and "toughness" are complex subjective terms that are difficult to translate exactly into mechanical parameters which can be measured objectively. When meat is cooked, muscle fibers decrease in diameter by up to 15% and collagen is gelatinized to varying degrees. The degraded collagen around muscle fasciculi traps a multitude of molten fat droplets which have been released from any marbling fat within the meat (Wang et al., 1954). Maximum tenderness is reached when the meat reaches a certain temperature, but this temperature differs between muscles and between animals, mostly in relation to the amount and strength of connective tissue. Beef semitendinosus, for example, becomes increasingly tender up to 67°C and then becomes tougher at higher temperatures (Sartorius and Child, 1938). Four major systems influence the texture of meat: (1) sarcomere length; (2) the cytoskeleton; (3) the connective tissue framework of meat; and (4) the breakdown of the muscle microstructure after slaughter.

Sarcomere Length

The relationship between sarcomere length and meat tenderness is now widely known in the meat industry, and the volume of scientific research on the topic is somewhat overwhelming. Marsh (1972) identified four discoveries which he thought were the most important.

1. At the onset of rigor mortis, carcass muscles may be stretched or contracted, depending largely on their position in the hanging carcass (Locker, 1959).
2. Relaxed muscles produce meat which is more tender than that from contracted muscles (Locker, 1960).
3. Rapid cooling before the start of rigor mortis causes muscles to shorten (*cold shortening*; Locker and Hagyard, 1963).
4. Freezing of meat before the completion of rigor mortis leads to extreme shortening when meat is thawed (*thaw shortening*).

The relationship between sarcomere length and the tenderness of cooked meat is shown in Figure 5-9. When sarcomeres in chicken breast muscle contract down to 1.3 μm, the tapered ends of the thick filaments penetrate the Z line (Hagopian, 1970). Changes in tenderness due to sarcomere length act both along and across the muscle fibers, and tensile strength and shear force measurements of meat toughness are quite closely related (Davey and Gilbert, 1977). Apart from the background effect of connective tissue, there are a number of other factors that influence meat tenderness.

1. Shortening causes a decrease in the number of muscle fibers transected per unit of muscle cross-sectional area.
2. Further shortening occurs as meat is cooked (Davey and Gilbert, 1975).

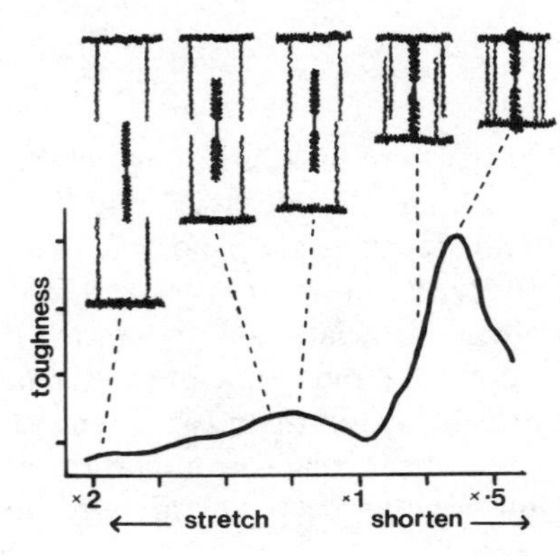

Figure 5-9 Relationship between meat toughness (resistance to shear) and sarcomere length. At length ×1, the muscle is neither stretched nor shortened. The sarcomeres above the graph show the relative longitudinal positions of thick and thin filaments at different degrees of sarcomere stretching and shortening. (Data from Marsh and Carse, 1974.)

3. There is an effect of pH on meat tenderness (Bouton et al., 1973a, 1973b).
4. The temperature at which meat sets in rigor mortis affects meat tenderness (Locker and Daines, 1975b).

Meat Texture

To meat graders, the texture of meat is a property that is related to the size and degree of separation of muscle fasciculi. The prominence of muscle fasciculi is related to the depth of the connective tissue that binds the muscle fibers into fasciculi (Schmitt et al., 1979). To food technologists, however, meat texture has a much broader meaning which includes those aspects of *rheology*, *microstructure*, and *sensory evaluation* which can be related to the phenomenon of meat tenderness. Rheology is the study of flowing systems, but the term is commonly applied to the study of food systems that are being deformed by an external force. Many methods have been invented for the measurement of meat texture, and some of them are shown in Figure 5-10. Some

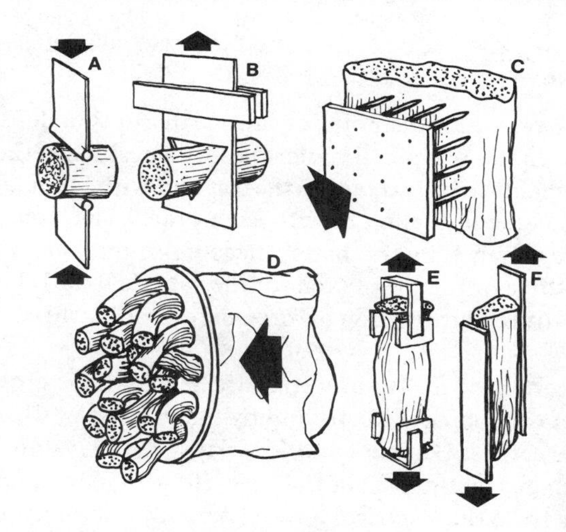

Figure 5-10 Some of the ways to measure meat texture: (A) blunt jaws; (B) Warner-Bratzler shear; (C) Armour-type tenderometer; (D) electric meat grinder method; (E) Instron test for tensile properties; (F) meat glued onto clamp.

methods simply measure the force that must be applied to a sample of meat before it is completely squashed, penetrated, severed, or pulled apart. Other methods enable a *force-deformation* curve to be plotted on a chart recorder. In this case, the test instrument may have a powerful electric motor which drives a crosshead mounted on a worm gear. The forces that act against the meat are measured continuously with a *strain gauge*, and they are plotted on the vertical axis of a force-deformation curve (Figure 5-11). Diagrams D and E in Figure 5-11 show force-deformation curves similar to those which might be produced by methods B and A, respectively, which are illustrated in Figure 5-10. The force-deformation curves of both methods share some common features:

1. There is a smooth buildup of force as the meat initially resists disruption.
2. Complete disruption may be preceded by partial breakdown of the microstructure.
3. There is a point of final disruption.

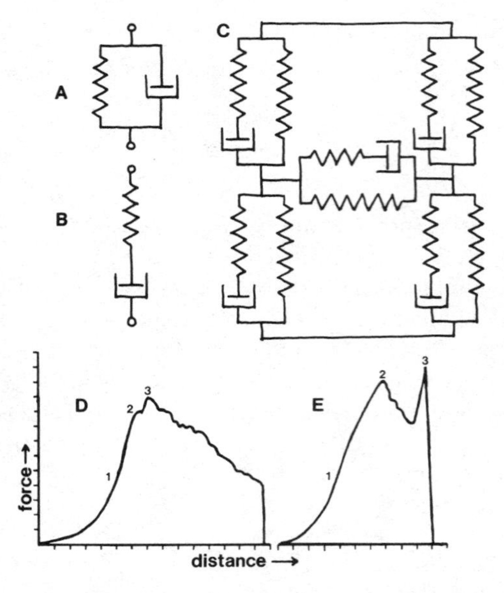

Figure 5-11 Models and measurements of meat texture: (A) Voigt-Kelvin model; (B) Maxwell model; (C) two-dimensional viscoelastic model; (D) Warner-Bratzler type of measurement; (E) Instron, compression by blunt jaws. (Data from deMan, 1976; Segars and Kapsalis, 1976; Pool and Klose, 1969; Rhodes et al., 1972.)

The direction taken by the curve after the point of sample disruption depends on the construction of the test unit. In the Warner-Bratzler shear, the residue of the meat core in the triangular opening remains to be severed. In the test unit with blunt jaws, the test is completed once the jaws come into contact with each other. The slope of the line which describes shear force versus time is a better measure of meat tenderness than either the maximum shear force or the time taken for the sample to fail (Hurwicz and Tischer, 1954). The quantitative interpretation of force-deformation curves is reviewed by Bourne (1976).

The rheological properties of meat cores are affected by their cross-sectional area, by the arrangement of their muscle fibers, and by the manner in which the meat cores are removed from the carcass (Kastner and Henrickson, 1969; Pool and Klose, 1969). The electric meat grinder method (D in Figure 5-10) provides an approximate measure of meat toughness from the electrical power consumption of a meat grinder. Tough meat slows the grinder and causes it to draw more power (Schoman et al., 1960). The tensile strength of meat can be a useful measurement of meat toughness, but one of the major problems that is encountered is the difficulty of securing the meat core in the apparatus. Jaws that hold the sample at each end cause secondary deformation patterns which complicate the overall measurement. Meat samples can be glued to the moving parts of the test unit in almost any configuration, but then it is difficult to determine which parts of the microstructure actually bear the load. In an extreme case (F in Figure 5-10), the connective tissue between muscle fibers may bear the load.

A number of mechanical models have been proposed to account for the *viscoelastic* properties of meat. Pure viscous properties may be represented diagrammatically by a piston in a cylinder (a dashpot). Forces that pull or push on the piston produce only a slow steady movement of the piston through the viscous oil in the cylinder. A pure spring, drawn diagrammatically as a zigzag line, reacts immediately to the forces acting on it, but it eventually reaches a point of maximum compression or extension and then stops behaving like a spring. The spring and the dashpot may be placed in parallel to make a *Voigt-Kelvin model,* or they may be placed in series to make a *Maxwell model* (Figure 5-11). The numerous other components and configurations found in food systems are described and explained by deMan (1976).

If meat were to behave like a pure spring, the start of the force-deformation relationship would be a straight line. This is not the case in practice (D and E in Figure 5-11) since the viscous components of meat give a curved shape to the start of the force-deformation relationship. An interesting property of meat is that when its microstructure is completely disrupted, the ultimate failure under load of each element in the microstructure is due to tensile failure. For example, at the microstructural level, the meat at the edge of a Warner-Bratzler blade is stretched across the blade until its tensile strength is overcome (Plate 8; Voisey and Larmond, 1974; Voisey, 1976). Unless meat is frozen solid, it cannot really be sheared in the same way that flakes may be knocked from a flint. Even when meat is compressed, it is the tensile strength of the microstructure that holds the sample together until the point of disruption. For example, diagram C in Figure 5-11 shows a two-dimensional model for meat under compression. As the meat is compressed, from top to bottom in the diagram, it starts to bulge sideways. The elements of the microstructure that resist this movement are shown as the horizontal components in the middle of the diagram. The meat core cannot be completely flattened until these microstructural components have failed in tensile strength.

Sarcomere length has a major effect on meat texture. Myosin molecule heads become locked onto the thin filaments when ATP is depleted after slaughter. The number of myosin molecules that can attach in this way is determined by sarcomere length. At a short sarcomere length, most of the myosin molecules can attach to actin, and the myofibril exhibits its greatest tensile strength. Within a few days after slaughter, however, *calcium-activated protease* weakens the Z line by releasing alpha-actinin (Dayton et al., 1976; Suzuki et al., 1975; Reville et al., 1976), and the thin filaments are pulled out of the Z line when the myofibril is stretched.

Connective Tissue Framework of Meat

The connective tissue content of muscle makes a major contribution to meat texture. Muscle fibers are embedded in a meshwork of connective tissue which is dominated by *collagen fibers.* This continuous meshwork is traditionally divided into three levels of organization. The *epimysium* covers the muscle surface, the *perimysium* binds individual muscle fibers into fasciculi (Plate 4), and the *endomysium* shrouds individual muscle fibers (Plate 5). The collagen fibers of the endomysium are very small, and are often called *reticular fibers.* They almost completely cover the surfaces of individual muscle fibers (Rowe, 1981).

Muscle fibers are anchored in the connective tissue framework of their muscle. The terminal thin filaments of myofibrils terminate in a dense material on the inner surface of the plasma membrane of the muscle fiber. The outer surface of the muscle fiber plasma membrane is secured in an amorphous basement membrane which also binds endomysial collagen fibers. At *myotendinous junctions*, the endomysium merges with the parallel collagen fibers of the tendon. The area of plasma membrane for the attachment of collagen fibrils outside, and myofibrils inside the muscle fiber is often increased by fingerlike projections and invaginations at the end of the muscle fiber (Korneliussen, 1973). At the muscle attachment, large numbers of muscle fibers may converge on a limited area where the muscle is attached to a tendon, to the periosteum, or to the perichondrium. When this occurs, there is no space for all the muscle fibers to terminate in the same plane at the end of the muscle. Myotendinous junctions then occur deep in the muscle, and only thin bundles of collagen fibrils from each muscle fiber extend into the crowded end of the muscle (Goss, 1944). In the large carcass muscles of meat animals, many muscle fibers terminate *intrafascicularly* within the belly of the muscle. Intrafascicularly terminating muscle fibers have tapered ends which are inserted into the endomysium of an adjacent muscle fiber (Swatland and Cassens, 1972).

The *sarcolemma* may also contribute to the directional or *anisotropic* properties of meat. In its broadest definition, the sarcolemma includes the plasma membrane of the muscle fiber together with an amorphous basement membrane in which reticular fibers are embedded on the muscle fiber surface (Kono et al., 1964; Zacks et al., 1973). The sarcolemma has sufficient longitudinal strength to bear the tension exerted across a region which has been experimentally deprived of myofibrils (Street and Ramsey, 1965). The sarcolemma is anisotropic, and is stiffer in its longitudinal direction than in its circumferential direction (Fields, 1970). In the past, some authors have used the term "sarcolemma" in a restricted manner as an alternative name for the plasma membrane of the muscle fiber. To avoid confusion, the term should be clearly defined when it is first used.

In the early days of scientific research on meat, the connective tissue content of meat was regarded as the sole cause of differences in tenderness between different muscles. Following the discovery of sarcomere length as a major factor affecting meat toughness (Locker, 1959, 1960), the contribution from intramuscular connective tissue came to be regarded as a background level of toughness determined solely by the amount of collagen and the degree of cross-linking. Thus events after slaughter that affected sarcomere length were not thought to affect the background connective tissue toughness. However, there is now some evidence that changes in muscle length can affect the arrangement of connective tissue fibers in the fibrous framework of a muscle.

Epimysial and perimysial collagen fibers are usually arranged in a crisscross lattice at an angle to the long axis of the muscle, and individual collagen fibers adopt a wavy or crimped shape when released from tension (Rowe, 1974). When tension is placed on a muscle while it is still in an extensible condition before the occurrence of rigor mortis, angular changes occur in the crisscross lattice. Collagen fibers that take up the tension become straight (Figure 5-12). Similar changes occur in the endomysial reticular fibers that surround individual muscle fibers (Swatland, 1975). In contracted muscles, endomysial fibers are mostly perpendicular to the long axes of their muscle fibers (Plate 7; Figure 5-13). In stretched muscles, endomysial fibers become realigned at an acute angle to the long axes of the muscle fibers (Plate 6). In living animals, these changes in connective tissue orientation probably help to protect the muscle against damage due to excessive stretching. Relationships between raw meat tenderness and

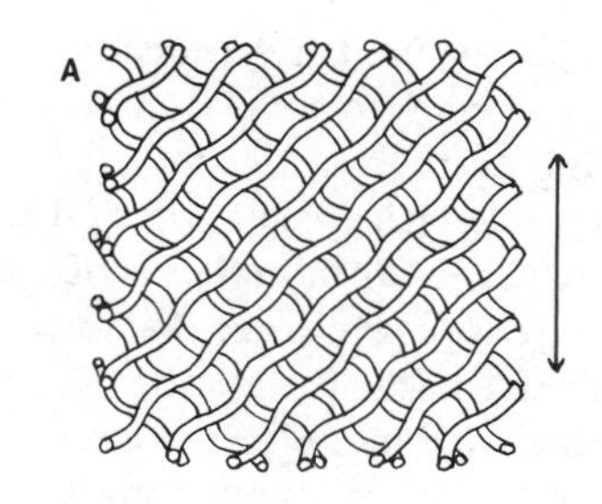

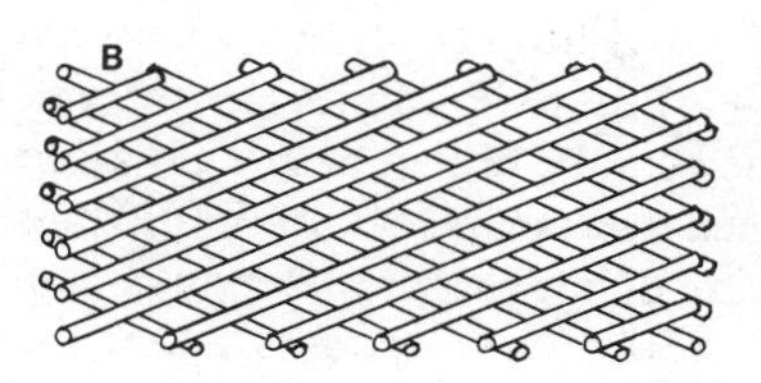

Figure 5-12 Diagram to illustrate changes in the arrangement of perimysial collagen fibers when a muscle contracts. The arrow indicates the direction of the longitudinal axis of the muscle: (A) muscle at resting length; (B) contracted muscle.

the degree of muscle contraction cannot, therefore, be completely explained solely by the effect of sarcomere length and filament overlap on tenderness (Dutson et al., 1976; Locker and Daines, 1976; Dransfield and Rhodes, 1976; Stanley and Swatland, 1976; Carroll et al., 1978; Swatland, 1978b).

If meat is pulled apart in a test instrument, the degree of slack in a connective tissue fiber may be important. If a collagen fiber is already under tension, it may be pulled apart more rapidly than if there is some series elasticity or slack to be taken up first. In a macroscopic analogy, if ropes of equal strength but different length are anchored together at their ends and are subjected to increasing tension, they will eventually fail one by one, starting with the shortest. Thus the force required to pull the system apart is equal to the tensile strength of one rope, rather than to the tensile strength of all the ropes combined, as would be the case if the ropes were equal in length and sharing the load. A further complication is added to the properties of the connective tissue framework of meat if the meat is cooked to the point at which collagen fibers start to shorten. Here, too, the orientation of the fibers may be important (Locker and Daines, 1975a).

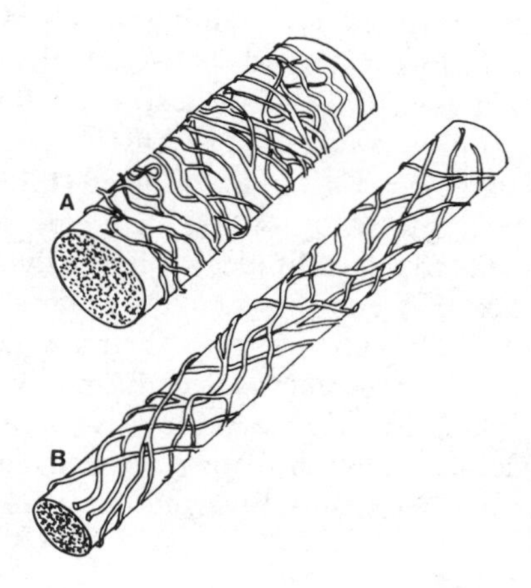

Figure 5-13 Change in the orientation of endomysial reticular fibers between a muscle fiber at contracted length (A) and at stretched length (B).

When tensile strength is measured dynamically, parallel-fibered systems may show a steep early rise in their force-deformation curves. Lattice meshworks of fibers show a shallow rise, depending on their lattice angles (Viidik, 1980). Thus meat that has set in rigor at a contracted length (Figures 5-12B and 5-13A) tends to show a shallow rise when tested along the longitudinal axis of the muscle fibers. In raw meat with a short sarcomere length, tensile forces are first engaged by the myofibrils. When these are broken apart, connective tissue fibers take the strain, change their configuration, and then fail. Thus meat with a short sarcomere length may have less tensile strength than an equivalent meat sample with long sarcomeres, even though short sarcomeres allow greater cross-linking strength between thick and thin filaments (Stanley and Swatland, 1976). Individual muscle fibers in meat are quite extensible, but the length to which they can be pulled before breaking is reduced when the meat has been aged or conditioned by a period of storage after slaughter (Wang et al., 1956).

The strength or texture of a meat sample may depend on the direction in which measurements are made. Anisotropy certainly exists in radial versus longitudinal directions relative to the muscle fiber axes, but it may also occur in different radial planes. In the straplike bovine sternomandibularis muscle, radial anisotropy may be due to relatively flat muscle fasciculi. If the fasciculi are flat, the perimysium may form layers whose united strength depends on the direction in which they are transected (Locker and Daines, 1976). Further information on the texture and rheology of meat is reviewed by deMan et al. (1976) and by Stanley (1976).

In the early years of scientific research on meat, when connective tissue content was regarded as the only major biological factor affecting meat tenderness, the increase in tenderness that occurred when beef was aged or conditioned after slaughter was rather puzzling. Early investigators naturally searched for chemical evidence of collagen degradation, but none was detected (Ramsbottom and Strandine, 1949; Wierbicki et al., 1954). Since then, attention has shifted to the myofibrillar component of meat toughness, and the problem of postmortem collagen degradation has not yet been properly settled. Wang (1949) found some histological changes which he attributed to the postmortem degradation of collagen, and his conclusions have been incorporated into textbooks (Birkner and Auerbach, 1960; Price and Schweigert, 1971). The collagen of aged beef loses its ability to bind a red histological stain, acid fuchsin. Wang (1949) measured the loss in staining with a microscope photometer, probably one that was equipped with a photoelectric cell primarily sensitive to green light. Arylmethane dyes such as acid fuchsin are strongly bound to collagen fibers which, in the living animal, have been subjected to tension (tendons or stretched skin). Collagen fibers from slack tissues are only lightly stained (Flint et al., 1975). Thus the changes in staining observed by Wang (1949) in conditioned beef might have resembled those that occur due to a lack of tension in the living animal.

There is some evidence that enzymes capable of attacking collagen exist in muscle and, in fact, it would be difficult to explain their absence, since the connective tissue framework of muscle must be remodeled during growth and after injury. Collagen degradation is initiated by *collagenases* but is completed by other enzymes (Gay and Miller, 1978). The degradation of bovine collagen by enzymes from the spleen has been studied experimentally (Etherington, 1977), and collagenases in muscle have been identified both biochemically (Canonico and Bird, 1970; Etherington, 1972) and histochemically (Montfort and Perez-Tamayo, 1975). Collagen also has a known turnover rate in living muscle (Gay and Miller, 1978; Bailey and Robins, 1976).

Kopp and Valin (1981) suggest that lysosomal enzymes in meat do not attack collagen because of their limited release or because of proteinase inhibitors. However, Stanley (1976) reported biochemical evidence of the degradation of collagen during conditioning. Wu et al. (1981) concluded that *lysosomal glycosidases* are involved in the postmortem degradation of collagen, and that their activity in enhanced when beef samples are conditioned at 37° C. It is difficult to use extrinsic collagenase activity for the measurement of the degree of cross-linking in collagen because a high fat content in the meat may protect collagen from digestion (Goll et al., 1964).

In living animals, healthy, active muscles are poorly endowed with lysosomes, perhaps because of the risk of autolysis during muscle exertion. In noncontractile tissues, degradative enzymes are used in the removal of structural proteins during cellular repair and maintenance, and they are safely stored in *lysosomes* until they are required. Collagenases may be stored in an inactive form, or they may only be synthesized in response to appropriate stimuli (Gay and Miller, 1978). If collagen fibers in meat are degraded during conditioning, the origin of the enzymes involved is difficult to explain. In diseased or bruised muscles, invasive cells from the blood may be involved (Hamdy et al., 1961; Tappel, 1965), but in properly exsanguinated meat without prior disease or bruising, this source is probably minimal. Degradative enzymes derived from the sarcoplasmic reticulum or from membrane-bounded structures within muscle fibers (Trout et al., 1979) might diffuse outward to reach distant collagen fibers. Sites of *acid phosphatase* and *protease* activity can be found within muscle fibers (Dutson et al., 1971; Stauber and Ong, 1982), and it has been suggested that another lysosomal enzyme, *glucuronidase*, diffuses outward from muscle fibers to attack collagen (Dutson and Lawrie, 1974). Although fibroblasts are near the fibers that they produce, they do not have many lysosomes. Intramuscular *mast cells* (Plate 9), however, are a rich source of lysosomes, and most of them are located in the perimysium (Swatland, 1978a). The *alkaline proteases* found in skeletal muscle may originate from mast cells (Bird et al., 1980), and proteolytic activity is reduced after experimental depletion of mast cell granules (Clark et al., 1980).

Meat Color

When meat is exposed to air, purple-red *myoglobin* absorbs oxygen and is converted to bright red *oxymyoglobin*. After prolonged exposure, oxymyoglobin is chemically oxidized to brown *metmyoglobin*. Green pigments due to the formation of *sulfmyoglobin* are rarely seen in the normal course of meat handling. Sulfmyoglobin may be formed if bacteria are able to produce hydrogen sulfide, which can then act on myoglobin (Barnes and Shrimpton, 1957). There are several different approaches to the measurement of meat color.

1. The overall concentration of isolated myoglobin pigments in solution can be measured by *absorbance spectrophotometry* in order to study differences

between red and white muscles, between species, between animals at different ages, and between various degrees of atmospheric exposure. Some typical values for overall myoglobin concentration are shown in Table 5-1.

2. The relative amounts of myoglobin, oxymyoglobin, and metmyoglobin can also be measured in situ by macroscopic or fiber optic *reflectance spectrophotometry*.
3. The color of the meat can be judged subjectively in terms related to human visual perception.

The derivatives of myoglobin in meat differ in their spectra of absorbance and reflectance. In other words, they absorb or reflect different amounts of light at different wavelengths (colors). Given a sample with a mixture of myoglobin derivatives, absorbance or reflectance may be measured at two different wavelengths. The ratio of the two measurements can be used to calculate the relative amounts of myoglobin derivatives. An *isobestic point* occurs when two or more spectra intersect to give the same value at the same wavelength. An isobestic point for myoglobin, oxymyoglobin, and metmyoglobin is at a wavelength of 525 nm. An absorbance peak for metmyoglobin is at 630 nm. Thus the ratio of measurements at 630 nm (mostly metmyoglobin) to measurements at 525 nm (sum of all three myoglobins) contains information on the amount of metmyoglobin as a fraction of the total myoglobins. The principles of this approach and other examples are reviewed by Hunt (1980).

There are three words which have special meanings in relation to the human perception of color. The general type of color, such as red, blue, or green, is called a *hue*. If, for the sake of explanation, red paint was mixed in increasing quantities into a pot of dull white paint, the paint would change through pale pink to dark red, yet the hue (red) is unchanged. The property of color that changes in this example is the intensity of the color, and this is called the *chroma*. If the demonstration had started with bright white paint instead of dull white, the final product would be brighter and

TABLE 5-1. OVERALL CONCENTRATION OF MYOGLOBIN IN DIFFERENT TYPES OF MEAT

Type	Age	Muscle	Myoglobin (per cent wet weight)
Veal	12 days	Longissimus dorsi	0.07
Steer	3 years	Longissimus dorsi	0.46
Mutton		Longissimus dorsi	0.25
Pork	5 months	Longissimus dorsi	0.030
Pork	6 months	Longissimus dorsi	0.038
Pork	7 months	Longissimus dorsi	0.044
Pork		Semitendinosus, pale part	0.149
Pork		Semitendinosus, dark part	0.4
Pork	1 year	Psoas, normal	0.210
Pork	1 year	Psoas, immobile	0.156
Chicken		Gastrocnemius	0.045
Chicken		Pectoralis	0.005

Sources: Lawrie, (1950, 1974) and Beecher et al. (1968).

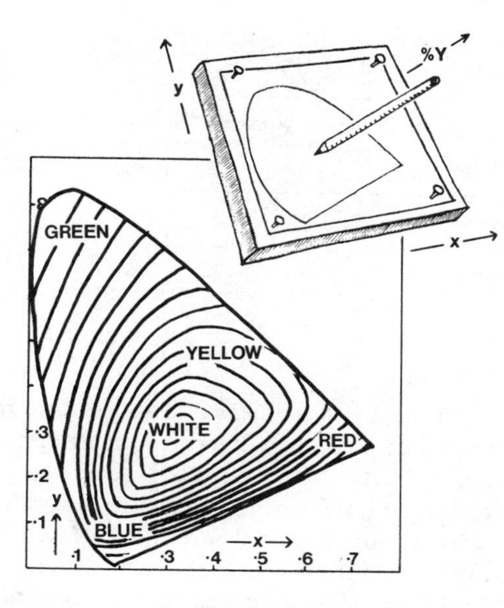

Figure 5-14 CIE system of color measurement. Hue and chroma are given by chromaticity coordinates *x* and *y* in the lower diagram. The upper part of the diagram shows a three-dimensional view of a pencil, perpendicular to the plane of the *x,y* graph pinned on a board. The height along the pencil measures luminosity as per cent *Y*.

would have a greater *luminosity*. There are several different systems for the measurement of color in these terms, and the choice of which system to use depends largely on what equipment is readily available.

In the method recommended by the International Commission on Illumination (CIE), the primary hues, red, green, and blue, are added or subtracted from each other to match any color. By a mathematical manipulation, it is possible to specify both hue and chroma in the CIE system by means of a single pair of *chromaticity coordinates* called *x* and *y* (Figure 5-14). Changes in hue follow the contour lines shown in Figure 5-14, while changes in chroma radiate from the central position of white. Luminosity is specified by a third coordinate relative to the plane of the chromaticity coordinates. To illustrate this dimension in Figure 5-14, the graph of the chromaticity coordinates *x* and *y* is shown pinned to a drawing board, and a pencil is held perpendicular to the board. Luminosity is measured at a point along the pencil, and this dimension is called *per cent Y*. In meat with a lot of marbling fat, it might be expected that measurements made by reflectance spectrophotometry would be biased by light reflected from marbling fat. However, this source of error appears to be quite small in actual practice (Tuma et al., 1962). Similarly, on the surface of freshly cut meat, differences in the wetness of the surface have only a small effect on luminosity (Swatland, 1982b). Details of the CIE system and its advantages are given by Francis and Clydesdale (1975). To minimize the ambiguity of published scientific data relating to meat color, it has been proposed that all results should now be reported using the CIE system (Boccard et al., 1981). Colorimetric data from other systems can be converted to the CIE system quite easily (computer programs are given by Francis and Clydesdale, 1975). Table 5-2 gives some typical CIE data relating to pork color. Chromaticity coordinates *x* and *y* are only slightly changed in PSE pork, probably in relation to the faster formation of metmyoglobin (Bembers and Satterlee, 1975). Large differences occur in luminosity.

TABLE 5-2. CIE MEASUREMENTS OF PORK COLOR (LONGISSIMUS DORSI)

	Typical CIE values		
Condition of the meat	x	y	Per cent Y
Extreme PSE	0.36	0.35	40
Normal	0.37	0.34	30
Extreme DFD	0.38	0.33	20

Meat pH

By titration, the total acidity of meat can be measured as the amount of a known concentration of sodium hydroxide that is required to neutralize a meat slurry containing a phenolphthalein indicator. However, the acidity of meat is more commonly measured by its hydrogen ion concentration. The pH scale is an inverted scale—the pH decreases as hydrogen ions become more concentrated. The pH scale is also a logarithmic scale—as pH drops from 7 to 6, the hydrogen ion concentration increases tenfold, from 10^{-7} to 10^{-6} M. Thus the characteristic shape of a postmortem pH curve in meat (Figure 5-15) is partly due to the logarithmic nature of the transformation

$$[H^+] = 10^{-pH}$$

$$pH = -\log[H^+]$$

where $[H^+]$ = hydrogen ion concentration.

Meat contains *buffers* (due to the presence of ATP and phosphate) which can capture hydrogen ions, so that the pH may give a poor measure of the total acidity. Dipeptides such as *carnosine* and *anserine* in meat also contribute to buffering capacity (Davey, 1960; Crush, 1970; Winnick et al., 1963; McManus, 1960; Parker, 1966). The presence of carnosine increases the extent of muscle contraction, at least in

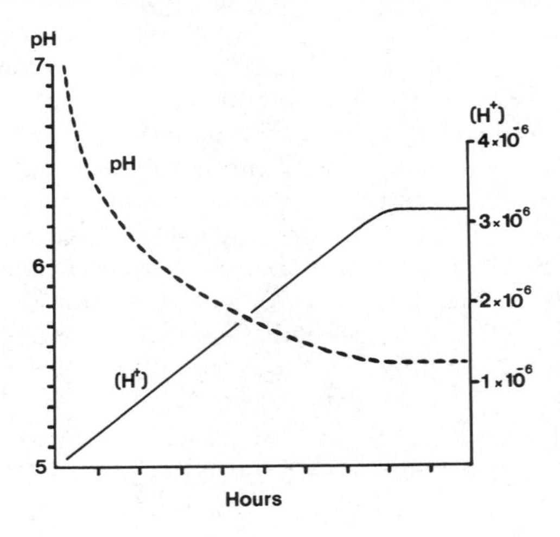

Figure 5-15 Relationship between pH and hydrogen ion concentration (H^+) in a postmortem pH curve.

model systems (Bowen, 1965). The pH declines after death because of the formation of *lactate* (lactic acid). But to produce the same drop in pH, the pork longissimus dorsi requires 33 per cent more lactate than the beef longissimus dorsi (Bendall, 1973).

If the glass electrode of a pH meter is pushed into a meat sample, it may be used to measure the pH of the meat in the immediate vicinity. This value may not represent the pH in the bulk of the sample. For example, lactate formation may be slower on the surface of meat exposed to atmospheric oxygen. Glass pH electrodes, no matter how carefully they are used, can sometimes give erroneous readings when they are pushed into a piece of meat. Very often the pH reading varies from one minute to another and it may never settle down to give a constant reading, particularly if the meat has a high pH and is relatively dry.

To avoid these problems, the pH of meat is often measured after the meat sample has been macerated in water. However, this creates two new problems: (1) lactate formation may be accelerated by cellular disruption; and (2) the buffers in the meat have been diluted and their properties have been slightly changed. Thus *sodium iodoacetate* (5 mM) is needed to prevent further lactate formation, and the ionic strength of the meat sample must be maintained by using 150 mM *potassium chloride* instead of water. The pH of meat is also affected by temperature. It decreases by about 0.15 when the meat is warmed from 20° C to 38° C, and it increases by about 0.2 when the meat is cooled from 20° C to 0° C (Bendall, 1973).

If the pH of meat drops to the point at which a meat protein bears no net charge, the protein exhibits its lowest solubility and water-binding capacity. This is called the *isoelectric point* of the protein. the isoelectric point of muscle proteins is near a pH of 5.5. The bulk of commerical meat gets very close to the isoelectric points of its various proteins, and this accounts for the ease with which meat will release water by evaporation or as drip loss. Young animals tend to produce meat with a higher pH than older animals (Tuma et al., 1963).

Mechanically Deboned Meat

The meat that remains on bones after they have been removed from the carcass can be retrieved by grinding the bones and meat fragments, and then subjecting them to great pressure behind a metal screen or sieve. Residual meat fragments are pushed through the screen and are collected as a product known as *mechanically deboned meat.* Mechanical deboning enables the recovery of meat from bones such as vertebrae which are difficult to clean manually. However, bone particles and fat are also extruded through the screen, together with the meat particles. With a hole size of 0.46 mm in the screen, Field et al. (1977) found that bone particles from beef neck muscles ranged in size from 0.08 to 0.11 mm. Bone particles of this size are probably dissolved in the acidic conditions found in the human stomach. Residual calcium content is normally maintained at relatively low levels in mechanically deboned meat (Field et al., 1976). Bone composition in meat animals was surveyed by Field et al. (1974), and bone particle size has been measured in mechanically deboned poultry meat (Froning, 1978). Mechanically deboned meat is generally used as a supplement in processed meat products.

Properties of Meat in Processed Products

Large amounts of meat are sold or stored in a frozen condition. The structure of meat is disrupted by the formation of ice crystals. Ice crystals are small and numerous when the meat is rapidly frozen, and they are large and less numerous when the meat is frozen slowly. In industrial conditions, where meat is frozen in large blocks, the outside freezes first and ice crystals are formed both within and between the muscle fibers. Deep in the meat, ice crystal formation is slower. The fluid inside the muscle fibers becomes concentrated and water is lost from the fibers to form ice crystals between the fibers (Menegalli and Calvelo, 1979; Bevilacqua et al., 1979).

Large amounts of meat are used to make processed meat products such as sausages. Although there are countless numbers of recipes, only a few different basic procedures are involved. Since ancient times, meat has been preserved or *cured* by treating it with *sodium chloride*. However, crude salt contains *sodium nitrate*, and some of this may be converted to *sodium nitrite* by bacterial action in the curing brine. Nitrite gives rise to *nitrous acid*, and then to *nitric oxide*. Nitric oxide reacts with the myoglobin of meat, and the product of this reaction creates the characteristic pink color of cured meat when cooked, *nitrosyl hemochrome*. Nitrite also contributes to the taste of meat products and lengthens their storage life.

The presence of sodium nitrite in a meat product prevents certain aspects of the growth and division of bacteria such as *Clostridium perfringens* and *Clostridium botulinum*. The first of these is a common cause of food poisoning, while the latter is the cause of a rare but often fatal form of food poisoning called *botulism*. The uncontrolled bacterial production of nitrite is undesirable, so that sodium nitrate is now excluded from meat products and sodium nitrite is added directly in small quantities (<125 parts per million). Processed meats contribute only several percent of the average consumer's intake of nitrite, but the nitrite levels in meat are kept as low as possible. Relatively high nitrite levels are found in some vegetables and even in human saliva, possibly as a natural defense against Clostridia.

Nitrites contribute to the formation of *nitrosamines* by adding a *nitroso* group (—N=O) to nitrogen atoms present in the structure of the meat. Nitrosamines occur in many substances in everyday use, but they are carcinogenic. Nitrosamine levels in cooked bacon can be minimized by the inclusion of *sodium ascorbate*. Without nitrite in processed meat products, the risk of botulism is increased. We are faced with three choices: (1) to stop eating meat products processed with nitrite; (2) to invent some other form of protection against botulism; or (3) to accept the risk to benefit ratio as it is at present. A combination of the last two choices suits most people. The extensive literature on this topic is reviewed by Lechowich et al. (1978). The properties of nitrite-free processed meats are reviewed by Buege (1980). There are slight changes in taste, and large changes in color (gray instead of pink).

Among the other common curing ingredients are (1) *sugar*, to modify the taste of the product and to reduce the availability of water for bacterial growth; (2) *ascorbic acid*, to maintain the color of the product; and (3) alkaline *phosphates*, to increase the water-binding capacity of the product. Curing ingredients are now usually injected with a battery of hollow needles or through major arteries. Some meat products are *smoked* with smoke from hickory or oak sawdust on a heating element. This modifies

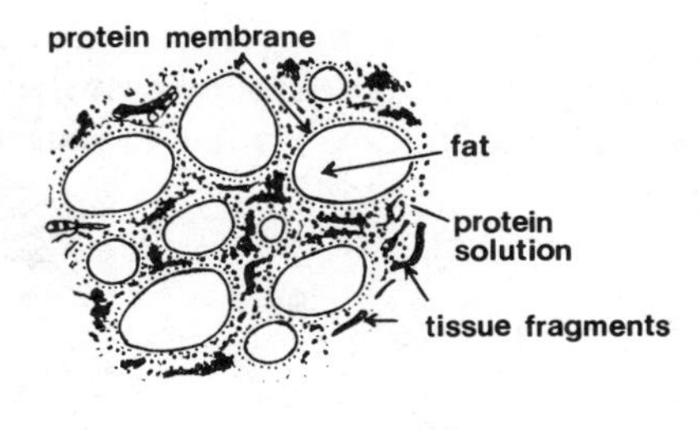

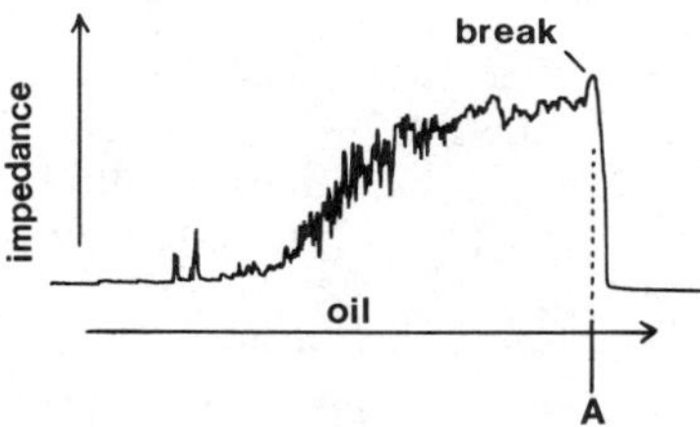

Figure 5-16 Structure of a meat emulsion and the measurement of emulsifying capacity from the amount of oil added to reach point A.

the taste of the product, retards *oxidative rancidity* of the fat, and tends to retard bacterial spoilage.

The structure of meat products such as frankfurters, bologna, and liver sausage resembles an *emulsion*. Fat droplets bounded by a protein membrane are trapped in a meshwork of myofibrillar proteins in water (Figure 5-16). The fat forms the *discontinuous phase* of the emulsion, and the aqueous proteins form the *continuous phase*. Actin and myosin become soluble in dilute salt solutions, so that the salt in the recipe strengthens the continuous phase of the emulsion. The continuous phase forms a gel-type matrix which, unlike oil-water emulsions. traps fat droplets and allows them to coalesce (Lee et al., 1981). Thus the microstructure of meat emulsions is usually more irregular than shown in Figure 5-16.

The *emulsifying capacity* is the ability of proteins and water to bind fat droplets in an emulsion. Fat droplets in sausages vary in diameter from 0.1 to 150 μm, depending on the type of product (Cassens et al., 1977). When the emulsifying capacity is exceeded, the fat separates into large masses in the product. These may rise to the top of the product if they are free to do so. When the product is cooked, masses of unemulsified fat detract from the value of the product. The emulsifying capacity of meat proteins is measured experimentally in the laboratory by slowly adding vegetable oil to a protein and saline (1 M NaCl) mixture in a blender. The *electrical resistance* or *impedance* of the mixture increases as the low-resistance pathway of the proteins in saline is progressively interrupted by emulsified oil droplets. When the emulsion breaks, almost all the oil rises to the top of the blender and the electrodes are again united by a low-resistance pathway if they are at the bottom of the blender (Figure 5-16). With other electrode configurations, emulsion breakdown may result in a rapid increase in resistance if the unemulsified oil then isolates the electrodes (Webb et al., 1970).

The emulsifying properties of myosin are far superior to those of actin or the sarcoplasmic proteins (Tsai et al., 1972a). Connective tissue adds nothing to the emulsifying capacity of a protein mixture and, in excess (>25 per cent of protein), may lead to pockets of *gelatin* in the final product when cooked. The amount of collagen in a processed meat product can be determined from the hydroxyproline content or by *computer image analysis* of stained collagen in microscopic sections (Hildebrandt et al., 1977). *Extenders* are proteins that are added to increase the emulsifying capacity. *Fillers* such as starch may be added to improve cooking yields and to strengthen the product for slicing.

REFERENCES

BAILEY, A. J., and ROBINS, S. P. 1976. *Sci. Prog. (Oxf.)* 63:419.

BARDEN, J. A., and MASON, P. 1978. *Science* 199:1212.

BARNES, E. M., and SHRIMPTON, D. H. 1957. *J. Appl. Bacteriol.* 20:273.

BEECHER, G. R., KASTENSCHMIDT, L. L., CASSENS, R. G., HOEKSTRA, W. G., and BRISKEY, E. J. 1968. *J. Food Sci.* 33:84.

BEMBERS, M., and SATTERLEE, L. D. 1975. *J. Food Sci.* 40:40.

BENDALL, J. R. 1969. *Muscles, Molecules and Movement.* Heinemann Educational Books, London.

BENDALL, J. R. 1973. Postmortem changes in muscle. In G. H. Bourne (ed.), *The Structure and Function of Muscle*, Vol. 2, P. 2, pp. 244–309. Academic Press, New York.

BEVILACQUA, A., ZARITSKY, N. E., and CALVELO, A. 1979. *J. Food Technol.* 14:237.

BIRKNER, M. L., and AUERBACH, E. 1960. *The Science of Meat and Meat Products.* W. H. Freeman and Company, Publishers, San Francisco.

BIRD, J. W. C., CARTER, J. H., TRIEMER, R. E., BROOKS, R. M., and SPANIER, A. M. 1980. *Fed. Proc.* 39:20.

BOCCARD, R., BUCHTER, L., CASTEELS, E., COSENTINO, E., DRANSFIELD, E., HOOD, D. E., JOSEPH, R. L., MACDOUGALL, D. B., RHODES, D. N., SCHON, I., TINBERGEN, B. J., and TOURAILLE, C. 1981. *Livestock Prod. Sci.* 8:385.

BOURNE, M. C. 1976. Interpretation of force curves from instrumental measurements. In J. M. deMan, P. W. Voisey, V. F. Rasper, and D. W. Stanley (eds.), *Rheology and Texture in Food Quality*, pp. 244–274. AVI Publishing Company, Westport, Conn.

BOUTON, P. E., CARROLL, F. D., FISHER, A. L., HARRIS, P. V., and SHORTHOSE, W. R. 1973a. *J. Food Sci.* 38:816.

BOUTON, P. E., CARROLL, F. D., HARRIS, P. V., and SHORTHOSE, W. R. 1973b. *J. Food Sci.* 38:404.

BOWEN, W. J. 1965. *Arch. Biochem. Biophys.* 112:436.

BUEGE, D. R. 1980. *Proc. Reciprocal Meat Conf.* 33:122.

CADWELL, J. J. S., and CASWELL, A. H. 1982. *J. Cell Biol.* 93:543.

CANONICO, P. G., and BIRD, J. W. C. 1970. *J. Cell Biol.* 45:321.

CARROLL, R. J., RORER, F. P., JONES, S. B., and CAVANAUGH, J. R. 1978. *J. Food Sci.* 43:1181.

CASSENS, R. G., SCHMIDT, R., TERRELL, R., and BORCHERT, L. L. 1977. *Microscopic Structure of Commercial Sausage.* Univ. Wis., Coll. Agric. Life Sci. Res. Div., Res. Rep. R2878.

CHOWRASHI, P. K., and PEPE, F. A. 1982. *J. Cell Biol.* 94:565.

CLARK, M. G., BEINLICH, C. J., MCKEE, E. E., LINS, J. A., and MORGAN, H. E. 1980. *Fed. Proc.* 39:26.

COHEN, C. 1975. *Sci. Am.* 233:36.

CRUSH, K. G. 1970. *Comp. Biochem. Physiol.* 34:3.

DAVEY, C. L. 1960. *Arch. Biochem. Biophys.* 89:296.

DAVEY, C. L., and GILBERT, K. V. 1975. *J. Food Technol.* 10:333.

DAVEY, C. L., and GILBERT, K. V. 1977. *Meat Sci.* 1:49.

DAYTON, W. R., GOLL, D. E., ZEECE, M. G., ROBSON, R. M., and REVILLE, W. J. 1976. *Biochemistry* 15:2150.

DEMAN, J. M. 1976. Mechanical properties of foods. In J. M. deMan, P. W. Voisey, V. F. Rasper, and D. W. Stanley (eds.), *Rheology and Texture in Food Quality*, pp. 8–27. AVI Publishing Company, Westport, Conn.

DEMPSEY, E. W., WISLOCKI, G. B., and SINGER, M. 1946. *Anat. Rec.* 96:221.

DRANSFIELD, E., and RHODES, D. N. 1976. *J. Sci. Food Agric.* 27:483.

DUTSON, T. R., and LAWRIE, R. A. 1974. *J. Food Technol.* 9:43.

DUTSON, T. R., PEARSON, A. M., MERKEL, R. A., KOCH, D. E., and WEATHERSPOON, J. B. 1971. *J. Anim. Sci.* 32:233.

DUTSON, T. R., HOSTETLER, L., and CARPENTER, Z. L. 1976. *J. Food Sci.* 41:863.

EBASHI, S., and NONOMURA, Y. 1973. Proteins of the myofibril. In G. H. Bourne (ed.), *The Structure and Function of Muscle*, 2nd ed., Vol. 3, pp. 285–362. Academic Press, New York.

EISENBERG, B., and EISENBERG, R. S. 1968. *Science* 160:1243.

EISENBERG, B. R., and EISENBERG, R. S. 1982. *J. Gen. Physiol.* 79:1.

EPPENBERGER, H. M., PERRIARD, J.-C., Rosenberg, U. B., and STREHLER, E. E. 1981. *J. Cell Biol.* 89:185.

ETHERINGTON, D. J. 1972. *Biochem. J.* 127:685.

ETHERINGTON, D. J. 1977. *Connect. Tissue Res.* 5:135.

FIELD, R. A., RILEY, M. L., MELLO, F. C., CORBRIDGE, M. H., and KOTULA, A. W. 1974. *J. Anim. Sci.* 39:493.

FIELD, R. A., KRUGGEL, W. G., and RILEY, M. L. 1976. *J. Anim. Sci.* 43:755.

FIELD, R. A., OLSON-WOMACK, S. L., and KRUGGEL, W. G. 1977. *J. Food Sci.* 42:1406.

FIELDS, R. W. 1970. *Biophys. J.* 10:462.

FLINT, M. H., LYONS, M. F., MEANEY, M. F., and WILLIAMS, D. E. 1975. *Histochem. J.* 7:529.

FRANCIS, F. J., and CLYDESDALE, F. M. 1975. *Food Colorimetry. Theory and Applications.* AVI Publishing Company, Westport, Conn.

FRONING, G. W. 1978. *Poult. Sci.* 57:1137.

GAY, S., and MILLER, E. J. 1978. *Collagen in the Physiology and Pathology of Connective Tissue*. Gustav Fischer Verlag, Stuttgart.

GOLL, D. E., HOEKSTRA, W. G., and BRAY, R. W. 1964. *J. Food Sci.* 29:608.

GOSS, C. M. 1944. *Am. J. Anat.* 74:259.

GRANGER, B. L., and LAZARIDES, E. 1979. *Cell* 18:1053.

HAGOPIAN, M. 1970. *J. Cell Biol.* 47:790.

HAMDY, M. K., MAY, K. N., and POWERS, J. J. 1961. *Proc. Soc. Exp. Biol. Med.* 108:185.

HEIZMANN, C. W., and HAUPTLE, M.-T. 1977. *Eur. J. Biochem.* 80:443.

HILDEBRANDT, G., KONIGSMANN, R., KRETSCHMER, F.-J., and RENNER, H. VON. 1977. *Fleischwirtschaft* 57:689.

HOYLE, G. 1967. *Am. Zool.* 7:435.

HOYLE, G. 1970. *Sci. Am.* 222:84.

HUNT, M. C. 1980. *Proc. Reciprocal Meat Conf.* 33:41.

HURWICZ, H., and TISCHER, R. G. 1954. *Food Technol.* 8:391.

HUXLEY, H. E. 1965. *Sci. Am.* 213:18.

HUXLEY, H. E. 1972. Molecular basis of contraction in cross-striated muscles. In G. H. Bourne (ed.), *The Structure and Function of Muscle*, 2nd ed., Vol. 1, pp. 302–387. Academic Press, New York.

INESI, G., and MALAN, N. 1976. *Life Sci.* 18:773.

KASTNER, C. L., and HENRICKSON, R. L. 1969. *J. Food Sci.* 34:603.

KONO, T., KAKUMA, F., HOMMA, M., and FUKUDA, S. 1964. *Biochem. Biophys. Acta* 88:155.

KOPP, J., and VALIN, C. 1981. *Meat Sci.* 5:319.

KORNELIUSSEN, H. 1973. *Z. Anat. Entwicklungsgesch.* 142:91.

KURODA, M., and MARUYAMA, K. 1976. *J. Biochem. (Tokyo)* 79:249.

LAWRIE, R. A. 1950. *J. Agric. Sci., Camb.* 40:356.

LAWRIE, R. A. 1974. *Meat Science.* Pergamon Press, Oxford.

LECHOWICH, R. V., CASSENS, R. G., ISSENBERG, P., PADBERG, D. I., SEBRANEK, J. G., SPENCER, J. V., and TERRELL, R. M. 1978. *Nitrite in Meat Curing: Risks and Benefits* Rep. 74. Council for Agricultural Science and Technology, Ames, Iowa.
LEE, C. M., CARROLL, R. J., and ABDOLLAHI, A. 1981. *J. Food Sci.* 46:1789.
LE GROS, CLARK, W. E., 1971. *The Tissues of the Body*. Clarendon Press, Oxford.
LOCKER, R. H. 1959. *J. Biochem. Biophys. Cytol.* 6:419.
LOCKER, R. H. 1960. *Food Res.* 25:304.
LOCKER, R. H., and DAINES, G. J. 1975a. *J. Sci. Food Agric.* 26:1711.
LOCKER, R. H., and DAINES, G. J. 1975b. *J. Sci. Food Agric.* 26:1721.
LOCKER, R. H., and DAINES, G. J. 1976. *J. Sci. Food Agric.* 27:186.
LOCKER, R. H., and HAGYARD, C. J. 1963. *J. Sci. Food Agric.* 14:787.
LOCKER, R. H., and LEET, N. G. 1975. *J. Ultrastruct. Res.* 52:64.
LOCKER, R. H., and LEET, N. G. 1976. *J. Ultrastruct. Res.* 56:31.
LOCKER, R. H. and WILD, D. J. C. 1982. *Meat Sci.* 7:189.
MANI, R. S., HERASYMOWYCH, O. S., and KAY, C. M. 1980. *Int. J. Biochem.* 12:333.
MARSH, B. B. 1972. Post-mortem muscle shortening and meat tenderness. *Proc. Meat Ind. Res. Conf.*, American Meat Industry Foundation, Chicago, pp. 109–124.
MARSH, B. B., and CARSE, W. A. 1974. *J. Food Technol.* 9:129.
MACFARLANE, J. J., and MORTON, D. J. 1978. *Meat Sci.* 2:281.
MARGARIA, R. 1972. *Sci. Am.* 226:84.
MARUYAMA, K. 1976. *J. Biochem.* 80:405.
MARUYAMA, K., and SHIMADA, Y. 1978. *Tissue Cell* 10:741.
MARUYAMA, K., NATORI, R., and NONOMURA, Y. 1976. *Nature (Lond.)* 262:58.
MARUYAMA, K., KIMURA, S., ISHII, T., KURODA, M., OHASHI, K., and MURAMATSU, S. 1977a. *J. Biochem.* 81:215.
MARUYAMA, K., KUNITOMO, S., KIMURA, S., and OHASHI, K. 1977b. *J. Biochem.* 81:243.
MARUYAMA, K., CAGE, P. E., and BELL, J. L. 1978. *Comp. Biochem. Physiol.* 61A:623.
MARUYAMA, K., KIMURA, S., OHASHI, K., and KUWANO, Y. 1981. *J. Biochem.* 89:701.
MCMANUS, R. 1960. *J. Biol. Chem.* 235:1398.
MENEGALLI, F. C., and CALVELO, A. 1979. *Meat Sci.* 3:179.
MONTFORT, I., and PEREZ-TAMAYO, R. 1975. *J. Histochem. Cytochem.* 23:910.
MUHL, Z. F. 1982. *J. Morphol.* 173:285.
MURRAY, J. M., and WEBER, A. 1974. *Sci. Am.* 230:58.
OFFER, G., MOOS, C., and STARR, R. 1973. *J. Mol. Biol.* 74:653.
OHASHI, K., KIMURA, S., DEGUCHI, K., and MARUYAMA, K. 1977a. *J. Biochem.* 81:233.
OHASHI, K., MASAKI, T., and MARUYAMA, K. 1977b. *J. Biochem.* 81:237.
OHASHI, K., FISCHMAN, D. A., OBINATA, T., and MARUYAMA, K. 1981. *Biomed. Res. (Tokyo)* 2:330.
PARKER, C. J. 1966. *Anal. Chem.* 38:1359.
PETTE, D. 1975. *Acta Histochem., Suppl.* 14S:47.
POOL, M. F., and KLOSE, A. A. 1969. *J. Food Sci.* 34:524.
PORTER, K. R., and FRANZINI-ARMSTRONG, C. 1965. *Sci. Am.* 212:72.
PORZIO, M. A., PEARSON, A. M., and CORNFORTH, D. P. 1979. *Meat Sci.* 3:31.
PRICE, J. F., and SCHWEIGERT, B. S. 1971. *The Science of Meat and Meat Products*. W. H. Freeman and Company, Publishers, San Francisco.
RAMSBOTTOM, J. M., and STRANDINE, E. J. 1949. *J. Anim. Sci.* 8:398.
REVILLE, W. J., GOLL, D. E., STROMER, M. H., ROBSON, R. M., and DAYTON, W. R. 1976. *J. Cell Biol.* 70:1.
RHODES, D. N., JONES, R. C. D., CHRYSTALL, B. B., and HARRIES, J. M. 1972. *J. Texture Stud.* 3:298.
ROBSON, R. M., STROMER, M. H., HUIATT, T. W., O'SHEA, J. M., HARTZER, M. K., RICHARDSON, F. L., and RATHBUN, W. E. 1980. *Int. Meat Res. Congr., Colo.* 26(1):22.

ROWE, R. W. D. 1974. *J. Food Technol.* 9:501.
ROWE, R. W. D. 1978. *Meat Sci.* 2:275.
ROWE, R. W. D. 1981. *Tissue Cell* 13:681.
SARTORIUS, M. J., and CHILD, A. M. 1938. *Food Res.* 3:619.
SCHMITT, O., DEGAS, T., PEROT, P. LANGLOIS, M. R., and DUMONT, B. L. 1979. *Ann. Biol. Anim. Biochem. Biophys.* 19(1A):1.
SCHOMAN, C. M., BELL, J., and BALL, C. O. 1960. *Food Technol.* 14:581.
SCOPES, R. K. 1970. Characterization and study of sarcoplasmic proteins. In E. J. Briskey, R. G. Cassens, and B. B. Marsh (eds.), *The Physiology and Biochemistry of Muscle as a Food, 2*, pp. 471–492. University of Wisconsin Press, Madison.
SEGARS, R. A., and KAPSALIS, J. G. 1976. *J. Texture Stud.* 7:129.
SOMLYO, A. V. 1979. *J. Cell Biol.* 80:743.
SQUIRE, J., EDMAN, A.-C., FREUNDLICH, A., HARFORD, J., and SJOSTROM, M. 1982. *J. Microsc.* 125:215.
STANLEY, D. W. 1976. The texture of meat and its measurement. In J. M. deMan, P. W. Voisey, V. F. Rasper, and D. W. Stanley (eds.), *Rheology and Texture in Food Quality* pp. 405–426. AVI Publishing Company, Westport, Conn.
STANLEY, D. W., and SWATLAND, H. J. 1976. *J. Texture Stud.* 7:65.
STAUBER, W. T., and ONG, S.-H. 1982. *Histochem. J.* 14:585.
STEWART, M., MORTON, D. J., and CLARKE, F. M. 1980. *Biochem. J.* 186:99.
STREET, S. F., and RAMSEY, R. W. 1965. *Science* 149:1379.
SUGI, H., and OCHI, R. 1967. *J. Gen. Physiol.* 50:2167.
SUZUKI, A., GOLL, D. E., STROMER, M. H., SINGH, I., and TEMPLE, J. 1973. *Biochim. Biophys. Acta* 295:188.
SUZUKI, A., NONAMI, Y., and GOLL, D. E. 1975. *Agric. Biol. Chem.* 39:1461.
SWATLAND, H. J. 1975. *J. Anim. Sci.* 41:78.
SWATLAND, H. J. 1978a. *Zentralbl. Veterinaermed. A* 25:556.
SWATLAND, H. J. 1978b. *Can. Inst. Food Sci. Technol. J.* 11:204.
SWATLAND, H. J. 1982a. *Can. J. Anim. Sci.* 62:15.
SWATLAND, H. J. 1982b. *J. Food Sci.* 47:1940.
SWATLAND, H. J. and CASSENS, R. G. 1972. *J. Anim. Sci.* 35:336.
TAPPEL, A. L. 1965. Lysosomes: Enzymes and catabolic reactions. In E. J. Briskey, R. G. Cassens, and J. C. Trautmant (eds.), *The Physiology and Biochemistry of Muscle as a Food*, pp. 237–250. University of Wisconsin Press, Madison.
TRINICK, J., and LOWEY, S. 1977. *J. Mol. Biol.* 113:343.
TROUT, J. J,. STAUBER, W. T., and SCHOTTELIUS, B. A. 1979. *Histochem. J.* 11:223.
TSAI, R., CASSENS, R. G., and BRISKEY, E. J. 1972a. *J. Food Sci.* 37:286.
TSAI, R., CASSENS, R. G., BRISKEY, E. J., and GREASER, M. L. 1972b. *J. Food Sci.* 37:612.
TUMA, H. J., HENRICKSON, R. L., STEPHENS, D. F., and MOORE, R. 1962. *J. Anim. Sci.* 21:848.
TUMA, H. J., HENRICKSON, R. L., ODELL, G. V., and STEPHENS, D. F. 1963. *J. Anim. Sci.* 22:354.
TUME, R. K. 1979. *Aust. J. Biol. Sci.* 32:177.
ULLRICK, W. C., TOSELLI, P. A., CHASE, D., and DASSE, K. 1977. *J. Ultrastruct. Res.* 60:263.
VIIDIK, A. 1980. Interdependence between structure and function in collagenous tissues. In A. Viidik and J. Vuust (eds.), *Biology of Collagen*, pp. 257–280. Academic Press, New York.
VOISEY, P. W. 1976. *J. Texture Stud.* 7:11.
VOISEY, P. W., and LARMOND, E. 1974. *Can. Inst. Food. Sci. Technol. J.* 7:243.
WALKER, S. M., and SCHRODT, G. R. 1966. *Nature* (*Lond.*) 211:935.
WALLIMANN, T., PELLONI, G., TURNER, D. C., and EPPENBERGER, H. M. 1978. *Proc. Natl. Acad. Sci. USA* 75:4296.
WALSH, T. P., WINZOR, D. J., CLARKE, F. M., MASTERS, C. J., and MORTON, D. J. 1980. *Biochem. J.* 186:89.
WANG, H. 1949. *Anat. Rec.* 105:537.

WANG, H., RASCH, E., BATES, V., BEARD, F. J., PIERCE, J. C., and HANKINS, O. G. 1954. *Food Res.* 19:314.
WANG, H., DOTY, D. M., BEARD, F. J., PIERCE, J. C., and HANKINS, O. G. 1956. *J. Anim. Sci.* 15:97.
WANSON, J.-C., and DROCHMANS, P. 1972. *J. Cell Biol.* 54:206.
WEBB, N. B., IVEY, F. J., CRAIG, H. B., JONES, V. A., and MONROE, R. J. 1970. *J. Food Sci.* 35:501.
WIERBICKI, E., KUNKLE, L. E., CAHILL, V. R., and DEATHERAGE, F. E. 1954. *Food Technol.* 8:506.
WINNICK, R. E., MOIKEHA, S., and WINNICK, T. 1963. *J. Biol. Chem.* 238:3645.
WU, J. J., DUTSON, T. R., and CARPENTER, Z. L. 1981. *J. Food Sci.* 46:1132.
ZACKS, S. I., SHEFF, M. F., and SAITO, A. 1973. *J. Histochem. Cytochem.* 21:703.

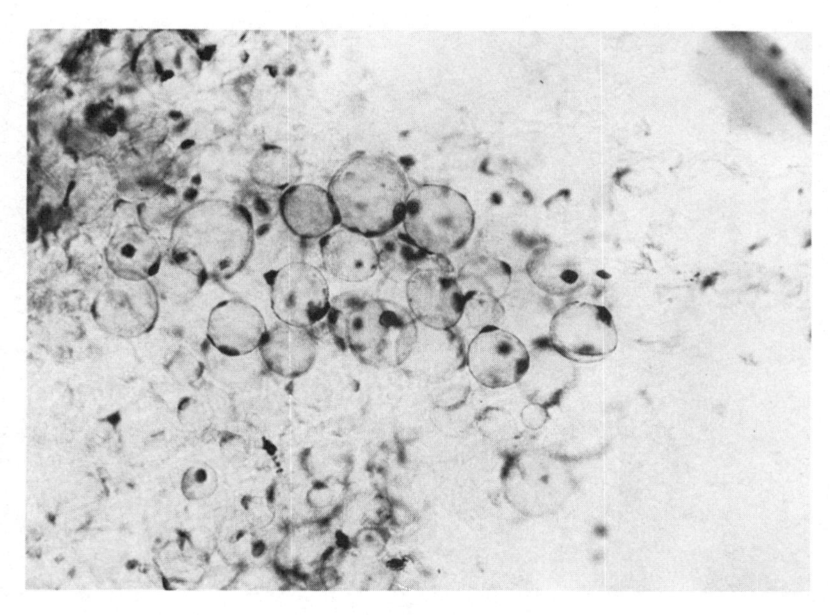

Plate 1 Isolated adipose cells with their nuclei stained.

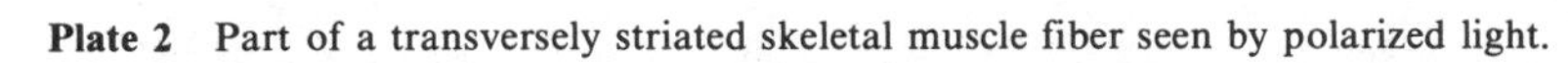

Plate 2 Part of a transversely striated skeletal muscle fiber seen by polarized light.

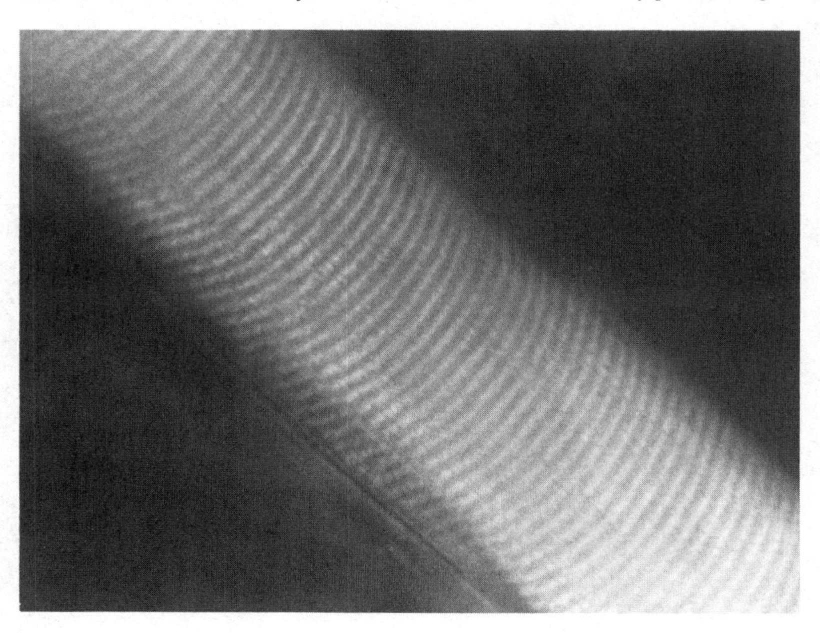

Plate 3 Tapered intrafascicularity terminating (ift) muscle fiber in a thick frozen section of beef stained with methylene blue.

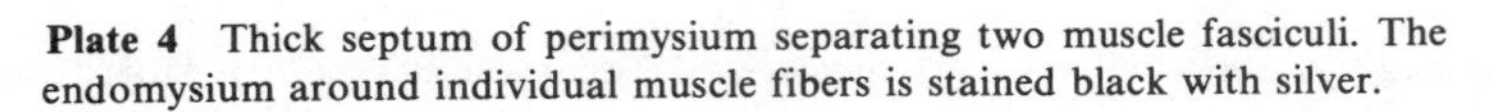

Plate 4 Thick septum of perimysium separating two muscle fasciculi. The endomysium around individual muscle fibers is stained black with silver.

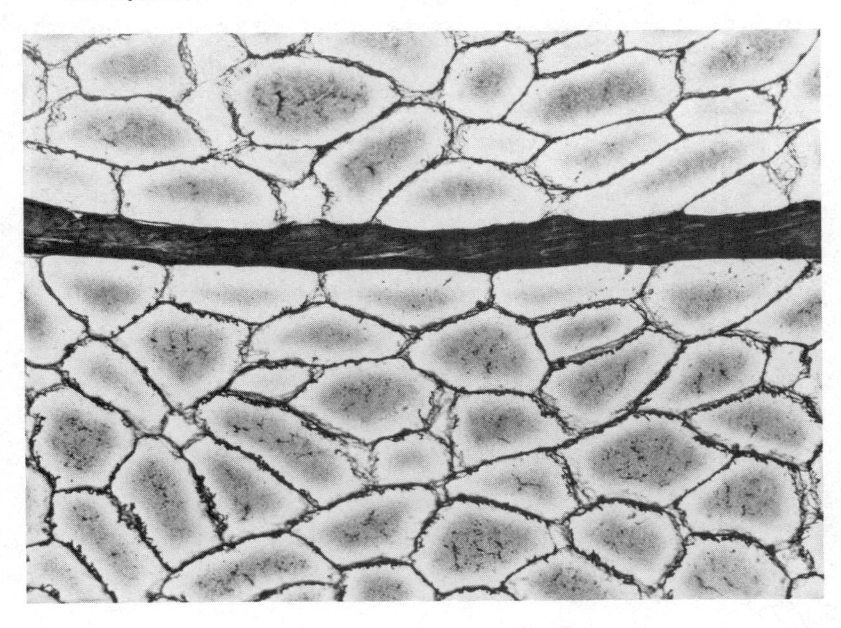

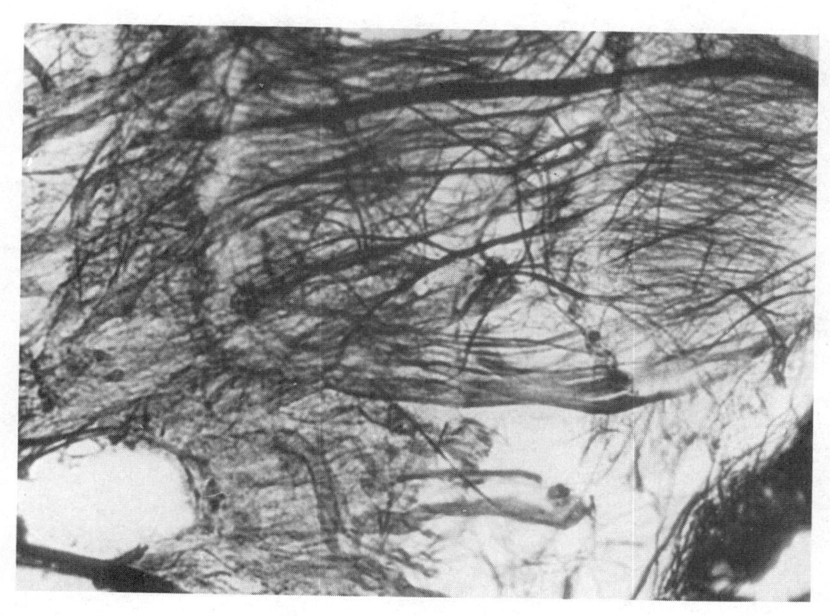

Plate 5 Endomysial reticular fibers on the surface of a muscle fiber.

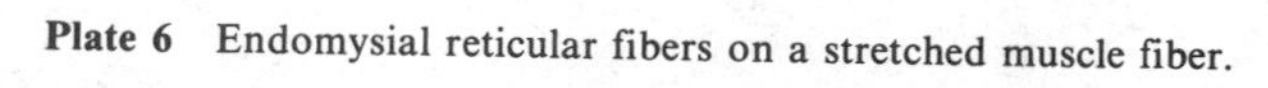

Plate 6 Endomysial reticular fibers on a stretched muscle fiber.

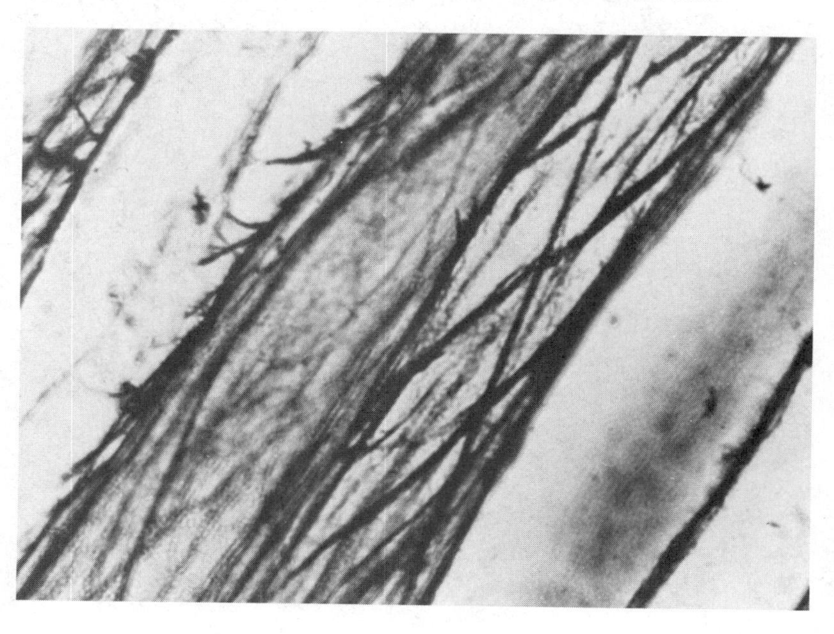

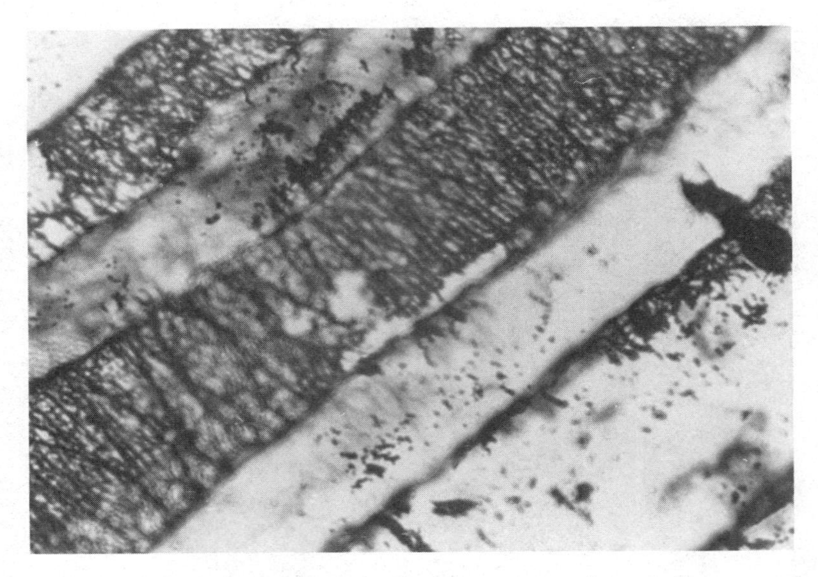

Plate 7 Endomysial reticular fibers on a contracted muscle fiber.

Plate 8 Connective tissue framework of a meat sample that has been compressed in front of the blade of a knife.

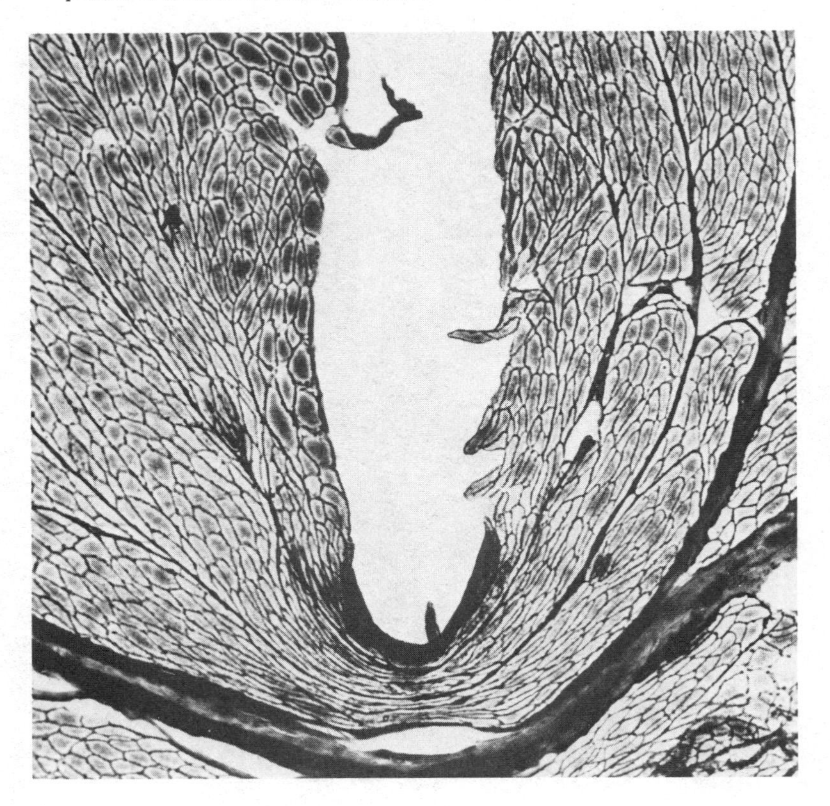

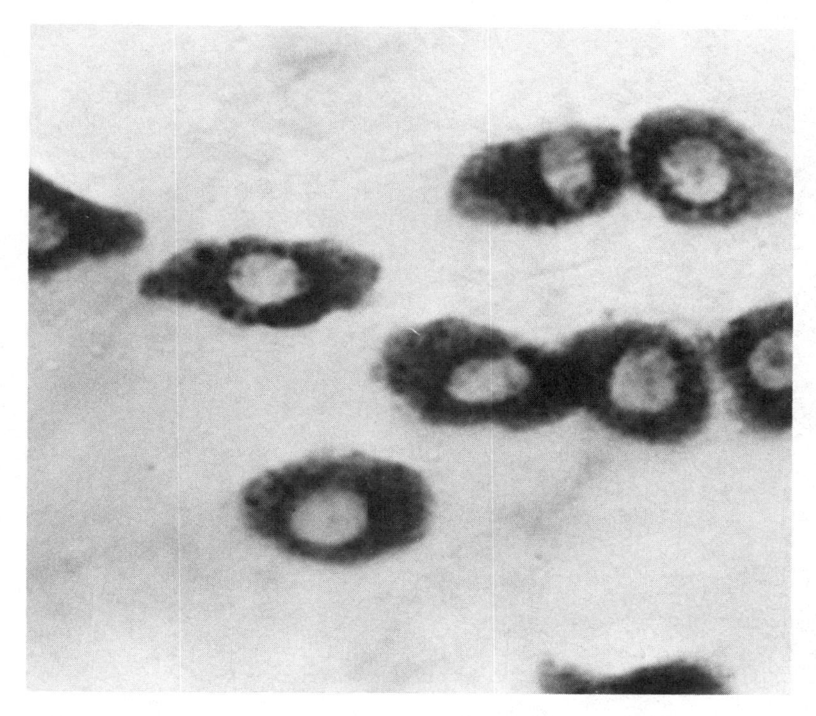

Plate 9 Mast cells in bovine epimysium exhibiting strong cytoplasmic reactions for acid phosphatase.

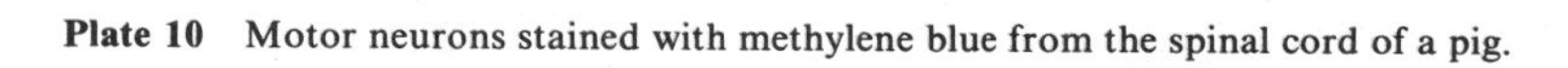

Plate 10 Motor neurons stained with methylene blue from the spinal cord of a pig.

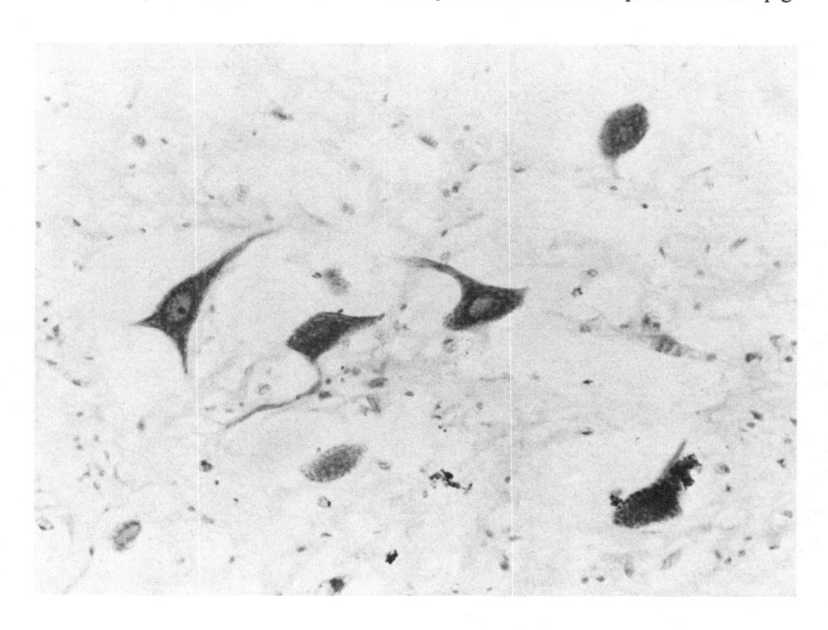

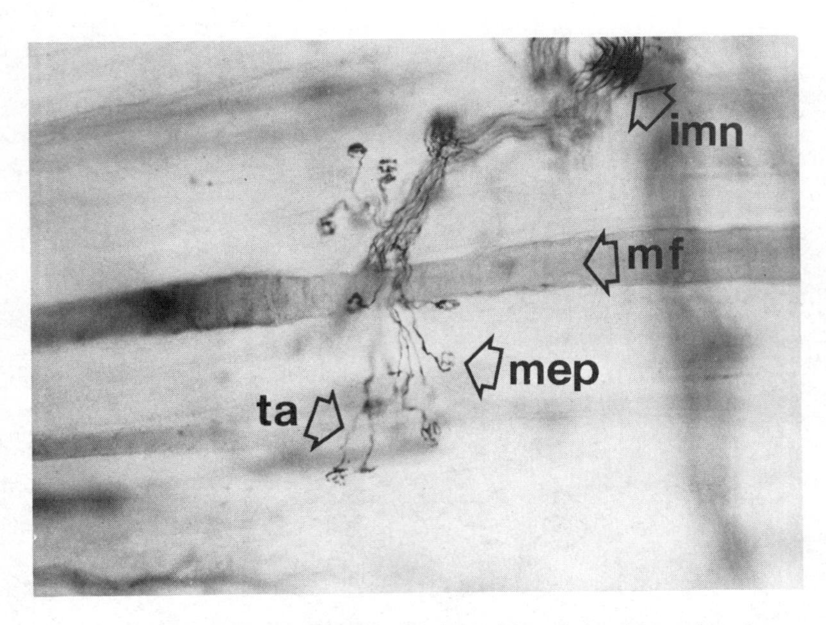

Plate 11 Thick frozen section of beef stained with methylene blue to show the innervation of muscle fibers: (imn) intramuscular nerve; (mf) a muscle fiber; (mep) motor end plate; (ta) terminal axon. Most of the muscle fibers are unstained.

Plate 12 Acetylcholinesterase activity in the innervation zones across a muscle fasciculus. Each small round area of reaction product indicates the position of a neuromuscular junction.

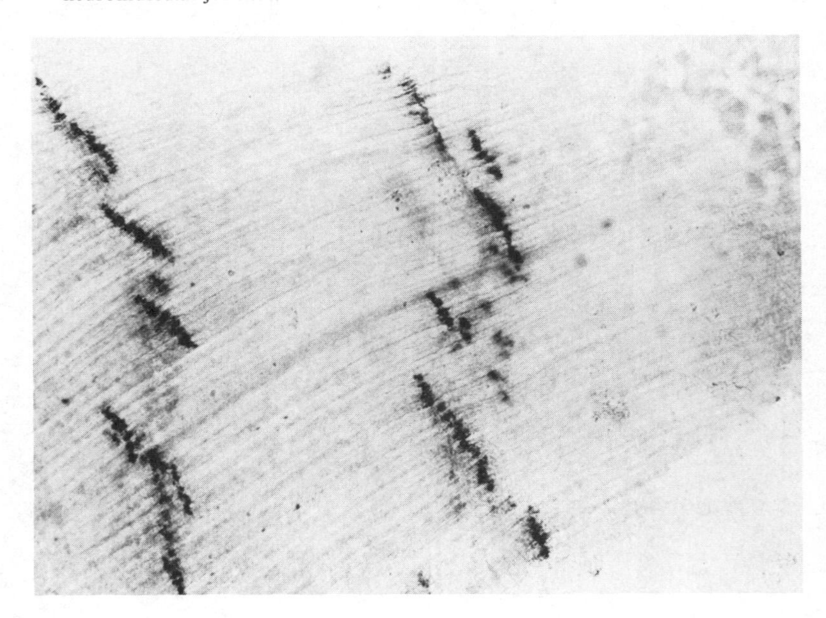

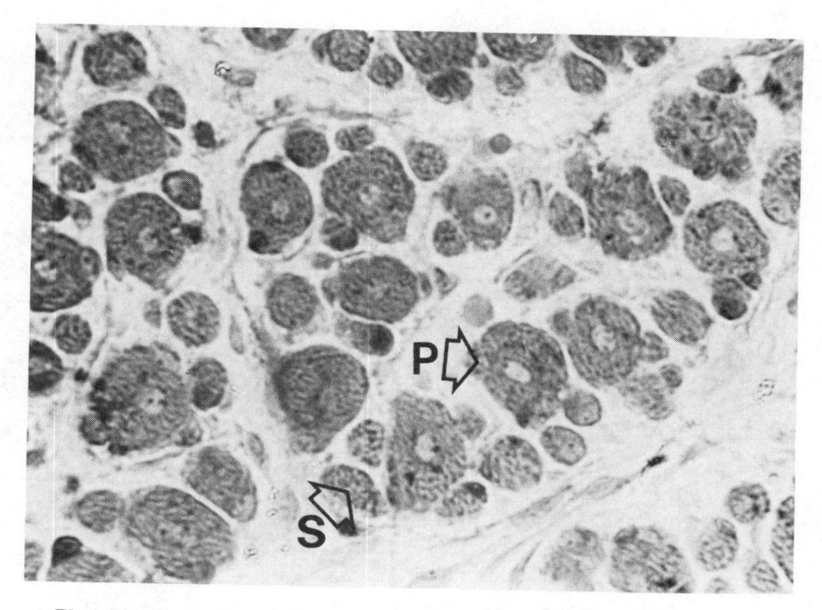

Plate 13 Transverse section of muscle from a fetal pig: (P) primary fiber or classical myotube; (S) secondary fiber.

Plate 14 Longitudinal section of muscle from a fetal pig: (P) primary fiber with axial nuclei; (S) secondary fiber being detached from the surface of the primary fiber.

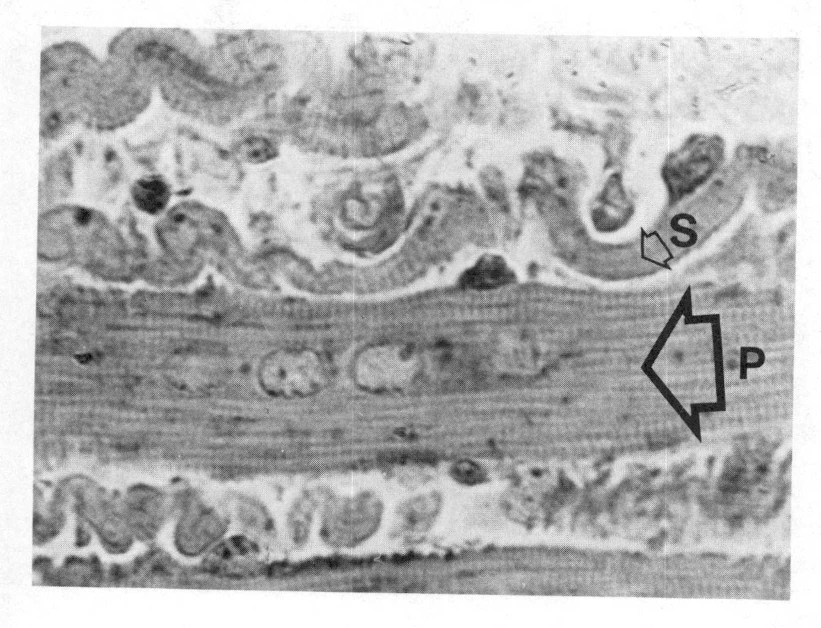

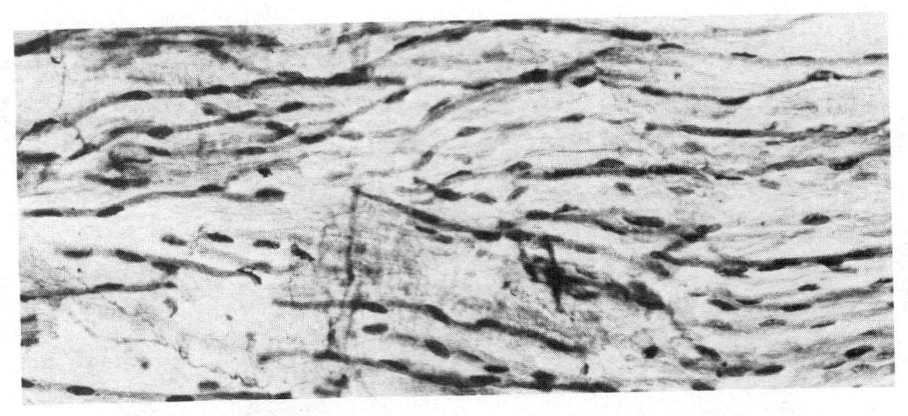

Plate 15 Large numbers of secondary fibers in fetal pig muscle.

Plate 16 Serial sections of beef muscle with the fiber types classified as white (w); intermediate (i); red (r); and transitional (*).

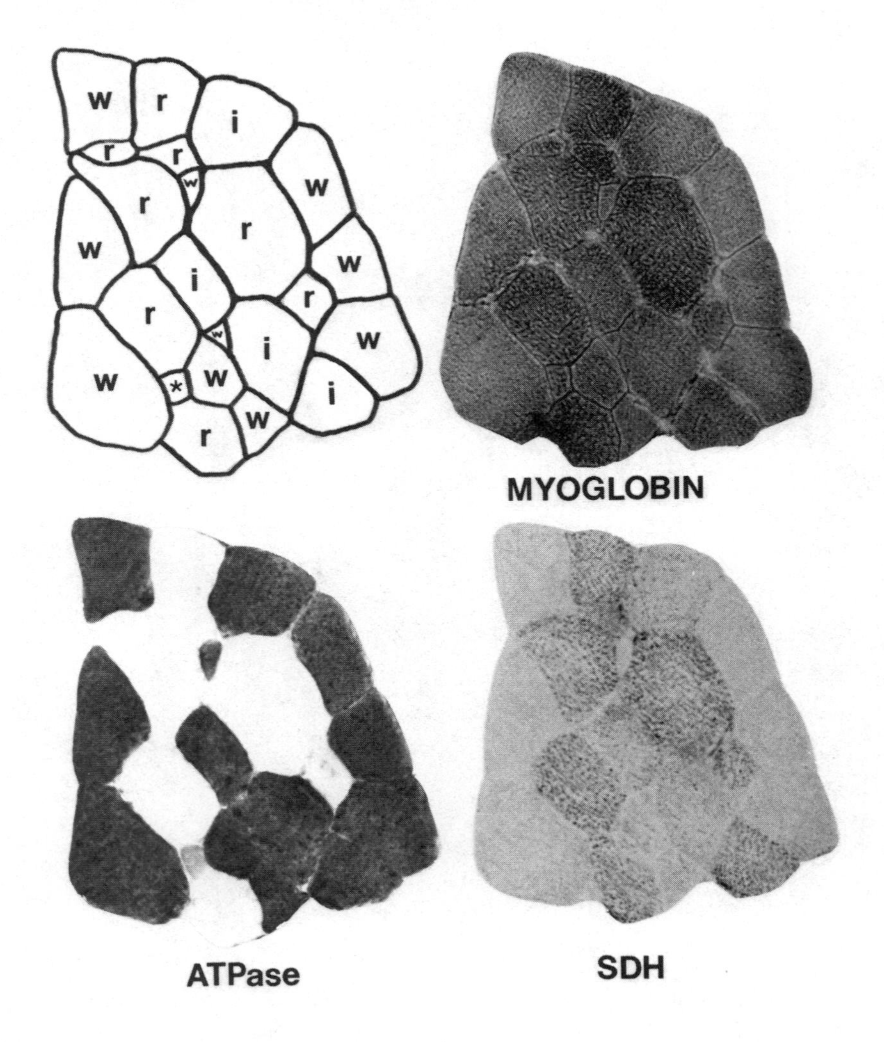

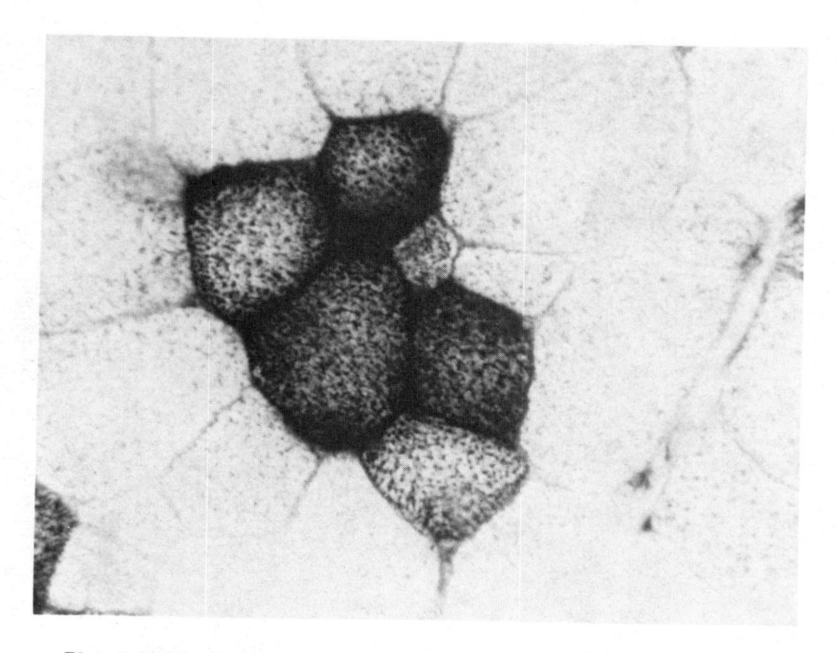

Plate 17 Characteristic feature of porcine muscle: the grouping of aerobic fibers in the center of a fasciculus.

Plate 18 Supracoracoideus muscle of a duck. The fibers with weak ATPase have the largest diameters.

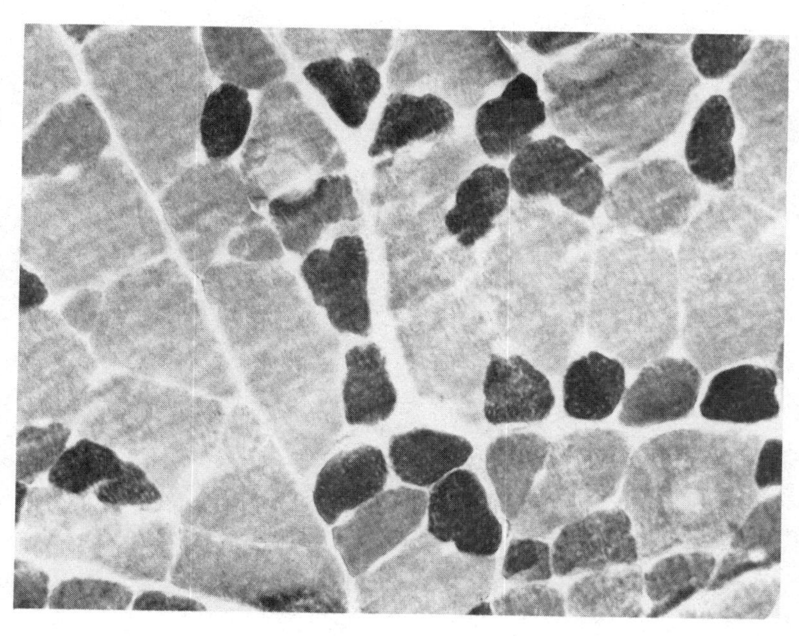

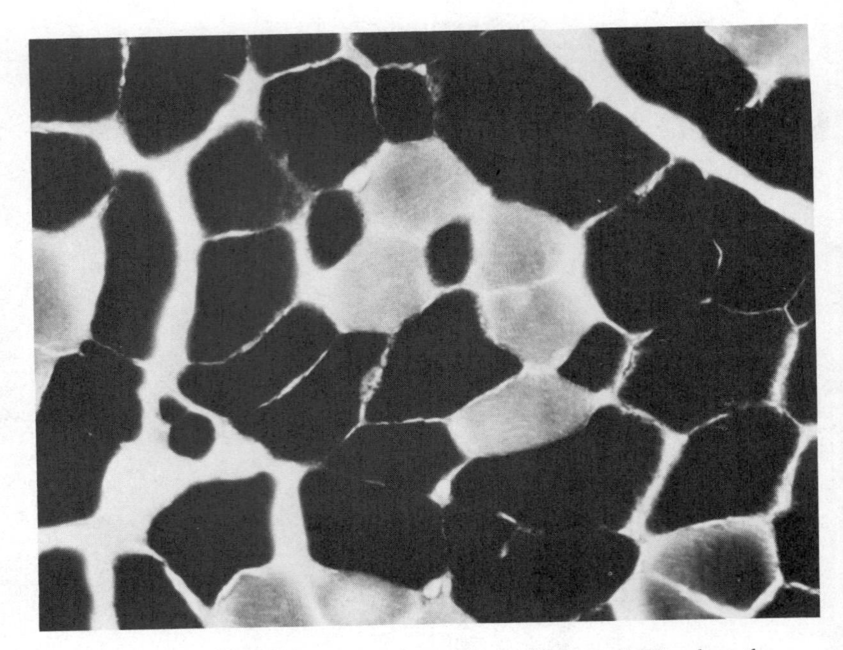

Plate 19 Sartorius muscle of a duck. The fibers with strong ATPase have the largest diameters.

Plate 20 Red muscle (vastus medialis) reacted for SDH from a young pork carcass.

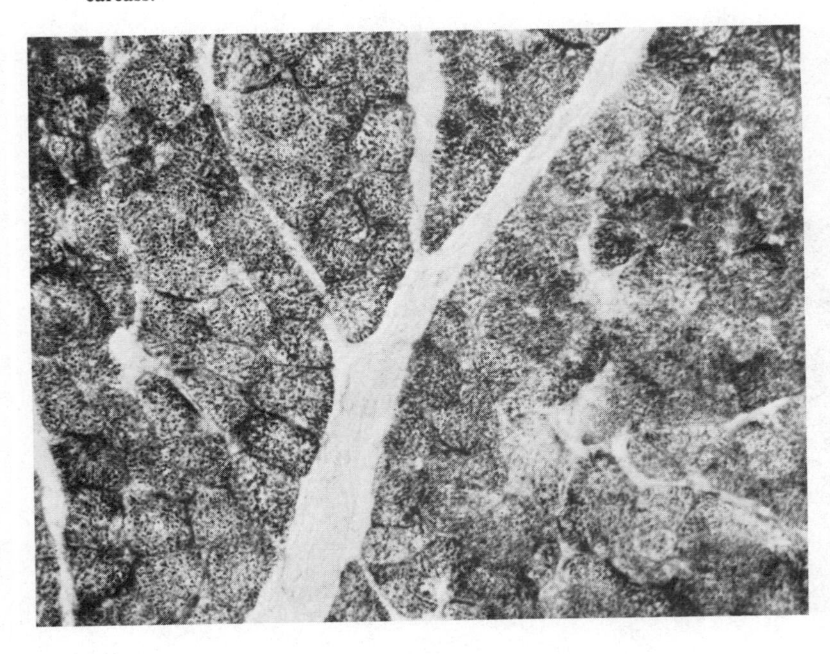

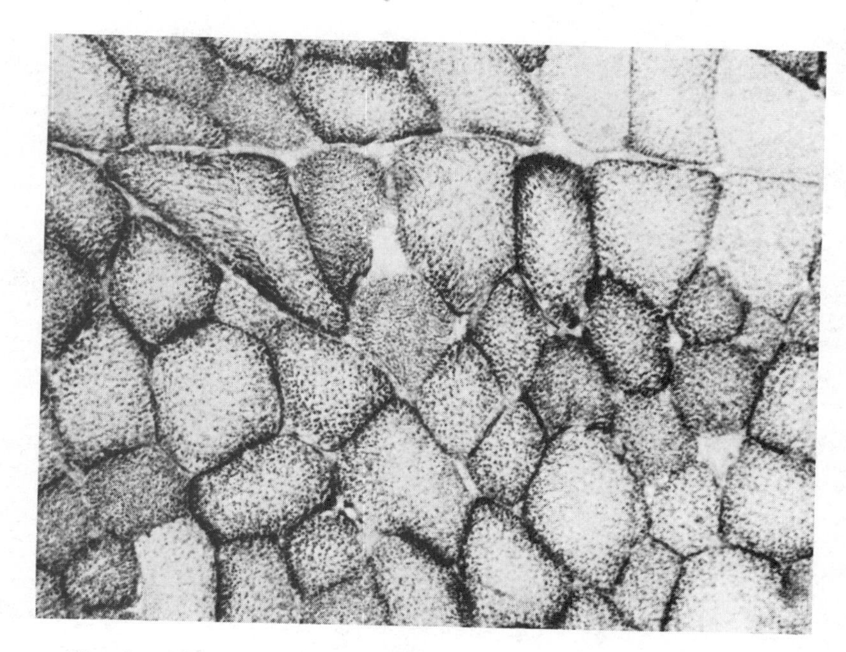

Plate 21 Red muscle (vastus medialis) reacted for SDH from a slaughter-weight pork carcass, showing a reduction in the concentration of mitochondria and different patterns of mitochondrial distribution.

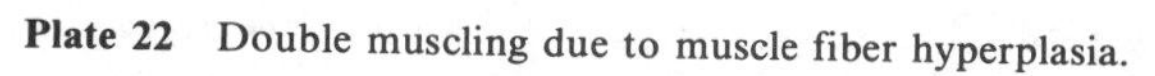

Plate 22 Double muscling due to muscle fiber hyperplasia.

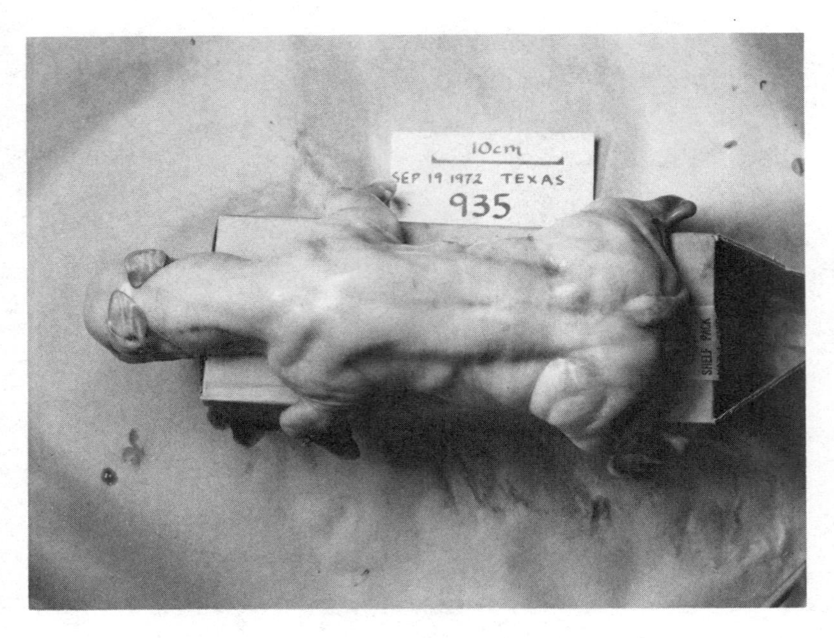

Plate 23 Muscle enlargements in a double-muscled fetus.

Plate 24 Section of bovine sternomandibularis muscle immediately after exsanguination. All fiber types contain an equal amount of glycogen, which is stained by the PAS reaction.

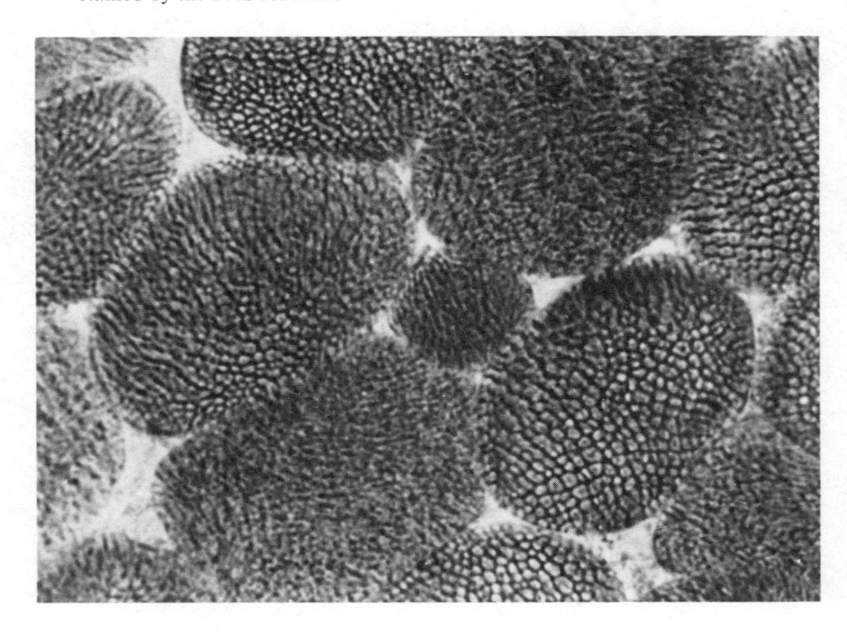

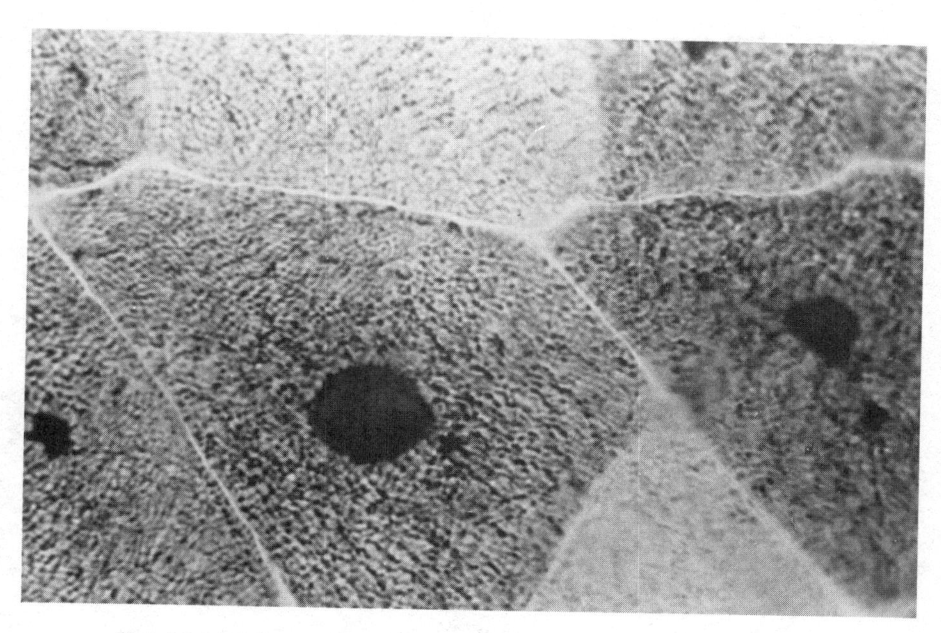

Plate 25 Axial cores of glycogen in the muscle fibers of a heavy pork carcass.

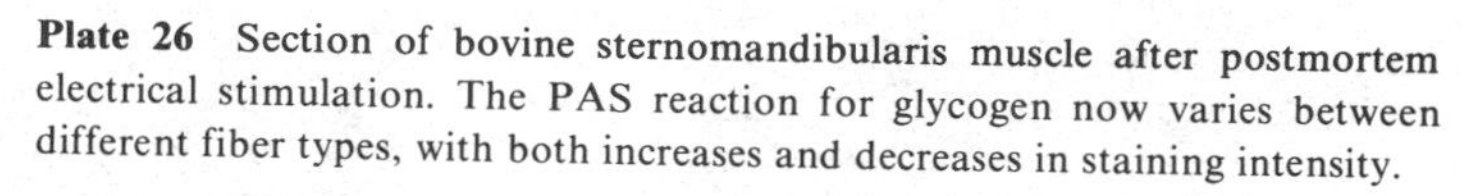

Plate 26 Section of bovine sternomandibularis muscle after postmortem electrical stimulation. The PAS reaction for glycogen now varies between different fiber types, with both increases and decreases in staining intensity.

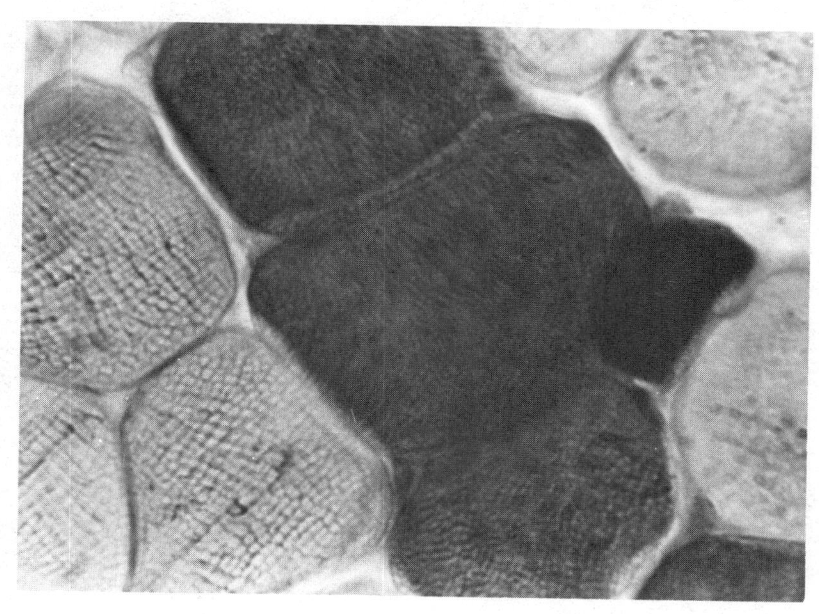

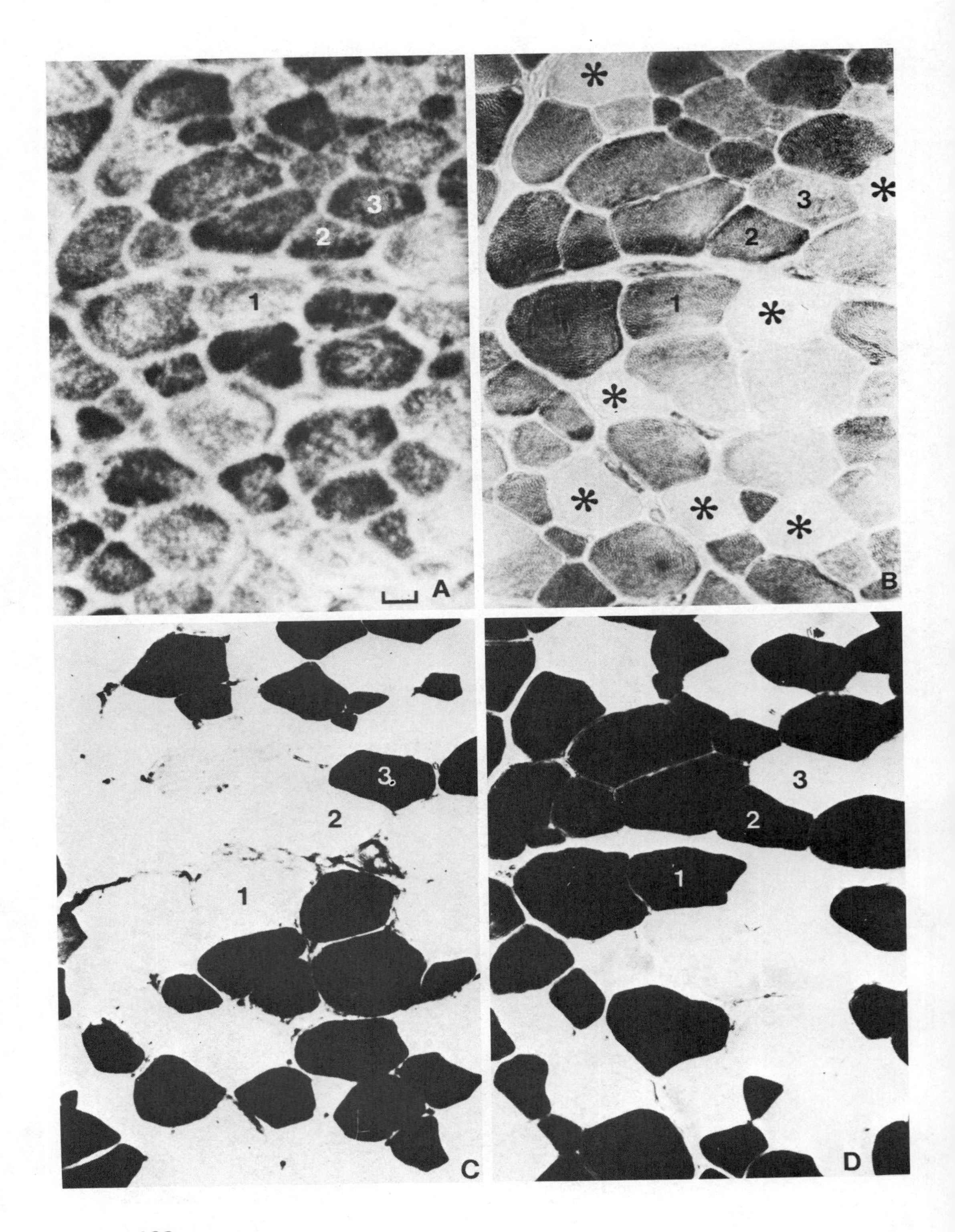
A
B
C
D
1
2
3
1
2
3
1
2
3
1
2
3

Plate 28 Electron micrograph of a longitudinal section through the edge of a muscle fiber. A transverse tubule (t) runs into the interior of the fiber from its surface. Mitochondria (m) are located between the fibrils. (Courtesy of T. R. Dutson. Micrograph prepared by M. W. Orcutt.)

Plate 27 (*opposite*) Serial sections of bovine sternomandibularis muscle demonstrating glycogen depletion due to intrinsic motor unit activity. Section A is reacted for aerobic enzymes, section B is stained for glycogen, section C is reacted for the acid-stable ATPase of slow fibers, and section D is reacted for the alkali-stable ATPase of fast fibers. Fiber 1 is a white fiber, fiber 2 is an intermediate fiber, and fiber 3 is a red fiber. The asterisks indicate the fibers that have depleted their glycogen because of intrinsic motor unit activity. (From Swatland, 1976, Can. Inst. Food Sci. Technol. J. 9:177).

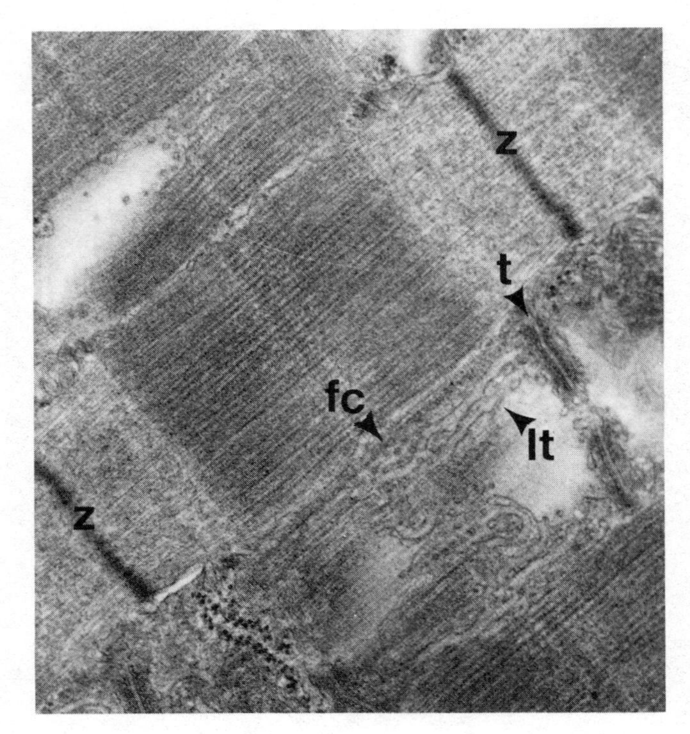

Plate 29 Electron micrograph of a longitudinal section through myofibrils. One fibril has been sectioned tangentially so that the membrane systems around it are visible. A transverse tubule (t) runs round the fibril at the level of the A-I junction. The sarcoplasmic reticulum around the fibril has some regional variations in structure. At the midpoint between two transverse tubules it appears as a fenestrated collar (fc), while extending up to the fenestrated collar on each side are longitudinal tubules (lt) of the sarcoplasmic reticulum. (Courtesy of T. R. Dutson. Micrograph prepared by M. W. Orcutt.)

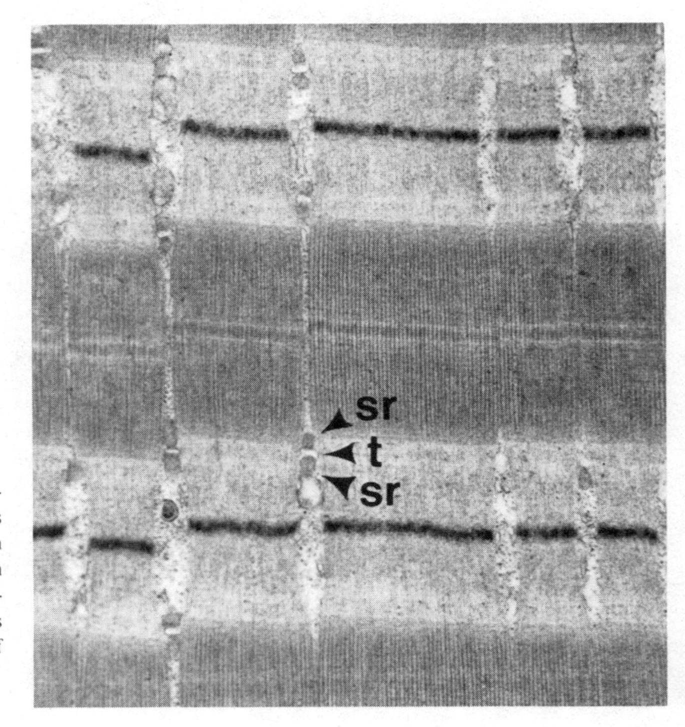

Plate 30 Electron micrograph of a longitudinal section through myofibrils. In this section, the transverse tubules (t) have been cut in cross section and they can be seen sandwiched between parts of the sarcoplasmic reticulum (sr). The three structures together are called a triad. (Courtesy of T. R. Dutson. Micrograph prepared by M. W. Orcutt.)

6

Muscle Fiber Differentiation and Neuromuscular Relationships

INTRODUCTION

In the live meat animal, skeletal muscles exist as a great mass of multinucleate cells which are precisely arranged and intimately related to the nerve cells that regulate their contraction and growth. The biological limits to protein synthesis and efficient meat production are hidden among the complex mechanisms that regulate the numbers, sizes, arrangement, and functional types of muscle fibers. Contemporary knowledge of these mechanisms has been derived from a number of different techniques, each with its own disadvantages and advantages. Electron microscopy gives highly magnified images loaded with fine detail, and has settled many of the outstanding problems left unresolved by light microscopy. The price of this advantage has been that tissue sections have become so thin that extensive reconstructions from serial sections are needed to determine the three-dimensional relationships of cells. Early microscopists peered through a murky sea of unstained or capriciously stained tissue looking for cytological features at the limit of resolution. They may have missed most of the fine details that we know today but some, like Schwann (1839) and Veratti (1902), saw and reported things that we are still rediscovering.

The morphological analysis of fixed and sectioned tissues may miss many of the biochemical subtleties of cell metabolism. The biochemical analysis of tissue homogenates, however, often sacrifices information relating to the activities of individual cells. When cultured in vitro, morphologically identical cells may show markedly different behavior patterns (Bellairs et al., 1980). Cultured cells, however, are usually reared in an artificial environment, both biochemically and spatially. The in vivo formation of muscle, *myogenesis*, occurs in a three-dimensional system, not spread over a two-dimensional surface. Thus in considering what is known about muscle fiber differentiation and neuromuscular relationships, it is rather important to note how each item of information is obtained, and to what extent the findings may reasonably be extrapolated to the situation in meat animals.

SOMITES

The prenatal origin of striated skeletal muscles is from *myoblast* cells. The cells that give rise to myoblasts may be called *premyoblasts* (Boyd, 1960) or presumptive myoblasts. The origin of premyoblasts is rather difficult to determine since premyoblasts are difficult to distinguish morphologically from other types of stem cells which are destined to give rise to other types of tissues. Figure 6-1 shows a greatly simplified plan of a transverse section through an embryo. The bottom of the figure, in particular, bears little or no resemblence to either mammalian or avian embryos but is included to indicate the general relationships of the *somites* to the *endoderm* (gut), to the *coelom* (body cavity) and to the *parietal lateral plate mesoderm* (ventral body wall) which will develop later. Each somite is composed of three zones around the *myocoele* cavity: the *sclerotome*, the *dermatome*, and the *myotome*. In bovine embryos, the cells of the dermatome and myotome differ in their arrangement so that the myocoele cavity may be merely an artifact produced by the histological processing of embryos (Haldiman, 1981).

Seen in a whole embryo, each somite appears as a cube of tissue, with left and right somites forming pairs in an anterior-to-posterior sequence. The *ectoderm* eventually forms the epidermis of the skin while the underlying connective tissue dermis is formed from the dermatome. The sclerotome is composed of loosely packed and morphologically undifferentiated *mesenchyme* cells. The mesenchyme surrounds the *notochord* and *neural tube* and later differentiates to form cartilagenous precursors of the vertebrae. In farm animals, myotomes become elongated from anterior to posterior, and this obscures the original cuboidal shape of each myotome. Extensive changes in the shape and orientation of somitic cells occur during the development of the somites (Youn and Malacinski, 1981).

Premyoblasts give rise to myoblasts which fuse to form muscle fiber precursors called *myotubes*. The axial muscles of the body, including the tongue and extraocular

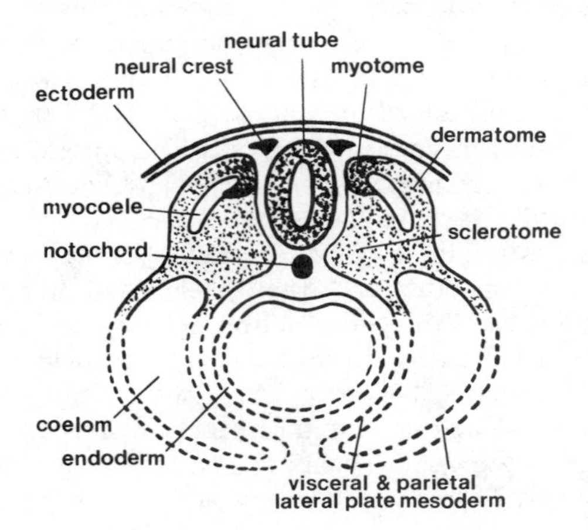

Figure 6-1 Simplified diagram of a transverse section of an embryo to show the general location of parts.

muscles, are derived from the somitic mesoderm of the myotomes. In many animals, the *limb buds* of the embryo become filled by mesenchyme cells which have broken away from the parietal layer of the lateral plate mesoderm. Thus Fischman (1972) reiterated the conclusion that the origin of limb muscles is from the differentiation in situ of mesenchyme derived from lateral plate mesoderm. In the chick wing and leg, however, myogenic stem cells may be derived from the somitic myotome (Jacob et al., 1978; Le Douarin, 1973; Chevallier, 1978; Chevallier et al., 1978; Christ et al., 1977). However, if wing buds are experimentally isolated before being invaded by somitic cells, their own mesoderm can differentiate into muscle (McLachlan and Hornbruch, 1979). Thus, under some conditions, the versatility of somatopleural cells enables them to give rise to skeletal muscle as well as to connective tissue in the limb (Mauger et al., 1980). The optimum growth and differentiation of mouse limb buds is also dependent on a cellular contribution from adjacent somites (Agnish and Kochhar, 1977). In mice, at least two precursor stem cells contribute to each somite (Gearhart and Mintz, 1972).

LIMB BUDS

Limb bud formation is regulated by a dialogue between ectoderm and mesoderm (Saunders, 1967): (1) the ectoderm makes an *apical ridge* in response to a message from the mesoderm; (2) the mesoderm grows to form parts of the limb in response to a message from the ectoderm; and (3) the mesoderm instructs the ectoderm to remain thick and to keep inducing mesodermal growth. The initial amount of premuscular mesoderm may be controlled by ectodermal factors (Milaire, 1967).

The mesoderm of the limb bud separates into muscle-forming and cartilage-forming regions. The *chondrogenic* regions which give rise to cartilage can be detected because they synthesize extracellular *proteoglycan* (Cioffi et al., 1980). Molecular differentiation is preceded by differential vascularization so that vascular growth might be involved in establishing metabolic gradients in the developing limb (Caplan and Koutroupas, 1973). Myosin can be detected with antiserum before myoblasts can be seen in the limb bud mesoderm (Medoff and Zwilling, 1972).

Very little is known about the factors that regulate the quantitative distribution of premuscle mesoderm, and nothing is known about the role of such factors in the regulation of the postnatal potential for muscle growth in meat animals. For example, the maximum postnatal number of muscle fibers might be predetermined by the number of stem cells and by the number of times that their offspring divide. Alternatively, both stem cell numbers and the number of divisions might be flexibly regulated by nutritional factors.

The cleavage of premuscle mesoderm in the limb buds of embryonic chicks (Romer, 1927), lizards (Romer, 1942, 1944), and mice (Lance-Jones, 1979) has been examined to establish muscle homologies as a basis for comparative anatomy. The evolutionary origin of tetrapod limb muscles from the dorsal and ventral muscle masses of ancestral fish fins is reflected in a primary cleavage of premuscle limb bud mesoderm into dorsal and ventral masses. Dorsal and ventral masses undergo a further sequence of cleavages (two in the lizard and five in mice) to form the individual muscles which are found postnatally. These events are completed quite early in development; by 13.5 days in mice (Lance-Jones, 1979), by 8 days in chicks (Romer, 1927), and by 17 mm crown-rump length (approximately 27 days gestation) in pigs (Carey, 1922).

MITOSIS

The mesodermal cells of somites and limb buds undergo frequent mitosis. The peak of mitotic activity in the limb buds of the chick embryo is at about 5 days incubation (Bloom and Buss, 1968). Dividing cells are rounded in shape and are locked into a mitotic cycle (Bischoff, 1970):

G_1 (gap one) or rest after the last mitosis (2.0 hours)

↓

S, synthesis of new DNA (4.3 hours)

↓

G_2 (gap two) or rest after DNA synthesis (2.4 hours)

↓

M, mitosis (0.8 hour)

↓

return to G_1 or become a myoblast (5–7 hours)

The escape from this cycle, when a stem cell becomes a postmitotic myoblast, appears to be irreversible. The cycle preceding a cell's escape has been termed the *quantal division*. The number of times that a clone of cells remains locked into the mitotic cycle might have a profound importance on myoblast numbers: just one extra cycle by all cells might double the number of myoblasts and give rise to extra muscle fibers (*hyperplasia*).

Another way of looking at this system of cell proliferation is to consider cells at the escape point in their mitotic cycle. Both the daughter cells produced by mitosis may stay in the cycle, both may escape to become myoblasts, or one may stay in and one may escape. With a population of cells, the percentage of escaping cells starts at 0 per cent in very young embryos, before the appearance of any myoblasts, and then increases toward, but never reaches 100 per cent (some stem cells remain as *satellite cells*, a source of muscle nuclei during growth and regeneration). Cell populations containing mixtures of premyoblast stem cells, mononucleate myoblasts, and fused myoblasts can be sorted with *arabinocytidine*. This prevents the formation of new myoblasts but does allow cell fusion. In cultures from 11-day chick embryos, about 20 per cent of cells are myoblasts, but the percentage is lower in younger embryos (Turner, 1978). Another way of sorting cells is to determine what percentage can be cloned to give rise to myoblasts capable of fusion. Chick leg bud mesoderm at 72 hours incubation contains 0 per cent, at 80 hours it contains 10 per cent, and at 6 days it reaches 60 per cent (Bonner and Hauschka, 1974). In human limb buds, comparable values are 14 per cent at 36 days, with a 90 per cent plateau from 100 to 172 days (Hauschka, 1974).

Bischoff (1970) and Buckley and Konigsberg (1974) suggested that another factor controlling cell proliferation might be the duration of the mitotic cycle, possibly by a variation of the duration of G_1. Cells that have escaped from the mitotic cycle to become myoblasts eventually fuse together, but the fusion of cells eventually becomes less frequent, as if inhibited. A slightly different model of the mitotic cycle with an escape point in late G_1 was proposed by Zalin (1979). Cells in G_1 may respond to *prostaglandin* E_1 with a transient increase in intracellular cyclic adenosine monophosphate (AMP) (Zalin, 1977). This may activate protein kinase (Zalin and Montague, 1974) and the onset of myoblast fusion. Konigsberg et al. (1978) also put the escape point in G_1. As discussed later, the nervous system exerts some regulation over muscle development, and its control over myoblast proliferation is probably achieved by varying the duration of G_1 rather than G_2 (Maden, 1978). Because of the importance of G_1 in the regulation of cell numbers, it is interesting to note that the G_1-S boundary is the point at which the cell synthesizes *calmodulin* (Means et al., 1982). Calmodulin is a protein that binds calcium ions and which is thought to be involved together with cyclic AMP in the regulation of many aspects of cell metabolism, growth, and division.

MYOBLASTS

The morphological features of postmitotic cells and of myoblasts prior to fusion are not unlike those of other types of precursor cells in the embryo. RNA synthesis dominates cell activity and results in a large ovoid nucleus, prominent *nucleoli* (which vary in number between species), diffuse *chromatin*, and many *ribosomes* (Fischman, 1970). Myoblasts are bipolar spindle-shaped cells (Figure 6-2), whereas fibroblasts tend to be triangular in shape. Myoblasts may form *tight junctions* where they are in contact with each other, usually at the tips of their elongated cytoplasmic extensions.

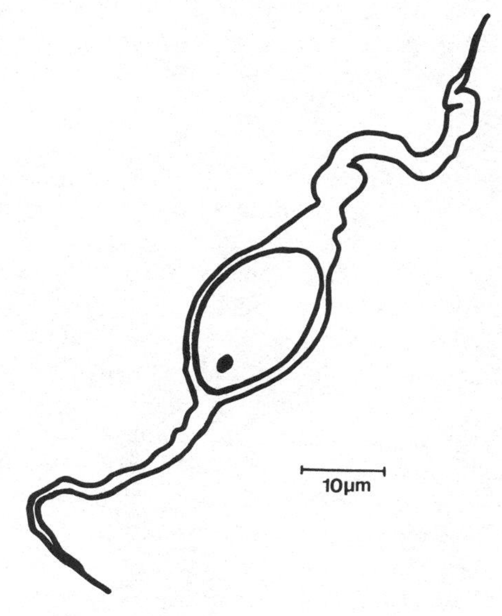

Figure 6-2 Myoblast from a fetal pig.

Protein Synthesis in Myoblasts

The synthesis of myosin is considered to be an exclusive feature of postmitotic cells, although a low level of myosin synthesis may be found in exponentially dividing clones of myoblasts (Schubert et al., 1973). A low concentration of actin and myosin may be found in many types of nonmuscle cells which are mobile or which are capable of changing their shape (Durham, 1974). The myosin which is synthesized by mitotic cells is like that of nonmuscle cells but, after the quantal mitosis, different genes are activated and these genes then make muscle-type myosin (Chi et al., 1975).

Early histological work on the formation of *myofibrils* in mononucleate myoblasts (Bardeen, 1900) has been endorsed by more recent in vitro studies. These more recent studies show that the formation of fibrils is accompanied by the formation of *T tubules* and *sarcoplasmic reticulum* (Vertel and Fischman, 1976; Moss and Strohman, 1976; Trotter and Nameroff, 1976; Levis et al., 1974; Cantini et al., 1979).

The rate of synthesis of muscle proteins is probably regulated by the levels of RNA (Daubas et al., 1981). Myoblasts develop and store a large amount of messenger RNA, and the coordinated synthesis of myofibrillar proteins proceeds very rapidly once cells have fused (Devlin and Emerson, 1979). The nucleic acid systems which are responsible for ribosomal protein synthesis are activated after the quantal mitosis, ready for the mass production of myofibrillar proteins after cell fusion (Doetschman et al., 1980; Zevin-Sonkin and Yaffe, 1980). The great concentration of ribosomes held in readiness for this activity is soon diluted by the resulting proteins. This causes a decrease in *basophilia* when myoblasts are observed by light microscopy. Stored ribosomes become arranged to form a helical pattern with 70 to 75 units; each set of units is called a *polyribosome* (Allen and Pepe, 1965). Other types of protein whose role is not understood may be produced by myoblasts up to the time of fusion (Garrels, 1979). Like the fibrillar proteins, the characteristic muscle enzyme *creatine phosphokinase* can also be found before myoblasts have fused (Keller and Nameroff, 1974). Although myoblasts no longer synthesize nuclear DNA after their quantal mitosis, they exhibit a fourfold increase in mitochondrial DNA which is associated with mitochondrial proliferation (Brunk, 1981).

Myoblast Fusion

Myoblasts fuse with each other to form multinucleate cells, which then give rise to multinucleate skeletal muscle fibers. Fusion is initiated at a single site between two myoblasts. The pore that is formed to link adjacent cells grows in size and leaves no trace of the intervening membranes (Lipton and Konigsberg, 1972). The cytoskeleton within myoblasts undergoes extensive remodeling at the time that myoblasts fuse to form myotubes (Fulton et al., 1981). The formation of myofibrils proceeds rapidly once fusion has occurred. Myofibrils accumulate under the cell membrane. Nuclei thereby become restricted to an axial core of sarcoplasm which is surrounded by myofibrils that are arranged to form a hollow cylinder or tube. At this stage, the whole cellular structure can be called a *myotube* because its structure is dominated by the hollow cylinder of myofibrils. Myoblasts do not normally fuse with other types of cells but, experimentally, myoblasts of one species can be induced to fuse with myoblasts of another species (Yaffe and Feldman, 1965). The surface morphology of myoblasts seen by scanning electron microscopy appears to differ between species (Dupont et al., 1979).

Myoblast fusion has been observed by time-lapse cinematography (Bachmann, 1980). Some myoblasts move to suitable positions prior to fusion, while in other cases, this may be unnecessary because aggregates of dividing cells have kept in contact with fused cells. Fusion is preceded by a period of cell-to-cell recognition in which the cells can still be chemically dispersed with ethylenediaminetetraacetic acid (EDTA). Recognition is followed by a period of adhesion in which *trypsin* must be added experimentally in order to disperse the cells. Finally, after membrane fusion, fused cells cannot be experimentally dispersed (Knudsen and Horwitz, 1977). Cultured myoblasts fuse when their numbers reach a certain density, perhaps in response to a high-molecular-weight chemical signal (Konigsberg, 1971). Within the myoblast, an increase in the level of adenosine 3′,5′-monophosphate (cylic AMP) initiates the events that lead to fusion (Curtis and Zalin, 1981). Myoblasts have surface antigens which are probably involved in cell-to-cell recognition (Friedlander and Fischman, 1977). During the process of myoblast differentiation, there are changes in the distribution of small particles on the inside of the cell membrane (Furcht and Wendelschafer-Crabb, 1978). Myoblast fusion is triggered by calcium ions but is inhibited by magnesium and potassium ions (Schudt et al., 1973).

Formation of Myofibrils

The synthesis of all the major proteins of the myofibril is simultaneous (Masaki and Yoshizaki, 1972), so that earlier morphological reports to the contrary may have been due to the detection of thin filaments formed from the ubiquitous type of actin rather than from the muscle type of actin. Fischman (1970) lists a number of different-sized filaments that may occur in myogenic cells. The myosin and actin of developing myofibrils appear as 15- to 16-nm-diameter and 5- to 6-nm-diameter filaments, respectively. The diameter of unincorporated myosin filaments is similar to those which have already joined a myofibril. The filaments of the Z line are 5 to 6 nm in diameter. Filaments with diameters of 5 to 6 nm also occur below the cell membrane, but these are probably the nonmuscle type of actin. Microtubules, with diameters from 22 to 25 nm, may be found in the axial core of myotubes. Microtubule subunits with a diameter of 10 μm may also be found. Filaments with diameters from 5 to 10 nm may be found at the myotendon junction.

Galey (1970) outlined three main schools of thought on how separate filaments might become assembled to form myofibrils. One school claimed to have seen Z-line material associated with the first aggregates of filaments and with rudimentary A and I bands. A second school claimed that filaments did not initially aggregate in register so as to form A and I bands. A third school considered that Z-line material was not necessary for filaments to become grouped in their cross-sectional hexagonal pattern. A fourth school might have been added, based on the idea that sarcomere formation is initiated by regularly repeating elements of the sarcoplasmic reticulum or transverse tubular system: this now appears unlikely (Warren, 1973), although there is evidence that Z lines are preceded by tubules in chick muscle (Allen, 1973). Much of the work in this area, even though relatively recent by some standards, must be reconsidered to take into account the respective contributions of *alpha-actinin, desmin*, and *vimentin*. At present, it appears that the regular structure and arrangement of sarcomeres proceeds in two stages. The initial formation of Z lines containing alpha-actinin is followed by the appearance of desmin and vimentin at the time that the lateral alignment of Z lines is established (Gard and Lazarides, 1980). *Filamin*, a protein normally found in smooth muscle, may also make a transient appearance when Z lines are being assembled (Gomer and Lazarides, 1981).

DETERMINATION OF MUSCLE FIBER ARRANGEMENT

The first myotubes formed in each embryonic muscle are involved in establishing the future arrangement of muscle fibers, as well as in establishing the approximate size and anatomical location of the muscle. Little is known about any of these three factors. The early development of muscle is closely related to skeletal and neural development. Nerves, muscles, and the skeleton interact so closely that it may be difficult to isolate causes from their effects. Thus a disorder in one element may soon modify either or both of the other elements. This may lead to a self-perpetuating defect in neuromuscular or skeletomuscular development, such as *congenital articular rigidity* (Chapter 2). The major nerve trunks grow into a limb bud by following the connective tissue framework of the bud, but developing muscles are necessary to invoke the formation of side branches to the muscle (Lewis et al., 1981).

Muscles may be attached to either the shaft (*diaphysis*) or the knob (*epiphysis*) of a bone. The longitudinal growth of bones occurs at cartilagenous *epiphyseal* plates. One of these plates is located between each epiphysis and its diaphysis. Thus, to retain their positions relative to each other during epiphyseal plate growth, some muscle attachments must migrate over the bone surface. Postnatally, muscle migrations are regulated by the bone (Grant and Hawes, 1977; Grant, 1978). Carey (1922) argued that muscle development in the limbs of fetal pigs was shaped by a dynamic interaction between linear skeletal growth and the resistance of muscles to stretching. The nervous system appears to have no direct part in the determination of myotube alignment (Read et al., 1971).

If muscle stretching really does shape muscle growth, the determination of muscle fiber arrangement might be explained by the *contact guidance theory* which Weiss (1955) elaborated to explain how nerve cells invade developing tissues. Myotubes and myoblasts might be guided by a matrix of very fine connective tissue fibers (Bardeen, 1900; Warren, 1981). Jacob et al. (1979) found that migrating myogenic stem cells in chick embryos branched into *filopodia* at their leading edges, and that stem cells followed the alignment of fine connective tissue fibers. In vitro, myoblasts only develop a parallel alignment if they are cultured on a type of collagen that forms distinct collagen fibers (John and Lawson, 1980). Myotubes might also be pulled into alignment by their already anchored ends, or fusing myoblasts might align themselves to follow the dominant directions of a stretched matrix. Yoshizato et al. (1981) suggested that the chances of myoblast fusion are increased when myoblasts become aligned on parallel collagen fibers. The arrangement of fibers at an angle to a muscle tendon is more difficult to explain. Perhaps the tensile forces that shape the connective tissue matrix of a *pennate* muscle are transmitted by intramuscular tendons. Another possibility is that the arrangement of myoblasts is influenced by the orientation of electrical fields. The regulation of animal and plant growth by bioelectric fields has been known for many years (Lund, 1947). Hinkle et al. (1981) found that cultured myoblasts become arranged with their long axes perpendicular to electric fields of 36 to 170 mV/cm. Muscle fibers also possess a magnetic asymmetry relative to their long axis (Arnold et al., 1958).

Intracellularly, the development of the parallel arrangement of myofibrils is dependent on the proper attachment of the whole cell (Puri et al., 1980). New aggregates of thick and thin filaments appear first at the periphery of cells (Kilarski and

Jakubowska, 1979) so that the longitudinal orientation of filaments might simply follow the direction of membrane stretching.

Many of the early histologists who studied myogenesis were impressed by the widespread evidence of cellular degeneration (*retrograde metamorphosis*) which they found in developing muscles (Achaval et al., 1966). Lysosomal systems capable of causing degeneration are well developed even in myoblasts (Bird et al., 1981). More recently, there has been a trend to dismiss degenerative phenomena during myogenesis as being a mere consequence of localized myotube *contracture* (a sustained and destructive contraction). An alternative viewpoint is to regard contracture followed by degeneration as a means of eliminating fibers which have failed to align themselves correctly within a muscle (Bardeen, 1900). Experimental proof for this was furnished by Nakai (1965), who cultured *intercostal* muscles between pieces of ribs. When the muscles were kept stretched by slow separation of the ribs, muscle fibers continued to develop but, when they were not stretched, the fibers degenerated. The passive stretching of myotubes activates the sodium ion pump of their membranes, and this is followed by increases in amino acid uptake and protein synthesis (Vandenburgh and Kaufman, 1981).

During the determination of fiber arrangement, any loss of cells by retrograde metamorphosis might affect muscle size. An experimental model is provided by the development of the striated *extrinsic ocular* muscles, which rotate the eyeball (Twitty, 1955). Transplantation of developing eyes from a large-eyed amphibian species to a small-eyed species results in an enlargement of the host's extrinsic ocular muscles. The cause of muscle enlargement is an increase in fiber numbers, *hyperplasia*. This is consistent with a stretch-induced reduction of cell loss, but other explanations are also possible.

MASS PRODUCTION OF MUSCLE FIBERS

In the descriptions of myogenesis given by many histology textbooks, we may be led to believe that all the skeletal muscle fibers of newborn animals are derived from myotubes by the radial migration of nuclei and a disruption of the tubular arrangement of myofibrils. In pigs, however, Schwann (1839) discovered that two types of muscle fiber precursors exist in the fetus: one type has a tubular arrangement of its myofibrils and can be called a *myotube*, while the other type lacks a tubular appearance and cannot reasonably be called a myotube. As yet, there is no generally agreed name for this second type of muscle fiber precursor. Here it is called a *secondary fetal muscle fiber*, shortened to *secondary fiber* (Plates 13, 14, and 15). The classical myotube may then be called a *primary fiber* or *primary myotube*. Unfortunately, this is the reverse of Schwann's original terminology since he worked backward from late to early embryos. The duality of muscle fiber precursors was almost completely ignored until Couteaux (1941) published his classic monograph on myogenesis. Since then, this knowledge has reappeared in a number of studies (Kelly and Zacks, 1969a; Kikuchi, 1971; Swatland, 1971; Thurley, 1972; Ashmore et al., 1972).

As yet, there is no general agreement on the histological significance of secondary fibers. The viewpoint taken here (Figure 6-3) is that secondary fiber formation is a process for the rapid mass production of relatively large numbers of new muscle fibers

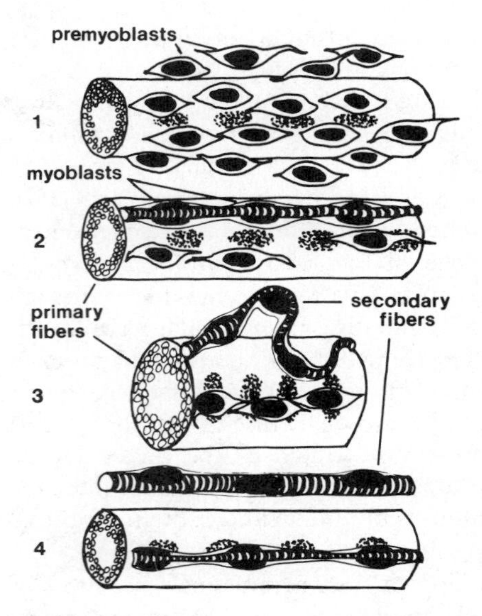

Figure 6-3 Stages in the mass production of secondary fetal muscle fibers: (1) dividing premyoblasts surround a primary fiber or myotube which has axial nuclei; (2) a secondary fiber is produced by myoblasts on the surface of a primary myotube; (3) the secondary fiber is detached because of contraction of the primary myotube; (4) the secondary fiber is now free and a new secondary fiber starts to form on the primary myotube. (From Swatland and Cassens, 1973.)

(Plate 15), taking advantage of two factors: first, that the general features of muscle architecture have already been established by the arrangement of primary myotubes, and second, that the surfaces of gently contracting myotubes provide an ideal site for myoblast contact and fusion (Swatland and Cassens, 1973). Myoblasts are commonly found clinging to myotubes. As myotubes contract, these surface myoblasts probably bump against each other. The dimensional changes that occur on the myotube surface, as it decreases in length and increases in radius, favor myoblast contact along the length of the myotube. Thus the long axis of fused myoblasts will follow that of their supporting myotube. With an in vitro model created by stretching the substrate of cultured avian myoblasts, the optimal rate of stretching for maximum myoblast alignment was found to be 0.2 mm/hour (Vandenburgh, 1982).

Strings of fused myoblasts which have been assembled on a primary myotube are soon reinforced continuously along their length by new myofibrils. New myofibrils are grouped as a solid core, mostly located away from the supporting myotube. Most secondary fibers formed in this way retain a more or less axial core of myofibrils rather than having a tubular arrangement of their myofibrils. Secondary fibers adhere to their primary myotubes by means of pseudopodial processes that project into invaginations on the surface of the primary myotube (Campion et al., 1981). Further contractions of the supporting myotube (as indicated by their short sarcomeres) do not spread to secondary fibers (as indicated by their long sarcomeres). Once a secondary fiber has acquired a substantial core of fibrils, differences in length due to contraction probably create a shear force between the secondary fiber and its supporting myotube. This leads to the separation of secondary fibers from their supporting myotubes, and accounts for the fact that secondary fibers are often sinuously folded when their myotubes are contracted.

The morphological categorization of prenatal muscle fibers into primary myotubes and secondary fibers is a general principle which, at best, can only account for a majority of cases. At worst, it does not take long to find a few fibers with features which are intermediate between those of primary and secondary fetal fibers. Schwann

(1839) was the first to notice these transitional cases. Why should early fibers have a tubular structure and why should later fibers be different? In the absence of any experimental evidence, the best approach to answer these questions is by deduction from the working hypothesis of secondary fiber mass production.

Early in muscle development, strings of myoblasts start to produce filaments below their plasma membrane (Tomanek and Colling-Saltin, 1977). If sustained by tension from successful terminal attachments, the continuous peripheral formation of new filaments gives rise to a complete tube of myofibrils below the plasma membrane. This radial symmetry provides the maximum surface area for mass production of secondary fibers. As secondary fibers start to acquire fibrils (fibrils which do not contract with those of the supporting myotube), shear forces develop between supporting myotubes and secondary fibers. Further proliferation of fibrils in secondary fibers leads to secondary fiber separation.

Once a secondary fiber has separated, its core of fibrils becomes crescent-shaped in transverse section (Figure 6-4) as the secondary fiber starts to utilize the free space below the membrane which was originally apposed to the supporting myotube. The

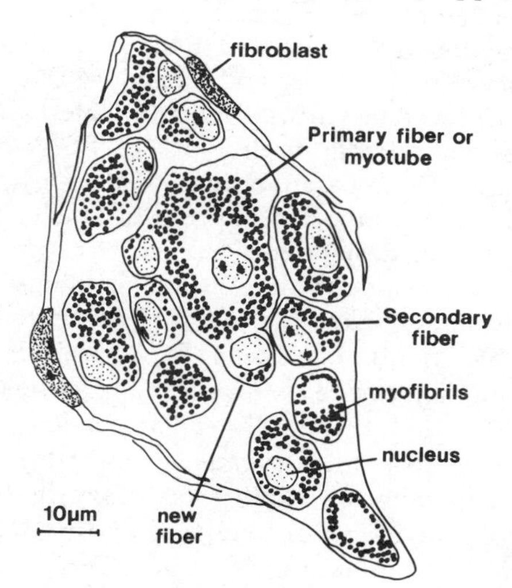

Figure 6-4 Transverse section of a developing muscle fasciculus in a fetal pig at 60 days gestation.

oldest secondary fibers, which are now pushed away from their parent myotubes by younger secondary fibers, begin to assume a tubular structure. Perhaps they can now support production of further secondary fibers themselves. By now, however, secondary fiber production is slowing down as fetal development nears completion. For reasons as yet unknown, the axial nuclei of myotubes move, or are pushed to a peripheral position below the cell membrane, and the morphological distinction between primary myotubes and secondary fibers is obscured.

The radial dimensions of primary and secondary fibers are difficult to measure histologically. Using linear measurements, primary myotubes appear to become smaller during postnatal development (Figure 6-5). Secondary fibers remain small until just before birth. However, the decrease in mean size of primary myotubes may be related to their less frequent contraction or to the detachment of secondary fibers. Similarly, the mean size of secondary fibers is biased by the constant addition of newly formed fibers with a small diameter.

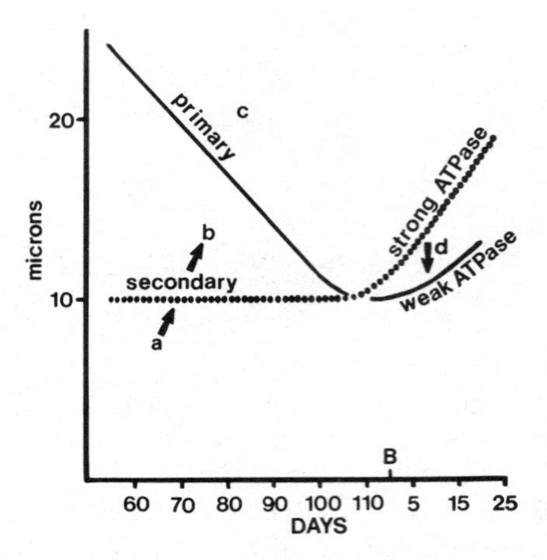

Figure 6-5 Radial growth of muscle fibers before and after birth (B) in pigs. Fiber size is indicated by mean minimum diameters (microns). The mean diameter of secondary fibers stays constant as new small fibers are formed (a) and as older fibers continue to grow (b). The mean diameter of primary fibers decreases, partly as a result of the detachment of secondary fibers and partly because of a decrease in the extent of fiber contraction (c). After birth, some fibers change from strong to weak ATPase activity (d). (Data from Swatland, 1973b, 1975a.)

It is difficult to estimate the numbers of secondary fetal fibers that are formed by mass production. The fibers that can be counted in a muscle cross section (*apparent number*) comprise only a fraction of those present in the whole muscle (*real number*). The apparent numbers of primary myotubes and secondary fibers indicate only the ratio of primary myotubes to secondary fibers. Even the interpretation of this ratio rests on the unproved assumption that both types of fibers maintain equal or constantly proportional lengths. The apparent ratio of primary myotubes to secondary fibers is difficult to determine toward the end of prenatal development. A few of the oldest secondary fibers may just be developing a tubular structure at about the same time that primary myotubes are just starting to lose their tubular structure.

The ratio of tubular to nontubular fibers describes a curve that starts with all tubular fibers and ends with all nontubular fibers (Figure 6-6). From this curve, the start of secondary fiber production can be estimated at approximately 55 days gestation in pigs (Thurley, 1972; Swatland, 1973b). The end of secondary fiber production is more difficult to pinpoint: subjective estimates range from 70 days (Swatland, 1973b) to 100 days (Thurley, 1972), depending on whether secondary fibers are counted while they are still on their supporting myotubes or when they are first detached. One way to interpret the curve of the ratio of tubular to nontubular fibers is to regard the rapid change in ratio from 55 to 70 days as a consequence of secondary fiber mass production, and to regard the slow change in ratio from 70 to 110 days as a consequence of nuclear migration. Thus at 70 days in porcine muscle, it appears that each primary myotube has supported the production of approximately five secondary fibers. In fetal calves, secondary fiber formation is completed by 20 cm crown-rump length

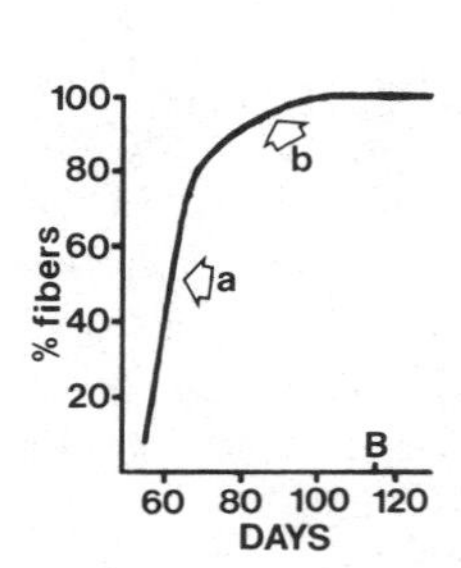

Figure 6-6 Change in the percentage of muscle fibers with axial myofibrils and peripheral nuclei in pigs before and after birth (B). The time scale is in days of gestation. Initially, the percentage increases due to the formation of secondary fibers (a). Just before birth, no new secondary fibers are formed, but primary myotubes are changing their internal structure (b). (Data from Swatland, 1973b).

(Stickland, 1978) at about 205 days gestation. In chick embryos, primary myotubes are all formed by approximately 11 days and the mass production of secondary fibers occurs up to 16 days of incubation (Kikuchi et al., 1972).

There are many quantitative aspects of fiber number, length, diameter, and arrangement to be considered before the cellular factors that determine muscle mass in meat animals can be considered. These are postponed until Chapter 7. The remainder of this chapter is concerned with the qualitative histochemical and physiological differences between muscle fibers, and with the way in which these differences are regulated by the nervous system.

INNERVATION OF MUSCLE FIBERS

The cell bodies of the *motor neurons* that innervate carcass muscle fibers are located in the *ventral horn* of the spinal cord (Plate 10; Figures 1-18 and 6-7). Basic dyes for light microscopy show that the *perikaryon* (cell body) is packed with *chromophilic* or *Nissl granules*. Electron microscopy reveals that each granule is an aggregate of *cisternae* belonging to the *rough endoplasmic reticulum*. Silver stains for light microscopy show that Nissl granules are set in a tightly woven meshwork of *neurofibrils*. Electron microscopy shows each neurofibril to be an aggregate of 10-nm-diameter *neurofilaments*. Interspersed among the neurofibrils are a few 20-nm-diameter *neurotubules* or *microtubules*. Small mitochondria are found throughout the perikaryon and its rootlike extensions (*dendrites*). The single long *axon*, which links each motor neuron to its group of muscle fibers (*motor unit*), usually departs from the perikaryon at a small *axon hillock* devoid of Nissl granules. The *axoplasm* of the axon contains tubules of endoplasmic reticulum, slender mitochondria, neurotubules, and neurofilaments. The nucleus of the motor neuron has a conspicuous *nucleolus*. Within the spinal cord, the axons of some motor neurons give rise to a few small *collateral*

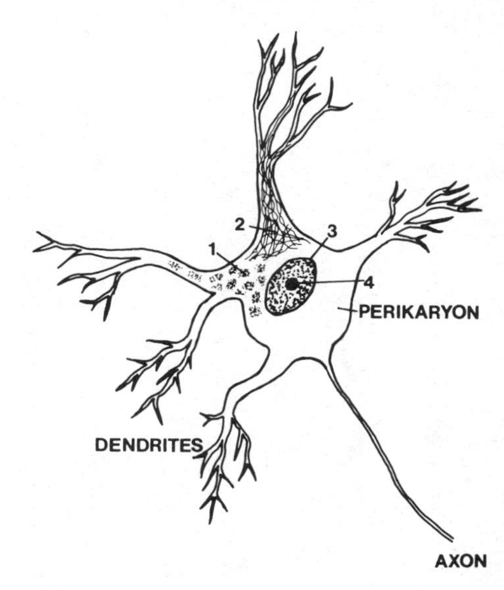

Figure 6-7 Representative parts of a motor neuron: (1) chromophilic or Nissl granules stained with basic dyes; (2) neurofibrils stained with silver; (3) nucleus; (4) nucleolus. (From Swatland, 1971.)

branches which convey feedback information to the central nervous system (Cullheim and Kellerth, 1978a, 1978b).

Motor axons in ventral nerve roots and in peripheral nerves are insulated by a *myelin sheath* formed by *Schwann cells*. The Schwann cell is tightly rolled around the axon many times so that concentric layers of Schwann cell membranes form a thick layer of insulation. In the depth of the insulating layer, *major dense lines* (3 nm) are formed by the apposition of the cytoplasmic faces of unit membranes, while faint *intraperiod lines* are formed by apposition of outer faces. Major dense lines are 12 nm apart. Schwann cells migrate along axons and become evenly distributed. They are separated by *nodes of Ranvier* where the myelin leaves a narrow bare ring of axon. The positions of the nodes of Ranvier are determined by the axon (Wiley-Livingston and Ellisman, 1980). Action potentials jump from one node to another so that the velocity of neural conduction is increased.

Axons in peripheral nerves are bound together by an *endoneurium* composed of fine collagen fibers, fibroblasts, and fixed macrophages. Bundles of axons are delineated by a more substantial *perineurium* to form fascicles. Fascicles are bound together by *epineurium* to form a complete nerve with a blood supply and often with protective adipose cells (Figure 6-8). The smallest-diameter axons (0.3 to 0.4 μm) in peripheral nerves are not myelinated, and are mainly *postganglionic sympathetic* fibers and *afferent* fibers of the peripheral nervous system which enter the dorsal roots of the spinal cord (Elfvin, 1968). The smallest myelinated axons (up to 3 μm in diameter) are *efferent preganglionic* fibers of the autonomic nervous system. Myelinated *somatic sensory* and *motor* axons range in diameter from 1 to 22 μm. In nonmyelinated axons, conduction velocity is proportional to the square root of axon diameter. In myelinated axons, conduction velocity is proportional to diameter.

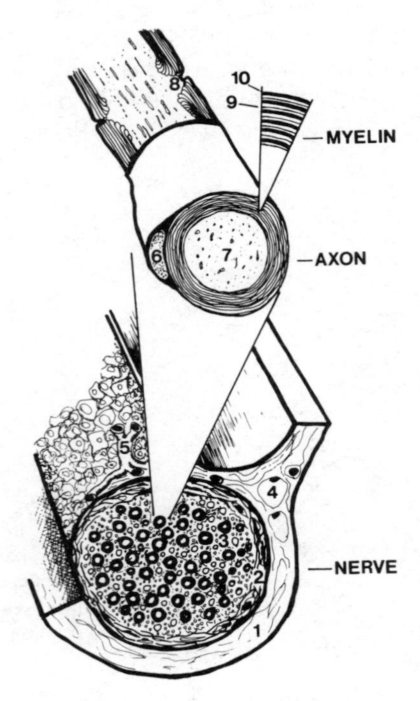

Figure 6-8 Structure of a peripheral nerve: (1) epineurium; (2) perineurium; (3) endoneurium; (4) adipose cell; (5) blood vessel; (6) Schwann cell nucleus; (7) axoplasm; (8) node of Ranvier; (9) intraperiod line; (10) major dense line. (From Swatland, 1971.)

When an axon enters a muscle, it branches so as to innervate all the scattered fibers of its motor unit (Plate 11). In simple muscles, the zone of muscle fiber innervation may be in an equatorial plane. In the muscles of meat animals the situation is more complex and, in pennate muscles, even short fasciculi in which all muscle fibers run from one end to the other may have two innervation bands (Swatland and Cassens, 1971). The pattern of intramuscular nerve branching in pennate muscles may follow the regional histochemical differentiation of muscle fibers (Galvas and Gonyea, 1980).

The final axonal branch which innervates a muscle fiber is called a *terminal axon*. The *functional terminal innervation ratio* was defined by Coers and Woolf (1959) as the mean number of muscle fibers innervated by the axonal branches radiating from an intramuscular nerve. The *absolute terminal innervation ratio* gives the mean number of *neuromuscular junctions* or *motor end plates* innervated by the axonal branches which radiate from an intramuscular nerve. There is a considerable subjective element in deciding how far back into the intramuscular nerve that branch points are to be recorded, but if standardized, the distal sample of the total degree of axonal branching is a useful measure of changes in the total amount of axonal branching (Coers et al., 1973). For example, if a motor neuron is lost by disease, surviving neurons may take over the denervated motor unit, and the functional terminal innervation ratio and absolute terminal innervation ratio are increased. This response is probably localized near denervated fibers and thus shows up well in terminal innervation ratios. Competition between surviving axons may lead to multiple innervation of recently denervated fibers, at least temporarily, so that the absolute terminal innervation ratio may show a greater increase than the functional terminal innervation ratio. In calves with hereditary congenital articular rigidity associated with cleft palate, terminal axons have more branches than normal (Rieger et al., 1979).

In muscles from healthy pigs, the functional terminal innervation ratio ranges from approximately 1.00 to 1.04. The absolute terminal innervation ratio ranges from approximately 1.01 to 1.13 because some muscle fibers have double motor end plates (Swatland and Cassens, 1971). Barker and Ip (1966) suggested that motor end plates might have a limited life span, and that double motor end plates might be a phase in the physiological replacement of motor end plates. This idea was rejected by Tuffery (1971), who considered that motor end plate growth and degeneration in normal muscle were separate and unrelated phenomena. However, the topic is far from settled. There is evidence of physiological regeneration of neuromuscular junctions among invertebrate animals (Govind and Pearce, 1981), and it has been suggested that ultrastructural remodeling of neuromuscular junctions in vertebrates may occur in response to developmental changes in muscle activity patterns (Cardasis and Padykula, 1981).

If peripheral nerves are crushed (not too close to the spinal cord), axons regenerate from their proximal stump and follow the distal connective tissue framework of the nerve back to their original muscle. Regenerating axons grow at a rate of approximately 1 mm/day (Palmer, 1976). Once the axons have returned to their muscle, the contact guidance of axons is probably supplemented by directional growth in response to chemical stimuli. Axonal branches are formed as axons compete to reinnervate muscle fibers (Edds, 1953). Beermann et al. (1977) crushed the sciatic nerves of pigs, and the mean functional terminal innervation ratio and absolute terminal innervation ratio increased to approximately 1.66 and 1.81, respectively, in deep parts of the semitendinosus muscle. The degeneration of axons distal to a damaged region is called *Wallerian degeneration*. The perikarya of regenerating neurons increase their rate of protein synthesis, with an accompanying change (*chromatolysis*) in their Nissl granules. Wallerian degeneration and *neurogenic muscle atrophy* can occur as a result of *dystocia* (difficult calving) in Charolais and in Simmental cattle (Tryphonas et al., 1974).

As a postscript to this introduction to the innervation of skeletal muscle fibers, it may be of interest to note that the idea of an autonomic innervation of skeletal muscle fibers has been revived (Barker and Saito, 1980). The idea was proposed in the 1920s and rejected in the 1930s (Youmans, 1967). The agricultural significance of this point relates to the possible involvement of adrenergic systems during the conversion of muscles to meat (Gregory, 1981). This general topic is considered in Chapter 9.

Motor Units

Each motor neuron innervates a number of muscle fibers in the same muscle. The group of muscle fibers innervated by a single neuron is called a *motor unit*. The fibers of each motor unit exhibit similar metabolic features (Nemeth et al., 1981). In short muscles, individual muscle fibers may extend without interruption from one end of a fasciculus to the other. But in muscles with long fasciculi, many fibers have tapered endings which do not reach to the end of the fasciculus. These are called *intrafascicularly terminating fibers* (Le Gros Clark, 1945). Tapered endings are anchored in the *endomysial* connective tissue around an adjacent normal-diameter fiber. Intrafascicularly terminating fibers create quite a problem in studies on muscle growth and innervation (Chapter 7). Contraction of an intrafascicularly terminating fiber may stretch the fiber on which it is anchored. Not only might the *series elasticity* of the passive fiber change the result of a contraction by an intrafascicularly terminating fiber, but the response of the stretched fiber will also be changed if, in turn, it is stimulated to contract.

Most muscle fibers have a motor end plate about halfway along their length. Neuromuscular junctions are established very early in development, and a considerable length of new material is subsequently added to each end of the fiber so that the motor end plate is left in the middle. Many muscles have relays of fasciculi, each with a zone of motor end plates. Jarcho et al. (1952) and Hunt and Kuffler (1954) found that muscle fibers could extend through two innervation zones with an end plate in each. In this case, however, it is difficult to prove that both end plates did not originate from a single axon which had branched before reaching the muscle. However, multiple innervation has also been reported in a muscle located along the length of the body and innervated by several different ventral roots from the spinal cord (Garven, 1925).

When a muscle is innervated by two nerves (which we will call nerve a and nerve b), it is possible to measure the tension (T) generated by stimulating each nerve separately (Ta followed by Tb). The sum of tensions generated by separate stimulation (Ta + Tb) may exceed the tension generated by stimulating both nerves simultaneously (Ta + b). Muscles where Ta + Tb > Ta + b have been claimed as evidence of multiple innervation of muscle fibers, arguing on the grounds that a multiply innervated fiber might contribute separately to both Ta and Tb but would not double its tension for Ta + b (Walker, 1961). Katz (1926) argued that tensions Ta and Tb might not be additive because of fascicular arrangement and leverage, and he demonstrated a case where Ta + Tb > Ta + b but heat production (H) resulting from contraction was Ha + Hb = Ha + b. The consensus at the present time is that multineuronal innervation of mammalian muscle fibers may occur in immature muscles which are still establishing appropriate neuromuscular connections. This condition, however, persists for only a short while after birth and is then lost (Redfern, 1970; Bagust et al., 1973; O'Brien et al., 1978; Miyata and Yoshioka, 1980). By experimental manipulation, mature muscle fibers can be induced to accept extra innervation (Frank et al., 1974), just as they will if their primary innervation is chemically blocked (Jansen et al., 1973). Multiple innervation is common in amphibians and fish (Bone, 1972; Slack and Docherty, 1978; Angaut-Petit and Mallart, 1979).

It is generally considered that the number of fibers in a motor unit is related to the precision with which a muscle must contract to perform its normal function. Motor units are thought to be small in precise muscles. From this it might be deduced that motor units are probably quite large in the major carcass muscles of meat animals since their movements are not precisely controlled. In most muscles, the fibers of a motor unit are widely scattered, both across the muscle belly and along its length (Van Harreveld, 1947a). Clark (1931) examined the sizes of motor units in two small muscles of the cat's limb, carefully taking into account the location of intrafascicularly terminating fibers, and it was found that motor units contained approximately 120 to 165 muscle fibers. Van Harreveld (1947b) found motor units with approximately 100 to 125 fibers in the sartorius muscles of rabbits. Unfortunately, there is little or no readily available information on motor unit sizes in meat animals. At a guess, 100 to 200 fibers might be the lower limit of motor unit size.

In double-muscled cattle with greatly increased real and apparent numbers of muscle fibers, the functional terminal innervation ratio is increased from approximately 1.02 to 1.18 (Swatland, 1973a) or to 1.11 (Novakofski et al., 1981). This finding suggests that motor units are larger in double-muscled cattle and that the increased numbers of muscle fibers are not matched by a completely adequate increase in numbers of motor neurons. Breed and age differences may occur in these innervation ratios, and although the values for heterozygous animals are intermediate between those of homozygous normal and homozygous double-muscled animals, they are of no practical value in the identification of latent heterozygous carriers (Novakofski et al., 1981).

Motor End Plates

At the neuromuscular junction, the terminal axon forms a *terminal ramification* of small branches which often have bulbous ends. The terminal ramification sits on a mound of muscle fiber sarcoplasm called the *Doyère hillock* (Figure 6-9). The terminal ramification and the Doyère hillock constitute a motor end plate. The motor end plate is covered by *perineural epithelial cells* instead of Schwann cells (Shanthaveerappa and Bourne, 1967). At the point of contact between the terminal ramification and the muscle fiber surface, the muscle fiber (covered by its basement membrane) is corrugated to form *synaptic clefts*. When an action potential arrives at a motor end plate, *synaptic vesicles* in the axoplasm (Korneliussen, 1972) fuse with the axonal membrane, and each vesicle releases a *quantal amount of acetylcholine* into the

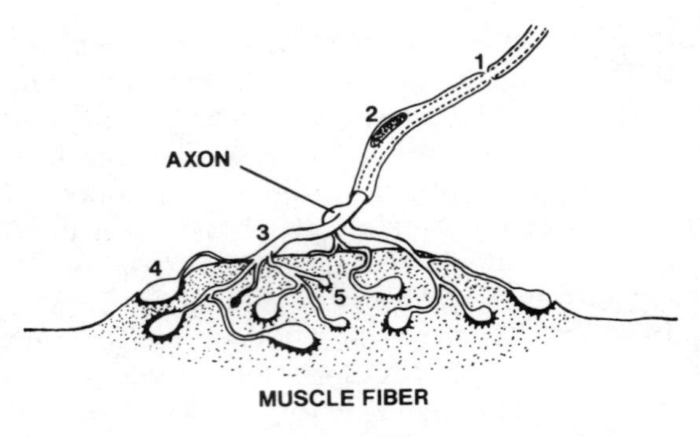

Figure 6-9 Neuromuscular junction or motor end plate: (1) node of Ranvier; (2) Schwann cell; (3) terminal ramification of axon; (4) axonal swelling in a synaptic trough on the muscle fiber surface; (5) Doyère hillock.

synaptic cleft. Acetylcholine diffuses across the synapse, from the axonal membrane to the muscle fiber membrane, and causes a change in the ionic permeability of the postsynaptic area of the muscle fiber membrane. This causes the *end plate potential* to change. If the change is of sufficient magnitude, it initiates an action potential on the surface of the muscle fiber. Acetylcholine is rapidly destroyed by *acetylcholinesterase* so that the muscle fiber can relax (Plate 12). In the brief instant of 1 second, this whole sequence of events at the neuromuscular junction can occur several hundred times.

It is relatively easy to measure the maximum diameter of the spread of a terminal ramification, but this may be a poor measure of the neuromuscular contact area, the most meaningful dimension in physiological terms (Salpeter and Eldefrawi, 1973). Similarly, it is difficult to select a dimension of muscle fiber size with which motor end plate dimensions can be meaningfully correlated without encountering the curvilinear relationships that exist between different dimensions (x versus x^2 versus x^3). The determination of muscle fiber surface area requires a knowledge of fiber length, and this is usually obscured by intrafascicularly terminating fibers. The nature of the relationships between motor end plate dimensions and muscle fiber dimensions depends on the source of variation in muscle fiber size; we may take animals of a set age and look at different-size fibers or, alternatively, we may examine muscle fibers at different stages during their growth. Despite the multiplicity of dimensions and bases of comparison, it is generally found that the growth of motor end plates is correlated with the growth of muscle fibers until motor end plates reach a certain size, after which, muscle fiber growth is no longer accompanied by motor end plate growth (Figure 6-10; Harris, 1954; Coers, 1955; Anzenbacher and Zenker, 1963; Gruber, 1966; Flamm, 1968; Granbacher, 1971; Swatland and Cassens, 1972a; Ip, 1974; Kordylewski, 1979b). Nystrom (1968b), however, thought that motor end plate diameters in cat muscles might always retain a 1:2 proportionality to muscle fiber diameters.

In large muscles, motor end plates are sometimes difficult to locate. Innervation zones can be mapped by using an electrical stimulator to find regions with a low

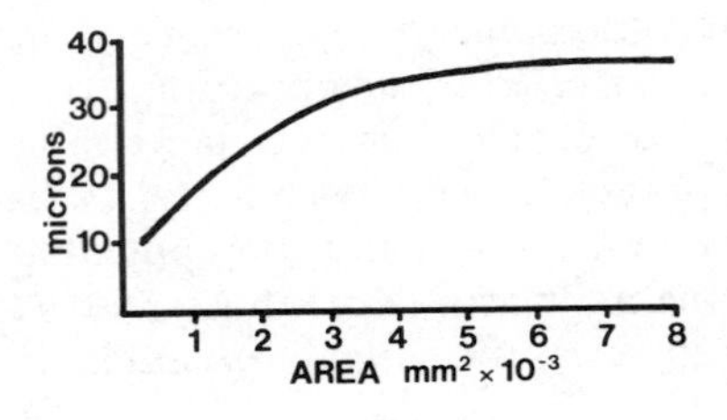

Figure 6-10 Relationship between motor end plate maximum diameter (microns) and muscle fiber cross-sectional area during growth. (Data from Swatland and Cassens, 1972a.)

threshold to excitation. As a muscle is frozen, motor end plates can be located by the way in which they form a focal point for ice crystal formation (England, 1970). The terminal ramification can be stained with either *methylene blue* or *silver*. Staining with methylene blue is probably dependent on the integrity of a membrane-associated ATPase (Kiernan, 1974), and usually requires fresh tissue. The precipitation of metallic silver on neurofilaments and on coarser elements of axonal structure can be undertaken on either frozen (Toop, 1976) or fixed tissue. The corrugated postsynaptic region can be demonstrated histochemically with reactions for acetylcholinesterase (Pearse, 1972), or with fluorescent labels for acetylcholine receptors (Anderson and Cohen, 1974). The proteolytic activity of motor end plates is sufficient to digest the different colored layers of gelatin in a photographic film (Poberai and Savay, 1976).

Actinlike filaments are probably involved in the development and maintenance of the morphology of the neuromuscular junction (Hall et al., 1981). Motor end plate morphology exhibits considerable variation within an individual muscle (Figure 6-11).

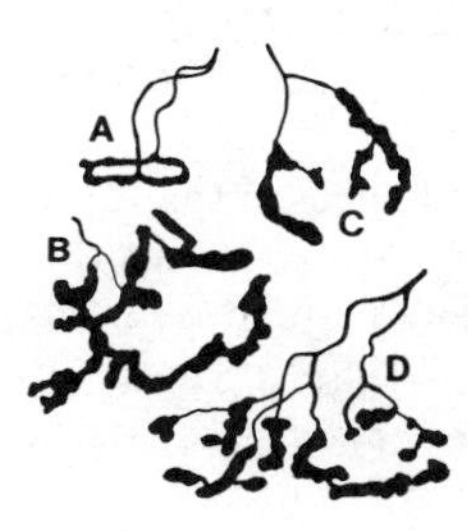

Figure 6-11 Motor end plate morphology in pigs: (A) simple ring; (B) complex ring; (C) simple arborescent type; (D) complex arborescent type. (From Swatland and Cassens, 1974.)

Some muscle fibers have compact motor end plates, while others have irregular terminal ramifications with exploratory sprouts. Motor end plates become very irregular in old animals (Gutmann and Hanzlikova, 1965). Complex motor end plates and axonal sprouting are conspicuous in muscles from stress-susceptible pigs (Swatland and Cassens, 1972a). In meat animals, the synaptic area often forms a ring which is innervated by a simplified terminal ramification (Swatland and Cassens, 1971; Bradley, 1977). This has also been observed in motor end plates of young rats (Cole, 1957).

The types of mammalian neuromuscular junctions discussed so far are in a morphological category known as the *en plaque* type. In poultry, another morphological type, known as *en grappe*, occurs on some muscle fibers; terminal axons run irregularly along the muscle fiber so that the postsynaptic area is scattered and is poorly developed relative to en plaque end plates (Hodges, 1974). Avian muscle fibers with en grappe innervation contract more slowly than fibers with en plaque endings. En grappe endings can be found in the extraocular muscles of some mammals (Namba et al., 1968; Zenker and Anzenbacher, 1964; Pilar and Hess, 1966), and in the small ear muscles of cats (Fernand and Hess, 1969), but they do not normally occur in the carcass muscles of cattle, sheep, and pigs.

PHYSIOLOGICAL DIFFERENTIATION OF MUSCLE FIBERS

The formation of muscle fibers from mesodermal cells, through a series of transitional cell types (premyoblast, myoblast, and myotube or secondary fiber), is a classical example of *cellular differentiation*. Cellular differentiation leads to an efficient and mutually advantageous division of labor among the tissues and organs of the body. In skeletal muscles, differentiation continues after the fibers have been formed and after they have reached a functional state. Physiological differentiation follows cellular differentiation, and creates populations of fast and slow fibers with appropriate sources of energy for contraction, either *aerobic* (using blood-borne oxygen for complete oxidation of substrates) or *anaerobic* (incomplete oxidation of carbohydrates without need for oxygen).

Certain muscles of the carcass are particularly dark or red. This color difference is due to a red pigment, *myoglobin*, in the sarcoplasm of muscle fibers. *Hemoglobin*, the pigment of red blood cells, brings oxygen to *capillaries* on the muscle fiber surface. From here, the transport of oxygen to the interior of the fiber is facilitated by myoglobin. Thus fibers which are specialized for aerobic metabolism develop a high myoglobin concentration. The dominant work of some muscles is to maintain a standing posture or else to contract slowly during locomotion, chewing, or breathing. Such muscles tend to contain a high proportion of slow-contracting and fatigue-resistant fibers with a high myoglobin concentration. These are the dark red muscles of the carcass.

The early scientific literature on the topic of red and white muscles was reviewed by Needham (1926) and by Denny-Brown (1929), starting with a report of red and white rabbit muscle by Lorenzini in 1678, and including the classical studies of Ranvier published between 1873 and 1889. Ranvier found that dark red muscles (1) contract slowly; (2) that they develop *tetanus* (a sustained contraction resulting from the fusion of rapid twitches) at lower rates of stimulation; (3) that they have relatively more sarcoplasm; (4) that they have more distinct *longitudinal* striations; and (5) that they are more resistant to fatigue. Longitudinal striations are caused by abundant sarcoplasm between the longitudinally arranged fibrils. In transverse sections of muscle fibers, differences in myofibrillar size, in the regularity of myofibrillar arrangement, and in the degree of myofibrillar separation can create two distinct patterns, *felderstruktur* in slow fibers and *fibrillenstruktur* in fast fibers.

In frogs (Peachey and Huxley, 1962; Page, 1965), electron microscopy shows that felderstruktur fibers have large ribbonlike myofibrils which are irregularly shaped and often fused. Fibrillenstruktur fibers have small compact myofibrils which are neatly separated by sarcoplasm. Felderstruktur fibers lack an M band, they have an irregular Z line, and they have no *triads* of *sarcoplasmic reticulum*. Fibrillenstruktur fibers have an M band, thick Z lines, and triads. In frogs, felderstruktur fibers slowly develop tension in response to repeated neural stimuli, they have relatively low resting potentials (60 mV instead of 90 mV), and they are activated by slowly spreading or *graded potentials* rather than by rapidly propagated *action potentials* (Katz, 1966). In frog muscle, the two types of muscle fibers differ in their responses to changes in their ionic environment (Irwin and Hein, 1966). In frog muscle, felderstruktur fibers have weak *myofibrillar* ATPase activity, whereas fibrillenstruktur fibers have strong ATPase activity (Engel and Irwin, 1967). A similar physiological differentiation in fiber ultrastructure, physiology, and myofibrillar ATPase is found in other amphibians (Kordylewski, 1979a; Watanabe et al., 1980) and in fish (Kilarski and Bigaj, 1969; Nag, 1972; Korneliussen and Nicolaysen, 1973; Pool et al., 1976; Raamsdonk et al., 1978).

In meat animals, muscle fibers are also differentiated into fast- and slow-

TABLE 6-1. PHYSIOLOGICAL DIFFERENTIATION OF MUSCLE FIBERS IN VERTEBRATES

Features	Very slow fiber of lower vertebrates	Slow fiber of birds and mammals	Fast fiber of birds and mammals
Type of contraction	Very slow contracture	Slow twitch	Fast twitch
Membrane response	Slow graded potential	Slow action potential	Fast action potential
Neuromuscular junction	En grappe	En grappe in birds but en plaque in mammals	En plaque
Myofibrillar ATPase	Weak	Weak	Strong
Myofibrillar arrangement	Felderstruktur	Intermediate	Fibrillenstruktur

contracting types, with appropriate differences in ultrastructure and myofibrillar ATPase activity. There is, however, an important difference between the condition found in the carcass muscles of farm mammals and that found in lower vertebrates: both the fast- and the slow-contracting types of muscle fibers in farm mammals respond to neural stimulation by developing propagated action potentials. Thus in this important feature, both the fiber types of mammals resemble the fast-contracting fiber type of lower vertebrates. In the carcass muscles of farm mammals, there are no fibers that develop slowly spreading graded potentials like those of felderstruktur fibers in lower vertebrates (Table 6-1). It should, perhaps, be admitted that the statement above is obtained by deduction rather than by direct observation. Thus the statement is made on the basis of a similarity between farm mammals and laboratory mammals, and on evidence which shows that both types of fibers in laboratory mammals resemble the fast-contracting fiber type of lower vertebrates. While the statement probably holds true for the carcass muscles of interest to us in meat production, there is some doubt as to the situation in specialized small muscles, such as the extraocular muscles. Of more serious consequence in agriculture, there is some confusion over the condition in poultry, where as mentioned earlier, the slow fibers are multiply innervated with en grappe endings.

In lower vertebrates, there is usually a clear distinction between slow fibers (with graded potentials and en grappe endings) and fast fibers (with action potentials and en plaque endings). In carcass muscles of cattle, sheep, and pigs, the situation is different, and both slow fibers and fast fibers have action potentials and en plaque endings (motor end plates). In poultry, the fast fibers present no problem—they have en plaque endings and they develop action potentials. However, the slow fibers of poultry have en grappe endings but they contract with a twitch, not with a very slow contracture (Table 6-1). Do they have graded potentials without action potentials, or do they have action potentials without graded potentials? Hodges (1974) concluded that felderstruktur fibers in poultry do not generate action potentials. Ginsborg (1960a, 1960b), however, produced experimental evidence that they could. Bowman and

Marshall (1971) managed to reconcile a number of such conflicting reports with the idea that slow fibers with en grappe endings in poultry might respond to a single stimulus with a graded potential, but that they might also respond to a series of impulses by developing action potentials. Perhaps the same thing is true of the specialized mammalian muscles which have en grappe endings (Bach-y-rita and Ito, 1966). In chicken muscles, fibers with en plaque innervation have strong alkaline ATPase activity and fibers with en grappe endings have weak alkaline ATPase activity (Toutant et al., 1981). However, further subtypes can also be distinguished on the basis of the pH stability of ATPase and on aerobic enzyme activity. The evolutionary development of fiber types is beyond the scope of the present work; some interesting ideas are discussed by Coers (1955).

We now leave the problems of fiber types in lower vertebrates and progress to the equally complex problem of physiological differentiation between muscle fibers in meat animals. In the earlier part of this chapter, myogenesis was considered in relation to the factors that might ultimately set limits on the muscularity of meat animals. It might be useful at this point to explain briefly the agricultural importance of the physiological differentiation of muscle fibers (Cassens and Cooper, 1971). In growing animals, red and white muscles exhibit differences in protein metabolism, with greater protein turnover rates in red muscles, possibly as a result of their higher concentration of mitochondria (Dadoune et al., 1978). Another point is that the physiological types of muscle fibers behave differently once an animal has been slaughtered and its muscles are being converted to meat (Chapter 9). Last, but not least, there are factors such as myoglobin and lipid concentration and connective tissue content which are a direct consequence of differences in the work patterns of living muscles. These factors have a major effect on the salability of meat. In beef, for example, Calkins et al. (1981) found that the size and frequency of different histochemical types of muscle fibers were correlated with meat toughness and marbling. In lamb, fiber-type composition is correlated with the juiciness and flavor intensity of the meat (Valin et al., 1982). Pork hams which exhibit a conspicuous range from pale to dark red muscles, an appearance called "two-toning," may become a problem if consumers prefer an even color distribution. Two-toning is due to an increase in the difference in the myoglobin concentration between red and white muscles. This may be a genetic trait (Wilson et al., 1959).

At first sight, historically speaking, it appeared that the relationship between fast and slow fibers in meat animals was quite simple. From the time of Ranvier onward, it had been known that fast fibers were usually white, while slow fibers were usually red. When redness was found to be due to myoglobin, and when myoglobin was found to be correlated with aerobic metabolism, this explained the relationship between redness and speed of contraction. The pale or white fibers, with a low aerobic potential, were found to be well endowed with glycolytic enzymes, which enabled them to obtain energy rapidly by the incomplete oxidation of glycogen (Dubowitz and Pearse, 1960). This explained why white fibers soon became fatigued once their glycogen stores were depleted and why they had to wait for the removal of lactate by the circulatory system. At the extremes of the range in physiological differentiation (fast white fibers versus slow red fibers) these discoveries are still valid. The problem, as we see it now, is that there are also fibers with a fast contraction speed and a dual energy supply. In other words, some fast fibers have both aerobic and anaerobic capabilities. The discovery of these fibers coincided in a most confusing way with a growing awareness that mammalian and avian slow red fibers were rather different from the slow red fibers with en gappe endings and graded membrane potentials which were known in lower vertebrates. It is difficult to write a research report on muscle fiber types without giving them names. Unfortunately, everybody seemed to use different names, and the numbers of fiber types that were recognized tended to be a function of the number of histochemical techniques which were used to identify them.

In human pathology, the dominant system for naming fiber types is based on the Type I (slow red) versus Type II (fast) nomenclature which was advocated by Brooke and Kaiser (1970) and by Engel (1974). Subtypes of I and II are given alphabetic designations to account for dual-purpose fibers (Askanas and Engel, 1975; Tunell and Hart, 1977; Mabuchi and Sreter, 1980b). A table of synonyms was prepared by Close (1972), who reviewed the correlation between contractile and histochemical features of muscle fiber types. The Type I–Type II system is also advocated by Khan (1976; 1977), with the major subtypes as Type I red, Type II red, and Type II white (the Type II red is the fast fiber with a dual aerobic-anaerobic energy system). Type I white fibers are rare (Khan, 1979). Thus the first feature on which fiber types are usually categorized is their possession of strong (Type II) or weak (Type I) alkaline myofibrillar ATPase. Next, muscle fibers are secondarily subdivided on the basis of their aerobic potential.

Unfortunately, although ATPase techniques are now mostly quite reliable, there are many different variations of the most common method. Changes in pH or modifications of the ATPase incubation medium can cause reaction intensities which are intermediate between the all-or-nothing criteria of strong versus weak ATPase activity. As an extreme case, reaction intensities are totally reversed if histological sections of muscle are preincubated at a low pH (Guth and Samaha, 1969; Guth and Samaha, 1970). Fast fibers have acid-labile, alkali-stable ATPase. Slow fibers have acid-stable, alkali-labile ATPase.

The biochemical basis of the histochemical reaction for myofibrillar ATPase is difficult to investigate. When muscle samples are homogenized for biochemical analysis, it is difficult to start with a pure population of muscle fibers which are all of the same type, but it is relatively easy to remove sources of nonmyofibrillar ATPase activity. Histochemically, on the other hand, fibers can be studied individually, but their myofibrils are now surrounded by other cellular components which may exhibit nonmyofibrillar ATPase activity. Biochemically, fast muscles have stronger ATPase activity than slow muscle (Barany, 1967). Biochemically, myosin from fast muscles exhibits alkaline stability, while myosin from slow muscles exhibits acid stability (Samaha et al., 1970b).

The common histochemical technique for myofibrillar ATPase (Padykula and Herman, 1955a, 1955b) employs calcium ion activation, but mitochondria and the sarcoplasmic reticulum also contain an alkali-stable calcium-activated ATPase (Padykula and Gauthier, 1963; Khan et al., 1972, 1975; Samaha and Yunis, 1973). Secondary steps in the histochemical reaction for ATPase create a black final reaction product from the initial reaction product, inorganic phosphate, which is liberated enzymatically by ATPase. Unfortunately, there is an opportunity for any phosphate which has been liberated from nonmyofibrillar ATPase to bind to myofibrils and so to contribute to the myofibrillar final reaction product (Guth, 1973). On the other hand, some investigators have concluded that intermyofibrillar reaction products might be an artifact originating from myofibrillar ATPase (Schiaffino and Bormioli, 1973a).

With all these potential problems, it is reassuring to find that the calcium-activated method for myofibrillar ATPase can be verified independently with an alternative lead salt method (Meijer and Vossenberg, 1977), and that the calcium-activated method is fairly reliable postmortem (Eriksson et al., 1980), provided that postmortem glycolysis is not too rapid. In studies on individually isolated muscle fibers, Zeman and Wood (1980) confirmed the existence of the relationship between contraction speed and histochemical myofibrillar ATPase activity. Thus, at present, studies on the physiological differentiation of muscle fibers are now firmly based on the correlation between contraction speed and myofibrillar ATPase activity, and there is a reasonable degree of agreement between physiologists, biochemists, and histologists.

The published literature on muscle fiber types in different species of mammals is very extensive, and it includes humans (Johnson et al., 1973) and other primates (Beatty et al., 1966; Ariano et al., 1973), rodents (Yellin, 1972; Beatty et al., 1974; Sexton and Gerston, 1967; Khan, 1978), carnivores (Taylor et al., 1973; George and Ronald, 1973; Gunn, 1978; Cardinet et al., 1972; Hamilton et al., 1974), horses (Gunn, 1972, 1973, 1978; Snow and Guy, 1980), and rabbits (Khan et al., 1973; Bass et al., 1973). The comparative myology of birds is reviewed systematically by George and Berger (1966).

It is difficult to make generalizations and still do justice to a myriad of exceptions and variations. With that shortcoming in mind, two major taxonomic systems can be identified in the agricultural literature on muscle fiber types. The first to appear was that used by R. G. Cassens and his research associates during the late 1960s. At this time, it was only just becoming apparent that the slow red fibers of mammals were fundamentally different from the slow red fibers of lower vertebrates. Although the histochemical reaction for myofibrillar ATPase was available, it had not been widely used on meat animals, basically because the primary motive for research was myoglobin and meat color. At the time, there was little reason to suspect that all red fibers did not have weak myofibrillar ATPase activity. Thus the terms *red fiber* and *white fiber* appeared adequate. The application of histochemical fiber typing to the problem of pale, soft, exudative (PSE) pork (Chapter 9) soon yielded some contradictory findings which led to the recognition of *intermediate* fibers with an intermediate aerobic potential. Intermediate fibers were found to resemble white fibers in having strong myofibrillar ATPase activity (Plate 16; Table 6-2).

Until this point, most researchers thought that fiber types in mature animals

TABLE 6-2. HISTOCHEMICAL CHARACTERISTICS OF RED, INTERMEDIATE, AND WHITE FIBERS

Characteristics	Red	Intermediate	White
ATPase activity	Weak	Strong	Strong
Aerobic metabolism	Strong	Strong or intermediate	Weak
Anaerobic metabolism	Weak	Strong	Strong

constituted more or less distinct species of fibers. As it gradually became apparent that myofibrillar ATPase was a more basic criterion of muscle fiber histochemistry than aerobic potential or redness, C. R. Ashmore and his research associates developed a terminology which demoted redness to a secondary characteristic which, in one fiber type, was a continuous variable (Table 6-3).

TABLE 6-3. CHARACTERISTICS OF ALPHA AND BETA FIBERS

Characteristic	Alpha	Beta
ATPase activity	Strong	Weak
Aerobic metabolism	Range from strong (R) to weak (W)	Strong (R)

The advantage of this system was that it openly recognized as a subjective decision, the often difficult problem of categorizing a fiber with an in-between aerobic potential. This was particularly important in growth studies since fibers of very young animals all have a strong aerobic potential, and anaerobic fibers are slow to develop (Plates 20 and 21). In mature animals, the interrelationship between the two systems is straightforward (Table 6-4). Further subdivisions of these three basic types can be made. Suzuki (1971a, 1971b, 1972, 1976), and Suzuki and Tamate (1974), for example, have identified five histochemical types of muscle fibers in sheep and cattle.

TABLE 6-4. INTERRELATIONSHIPS OF THE MAJOR HISTOCHEMICAL FIBER TYPES

White =	alpha W
Intermediate =	alpha R
Red =	beta R

Many of the cellular features associated with aerobic and anaerobic metabolism in muscle fibers are fairly straightforward. Aerobic fibers are served by a more dense capillary meshwork than fibers with a poor aerobic potential; their sarcoplasm contains more mitochondria and more lipid droplets; and the enzymes involved in aerobic metabolism are more concentrated. Quantitatively, however, the range from dominantly aerobic to dominantly anaerobic metabolism is usually a continuous variable, and it is seldom broken into discontinuous steps. Spamer and Pette (1977, 1979) have demonstrated this in enzyme assays of individual muscle fibers using indicator enzymes for the Krebs cycle, beta oxidation of fatty acids and anaerobic glycolysis. In many cases, the aerobic-anaerobic potential of a fiber can also be related to differences in isoenzyme activity (Sawaki and Peter, 1972; McMillan and Wittum, 1971). Thus, as we begin to view metabolic differentiation as a continuous variable, we may start to think of fiber types as frequency peaks in a population (Schmalbruch and Kamieniecka, 1975). But what about myofibrillar ATPase? Is it really a discontinuous variable, strong versus weak, as implied in both taxonomic systems for histochemical fiber types in meat animals?

In chicken muscles that differ in their contraction speed and type of innervation, immunohistochemical studies on the distribution of fast and slow myosin support the idea that myofibrillar ATPase is a discontinuous variable; no evidence of in-between cases was found by Arndt and Pepe (1975). In rat muscle, the biochemical analysis of myosin from individual muscle fibers provides further support for discontinuity (Pette and Schnez, 1977). What, then, leads us to question the distinctness of fiber typing by myofibrillar ATPase? There are two things. First, by manipulating the pH of ATPase incubation media, it is possible to generate more than two staining responses, and these do not fit very well with the dual ATPase-aerobic systems of fiber type categorization (Nemeth et al., 1979; Nemeth and Pette, 1980; Suzuki and Cassens, 1980a). Second, there is evidence that the physiological differentiation of muscle fibers is a dynamic balance in the division of labor, and that the balance may change during growth or in response to a change in the work pattern of a muscle. Thus, to some researchers, the histochemical categorization of muscle fibers by any method, including myofibrillar ATPase, is merely a useful, but artificial subdivision of a continuously variable range (Billeter et al., 1980; Hintz et al., 1980). Guth and Yellin (1971) concluded that muscle fibers undergo a continual alteration throughout life as an adaptation to changing functional demands, and that "fiber type" merely reflects the constitution of a fiber at any particular time. From an agricultural viewpoint, this is particularly interesting since it suggests the existence of some degree of genetic or developmental plasticity in the fiber type continuum. In meat animals, this might be a vital link in relating muscle growth to meat quality. Before exploring this topic, however, it might be best to complete this review of neuromuscular relationships.

NERVE-MUSCLE INTERACTIONS

Motor End Plates

Slow red fibers are innervated by neurons with smaller perikarya than those that innervate fast pale fibers (Henneman and Olson, 1965; Sato et al., 1977). In rat muscles, Ogata and Murata (1969) found that red fibers had small motor end plates with few synaptic clefts, relative to those of white fibers. Padykula and Gauthier (1970) found that presynaptic components also varied between red and white fibers. White fibers were innervated over a larger synaptic area, they had a more extensive terminal ramification, and they had a greater concentration of synaptic vesicles. Korneliussen and Waerhaug (1973) and Waerhaug and Korneliussen (1974) found that terminal ramifications on white fibers were more slender, but more widely spread. Unfortunately, these studies do not correspond very well with other work (including some on pork muscle) in which motor end plates with larger diameters or synaptic areas have been found in red muscle (Coers, 1955; Nystrom, 1968a; Swatland and Cassens, 1972a; Dias, 1974). It is likely, however, that both the physiological differentiation and the size of muscle fibers are related to the morphology and size of neuromuscular junctions. Physiological differentiation and relative fiber size are both complex variables, and this may explain the discrepancy. Nerve terminals on muscle fibers with a fast speed of contraction are metabolically more active than those on slow fibers (Pickett, 1980).

Motor Units

The anatomical location of muscle fibers within a motor unit can be experimentally determined by stimulating individual motor axons of the ventral roots of the spinal cord, or by stimulating the major nerve trunks which radiate from the spinal cord. After exhaustive stimulation, the fibers of an active motor unit can be identified by their loss of glycogen. The active part of the muscle is frozen in liquid nitrogen so that frozen transverse sections can be histochemically tested for glycogen (Doyle and Mayer, 1969; Kugelber, 1973; Eisen et al., 1974; Burke et al., 1974; Frederick et al., 1978). The fibers of a single motor unit normally have identical or very similar histochemical characteristics, probably because these features are regulated in some way by the motor neuron. Sometimes, however, histochemically foreign muscle fibers appear in a stimulated motor unit (Edstrom and Kugelberg, 1968).

Although the fibers of a motor unit are scattered throughout their muscle, the distribution of different fiber types within a fasciculus may not be random. Type I (slow) fibers tend to be grouped centrally while Type II (fast) fibers tend to be peripheral (James, 1971, 1972). This pattern is particularly conspicuous in pork muscles (Moody and Cassens, 1968), even though the fibers of motor units are scattered through the muscle as in other species (Szentkuti and Cassens, 1979). When muscles are partially denervated, the reinnervation of denervated muscle fibers by surviving neurons gives rise to a localized group of fibers that all have the same histochemical characteristics. The functional terminal innervation ratio is increased, newly formed small motor end plates lower the mean motor end plate diameter, and muscle fibers assume transitional features as they come under the control of a new neuron (Morris and Raybould, 1971).

Development of Neuromuscular Junctions

Some of the early research on the formation of neuromuscular junctions is difficult to understand since early investigators held some ideas which we no longer accept. Some of the early investigators thought that the terminal ramification fused into the muscle fiber sarcoplasm; others thought that the multinucleate condition of muscle fibers was due to the incomplete separation of myoblasts after nuclear division; and others mistook secondary fiber formation for the longitudinal fission of myotubes. Despite these handicaps, many of the essential events concerning the development of neuromuscular junctions have been known for many years (Cuajunco, 1942). At an early stage in muscle development, thin, naked axons are found among the first myotubes to be formed. These axons branch frequently and make passing contact with lots of myotubes so that each myotube is touched by several different axons. As the postsynaptic structure of the neuromuscular junction begins to develop at the site of a future motor end plate, superfluous axons disappear and only one axon survives to innervate the future motor end plate. Secondary fibers may become innervated by axons which supply the parent myotube (Kelly and Zacks, 1969b). In early research, it was difficult to decide whether a muscle fiber attracted an axon to itself to form a neuromuscular junction, or whether the axon prompted the muscle fiber to form a neuromuscular junction that it could innervate: The latter alternative now appears most likely (Lentz, 1969; Veneroni and Murray, 1969; Juntunen, 1973). Motor end plate growth is very rapid just after birth (Swatland, 1975a).

With a few exceptions, the muscle fibers of a motor unit are all of the same histochemical type. Do muscle fibers select an appropriate motor neuron for their type, or do motor neurons dictate to the muscle fibers the type of physiological differentiation that they should develop? The second alternative seems more likely, at least after birth, since muscle fibers attempt to change their type of physiological differentiation if they are experimentally *cross-reinnervated* by a different type of motor neuron. In some cases, physiological differentiation becomes apparent during early development at about the same time that axonal distribution is being trimmed to the final pattern of survivors. Riley (1977) has shown that there is, however, no causal relationship between these two events. The possibility that degenerative events might shape neural networks by removing inappropriate connections has been known for a long time. Thus, in neuromuscular relationships, it was thought that the loss of multiple innervation during development might cause an appropriate loss of motor neurons in the spinal cord (Bagust et al., 1973). Holt and Sohal (1978) were unable to find any evidence of such an interaction.

Hughes (1968) provides an excellent survey of developmental interactions between nerves and muscles and between motor neurons and higher neural centers. Only a few general points need be summarized here on these topics. There is some evidence that the brain exerts an effect on muscle development, even to the point of affecting muscle fiber histochemistry. However, it is difficult to separate direct effects, which might result from regulatory systems, from indirect effects, which might result from changes in muscle work patterns. Even before birth, muscles are quite active at certain times. In the explanation of secondary fiber formation given earlier, attention was focused on the role of contraction in primary myotubes. Since myotubes are innervated early in development, the nervous system may exert its effect by influencing myotube contraction and secondary fiber formation. Karpati and Engel (1968) found that *cordotomy* (removal of a length of spinal cord between hind limb motor neurons and the brain) increased the frequency of occurrence of Type II fibers, but whether this was due to Type I loss or to the transformation of fibers from Type I to II could not be determined. Further evidence that the central nervous system affects muscle development comes from the fact that children with brain defects often have changes in the histochemistry of their muscle fibers (Fenichel, 1967, 1969; Hooshmand et al., 1971).

The anatomical distribution of carcass muscles is topographically represented in the circuitry of the spinal cord by groups of motor neurons. Each group of neurons in the spinal cord serves a single muscle. In chickens, groups of motor neurons located laterally in the spinal cord innervate muscles which are derived from the dorsal muscle mass, while more medial groups of neurons innervate muscles which are derived from the ventral muscle mass (Landmesser, 1978a). Neuronal group patterns are established before muscle cleavage occurs (Landmesser, 1978b). Groups of motor neurons tend to develop and to mature in an anterior to posterior sequence, just as limb muscles tend to develop in a proximal to distal sequence (Romanes, 1941). The possibility that this coincidence is responsible for the matching of anterior neuronal groups to proximal limb muscles has been rejected, at least in chicken muscle (Landmesser and Morris, 1975; Lance-Jones and Landmesser, 1980a, 1980b). Barron (1953) concluded that muscle fiber numbers were regulated by the numbers of motor neurons that developed. Hamburger (1934), however, thought that the number of muscle fibers was only one of several factors that regulated the number of neurons. MeLennan (1982) concludes that the number of "myotube clusters" in a muscle determines its capacity to support motor neurons during the period of neuronal cell death. The extent of muscle activity also affects neuronal numbers (Pittman and Oppenheim, 1979). The number of motor neurons becomes fixed at the time that integrated neuromuscular development is completed. After this time, no new neurons are added and numbers are stable, but neurons can be lost, particularly in old age (Tomlinson and Irving, 1977).

Trophic Effect of Nerves

Motor neurons exert a long-term regulation over the physiological and metabolic properties of the fibers in their motor unit. This is often called the *trophic effect* of nerve on muscle. The word "trophic" implies something of a nutritive effect, as if the nerve was feeding the muscle, but its current usage sometimes includes possible nonnutritive effects, such as the frequency patterns of nerve impulses to the muscle. The idea that nerves might have a trophic function is far from new, and it probably originates from ancient observations on the degenerative fate that overtakes many organs once they have been denervated (Haliburton, 1909). Trophic effects may be bidirectional since there are some retrograde trophic effects which travel from the muscle to the nerve (Czeh et al., 1978). For example, presynaptic terminal *boutons* on motor neuron perikarya are lost when axons are cut, and they are restored when neuromuscular contact is reestablished (Cull, 1974). Similarly, there are soluble fractions from skeletal muscle tissue that may promote growth and differentiation in the embryonic spinal cord (Hsu et al., 1982). The extensive scientific literature on neurotrophic effects is reviewed by Guth (1968, 1969), Harris (1974) and Gutmann (1976).

Denervation

The denervation of a muscle causes a decrease in its speed of contraction, a decrease in muscle fiber diameters, a loss of transverse striations, and eventually, disintegration of the muscle fibers (Gutmann and Zelena, 1962). Although the act of denervation deprives a muscle of any trophic substances that it might normally receive from its nerve, denervation also terminates the muscle's normal work pattern. Trophic deprivation greatly overshadows the relatively minor effects of inactivity, although over many years, inactivity alone can be lethal to a muscle. In immobilized muscles, the diameters of both major histochemical fiber types are diminished (Eisen et al., 1973). In denervated muscles, fibers with strong ATPase activity may start to reduce their diameter more rapidly than weak ATPase fibers (Guth et al., 1971). Denervated muscle fibers appear to contain abnormally large numbers of nuclei, although how much of this appearance is the result of mitosis and how much is an illusion caused by the shrinking medium between the nuclei is difficult to assess. In denervated muscles, the various proteins of the muscle fiber are all lost at about the same rate (Kohn, 1964). Mitochondrial volume is reduced, both in relative and absolute terms, but the sarcotubular system first increases then decreases in absolute volume (Stonnington and Engel, 1973).

In experimental situations, the results of denervation do not always follow a simple pattern. Mammalian half-diaphragms and avian slow muscles that have been denervated may become enlarged, perhaps as a result of passive stretching caused by remaining muscles (Miledi and Slater, 1969; Hikida and Bock, 1972). In the diaphragm of the rat, denervation leads to the accumulation of ribosomes and rough endoplasmic reticulum (Gauthier and Schaeffer, 1974).

Once a developing muscle fiber reaches a state of functional competence, the postsynaptic area of its membrane is far more sensitive to acetylcholine than the area of membrane beyond the neuromuscular junction. In denervated muscle fibers, sensitivity to acetylcholine gradually spreads from this focal point so that the fiber becomes *hypersensitive*. Hypersensitive fibers may even respond to the low levels of acetylcholine which are released into the cellular environment by parasympathetic or sympathetic nerves. Hypersensitive fibers respond to acetylcholine with a relatively slow contracture which is mediated by the release of calcium ions from the sarcoplasmic reticulum (Lullmann and Sunano, 1973).

Acetylcholine sensitivity is caused by a large number of discrete acetylcholine receptors. These can be individually labeled and counted because they bind *alpha-bungarotoxin* labeled with radioactive iodide. Bungarotoxin is a type of snake venom that binds to acetylcholine receptors to cause paralysis of the snake's prey. Acetylcholine receptors are found inside myoblasts prior to cell fusion (Teng and Fiszman, 1976). After cell fusion, the receptors appear on the surface membrane of the myoblast, and about 90 receptors are added each hour to each square micron of membrane (Hartzell and Fambrough, 1973). At first, acetylcholine receptors are widely distributed over the whole muscle fiber membrane, with slight accumulations over muscle fiber nuclei (Fischbach and Cohen, 1973). As the motor end plate develops, however, receptor density becomes greatly concentrated in the postsynaptic region of the neuromuscular junction. An increase in the chemosensitive area also occurs when the release of acetylcholine from nerve terminals is blocked by *botulinum toxin* (Thesleff, 1960).

Although the initial cellular differentiation of muscle fibers can be accomplished in the absence of motor neurons, muscle fibers do not survive long unless they are innervated (Zelena, 1962). Ultrastructural and histochemical aspects of the physiological differentiation of muscle fibers are lost if developing muscles are denervated (Shafiq et al., 1972; Hanzlikova and Schiaffino, 1973).

Crossed Reinnervation

The concept of neurotrophism was initially founded on a knowledge of the degenerative changes that occur in denervated muscles. In 1960, Buller et al. (1960a, 1960b) reported some experiments of a more positive nature in which it was shown that the speed of muscle contraction was neurally regulated. Transected axons are capable of growing back to reinnervate their motor units. In crossed reinnervation experiments, a nerve to a fast muscle and a nearby nerve to a slow muscle are both cut: their connections are then experimentally reversed so that the fast muscle's original nerve now reinnervates the slow muscle, and vice versa. The muscle that was originally fast now becomes slow. Depending on the species of animal, the muscle that was once slow becomes fully, or almost as fast as the original fast muscle. The nerves that normally innervate fast muscles might differ from those that normally innervate slow muscles in two general ways: (1) in the physical or chemical nature of the trophic factor which they deliver; and (2) in the frequency profile of the impulses which they normally convey. These two possibilities are as difficult to disentangle as the effects of denervation and disuse. There is independent evidence for both, but they might also be interdependent; for example, impulse frequency might influence the rate of delivery of trophic substance. The experimental tactics for the investigation of this problem are discussed by Guth (1969). Similar results to those of crossed reinnervation experiments may be obtained by the transplantation of muscles, so that a fast muscle regenerates in the original location of a slow muscle and vice versa (Gutmann and Carlson, 1975).

The physiological changes that are produced by experimental crossed reinnervation involve changes in gene expression as well as changes in the types of proteins synthesized to replace those lost by normal turnover (Samaha et al., 1970a). This may result in the transformation of muscle fibers from one histochemical type to another. Guth et al. (1970) classified fibers into a type with acid-labile ATPase and a type with alkali-labile ATPase. A third category was created for fibers with an intermediate pH lability. Acid-labile fibers could be converted to alkali-labile fibers, and vice versa. Intermediate fibers could be converted to either acid-labile or alkali-labile fibers. However, neither acid-labile nor alkali-labile fibers could be converted to the intermediate category. A certain number of fibers retain their original properties, despite contradictory reinnervation (Yellin, 1975). The completeness of histochemical conversion is enhanced if injured muscles are cross-reinnervated while they are regenerating (Riley, 1974). The extent of histochemical conversion in cross-reinnervated fibers is matched by changes in the overall contractile properties of a muscle (Crockett and Edgerton, 1975).

Trophic Substances

At present, the exact nature of the trophic substance or substances which are released by nerve terminals to regulate gene expression in muscle fibers is unknown. Trophic factors capable of influencing muscle development and growth are not limited to the motor neurons that innervate skeletal muscles. *Preganglionic* axons of the *sympathetic nervous system* can be induced to reinnervate skeletal muscle fibers, and their trophic action then resembles that of a natural nerve to a slow muscle (Ramirez and Luco, 1973). There is some evidence that trophic substances from nerves need not necessarily be released at a neuromuscular junction in order to achieve their effect. Lentz (1972) showed that motor end plate structure and cholinesterase activity could be sustained in a cultured muscle by nerve explants which did not directly innervate the muscle fibers. Fex and Sonesson (1970) thought that the grafting of a fast nerve into a normally innervated slow muscle might lead to changes in trophic control without any new motor end plates being formed. However, Crockett and Edgerton (1974) were unable to confirm the existence of such an effect. Supporting evidence for remote action by trophic substances comes from the fact that nervous tissue is not alone in producing trophic substances that act on muscle.

Generalized or systemic trophic factors which act on muscle can be measured with a *bioassay* using myoblast proliferation (Ozawa and Kohama, 1978). In chickens, the levels of systemic trophic factors that act on muscle are high in the serum but are low in the spinal cord. In embryos, high trophic levels accompany myogenic activity, with a peak at 12 days incubation and another peak just after hatching (Kohama and Ozawa, 1978). Studies on the incorporation of radioactive *thymidine* (Jabaily and Singer, 1978) show that substances in brain and liver tissue have a strong trophic effect on muscle. Trophic extracts from these tissues are heat and trypsin labile, and have high (>5000)- and low (<5000)-molecular-weight components which act together.

The background levels of the systemic trophic factors that stimulate muscle growth decline once the early formative stages of muscle development are complete. The level of trophic factors emanating from nervous tissue, however, remains high. Thus the control of muscle differentiation and growth is monopolized by the nervous system. Muscles can develop without innervation in embryos and fetuses because of the high background level of systemic trophic factors. Later, however, they become totally dependent on their innervation once the background trophic levels have declined. Embryonic chick brain extracts have a high (>300,000)-molecular-weight component which stimulates morphological development, acetylcholinesterase activity, and protein synthesis in muscle (Oh, 1975). Adult chicken sciatic nerve extract has a component with a molecular weight between 10,000 and 50,000 which has a comparable effect (Oh, 1976). Trophic protein from the sciatic nerves of chickens causes an increase in the uptake of labeled leucine by myotubes in vitro (Markelonis et al., 1980). The existence of this trophic factor from nerve tissue was confirmed by Popiela (1978). Pituitary and brain tissues from cattle contain a polypeptide which stimulates mitosis and delays fusion in myoblasts (Gospodarowicz et al., 1976). *Mast cells* may also secrete substances which facilitate mitosis in surrounding tissue cells (Franzen and Norrby, 1980).

Axoplasmic Flow

Trophic factors appear to be released from nerve terminals, often at a great distance from the perikaryon. Where are trophic substances produced in the neuron? Motor axons actively grow toward their muscles when neuromuscular relationships are first established. This is achieved by a great increase in the volume of axoplasm. The source of new axoplasm in the perikaryon is sufficient to allow repeated regeneration of transected axons, no matter how often they are cut back. If axons are constricted by a tight external ring or "bottleneck," the movement of axoplasm down the axon causes a bulge to develop proximal to the constriction. Distally, the axon becomes narrower due to a reduction of the axoplasmic flow (Weiss, 1955).

Axoplasmic flow in intact mature axons is currently regarded as a bidirectional process with both slow (1 mm/day) and fast (400 mm/day) streams (Gutmann, 1976). Slow transport probably accounts for the growth and regeneration of axons since the amount of material conveyed by slow transport greatly exceeds that carried by fast transport (Bradley and Jaros, 1973). Fast transport is channeled in some way by microtubules or vesicles in the axon, and is sensitive to oxygen and temperature levels. Fast axoplasmic transport can even be demonstrated in axoplasm that has been experimentally extruded from an axon (Brady et al., 1982). *Colchicine*, a substance that disrupts microtubules, interferes with axoplasmic flow and causes muscles to exhibit some of the features that normally accompany denervation (Hofmann and Peacock, 1973). With the use of *vinblastine* to disrupt axoplasmic flow, some symptoms of denervation can be induced, together with changes in muscle fiber enzyme levels (Kauffman et al., 1976). Slow transport carries mostly high-molecular-weight soluble proteins, whereas fast transport carries particles and a small amount of soluble protein (Sabri and Ochs, 1972). The speed of fast but not of slow axonal transport is increased by muscle *ischemia* or by administration of *5-hydroxytryptamine* (Wood and Boegman, 1975). Two additional streams of axoplasmic transport with speeds intermediate between those of the fast and slow streams have also been reported (Korr and Appeltauer, 1974). The evidence that at least some trophic substances are carried by the axoplasmic flow is quite convincing. For example, cholinesterase activity on the muscle fiber is regulated by a trophic substance from the axoplasmic flow (Younkin et al., 1978).

Activity Patterns

Three possibilities for the long-term regulation of muscle fibers by their neurons are: (1) that regulation is somehow related to the transmission of acetylcholine; (2) that regulation is due to transmission of a substance which is not acetylcholine; and (3) that regulation is not due to the transmission of a substance but to the pattern of muscle fiber activity. In controversies involving possibility 1 versus possibility 2, the main problem is the difficulty of establishing all the effects of any biologically active substances that are used in an experiment. For example, it is generally agreed that *botulinum toxin* prevents the release of acetylcholine, but experiments using this toxin are open to two interpretations, either to dwell on (Bray and Harris, 1975) or to pass over (Drachman and Johnston, 1975) the secondary effect of botulinum toxin on the release of substances carried by fast axoplasmic transport.

Experiments that attempt to deal solely with activity patterns are fraught with their own problems. One experimental approach is to deprive a group of motor neurons of their excitatory input by cutting the spinal cord between them and the brain, and by cutting dorsal roots so as to prevent any sensory input. The resulting disuse atrophy of muscle can be prevented by indirect electrical stimulation. Stimulation is most effective if the muscle can contract against a load: stimulation is least effective if the muscle's tendons are cut (Eccles, 1944). Riley and Allin (1973) found that, when isolated neurons and their motor units are stimulated with a slow (10-Hz) train of impulses, aerobic enzyme activity is increased and anaerobic enzyme activity is decreased. Rapid (50-Hz) impulse trains create the opposite response. However, apart from a few fibers with intermediate characteristics, ATPase activity is unchanged by these treatments. When the same experiment is undertaken in young animals in which physiological differentiation of muscle fibers is still being decided, the results are more complex. Gallego et al. (1978) found that neuronal isolation interferes with slow muscle development but that muscles also exert a retrograde influence on motor neuron development.

Electrical stimulation of either denervated or normally innervated muscles has also been used as a technique to study activity patterns. Denervated slow muscles retain their slow contraction speed when they are kept active by slow (10-Hz) trains of stimuli, but fast (100-Hz) trains cause slow muscles to acquire some of the physiological and histochemical properties of fast muscles (Lomo et al., 1974). When normally innervated fast muscles are exercised more frequently by means of an implanted electronic stimulator, they acquire structural characteristics, such as thick

Z lines and extra mitochondria, that are typical of slow muscles (Salmons et al., 1978). In an attempt to find a functional link between muscle activity and fiber hypertrophy in skeletal muscles, it has been suggested that the synthesis of myofibrillar proteins is enhanced by high levels of free *creatine* in active muscles. In adult muscle, however, the evidence of this is negligible (Hofmann et al., 1978).

A complicated set of responses is elicited when muscle activity patterns are modified by *tenotomy* (severing tendons) or by immobilization of a muscle (joint fixation or plaster cast). Tenotomy of a slow muscle may cause the development of fast muscle characteristics (Hikida, 1972). Immobilization of a muscle at a short length leads to a structural decrease in muscle length known as *myostatic contracture* (Ranson and Sams, 1928) and ultimately to degeneration of muscle fibers (Cooper, 1972), depending on whether the muscle is immobilized at a long or a short length (Ralston et al., 1952). Structural changes in the length of immobilized muscles involve changes in the number and rate of formation of sarcomeres (Williams and Goldspink, 1973; Muhl and Grimm, 1975). Structural shortening of sarcomeres in immobilized muscles may cause myofibrillar degeneration (Shear, 1978).

Motor Unit Volume

In animal agriculture, our primary interest is in understanding the factors that control muscle mass. However, now that we know that the development of muscle and nerve are intertwined, three volumetric compartments in the neuromuscular system can be envisioned:

1. perikaryon volume;
2. axonal volume;
3. muscle fiber volume.

If long-term regulation of muscle fiber metabolism is due to a trophic substance which is synthesized in compartment 1, travels through compartment 2, and arrives in compartment 3, what will happen to trophic regulation if compartment 3 is expanded by genetic and nutritional manipulation? Steinhauf et al. (1976) considered it to be a well-known fact that muscularity and meat quality are often inversely correlated. Is such a relationship caused by a change in the long-term trophic control of muscle metabolism by neurons?

Axonal volume is larger than might be thought at first sight. Not only are some axons very long, but all motor axons undergo extensive branching. In one case studied, the cross-sectional area of an axon before branching was approximately 60 μm^2 but the total cross-sectional area of all its branches amounted to 64,000 μm^2 (Zenker and Hohberg, 1973). Axonal diameter in mice may increase after exercise training on a treadmill (Samorajski and Rolsten, 1975).

Henneman and Olson (1965) proposed that motor neuron size is related to the physiological differentiation of muscle fibers because of a functional relationship between the size and the excitability of neurons. Neurons with a large perikaryon and a large diameter axon are less easily excited than small neurons. Thus motor units innervated by large neurons are used less frequently. This relationship may also have a retrograde component in which muscle activity may affect the size of motor neurons. In rat soleus muscles which have been overloaded by the denervation of synergistic muscles, the mean diameters of some motor axons are increased (Edds, 1950). Similarly, if some of the neurons in a group of motor neurons are destroyed, the survivors may increase their volume as they take over the extra muscle volume to be innervated (Cavanaugh, 1951; Stefanelli, 1950). Conversely, when motor unit volume is experimentally decreased, there is an increase in the release of acetylcholine at the remaining neuromuscular junctions (Herrera and Grinnell, 1980).

For nerves to muscles with mixed fiber populations, frequency histograms for axon diameters are usually bimodal with peaks at 8 and 15 μm; both large and small peaks contain *afferent* (sensory) and *efferent* (motor) axons (Young, 1950). Although nerves to red muscles with a postural function may have a unimodal distribution of small diameter axons, it is important to look at muscle position in the carcass. Axons to muscles located proximally in a limb may be slightly larger than axons to distal muscles. The axons of neurons that originate anteriorly in the spinal cord may have slightly larger diameters than those that originate posteriorly. Axon diameter shows no relationship to motor unit size, either numerically or volumetrically. Thus Young (1950) considered that there was no correlation between axon diameter and the type of physiological differentiation of fibers in a motor unit. This statement needs to be reevaluated using modern histochemical methods, but the problem of muscle position cannot be ignored.

ORIGIN AND TRANSFORMATION OF FIBER TYPES

Histochemical fiber types are important in meat animals because they influence the distribution of myoglobin and meat color. Histochemical fiber types also react differently during the conversion of muscles to meat, because they contain different levels of glycogen and anaerobic enzymes. Before it became known that fibers could change from one type to another, growth-related changes in fiber types were not adequately controlled in agricultural experiments on muscle fiber histochemistry. In experimental animals such as mice, differences in fiber-type ratios exist between sexes and between genetic lines (Vaughan et al., 1974). Similar variation may or may not exist in meat animals and poultry, but its nonexistence cannot simply be assumed uncritically.

The transformation of histochemical fiber types is quite important in meat animals, and it is worth taking a broad look at the situation in other animals. Many of the experiments undertaken on laboratory animals would be difficult to repeat in large meat animals. An important point to bear in mind, however, is that all the muscle fibers of meat animals are probably formed by the time of birth. Some fibers may be older than others, but relative differences in age are diminished as all the fibers grow older. This situation is in sharp contrast to the situation in lower vertebrates such as fish, where new muscle fibers are added after hatching. Thus, in white muscles, the red fibers may in fact be juvenile white fibers (Weatherley and Gill, 1981).

Comparative Survey

The central problem concerning the origin of histochemical fiber types is the relationship of primary myotubes and secondary fibers (defined by size, position, morphology, and time of formation) to histochemical fibers types (defined by ATPase activity and the ratio of aerobic to anaerobic enzymes). In human muscle, this relationship was first explored by Fenichel (1963). Primary myotubes and secondary fibers were categorized eponymously as *Wohlfart B and A fibers*, respectively. All primary myotubes were found to have weak ATPase activity, whereas secondary fibers had either weak or strong ATPase activity.

In newborn babies, 15% to 20% of muscle fibers do not yet exhibit histochemical differentiation (Colling-Saltin, 1978). In comparisons between species, however, the degree of fiber-type differentiation at birth must be considered in the context of the degree of relative maturity at birth; species whose young are born in a relatively immature state tend to be less advanced in the physiological differentiation of their muscle fibers (Dubowitz, 1963). Fiber-type ratios in adult human muscles can be changed by different types of exercise training (Jansson et al., 1978), as in other primates (Edgerton et al., 1972).

Dubowitz (1963) considered that the muscle fibers of rats were not histochemically differentiated at birth. Differentiation was thought to occur between 7 and 10 days after birth, with some variation between different muscles. A nutritional effect was suspected since the more mature neonates of smaller litters showed earlier differentiation. Howells and Jordan (1978) later found that perinatal undernutrition resulted in a long-lasting decrease in succinate dehydrogenase in fast muscle but not in slow muscle. Vitamin B_1 deficiency in rats also strikes preferentially at muscle fibers with strong ATPase and strong aerobic enzyme activity (Alvarado-Mallart and Przybyslawski, 1976). The physiological development of fast-contracting muscles in rats was examined by Drachman and Johnston (1973). Contraction speed is slow at birth and reaches adult speed at 21 days. Another finding that seemed to confirm the work of Dubowitz (1963) was that marked differences in muscle electrolytes occur in rats between 3 and 14 days after birth (Bergstrom et al., 1971).

Improvements in histochemical techniques called for a revision of some of these initial conclusions. Kelly and Schotland (1972) superseded the early work on rats by demonstrating histochemical differentiation in fetal muscles, and Fenichel's (1963) work on human muscle was confirmed with the discovery that primary myotubes had weak ATPase reactions while secondary fetal fibers developed either weak or strong reactions, sometimes switching from strong to weak reaction intensity. Kelly and Schotland (1972) suggested that secondary fibers which developed weak ATPase might share the innervation of primary myotubes, thus explaining why weak ATPase fibers tend to be located more centrally within their fasciculi. The basis for this suggestion comes from the fact that new muscle fibers are rapidly innervated by generations of new neurons whose axons invade the musculature. Primary fibers are located axially in their fasciculi, while older secondary fibers are pushed to the periphery by the development of even younger secondary fibers. Arguments have been presented for (Swatland and Cassens, 1973) and against (Beerman et al., 1978) the application of this idea to the even more sharply defined concentric arrangement of fiber types in pork muscles.

In a combined histochemical and physiological study of soleus muscles of rats from 5 to 34 weeks after birth, Kugelberg (1976) found a decrease (from 33% to 10%) in the percentage of Type II fibers and an increase in Type I fibers. Transitional stages between strong and weak ATPase activity were initiated by an increase in acid-stable ATPase, and were followed by a decrease in formaldehyde-stable and alkali-stable

ATPase. Three types of myosin can be identified by immunohistochemistry in rat muscles (Pierobon-Bormioli et al., 1981), and these correspond to the three main fiber types: Type I (slow twitch), Type IIA (fast red), and Type IIB (fast white). However, double reactivity occurs in some fibers so that some fibers appear with both Type I plus Type IIA reactivity, while others appear with both Type IIA plus Type IIB reactivity. In the face of evidence such as this, it is difficult to regard ATPase activity as a genetically fixed and stable feature of muscle fibers.

The effects of different types of exercise training have been examined in rats. *Treadmill* running causes a decrease in biochemically determined ATPase activity in fast-contracting aerobic muscle, but not in fast-contracting anaerobic muscle (Baldwin et al., 1975). *Lipoprotein lipase* activity and, presumably, the ability to take up and oxidize fatty acids is increased in all three major fiber types after treadmill training (Borensztajn et al., 1975). Muscle training on a treadmill causes aerobic muscles to become more like cardiac muscle in their enzymatic properties (Baldwin et al., 1973). Muscle training by endurance swimming in young but not mature rats may cause an increase in the percentage of strong ATPase fibers in slow-contracting muscles but not in fast-contracting muscles (Syrovy et al., 1972). Short-term sprint training causes the slow soleus muscle of the rat to acquire some of the properties of a fast muscle (Staudte et al., 1973). Anabolic derivatives of *testosterone* enhance the muscle response of female rats to *isometric* exercise training, but no effect is found in male rats (Exner et al., 1973a, 1973b). Isometric training by itself may cause an increase in the percentage of fibers with strong ATPase and strong anaerobic enzyme activity. This is due to a loss of fibers with strong ATPase and strong aerobic activity; the percentage of fibers with weak ATPase activity is unchanged (Muller, 1975).

Muscle *hypertrophy* or enlargement can be due to a number of different causes. The enlarged muscles of meat animals represent only one particular type of muscle enlargement. Nevertheless, the phenomenon of muscle enlargement in laboratory animals is of general interest if we wish to understand some of the limitations to muscularity that might exist in meat animals. Genetic selection for postweaning gain in rats leads to larger adults with larger muscles. The greater part of the extra muscle mass originates from the enlargement of fibers which are specialized for anaerobic metabolism rather than for aerobic metabolism (Swatland and Cassens, 1972b).

Compensatory hypertrophy is a common experimental model for muscle enlargement in laboratory animals. Compensatory hypertrophy may be induced by removing or inactivating certain muscles of the body, so as to overwork those that remain. Compensatory hypertrophy may involve neurotrophism, as well as myogenic factors such as stretch-induced changes in ionic metabolism (Gutmann et al., 1971; Hofmann, 1980). In fast muscles, compensatory hypertrophy differs from exercise-induced muscle enlargement; the former is associated with increased mitochondrial volume while the latter is associated with an increased myofibrillar mass (Seiden, 1976). Compensatory hypertrophy causes an increase in aerobic metabolism (Carlo et al., 1975), and muscle fiber enlargement is most marked in fibers with weak ATPase and strong aerobic enzyme activity (Yellin, 1974). In slow muscles which are undergoing hypertrophy, initial increases in sarcoplasmic proteins are followed by an increase in myofibrillar proteins (Mackova and Hnik, 1972). Perhaps the initial sarcoplasmic phase of enlargement in compensatory hypertrophy should not be regarded as true hypertrophy (Mackova and Hnik, 1973). Animal age has a marked effect on the degree of histochemical transformation which occurs during compensatory hypertrophy. In neonatal rats, transformation may lead to the disappearance of anaerobic fibers and to an increase in the percentage of fibers with weak ATPase. One or two months after birth, however, histochemical transformation is negligible (Schiaffino and Bormioli, 1973b).

In kittens, the physiological differentiation of muscle fibers appears after birth. The slow-contracting muscles of adult cats retain the slowness that is found in newborn kittens, while other muscles may increase their speed of contraction and reach adult fast muscle speeds by 6 or 7 weeks after birth (Hammarberg and Kellerth, 1975a). A peculiarity of the development of slow muscles, in cats and in other species, is that contraction speeds begin to get faster after birth, but then they slow down again. Thus a few weeks after birth, the contraction speed of slow muscles may be slightly faster than at any other time during the animal's life. This transient increase in contraction speed does not involve any other aspects of physiological differentiation such as half-relaxation time, susceptibility to fatigue, and duration of after-hyperpolarization (Hammarberg and Kellerth, 1975c).

In newborn kittens, the differentiation of phosphorylase, glycogen, lipids, and aerobic enzymes appears after the differentiation of ATPase, and is complete by 6 to 7 weeks after birth (Nystrom, 1968a). In adult cats, most motor units can be categorized as either (1) fast contracting and easily fatigued; (2) fast contracting and fatigue resistant; or (3) slow contracting and fatigue resistant (Hammarberg and Kellerth, 1975b). Within each fiber type, there is a considerable range in ultrastructure (Galvas et al., 1982).

In cats, muscle training by weight lifting causes an increase in the percentage of fibers with strong ATPase and strong anaerobic enzyme activity at the expense of fibers with strong ATPase and strong aerobic enzyme activity. The percentage of fibers with weak ATPase and strong aerobic enzyme activity is unchanged (Gonyea and Bonde-Petersen, 1978). In the slow soleus muscles of kittens, the percentage of fibers with strong ATPase activity decreases during normal development and is accelerated by compensatory hypertrophy (Tomanek, 1975; Wetzel et al., 1973). In adult cats, compensatory hypertrophy produces no change in fiber-type ratios (Walsh et al., 1978).

Wirsen and Larsson (1964) used differences in phosphorylase activity to identify primary and secondary types of muscle fibers in mice. Primary fibers exhibited a strong phosphorylase reaction, whereas secondary fibers did not. This condition is unlike that found in most other species. A possible explanation is that intrinsic glycogen is needed as a primer for amylose synthesis in the histochemical reaction for phosphorylase, and primary fibers might have been the first to acquire a substantial number of glycogen granules (Platzer, 1978). In mice, the histochemical differentiation of muscle fibers is not complete until 3 weeks after birth. During this period, mitochondria increase in number. After reaching a body weight of 20 to 25 g, however, no further increase takes place in succinate dehydrogenase activity (Goldspink, 1962). Thus adult fibers with a large diameter are considered to have diluted their fixed mitochondrial content in an expanded myofibrillar mass (Goldspink, 1969). In mice, compensatory hypertrophy causes an increase in myofibrillar mass (Rowe, 1969) and an enlargement of some, but not all muscle fibers (Rowe and Goldspink, 1968).

At birth, the muscle fibers of rabbits exhibit some histochemical differentiation of their ATPase activity. Small (probably secondary) fibers react strongly for ATPase, while large (probably primary) fibers vary from strong to moderate ATPase activity (Guth and Samaha, 1972). A major problem in young rabbits, as well as in the neonates of other mammals, is that histochemical reactions which are reliable indicators of biochemical ATPase activity in adults may be misleading in very young animals. Thus Guth and Samaha (1972) found that the histochemical reaction for ATPase overestimated the biochemical activity of ATPase so that the histochemical differentiation of fast fibers appeared to precede their functional or biochemical differentiation. The explanation of this paradox may be that rabbits (and probably most other species) have three forms of myosin: an embryonic type, a fast adult type, and a slow adult type. Biochemically, the embryonic type is almost as fast as the adult fast type, and it is also stable at a high pH (Sreter et al., 1975).

In newborn rabbits, red and white muscles receive a similar *blood flow*. By 3 to 4 weeks, the flow to white muscle has dropped to its low adult value. This aspect of physiological differentiation between muscles occurs in an anterior-to-posterior sequence (Wooten and Reis, 1972). Rabbit fiber types are transformed by crossed reinnervation (Jobsis and Meijer, 1973a, 1973b) and by changes in activity patterns (Pette et al., 1976), as in other species.

In some rodents, the temporal muscles of the jaw are larger in males than in females. This sex difference has been used to study the effect of *androgens* on muscle development (Gutmann and Hanzlikova, 1970). In males, temporal muscles are reduced in size when the animals are castrated. The administration of testosterone to castrated males restores the size of their temporal muscles. The treatment of females with androgens causes their muscles to switch from an aerobic to an anaerobic type of metabolism. This might be a direct effect of androgens on muscle RNA, but it might also be an indirect effect caused by differences in the degree of muscle activity between males and females. Maxwell et al. (1973) found that age-related transformations of muscle fiber histochemistry occurred at the same time as changes in *apparent numbers* of muscle fibers (fiber number seen in transverse section). This raises a difficult problem which we have not yet solved in meat animals. Do muscle fibers really change from one histochemical type to another during normal development, or do the percentages of fiber types change because of fiber losses or rearrangement?

The geographical range of meat animals extends from hot to cold climates, and little work has yet been undertaken on the long-term regulation of muscle physiology in relation to climate. Thus it is interesting to note that the muscles of awake and hibernating hamsters maintain their physiological and histochemical differentiation regardless of temperature (Vyskocil and Gutmann, 1977).

Pigs

The postnatal physiological differentiation of the fast anterior tibialis muscle in pigs was examined by Campion et al. (1973). The muscle is slow at birth, it reaches its fastest contraction time by 14 days, and then it slows down slightly, as in other species. Differences in glycolytic enzyme activity between the red trapezius muscle and the white longissimus dorsi muscle occur on a comparable time scale. An adult degree of differentiation is developed by 2 weeks after birth (Dalrymple et al., 1974). Some enzymes, however, are slightly slower in their rate of differentiation between red and white muscle. *Glutamate oxaloacetate transaminase* takes approximately 5 weeks, while glycogen phosphorylase reaches a maximum level of differentiation at 8 weeks. After 8 weeks, differences between red and white muscles may decline (Cooper et al., 1971). Both the red trapezius muscle and the white semimembranosus muscle contain high levels of glycogen at birth. After birth, glycogen levels decline rapidly to reach an adult level of intermuscular differentiation within a few weeks (Dalrymple et al., 1973).

The differences in red coloration between various pork muscles are related to the incidence of aerobic and anaerobic muscle fibers. An unusual feature of most pork muscles is that the tendency for aerobic fibers to be located centrally in their fasciculi is more extreme than in any other species yet identified (Plate 17). Thus the concentric arrangement of primary myotubes and secondary fetal fibers is preserved after birth. The reason it is well preserved in pigs, yet becomes muddled in other species, is unknown. In the longissimus dorsi muscle, fiber-type differentiation on the basis of aerobic enzyme activity is only slightly developed at birth, but becomes well developed by 2 weeks (Cassens et al., 1968). With reactions for phosphorylase and ATPase, Cooper et al. (1970) found an indistinct histochemical differentiation between fiber types at birth. Ashmore et al. (1973) had more success with the ATPase reaction, and were able to show that primary myotubes had weak ATPase while secondary fibers had strong ATPase, following the pattern first identified by Fenichel (1963) in human muscle. Using fluorescence microscopy, Beermann and Cassens (1977) found that secondary fibers in fetal pig muscle have a greater concentration of RNA than primary fibers.

The percentage of white fibers in pork muscles differs between breeds and is related to the extent to which the meat yield of a breed has been improved by selective breeding. The muscles of wild pigs are dominated by red fibers, whereas those of the most improved breeds are dominated by white fibers with a large diameter (Rahelic and Puac, 1981).

Ontogenetic transformations of muscle fiber histochemistry can be found soon after birth in pigs (Figure 6-5). In the sartorius muscle, there is a marked increase in the percentage of fibers with weak ATPase (with an acid-stable ATPase reaction) during the 10 days after birth (Swatland, 1975a). Neonatal increases in the percentages of fibers with weak ATPase activity (alkali-labile, acid-stable ATPase) have been found in masseter, trapezius, longissimus dorsi, rectus femoris, and vastus intermedius muscles (Suzuki and Cassens, 1980b). During growth to market weight, the longissimus dorsi muscle exhibits an increase in the percentage of fibers with weak aerobic enzyme activity (Van Den Hende et al., 1972), as well as an increase in the percentage of fibers with weak ATPase activity (Davies, 1972). The decrease in the percentage of fibers with strong aerobic activity (Figure 6-12) occurs in the fiber population that has strong ATPase activity (Figure 6-13; Swatland, 1977a). A

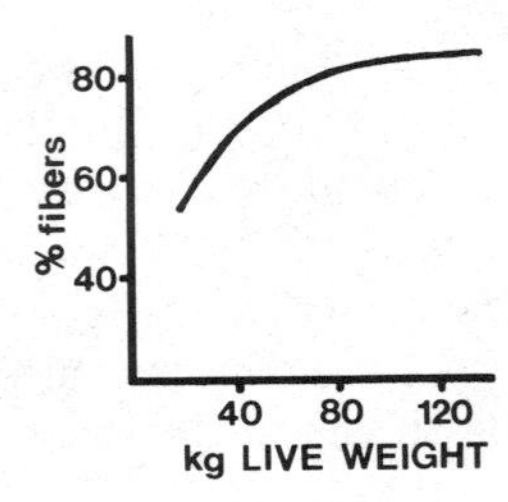

Figure 6-12 Change in the percentage of anaerobic muscle fibers in the longissimus dorsi muscles of pigs at different live weights. (Data from Swatland, 1975c.)

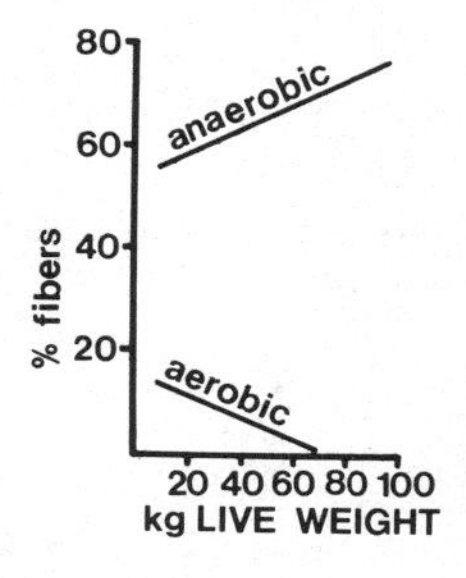

Figure 6-13 Transformation to anaerobic metabolism in muscle fibers with strong ATPase activity in the longissimus dorsi muscles of pigs at different live weights. (From Swatland, 1977a.)

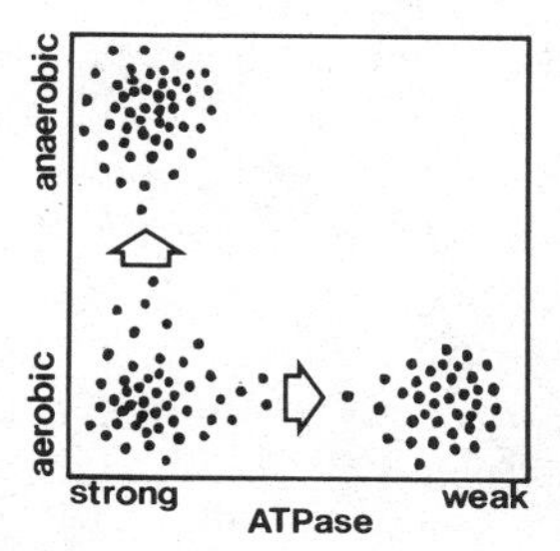

Figure 6-14 Scattergram to demonstrate the dominant postnatal transformations in the physiological differentiation of muscle fibers in the porcine longissimus dorsi. Each dot represents a muscle fiber which is described by its x,y position.

common pool of fibers with strong ATPase activity and strong aerobic enzyme activity provides the source of fibers that switch to weak aerobic activity or to weak ATPase activity (Figure 6-14; Swatland, 1977b). In pigs, as well as in cattle, fiber transformations are sometimes found in the reverse direction (from weak to strong ATPase). In pigs, this has been detected in the vastus medialis, where the percentage of fibers with strong ATPase may increase with animal live weight (Swatland, 1978).

Sheep and Goats

Secondary fetal fibers are formed between 60 and 100 days gestation in fetal lambs (Ashmore et al., 1972). By 70 days, primary myotubes exhibit weak ATPase activity, whereas secondary fibers exhibit strong ATPase activity. From birth to 5 years of age, sheep show a decrease in their percentage of fibers with strong ATPase activity (White et al., 1978; Sivachelvan and Davies, 1981). Using a histochemical reaction for succinate dehydrogenase, Beckett and Bourne (1960) were able to identify histochemical differences between the muscle fibers of fetal goats at 130 days gestation. The presence of primary myotubes and secondary fibers in fetal lambs was first reported by Quain (1856).

Cattle

Ommer (1971) found that the muscles of fetal calves were composed of small-diameter fibers (probably secondary fibers) located around large-diameter fibers (probably primary myotubes). After 200 days gestation, fibers exhibited differentiation on the basis of their succinate dehydrogenase activity. Ultrastructural studies have confirmed the existence of primary and secondary muscle fibers in fetal calves, and have shown that ultrastructural differentiation into fiber types is present by 8 months gestation (Russell and Oteruelo, 1981). Suzuki et al. (1976) found that a high plane of nutrition in steers grown to market weight caused fibers to switch from type A (strong ATPase and strong aerobic enzyme activity) to type B (strong ATPase and weak aerobic enzyme activity). In the longissimus dorsi and semimembranosus, Johnston et al. (1981) found that some alpha white fibers were replaced by beta red fibers. The opposite transformation was found in the biceps femoris and semitendinosus. Steers tend to have a higher percentage of beta red fibers and a lower percentage of alpha white fibers than do heifers (Johnston et al., 1981). Cows may lose fast-contracting white fibers if the plane of nutrition is inadequate during pregnancy (Reid et al., 1980).

Chickens

When chicks are hatched, their fast and slow muscles already exhibit physiological differentiation with regard to contraction speed (Melichna et al., 1974). The primary myotubes and secondary fibers that are formed in the embryo (Kikuchi, 1971) give rise, respectively, to fibers with weak ATPase and to fibers with strong ATPase activity (Ashmore and Doerr, 1971). The conversion of aerobic to anaerobic fibers in the pectoralis proceeds rapidly for about 3 days after hatching, then continues more slowly until 100 days (Kiessling, 1977). Kiessling (1978) examined the histochemistry of muscle fibers in hybrid crosses between guinea hens and laying hens. In the pectoralis, the hybrid had the large-diameter fibers of a guinea hen, but also the high proportion of anaerobic fibers typical of a laying hen. The fibers in the sartorius muscle of the hybrid resembled those in the sartorius of the laying hen.

There are three types of myosin in the chicken: (1) an embryonic type; (2) a fast adult type; and (3) a slow adult type. Cultured myoblasts that have been isolated from presumptive fast, and from presumptive slow embryonic muscles continue to produce the embryonic type of myosin (Nougues, 1980). Stockdale et al. (1981), however, isolated myoblasts from a future slow muscle and found that both fast and slow myosins were present. Both fiber types seen with the ATPase reaction are aerobic at hatching, but some fibers with strong ATPase reactions later become anaerobic (Ashmore and Doerr, 1971). As in mammals, there is a transformation of some muscle fibers from strong to weak ATPase activity during later muscle development (Melichna et al., 1974). Just before and after hatching, the anterior latissimus dorsi shows a transformation of some fibers from acid-stable to alkali-stable ATPase (Toutant et al., 1980). Fiber transformations from strong to weak ATPase activity (Sola et al., 1973) and from weak to strong aerobic activity (Holly et al., 1980) may occur in response to stretch-induced muscle enlargement.

The metabolic differentiation of muscle fibers is affected by a nonneural influence from the thymus (Cosmos et al., 1977). At first sight this may seem a strange association (between skeletal muscle and the thymus gland), but the thymus gland often contains a small number of striated muscle fibers. The supression of misplaced striated muscle fibers might be a function of the immune system. Erroneous attacks on properly located striated muscles fibers, such as those that occur in the disease myasthenia gravis, have been tentatively linked with disorders of the thymus (Kao and Drachman, 1977).

SENSORY INNERVATION OF MUSCLE

Carcass muscles are well endowed with both unspecialized sensory nerve endings (Stacey, 1969) and with complex sensory organs; as yet, neither have been implicated in any aspect of muscle growth in meat animals, but it seems unwise to ignore them altogether. *Neuromuscular spindles* are complex sensory organs. They are derived from normal muscle fibers which have been innervated by sensory axons rather than by motor axons. These modified muscle fibers in neuromuscular spindles are called *intrafusal* fibers, so as to distinguish them from normal muscle fibers, which may then be called *extrafusal* fibers. There are two basic types of intrafusal fibers. *Nuclear chain* fibers are rather like myotubes since they have axial nuclei along their length. *Nuclear bag* fibers are swollen at their midlength to accommodate a group of nuclei. Large-diameter sensory axons (group Ia afferents with diameters from 12 to 20 μm) wind around the intrafusal fibers at their midlength like coiled springs. Intrafusal fibers also bear simple sensory endings derived from small sensory axons (group II afferents, from 6 to 12 μm in diameter). The myofibrils of intrafusal fibers are stimulated to contract by *gamma motor neurons*.

Neuromuscular spindles are particularly important in the maintenance of body posture and in precise muscle movements. In calves, for example, the muscles that open the jaws have no neuromuscular spindles, but the muscles that close the jaws have over 1000 spindles on each side of the head (Kubota et al., 1980). Sensory endings are activated by being stretched. Their activity is relayed to the spinal cord via its dorsal roots. Sensory neurons may then interact with the *alpha motor neurons* which innervate extrafusal muscle fibers. It was once thought that the nervous system might set and maintain muscle length by adjusting the length of intrafusal fibers, which, in turn, would exercise a servomotor control over muscle length. This is now considered to be unlikely (Matthews, 1981). The overload cutoff switch for muscle contraction is the *Golgi tendon organ* (Schoultz and Swett, 1972). When it detects a dangerously high tension within the muscle, the Golgi tendon organ inhibits alpha motor neurons.

In rat muscles, the apparent number of intrafusal fibers may continue to increase for some time after birth (Marchand and Eldred, 1969). A similar increase occurs in pigs (Swatland, 1974). The capsules of neuromuscular spindles are partly composed of *elastin* fibers (Cooper and Gladden, 1974). Intrafusal fibers sometimes extend beyond their capsules to terminate in the same way as intrafascicularly terminating extrafusal fibers (Bridgeman et al., 1969). In pigs, intrafusal fibers exhibit little or no growth in diameter after birth, and they already exhibit histochemical differentiation of their ATPase activity at birth (Swatland, 1975b).

REFERENCES

ACHAVAL, M., PIANTELLI, A., and REBELLO, M. A. 1966. *Acta Neurol. Latinoam.* 12:153.

AGNISH, N. D., and KOCHHAR, D. M. 1977. *Dev. Biol.* 56:174.

ALLEN, E. R. 1973. *Z. Zellforsch.* 145:167.

ALLEN, E. R., and PEPE, F. A. 1965. *Am. J. Anat.* 116:115.

ALVARADO-MALLART, R. M., and PRZYBYSLAWSKI, J. 1976. *Ann. Histochim.* 21:129.

ANDERSON, M. J., and COHEN, M. W. 1974. *J. Physiol.* 237:385.

ANGAUT-PETIT, D., and MALLART, A. 1979. *J. Physiol.* 289:203.

ANZENBACHER, H., and ZENKER, W. 1963. *Z. Zellforsch.* 60:860.

ARIANO, M. A., ARMSTRONG, R. B., and EDGERTON, V. R. 1973. *J. Histochem. Cytochem.* 21:51.

ARNDT, I., and PEPE, F. A. 1975. *J. Histochem. Cytochem.* 23:159.

ARNOLD, W., STEELE, R., and MUELLER, H. 1958. *Proc. Natl. Acad. Sci. USA* 44:1.

Ashmore, C. R., and Doerr, L. 1971. *Exp. Neurol.* 30:431.
Ashmore, C. R., Robinson, D. W., Rattray, P., and Doerr, L. 1972. *Exp. Neurol.* 37:241.
Ashmore, C. R., Addis, P. B., and Doerr, L. 1973. *J. Anim. Sci.* 36:1088.
Askanas, V., and Engel, W. K. 1975. *Neurology* 25:879.
Bachmann, P. 1980. *Cell Tissue Res.* 206:431.
Bach-y-Rita, P., and Ito, F. 1966. *J. Gen. Physiol.* 49:1177.
Bagust, J., Lewis, D. M., and Westerman, R. A. 1973. *J. Physiol.* 229:241.
Baldwin, K. M., Winder, W. W., Terjung, R. L., and Holloszy, J. O. 1973. *Am. J. Physiol.* 225:962.
Baldwin, K. M., Winder, W. W., and Holloszy, J. O. 1975. *Am. J. Physiol.* 229:422.
Barany, M. 1967. *J. Gen. Physiol.* 50:197.
Bardeen, C. R. 1900. *Johns Hopkins Hosp. Rep.* 9:367.
Barker, D., and Ip, M. C. 1966. *Proc. R. Soc. B* 163:538.
Barker, D., and Saito, M. 1980. *J. Physiol.* 307:16P.
Barron, D. H. 1953. Some factors regulating the form and organisation of the motoneurones of the spinal cord. In J. L. Malcolm, J. A. B. Gray, and G. E. W. Wolstenholme (eds.), *The Spinal Cord.* Ciba Symposium, Little, Brown and Company, Boston.
Bass, A., Gutmann, E., Melichna, J., and Syrovy, I. 1973. *Physiol. Bohemoslov.* 22:477.
Beatty, C. H., Basinger, G. M., Dully, C. C., and Bocek, R. M. 1966. *J. Histochem. Cytochem.* 14:590.
Beatty, C. H., Curtis, S., Young, M. K., and Bocek, R. M. 1974. *Am. J. Physiol.* 227:268.
Beckett, E. B., and Bourne, G. H. 1960. Histochemistry of developing skeletal and cardiac muscle. In G. H. Bourne (ed.), *The Structure and Function of Muscle*, Vol. 1, Chap. 4. Academic Press, New York.
Beermann, D. H., and Cassens, R. G. 1977. *J. Histochem. Cytochem.* 25:439.
Beermann, D. H., Cassens, R. G., Couch, C. C., and Nagle, F. J. 1977. *J. Neurol. Sci.* 31:207.
Beermann, D. H., Cassens, R. G., and Hausman, G. J. 1978. *J. Anim. Sci.* 46:125.
Bellairs, R., Sanders, E. J., and Portch, P. A. 1980. *J. Embryol. Exp. Morphol.* 56:41.
Bergstrom, J., Boethius, J., and Hultman, E. 1971. *Acta Physiol. Scand.* 81:164.
Billeter, R., Weber, H., Lutz, H., Howald, H., Eppenberger, H. M., and Jenny, E. 1980. *Histochemistry* 65:249.
Bird, J. W. C., Roisen, F. J., Yorke, G., Lee, J. A., McElligott, M. A., Triemer, D. F., and St. John, A. 1981. *J. Histochem. Cytochem., Suppl. 29*, 3A:431.
Bischoff, R. 1970. The myogenic stem cell in development of skeletal muscle. In A. Mauro, S. A. Shafiq, and A. T. Milhorat (eds.), *Regeneration of Striated Muscle, and Myogenesis.* Excerpta Medica, Amsterdam.
Bloom, S. E., and Buss, E. G. 1968. *Poult. Sci.* 47:837.
Bone, Q. 1972. *J. Cell Sci.* 10:657.
Bonner, P. H., and Hauschka, S. D. 1974. *Dev. Biol.* 37:317.
Borensztajn, J., Rone, M. S., Babirak, S. P., McGarr, J. A., and Oscai, L. B. 1975. *Am. J. Physiol.* 229:394.
Bowman, W. C., and Marshall, I. G. 1971. Muscle. In D. J. Bell and B. M. Freeman (eds.), *Physiology and Biochemistry of the Domestic Fowl*, Vol. 2, pp. 707–737. Academic Press, New York.
Boyd, J. D. 1960. Development of striated muscle. In G. H. Bourne (ed.), *The Structure and Function of Muscle*, Vol. 1, pp. 63–85. Academic Press. New York.
Bradley, R. 1977. *Res. Vet. Sci.* 23:250.
Bradley, W. G., and Jaros, E. 1973. *Brain* 96:247.
Brady, S. T., Lasek, R. J., and Allen, R. D. 1982. *Science* 218:1129.
Bray, J. J., and Harris, A. J. 1975. *J. Physiol.* 253:53.
Bridgeman, C. F., Shumpert, E. E., and Eldred, E. 1969. *Anat. Rec.* 164:391.

BROOKE, M. H., and KAISER, K. K. 1970. *Arch. Neurol.* 23:369.
BRUNK, C. F. 1981. *Exp. Cell Res.* 136:305.
BUCKLEY, P. A., and KONIGSBERG, I. R. 1974. *Dev. Biol.* 37:193.
BULLER, A. J., ECCLES, J. C., and ECCLES, R. M. 1960a. *J. Physiol.* 150:399.
BULLER, A. J., ECCLES, J. C., and ECCLES, R. M. 1960b. *J. Physiol.* 150:417.
BURKE, R. E., LEVINE, D. N., ZAJAC, F. E., TSAIRIS, P., and ENGEL, W. K. 1971. *Science* 174:709.
BURKE, R. E., LEVINE, D. N., SALCMAN, M., and TSAIRIS, P. 1974. *J. Physiol.* 238:503.
CALKINS, C. R., DUTSON, T. R., SMITH, G. C., CARPENTER, Z. L., and DAVIES, G. W. 1981. *J. Food Sci.* 46:708.
CAMPION, D. R., CASSENS, R. G., and NAGLE, F. J. 1973. *Growth* 37:257.
CAMPION, D. R., FOWLER, S. P., HAUSMAN, G. J., and REAGAN, J. O. 1981. *Acta Anat.* 110:277.
CANTINI, M., SARTORE, S., VITADELLO, M., and SCHIAFFINO, S. 1979. *Cell Biol. Int. Rep.* 3:151.
CAPLAN, A. I., and KOUTROUPAS, S. 1973. *J. Embryol. Exp. Morphol.* 29:571.
CARDASIS, C. A., and PADYKULA, H. A. 1981. *Anat. Rec.* 200:41.
CARDINET, G. H., FEDDE, M. R., and TUNELL, G. L. 1972. *Lab. Invest.* 27:32.
CAREY, E. J. 1922. *J. Morphol.* 37:1.
CARLO, J. W., MAX, S. R., and RIFENBERICK, D. H. 1975. *Exp. Neurol.* 48:222.
CASSENS, R. G., and COOPER, C. C. 1971. *Adv. Food Res.* 19:1.
CASSENS, R. G., COOPER, C. C., MOODY, W. G., and BRISKEY, E. J. 1968. *J. Anim. Morphol. Physiol.* 15:135.
CAVANAUGH, M. W. 1951. *J. Comp. Neurol.* 94:181.
CHEVALLIER, A. 1978. *Wilhelm Roux's Arch.* 184:57.
CHEVALLIER, A., KIENY, M., and MAUGER, A. 1978. *J. Embryol. Exp. Morphol.* 43:263.
CHI, J. C., FELLINI, S. A., and HOLTZER, H. 1975. *Proc. Natl. Acad. Sci. USA* 72:4999.
CHRIST, B., JACOB, H. J., and JACOB, M. 1977. *Anat. Embryol.* 150:171.
CIOFFI, M., SEARLS, R. L., and HILFER, S. R. 1980. *J. Embryol. Exp. Morphol.* 55:195.
CLARK, D. A. 1931. *Am. J. Physiol.* 96:296.
CLOSE, R. I. 1972. *Physiol. Rev.* 52:129.
COERS, C. 1955. *Acta Neurol. Psychiatr. Belg.* 55:741.
COERS, C., and WOOLF, A. L. 1959. *The Innervation of Muscle: A Biopsy Study*. Charles C Thomas, Publisher, Springfield, Ill.
COERS, C., TELERMAN-TOPPET, N., and GERARD, J.-M. 1973. *Arch. Neurol.* 29:210.
COLE, W. V. 1957. *J. Comp. Neurol.* 108:445.
COLLING-SALTIN, A.-S. 1978. *J. Neurol. Sci.* 39:169.
COOPER, C. C., CASSENS, R. G., KASTENSCHMIDT, L. L., and BRISKEY, E. J. 1970. *Dev. Biol.* 23:169.
COOPER, C. C., CASSENS, R. G., KASTENSCHMIDT, L. L., and BRISKEY, E. J. 1971. *Pediatr. Res.* 5:281.
COOPER, R. R. 1972. *J. Bone Joint Surg.* 54A:919.
COOPER, S., and GLADDEN, M. H. 1974. *Q. J. Exp. Physiol.* 59:367.
COSMOS, E., PEREY, D. Y. E., BUTLER, J., and ALLARD, E. P. 1977. *Differentiation* 9:139.
COUTEAUX, R. 1941. *Bull. Biol. France et Belg.* 75:101.
CROCKETT, J. L., and EDGERTON, V. R. 1974. *Exp. Neurol.* 43:207.
CROCKETT, J. L., and EDGERTON, V. R. 1975. *J. Neurol. Sci.* 25:1.
CUAJUNCO, J. 1942. *Contrib. Embryol.* 30:129.
CULL, R. E. 1974. *Exp. Brain Res.* 20:307.
CULLHEIM, S., and KELLERTH, J.-O. 1978a. *J. Physiol.* 281:285.
CULLHEIM, S., and KELLERTH, J.-O. 1978b. *J. Physiol.* 281:301.
CURTIS, D. H., and ZALIN, R. J. 1981. *Science* 214:1355.
CZEH, G., GALLEGO, R., KUDO, N., and KUNO, M. 1978. *J. Physiol.* 281:239.
DADOUNE, J. P., TERQUEM, A., and ALFONSI, M. F. 1978. *Cell Tissue Res.* 193:269

Dalrymple, R. H., Kastenschmidt, L. L., and Cassens, R. G. 1973. *Growth* 37:19.
Dalrymple, R. H., Cassens, R. G., and Kastenschmidt, L. L. 1974. *J. Cell. Physiol.* 83:251.
Daubas, P., Caput, D., Buckingham, M., and Gros, F. 1981. *Dev. Biol.* 84:133.
Davies, A. S. 1972. *J. Anat.* 113:213.
Denny-Brown, D. E. 1929. *Proc. R. Soc. B* 104:371.
Devlin, R. B., and Emerson, C. P. 1979. *Dev. Biol.* 69:202.
Dias, P. L. R. 1974. *J. Anat.* 117:453.
Doetschman, T. C., Dym, H. P., Siegel, E. J., and Heywood, S. M. 1980. *Differentiation* 16:149.
Doyle, A. M., and Mayer, R. F. 1969. *Neurology*, 19:296.
Drachman, D. B., and Johnston, D. M. 1973. *J. Physiol.* 234:29.
Drachman, D. B., and Johnston, D. M. 1975. *J. Physiol.* 252:657.
Dubowitz, V. 1963. *Nature* (*Lond.*) 197:1215.
Dubowitz, V., and Pearse, A. G. E. 1960. *Nature* (*Lond.*) 185:701.
Dupont, L., Buckingham, M. E., and Gros, F. 1979. *Biol. Cell.* 34:1.
Durham, A. C. H. 1974. *Cell* 2:123.
Eccles, J. C. 1944. *J. Physiol.* 103:253.
Edds, M. V. 1950. *J. Comp. Neurol.* 93:258.
Edds, M. V. 1953. *Q. Rev. Biol.* 28:260.
Edgerton, V. R., Barnard, R. J., Peter, J. B., Gillespie, C. A., and Simpson, D. R. 1972. *Exp. Neurol.* 37:322.
Edstrom, L., and Kugelberg, E. 1968. *J. Neurol. Neurosurg. Psychiatry* 31:424.
Eisen, A. A., Carpenter, S., Karpati, G., and Bellavance, A. 1973. *J. Neurol. Sci.* 20:457.
Eisen, A., Karpati, G., Carpenter, S., and Danon, J. 1974. *Neurology* 24:878.
Elfvin, L. G. 1968. The structure and composition of motor, sensory and autonomic nerves and nerve fibers. In G. H. Bourne (ed.), *The Structure and Function of Nervous Tissue*, Vol. 1, Chap. 9. Academic Press, New York.
Engel, W. K. 1974. *Neurology* 24:344.
Engel, W. K., and Irwin, R. L. 1967. *Am. J. Physiol.* 213:511.
England, J. M. 1970. *J. Anat.* 106:311.
Eriksson, P.-O., Eriksson, A., Ringqvist, M., and Thornell, L.-E. 1980. *Histochemistry* 65:193.
Exner, G. U., Staudte, H. W., and Pette, D. 1973a. *Pfluegers Arch.* 345:1.
Exner, G. U., Staudte, H. W., and Pette, D. 1973b. *Pfluegers Arch.* 345:15.
Fenichel, G. M. 1963. *Neurology* 13:219.
Fenichel, G. M. 1967. *Dev. Med. Child Neurol.* 9:419.
Fenichel, G. M. 1969. *Arch. Neurol.* 20:644.
Fernand, V. S. V., and Hess, A. 1969. *J. Physiol.* 200:547.
Fex, S., and Sonesson, B. 1970. *Acta Anat.* 77:1.
Fischbach, G. D., and Cohen, S. A. 1973. *Dev. Biol.* 31:147.
Fischman, D. A. 1970. *Curr. Top. Dev. Biol.* 5:235.
Fischman, D. A. 1972. Development of striated muscle. In G. H. Bourne (ed.), *The Structure and Function of Muscle*, 2nd ed., Vol. 1, Pt. 1, pp. 75–148. Academic Press, New York.
Flamm, J. 1968. *Z. Anat. Entwicklungsgesch.* 127:359.
Frank, E., Jansen, J. K. S., Lomo, T., and Westgaard, R. 1974. *Nature* (*Lond.*) 247:375.
Franzen, L., and Norrby, K. 1980. *Cell Tissue Kinet.* 13:635.
Franzini-Armstrong, C. 1973. Membranous systems in muscle fibers. In G. H. Bourne (ed.), *The Structure and Function of Muscle*, 2nd ed., Vol. 2, Pt. 2, pp. 532–619. Academic Press, New York.
Frederick, E. C., Hamant, M. F., Rasmussen, S. A., Chan, A. K., and Goslow, G. E. 1978. *Experientia* 34:372.

FRIEDLANDER, M., and FISCHMAN, D. A. 1977. *J. Supramol. Struct.* 7:323.
FULTON, A. B., PRIVES, J., FARMER, S. R., and PENMAN, S. 1981. *J. Cell Biol.* 91:103.
FURCHT, L. T., and WENDELSCHAFER-CRABB, G. 1978. *Differentiation* 12:39.
GALEGO, R., HUIZAR, P., KUDO, N., and KUNO, M. 1978. *J. Physiol.* 281:253.
GALEY, F. 1970. Ultrastructural differentiation. In O. A. Schjeide and J. de Vellis (eds.), *Cell Differentiation.* Van Nostrand Reinhold Company, New York.
GALVAS, P. E., and GONYEA, W. J. 1980. *Am. J. Anat.* 159:147.
GALVAS, P. E., NEAVES, W. B., and GONYEA, W. J. 1982. *Anat. Rec.* 203:1.
GARD, D. L., and LAZARIDES, E. 1980. *Cell* 19:263.
GARRELS, J. I. 1979. *Dev. Biol.* 73:134.
GARVEN, H. S. D. 1925. *Brain* 48:380.
GAUTHIER, G. F., and SCHAEFFER, S. F. 1974. *J. Cell Sci.* 14:113.
GEARHART, J. D., and MINTZ, B. 1972. *Dev. Biol.* 29:27.
GEORGE, J. C., and BERGER, A. J. 1966. *Avian Myology.* Academic press, New York.
GEORGE, J. C., and RONALD, K. 1973. *Can. J. Zool.* 51:833.
GINSBORG, B. L. 1960a. *J. Physiol.* 150:707.
GINSBORG, B. L. 1960b. *J. Physiol.* 154:581.
GOLDSPINK, G. 1962. *Comp. Biochem. Physiol.* 7:157.
GOLDSPINK, G. 1969. *Life Sci.* 8:791.
GOMER, R. H., and LAZARIDES, E. 1981. *Cell* 23:524.
GONYEA, W., and BONDE-PETERSEN, F. 1978. *Exp. Neurol.* 59:75.
GOSPODAROWICZ, D., WESEMAN, J., MORAN, J. S., and LINDSTROM, J. 1976. *J. Cell Biol.* 70:395.
GOVIND, C. K., and PEARCE, J. 1981. *Science* 212:1522.
GRANBACHER, N. 1971. *Z. Anat. Entwicklungsgesch.* 135:76.
GRANT, P. G. 1978. *J. Anat.* 127:157.
GRANT, P. G., and HAWES, M. R. 1977. *J. Anat.* 123:361.
GREGORY, N. G. 1981. Neurological control of muscle metabolism and growth in stress sensitive pigs. In T. Froystein, E. Slinde, and N. Standal (eds.), *Porcine Stress and Meat Quality*, pp. 11–20. Agric. Food Res. Soc., As, Norway.
GRUBER, H. VON. 1966. *Acta Anat.* 64:628.
GUNN, H. M. 1972. *Equine Vet. J.* 4:144.
GUNN, H. M. 1973. *Equine Vet. J.* 5:77.
GUNN, H. M. 1978. *J. Anat.* 127:615.
GUTH, L. 1968. *Physiol. Rev.* 48:645.
GUTH, L. 1969. *Neurosci. Res. Prog. Bull.* 7:1.
GUTH, L. 1973. *Exp. Neurol.* 41:440.
GUTH, L., and SAMAHA, F. J. 1969. *Exp. Neurol.* 25:138.
GUTH, L., and SAMAHA, F. J. 1970. *Exp. Neurol.* 28:365.
GUTH, L., and SAMAHA, F. J. 1972. *Exp. Neurol.* 34:465.
GUTH, L., and YELLIN, H. 1971. *Exp. Neurol.* 31:277.
GUTH, L., SAMAHA, F. J., and ALBERS, R. W. 1970. *Exp. Neurol.* 26:126.
GUTH, L., DEMPSEY, P. J., and COOPER, T. 1971. *Exp. Neurol.* 32:478.
GUTMANN, E. 1976. *Annu. Rev. Physiol.* 38:177.
GUTMANN, E., and CARLSON, B. M. 1975. *Pfluegers Arch.* 353:227.
GUTMANN, E., and HANZLIKOVA, V. 1965. *Gerontologia* 11:12.
GUTMANN, E., and HANZLIKOVA, V. 1970. *Histochemie* 24:287.
GUTMANN, E., and ZELENA, J. 1962. Morphological changes in the denervated muscle. In E. Gutmann (ed.), *The Denervated Muscle*, Chap. 2. Publishing House of the Czechoslovak Academy of Sciences, Prague.
GUTMANN, E., SCHIAFFINO, S., and HANZLIKOVA, V. 1971. *Exp. Neurol.* 31:451.

HALDIMAN, J. T. 1981. *Zentralbl. Veterinaermed. C* 10:289.
HALIBURTON, W. D. 1909. Trophic nerves. In *Handbook of Physiology*, 9th ed., pp. 854–855. P. Blakiston's Son & Co., Philadelphia.
HALL, Z. W., LUBIT, B. W., and SCHWARTZ, J. H. 1981. *J. Cell Biol.* 90:789.
HAMBURGER, V. 1934. *J. Exp. Zool.* 68:449.
HAMILTON, M. J., HEGREBERG, G. A., and GORHAM, J. R. 1974. *Am. J. Vet. Res.* 35:1321.
HAMMARBERG, C., and KELLERTH, J.-O. 1975a. *Acta Physiol. Scand.* 95:166.
HAMMARBERG, C., and KELLERTH, J.-O. 1975b. *Acta Physiol. Scand.* 95:231.
HAMMARBERG, C., and KELLERTH, J.-O. 1975c. *Acta Physiol. Scand.* 95:243.
HANZLIKOVA, V., and SCHIAFFINO, S. 1973. *Z. Zellforsch.* 147:75.
HARRIS, A. J. 1974. *Annu. Rev. Physiol.* 36:251.
HARRIS, C. 1954. *Am. J. Pathol.* 30:501.
HARTZELL, H. C., and FAMBROUGH, D. M. 1973. *Dev. Biol.* 30:153.
HAUSCHKA, S. D. 1974. *Dev. Biol.* 37:345.
HENNEMAN, E., and OLSON, C. B. 1965. *J. Neurophysiol.* 28:581.
HERRERA, A. A., and GRINNELL, A. D. 1980. *Nature* (*Lond.*) 287:649.
HIKIDA, R. S. 1972. *Exp. Neurol.* 35:265.
HIKIDA, R. S., and BOCK, W. J. 1972. *Z. Zellforsch.* 128:1.
HINKLE, L., MCCAIG, C. D., and ROBINSON, K. R. 1981. *J. Physiol.* 314:121.
HINTZ, C. S., LOWRY, C. V., KAISER, K. K., MCKEE, D., and LOWRY, O. H. 1980. *Am. J. Physiol.* 239:C58.
HODGES, R. D. 1974. *The Histology of the Fowl.* Academic Press, New York.
HOFMANN, W. W. 1980. *J. Neurol. Sci.* 45:205.
HOFMANN, W. W., and PEACOCK, J. H. 1973. *Exp. Neurol.* 41:345.
HOFMANN, W. W., BUTTE, J., and LEON, H. A. 1978. *Am. J. Physiol.* 235:C199.
HOLLY, R. G., BARNETT, J. G., ASHMORE, C. R., TAYLOR, R. G., and MOLE, P. A. 1980. *Am. J. Physiol.* 238:C62.
HOLT, R. K., and SOHAL, G. S. 1978. *Am. J. Anat.* 151:313.
HOOSHMAND, H., MARTINEZ, A. J., and ROSENBLUM, W. I. 1971. *Arch. Neurol.* 24:561.
HOWELLS, K. F., and JORDAN, T. C. 1978. *Histochemistry* 58:97.
HUGHES, A. F. W. 1968. *Aspects of Neural Ontogeny.* Logos Press/Academic Press, London.
HUIZAR, P., KUNO, M., and MIYATA, Y. 1975. *J. Physiol.* 252:465.
HUNT, C. C., and KUFFLER, S. W. 1954. *J. Physiol.* 126:293.
HSU, L., NATYZAK, D., and TRUPIN, G. L. 1982. *J. Embryol. Exp. Morphol.* 71:83.
IP, M. C. 1974. *Anat. Rec.* 180:605.
IRWIN, R. L., and HEIN, M. M. 1966. *Am. J. Physiol.* 211:1117.
JABAILY, J., and SINGER, M. 1978. *Dev. Biol.* 64:189.
JACOB, M., CHRIST, B., and JACOB, H. J. 1978. *Anat. Embryol.* 153:179.
JACOB, M., CHRIST, B., and JACOB, H. J. 1979. *Anat. Embryol.* 157:291.
JAMES, N. T. 1971. *J. Anat.* 110:335.
JAMES, N. T. 1972. *J. Neurol. Sci.* 17:41.
JANSEN, J. K. S., LOMO, T., NICOLAYSEN, K., and WESTGAARD, R. H. 1973. *Science* 181:559.
JANSSON, E., SJODIN, B., and TESCH, P. 1978. *Acta Physiol. Scand.* 104:235.
JARCHO, L. W., EYZAGUIRE, C., BERMAN, B., and LILIENTHAL, J. L. 1952. *Am. J. Physiol.* 168:446.
JOBSIS, A. C., and MEIJER, A. E. F. H. 1973a. *Histochemie* 36:51.
JOBSIS, A. C., and MEIJER, A. E. F. H. 1973b. *Histochemie* 36:63.
JOHN, H. A., and LAWSON, H. 1980. *Cell Biol. Int. Rep.* 4:841.
JOHNSON, M. A., POLGAR, J., WEIGHTMAN, D., and APPLETON, D. 1973. *J. Neurol. Sci.* 18:111.
JOHNSTON, D. M., MOODY, W. G., BOLING, J. A., and BRADLEY, N. W. 1981. *J. Food Sci.* 46:1760.
JUNTUNEN, J. 1973. *Z. Anat. Entwicklungsgesch.* 143:1.

KAO, I., and DRACHMAN, D. B. 1977. *Science* 195:74.
KARPATI, G., and ENGEL, W. K. 1968. *Neurology* 18:681.
KATZ, B. 1966. *Nerve, Muscle and Synapse*, pp. 160–161. McGraw-Hill, New York.
KATZ, L. N. 1926. *Proc. R. Soc. B* 99:1.
KAUFFMAN, F. C., ALBUQUERQUE, E. X., WARNICK, J. E., and MAX, S. R. 1976. *Exp. Neurol.* 50:60.
KELLER, J. M., and NAMEROFF, M. 1974. *Differentiation* 2:19.
KELLY, A. M., and SCHOTLAND, D. L. The evolution of the "checkerboard" in a rat muscle. In B. Q. Banker, R. J. Przybylski, J. P. Van Der Meulen, and M. Victor (eds.), *Research in Muscle Development and the Muscle Spindle*, pp. 32–48. Excerpta Medica, Amsterdam.
KELLY, A. M., and ZACKS, S. I. 1969a. *J. Cell Biol.* 42:135.
KELLY, A. M., and ZACKS, S. I. 1969b. *J. Cell Biol.* 42:154.
KHAN, M. A. 1976. *Histochemistry* 50:9.
KHAN, M. A. 1977. *Cell. Mol. Biol.* 22:383.
KHAN, M. A. 1978. *Histochemistry* 55:129.
KHAN, M. A. 1979. *Histochem. J.* 11:321.
KHAN, M. A., PAPADIMITRIOU, J. M., HOLT, P. G., and KAKULAS, B. A. 1972. *Histochemie* 30:329.
KHAN, M. A., PAPADIMITRIOU, J. M., HOLT, P. G., and KAKULAS, B. A. 1973. *Histochemie* 36:173.
KHAN, M. A., PAPADIMITRIOU, J. M., and KAKULAS, B. A. 1975. *Histochemistry* 43:101.
KIERNAN, J. A. 1974. *Histochemistry* 40:51.
KIESSLING, K.-H. 1977. *Swed. J. Agric. Res.* 7:115.
KIESSLING, K.-H. 1978. *Swed. J. Agric. Res.* 8:55.
KIKUCHI, T. 1971. *Tohoku J. Agric. Res.* 22:1.
KIKUCHI, T., NAGATANI, T., and TAMATE, H. 1972. *Tohoku J. Agric. Res.* 23:149.
KILARSKI, W., and BIGAJ, J. 1969. *Z. Zellforsch.* 94:194.
KILARSKI, W., and JAKUBOWSKA, M. 1979. *Z. Mikrosk. Anat. Forsch.* 93:1159.
KNUDSEN, K. A., and HORWITZ, A. F. 1977. *Dev. Biol.* 58:328.
KOHAMA, K., and OZAWA, E. 1978. *Muscle Nerve* 1:236.
KOHN, R. R. 1964. *Am. J. Pathol.* 45:435.
KONIGSBERG, I. R. 1971. *Dev. Biol.* 26:133.
KONIGSBERG, I. R., SOLLMANN, P. A., and MIXTER, L. O. 1978. *Dev. Biol.* 63:11.
KORDYLEWSKI, L. 1979a. *Z. Mikrosk.-Anat. Forsch.* 93:225.
KORDYLEWSKI, L. 1979b. *Z. Mikrosk.-Anat. Forsch.* 93:1038.
KORNELIUSSEN, H. 1972. *J. Neurocytol.* 1:279.
KORNELIUSSEN, H., and NICOLAYSEN, K. 1973. *Z. Zellforsch.* 143:273.
KORNELIUSSEN, H., and WAERHAUG, O. 1973. *Z. Anat. Entwicklungsgesch.* 140:73.
KORR, I. M., and APPELTAUER, G. S. L. 1974. *Exp. Neurol.* 43:452.
KUBOTA, K., KOMATSU, S., NAKAMURA, M., and MASEGI, T. 1980. *Anat. Rec.* 197:413.
KUGELBERG, E. 1973. *J. Neurol. Sci.* 20:177.
KUGELBERG, E. 1976. *J. Neurol. Sci.* 27:269.
LANCE-JONES, C. 1979. *J. Morphol.* 162:275.
LANCE-JONES, C., and LANDMESSER, L. 1980a. *J. Physiol.* 302:559.
LANCE-JONES, C., and LANDMESSER, L. 1980b. *J. Physiol.* 302:581.
LANDMESSER, L. 1978a. *J. Physiol.* 284:371.
LANDMESSER, L. 1978b. *J. Physiol.* 284:391.
LANDMESSER, L., and MORRIS, D. G. 1975. *J. Physiol.* 249:301.
LE DOUARIN, N. M. 1973. *Dev. Biol.* 30:217.
LE GROS CLARK, W. E. 1945. *The Tissues of the Body*. 2nd ed., p. 113. Clarendon Press, Oxford.
LENTZ, T. L. 1969. *J. Cell Biol.* 42:431.
LENTZ, T. L. 1972. *J. Cell Biol.* 55:93.

LEVIS, A. G., FURLAN, D., and BIANCHI, V. 1974. *Acta Embryol. Exp. (Palermo)* 2:123.

LEWIS, J., CHEVALLIER, A., KIENY, M., and WOLPERT, L. 1981. *J. Embryol. Exp. Morphol.* 64:211.

LIPTON, B. H., and KONIGSBERG, I. R. 1972. *J. Cell Biol.* 53:348.

LOMO, T., WESTGAARD, R. H., and DAHL, H. A. 1974. *Proc. R. Soc. B* 187:99.

LULLMANN, H., and SUNANO, S. 1973. *Pfluegers Arch.* 342:271.

LUND, E. J. 1947. *Bioelectric Fields and Growth.* University of Texas Press, Austin.

MABUCHI, K., and SRETER, F. A. 1980b. *Muscle Nerve* 3:233.

MACKOVA, E., and HNIK, P. 1972. *Physiol. Bohemoslov.* 21:9.

MACKOVA, E., and HNIK, P. 1973. *Physiol. Bohemoslov.* 22:43.

MADEN, M. 1978. *J. Embryol. Exp. Morphol.* 48:169.

MARCHAND, E. R., and ELDRED, E. 1969. *Exp. Neurol.* 25:655.

MARKELONIS, G., OH, T. H., and DERR, D. 1980. *Exp. Neurol.* 70:598.

MASAKI, T., and YOSHIZAKI, C. 1972. *J. Biochem. (Tokyo)* 71:755.

MATTHEWS, P. B. C. 1981. *J. Physiol.* 320:1.

MAUGER, A., KIENY, M., and CHEVALLIER, A. 1980. *Arch. Anat. Microsc.* 69:175.

MAXWELL, L. C., FAULKNER, J. A., and LIEBERMAN, D. A. 1973. *Am. J. Physiol.* 224:356.

MCLACHLAN, J. C., and HORNBRUCH, A. 1979. *J. Embryol. Exp. Morphol.* 54:209.

MCLENNAN, I. S. 1982. *Dev. Biol.* 92:263.

MCMILLAN, P. J., and WITTUM, R. L. 1971. *J. Histochem. Cytochem.* 19:421.

MEANS, A. R., TASH, J. S., and CHAFOULEAS, J. G. 1982. *Physiol. Rev.* 62:1.

MEDOFF, J., and ZWILLING, E. 1972. *Dev. Biol.* 28:138.

MEIJER, A. E. F. H., and VOSSENBERG, R. P. M. 1977. *Histochemistry* 52:45.

MELICHNA, J., GUTMANN, E., and SYROVY, I. 1974. *Physiol. Bohemoslov.* 23:511.

MILAIRE, J. 1967. The contribution of histochemistry to our understanding of limb morphogenesis and some of its congenital deviations. In C. H. Frantz (ed.), *Normal and Abnormal Embryological Development*, pp. 27–75. Publ. 1497. National Research Council, Washington, D.C.

MILEDI, R., and SLATER, C. R. 1969. *Proc. R. Soc. B* 174:253.

MIYATA, Y., and YOSHIOKA, K. 1980. *J. Physiol.* 309:631.

MOODY, W. G., and CASSENS, R. G. 1968. *J. Anim. Sci.* 27:961.

MORRIS, C. J., and RAYBOULD, J. A. 1971. *J. Neurol. Sci.* 13:181.

MOSS, P. S., and STROHMAN, R. C. 1976. *Dev. Biol.* 48:431.

MUHL, Z. F., and GRIMM, A. F. 1975. *Experientia* 31:1053.

MULLER, W. 1975. *Cell Tissue Res.* 161:225.

NAG, A. C. 1972. *J. Cell Biol.* 55:42.

NAKAI, J. 1965. *Exp. Cell Res.* 40:307.

NAMBA, T., NAKAMURA, T., TAKAHASHI, A., and GROB, D. 1968. *J. Comp. Neurol.* 134:385.

NEEDHAM, D. M. 1926. *Physiol. Rev.* 6:1.

NEMETH, P. M., and PETTE, D. 1980. *J. Histochem. Cytochem.* 28:193.

NEMETH, P., HOFER, H.-W., and PETTE, D. 1979. *Histochemistry* 63:191.

NEMETH, P. M., PETTE, D., and VRBOVA, G. 1981. *J. Physiol.* 311:489.

NOUGUES, J. 1980. *C. R. Acad. Sci. Paris* 290:D223.

NOVAKOFSKI, J. E., KAUFFMAN, R. G., and CASSENS, R. G. 1981. *J. Anim. Sci.* 52:1430.

NYSTROM, B. 1968a. *Acta Neurol. Scand.* 44:295.

NYSTROM, B. 1968b. *Acta Neurol. Scand.* 44:363.

NYSTROM, B. 1968c. *Acta Neurol. Scand.* 44:405.

O'BRIEN, R. A. D., OSTBERG, A. J. C., and VRBOVA, G. 1978. *J. Physiol.* 282:571.

OGATA, T., and MURATA, F. 1969. *Tohoku J. Exp. Med.* 98:107.

OH, T. H. 1975. *Exp. Neurol.* 46:432.

OH, T. H. 1976. *Exp. Neurol.* 50:376.

OMMER, P. A. 1971. *Experientia* 27:173.

OZAWA, E., and KOHAMA, K. 1978. *Muscle Nerve* 1:230.
PADYKULA, H. A., and GAUTHIER, G. F. 1963. *J. Cell Biol.* 18:87.
PADYKULA, H. A., and GAUTHIER, G. F. 1970. *J. Cell Biol.* 46:27.
PADYKULA, H. A., and HERMAN, E. 1955a. *J. Histochem. Cytochem.* 3:161.
PADYKULA, H. A., and HERMAN, E. 1955b. *J. Histochem. Cytochem.* 3:170.
PAGE, S. 1965. *J. Cell Biol.* 26:477.
PALMER, A. C. 1976. *Introduction to Animal Neurology*, 2nd ed. Blackwell Scientific Publications, Oxford.
PEACHEY, L. D., and HUXLEY, A. F. 1962. *J. Cell Biol.* 13:177.
PEARSE, A. G. E. 1972. *Histochemistry. Theoretical and Applied*, Vol. 2. Churchill Livingstone, Edinburgh.
PETTE, D., and SCHNEZ, U. 1977. *Histochemistry* 54:97.
PETTE, D., MULLER, W., LEISNER, E., and VRBOVA, G. 1976. *Pfluegers Arch.* 364:103.
PICKETT, J. B. 1980. *Science* 210:927.
PIEROBON-BORMIOLI, S., SARTORE, S., LIBERA, L. D., VITADELLO, M., and SCHIAFFINO, S. 1981. *J. Histochem. Cytochem.* 29:1179.
PILAR, G., and HESS, A. 1966. *Anat. Rec.* 154:243.
PITTMAN, R., and OPPENHEIM, R. W. 1979. *J. Comp. Neurol.* 187:425.
PLATZER, A. C. 1978. *Anat. Rec.* 190:639.
POBERAI, M., and SAVAY, G. 1976. *Acta Histochem.* 57:44.
POOL, C. W., RAAMSDONK, W. VAN, DIEGENBACH, P. C., MIJZEN, P., SCHENKKAN, E. J., and STELT, A. VAN DER. 1976. *Acta Histochem.* 57:20.
POPIELA, H. 1978. *Exp. Neurol.* 62:405.
PURI, E. C., CARAVETTI, M., PERRIARD, J.-C., TURNER, D. C., and EPPENBERGER, H. M. 1980. *Proc. Natl. Acad. Sci. USA* 77:5297.
QUAIN, J. 1856. *Elements of Anatomy*, Vol. 1. Walton and Maberly, London.
RAAMSDONK, W. VAN, POOL, C. W., and KRONNIE, G. TE. 1978. *Anat. Embryol.* 153:137.
RAHELIC, S., and PUAC, S. 1981. *Meat Sci.* 5:439.
RALSTON, H. J., FEINSTEIN, B., and INMAN, V. T. 1952. *Fed. Proc.* 11:127.
RAMIREZ, B., and LUCO, J. V. 1973. *J. Neurobiol.* 4:525.
RANSON, S. W., and SAMS, C. F. 1928. *J. Neurol. Psychopathol.* 8:304.
READ, S. E., TAKEDA, M., and KIRKALDY-WILLIS, W. H. 1971. *Wilhelm Roux's Arch.* 167:187.
REDFERN, R. A. 1970. *J. Physiol.* 209:701.
REID, I. M., ROBERTS, C. J., and BAIRD, G. D. 1980. *J. Agric. Sci., Camb.* 94:239.
RIEGER, F., PINCON-RAYMOND, M., DREYFUS, P., GUITTARD, M., and FARDEAU, M. 1979. *Ann. Genet. Sel. Anim.* 11:371.
RILEY, D. A. 1974. *Am. J. Anat.* 140:609.
RILEY, D. A. 1977. *Exp. Neurol.* 56:400.
RILEY, D. A., and ALLIN, E. F. 1973. *Exp. Neurol.* 40:391.
ROMANES, G. J. 1941. *J. Anat.* 76:112.
ROMER, A. S. 1927. *J. Morphol.* 43:347.
ROMER, A. S. 1942. *J. Morphol.* 71:251.
ROMER, A. S. 1944. *J. Morphol.* 74:1.
ROWE, R. W. D. 1969. *Comp. Biochem. Physiol.* 28:1449.
ROWE, R. W. D., and GOLDSPINK, G. 1968. *Anat. Rec.* 161:69.
RUSSELL, R. G., and OTERUELO, F. T. 1981. *Anat. Embryol.* 162:403.
SABRI, M. I., and OCHS, S. 1972. *J. Neurobiol.* 4:145.
SALMONS, S., GALE, D. R., and SRETER, F. A. 1978. *J. Anat.* 127:17.
SALPETER, M. M., and ELDEFRAWI, M. E. 1973. *J. Histochem. Cytochem.* 21:769.
SAMAHA, F. J., and YUNIS, E. J. 1973. *Exp. Neurol.* 41:431.
SAMAHA, F. J., GUTH, L., and ALBERS, R. W. 1970a. *Exp. Neurol.* 27:276.
SAMAHA, F. J., GUTH, L., and ALBERS, R. W. 1970b. *J. Biol. Chem.* 245:219.

Samorajski, T., and Rolsten, C. 1975. *J. Comp. Neurol.* 159:553.
Sato, M., Mizuno, N., and Konishi, A. 1977. *J. Comp. Neurol.* 175:27.
Saunders, J. W. 1967. Control of growth patterns in limb development. In C. H. Frantz (ed.), *Normal and Abnormal Embryological Development*, pp. 16–26. Publ. 1497. National Research Council, Washington, D.C.
Sawaki, S., and Peter, J. B. 1972. *Exp. Neurol.* 35:421.
Schiaffino, S., and Bormioli, S. P. 1973a. *J. Histochem. Cytochem.* 21:142.
Schiaffino, S., and Bormioli, S. P. 1973b. *Exp. Neurol.* 40:126.
Schmalbruch, H., and Kamieniecka, Z. 1975. *J. Histochem. Cytochem.* 23:395.
Schoultz, T. W., and Swett, J. E. 1972. *J. Neurocytol.* 1:1.
Schubert, D., Tarikas, H., Humphreys, S., Heinemann, S., and Patrick, J. 1973. *Dev. Biol.* 33:18.
Schudt, C., Van der Bosch, J., and Pette, D. 1973. *FEBS Lett. (Amst.)* 32:296.
Schwann, T. 1839. *Mikroskopische Untersuchungen*, H. Smith (trans.). Sydenham Society London, 1847.
Seiden, D. 1976. *Am. J. Anat.* 145:459.
Sexton, A. W., and Gerston, J. W. 1967. *Science* 157:199.
Shafiq, S. A., Asiedu, S. A., and Milhorat, A. T. 1972. *Exp. Neurol.* 35:529.
Shanthaveerappa, T. R., and Bourne, G. H. 1967. *Int. Rev. Cytol.* 21:353.
Shear, C. R. 1978. *J. Cell Sci.* 29:297.
Sivachelvan, M. N., and Davies, A. S. 1981. *J. Anat.* 132:545.
Slack, J. R., and Docherty, J. R. 1978. *Cell Tissue Res.* 186:171.
Snow, D. H., and Guy, P. S. 1980. *Res. Vet. Sci.* 28:137.
Sola, O. M., Christensen, D. L., and Martin, A. W. 1973. *Exp. Neurol.* 41:76.
Spamer, C., and Pette, D. 1977. *Histochemistry* 52:201.
Spamer, C., and Pette, D. 1979. *Histochemistry* 60:9.
Sreter, F. A., Balint, M., and Gergely, J. 1975. *Dev. Biol.* 46:317.
Stacey, M. J. 1969. *J. Anat.* 105:231.
Staudte, H. W., Exner, G. U., and Pette, D. 1973. *Pfluegers Arch.* 344:159.
Stefanelli, A. 1950. Some comments on regeneration in the central nervous system. In P. Weiss (ed.), *Genetic Neurology*. University of Chicago Press, Chicago.
Steinhauf, D., Weniger, J. H., and Mader, H.-P. 1976. Observations on the apparent antagonism between meat producing capacity and meat quality in pigs. In D. Lister, D. N. Rhodes, V. R. Fowler, and M. F. Fuller (eds.), *Meat Animals: Growth and Productivity*, pp. 373–385. Plenum Press, New York.
Stickland, N. C. 1978. *Zentralbl. Veterinaermed. C* 7:193.
Stockdale, F. E., Baden, H., and Raman, N. 1981. *Dev. Biol.* 82:168.
Stonnington, H. H., and Engel, A. G. 1973. *Neurology* 23:714.
Suzuki, A. 1971a. *Jap. J. Zootech. Sci.* 42:39.
Suzuki, A. 1971b. *Jap. J. Zootech. Sci.* 42:463.
Suzuki, A. 1972. *Jap. J. Zootech. Sci.* 43:161.
Suzuki, A. 1976. *Jap. J. Zootech. Sci.* 47:95.
Suzuki, A., and Cassens, R. G. 1980a. *Histochem. J.* 12:687.
Suzuki, A., and Cassens, R. G. 1980b. *J. Anim. Sci.* 51:1449.
Suzuki, A., and Tamate, H. 1974. *Acta Histochem. Cytochem.* 7:319.
Suzuki, A., Tamate, H., and Okada, M. 1976. *Tohoku J. Agric. Res.* 27:20.
Swatland, H. J. 1971. *Proc. Reciprocal Meat Conf.* 24:400.
Swatland, H. J. 1973a. *J. Anim. Sci.* 36:355.
Swatland, H. J. 1973b. *J. Anim. Sci.* 37:536.
Swatland, H. J. 1974. *J. Anim. Sci.* 38:752.
Swatland, H. J. 1975a. *Res. Vet. Sci.* 18:253.
Swatland, H. J. 1975b. *Res. Vet. Sci.* 18:258.

SWATLAND, H. J. 1975c. *Histochem. J.* 7:459.
SWATLAND, H. J. 1977a. *Zentralbl. Veterinaermed.* A 24:248.
SWATLAND, H. J. 1977b. *Histochem. J.* 9:751.
SWATLAND, H. J. 1978. *Anim. Prod.* 27:229.
SWATLAND, H. J., and CASSENS, R. G. 1971. *J. Anim. Sci.* 33:750.
SWATLAND, H. J., and CASSENS, R. G. 1972a. *J. Comp. Pathol.* 82:229.
SWATLAND, H. J., and CASSENS, R. G. 1972b. *J. Anim. Sci.* 34:21.
SWATLAND, H. J., and CASSENS, R. G. 1973. *J. Anim. Sci.* 36:343.
SWATLAND, H. J., and CASSENS, R. G. 1974. *J. Anim. Sci.* 38:1092.
SWATLAND, H. J., and KIEFFER, N. M. 1974. *J. Anim. Sci.* 39:42.
SYROVY, I., GUTMANN, E., and MELICHNA, *J. 1972. Physiol. Bohemoslov.* 21:633.
SZENTKUTI, L., and CASSENS, R. G. 1979. *J. Anim. Sci.* 49:693.
TAYLOR, A., CODY, F. W. J., and BOSLEY, M. A. 1973. *Exp. Neurol.* 38:99.
TENG, N. H., and FISZMAN, M. Y. 1976. *J. Supramol. Struct.* 4:381.
THESLEFF, S. 1960. *Physiol. Rev.* 40:734.
THURLEY, D. C. 1972. *Br. Vet. J.* 128:355.
TOMANEK, R. J. 1975. *Dev. Biol.* 42:305.
TOMANEK, R. J., and COLLING-SALTIN, A.-S. 1977. *Am. J. Anat.* 149:227.
TOMLINSON, B. E., and IRVING, D. 1977. *J. Neurol. Sci.* 34:213.
TOOP, J. 1976. *Stain Technol.* 51:1.
TOUTANT, J. P., TOUTANT, M. N., RENAUD, D., and LE DOUARIN, G. H. 1980. *Cell Differ.* 9:305.
TOUTANT, J. P., ROUAUD, T., and LE DOUARIN, G. H. 1981. *Histochem. J.* 13:481.
TROTTER, J. A., and NAMEROFF, M. 1976. *Dev. Biol.* 49:548.
TRYPHONAS, L., HAMILTON, G. F., and RHODES, C. S. 1974. *J. Am. Vet. Med. Assoc.* 164:801.
TUFFERY, A. R. 1971. *J. Anat.* 110:221.
TUNELL, G. L., and HART, M. N. 1977. *Arch. Neurol.* 34:171.
TURNER, D. C. 1978. *Differentiation* 10:81.
TWITTY, V. 1955. Eye. In B. H. Willier, P. A. Weiss, and V. Hamburger (eds.), *Analysis of Development*, pp. 402–414.
VALIN, C., TOURAILLE, C., VIGNERON, P. and ASHMORE, C. R. 1982. *Meat Sci.* 6:257.
VANDENBURGH, H. H. 1982. *Dev. Biol.* 93:438.
VANDENBURGH, H. H., and KAUFMAN, S. 1981. *J. Cell. Physiol.* 109:205.
VAN DEN HENDE, C., MUYLLE, E., OYAERT, W., and DE ROOSE, P. 1972. *Zentralbl. Veterinaermed. A* 19:102.
VAN HARREVELD, A. 1947a. *Arch. Neerl. Physiol.* 28:408.
VAN HARREVELD, A. 1947b. *Am. J. Physiol.* 151:96.
VAUGHAN, H. S., AZIZ-ULLAH, G., GOLDSPINK, G., and NOWELL, N. W. 1974. *J. Histochem. Cytochem.* 22:155.
VENERONI, G., and MURRAY, M. R. 1969. *J. Embryol. Exp. Morphol.* 21:369.
VERATTI, E. 1902. *Mem. Ist. Lomb., Cl. Sci. Nat.* 19:87; translated in *J. Biophys. Biochem. Cytol., Suppl.* 10:1 (1961).
VERTEL, B. M., and FISCHMAN, D. A. 1976. *Dev. Biol.* 48:438.
VYSKOCIL, F., and GUTMANN, E. 1977. *J. Comp. Physiol.* 122:385.
WAERHAUG, O., and KORNELIUSSEN, H. 1974. *Z. Anat. Entwicklungsgesch.* 144:237.
WALKER, L. B. J. 1961. *Anat. Rec.* 139:1.
WALSH, J. V., BURKE, R. E., RYMER, W. Z., and TSAIRIS, P. 1978. *J. Neurophysiol.* 41:496.
WARREN, R. H. 1973. *Anat. Rec.* 177:225.
WARREN, R. H. 1981. *Tissue Cell* 13:773.
WATANABE, K., SASAKI, F., TAKAHAMA, H., and ISEKI, H. 1980. *J. Anat.* 130:83.
WEATHERLEY, A. H., and GILL, H. S. 1981. *Experientia* 37:1102.
WEISS, P. 1955. Nervous system. In B. H. Willier, P. A. Weiss, and V. Hamburger (eds.), *Analysis of Development*, pp. 346–401. W. B. Saunders Company, Philadelphia.

WETZEL, M. C., GERLACH, R. L., STERN, L. Z., and HANNAPEL, L. K. 1973. *Exp. Neurol.* 39:223.
WHITE, N. A., MCGAVIN, M. D., and SMITH, J. E. 1978. *Am. J. Vet. Res.* 39:1297.
WILEY-LIVINGSTON, C. A., and ELLISMAN, M. H. 1980. *Dev. Biol.* 79:334.
WILLIAMS, P. E., and GOLDSPINK, G. 1973. *J. Anat.* 116:45.
WILSON, G. D., GINGER, I. D., and SCHWEIGERT, B. S. 1959. *J. Anim. Sci.* 18:1080.
WIRSEN, C., and LARSSON, K. S. 1964. *J. Embryol. Exp. Morphol.* 12:759.
WOOD, P. L., and BOEGMAN, R. J. 1975. *Exp. Neurol.* 48:136.
WOOTEN, G. F., and REIS, D. J. 1972. *Int. J. Neurosci.* 3:155.
YAFFE, D., and FELDMAN, M. 1965. *Dev. Biol.* 11:300.
YELLIN, H. 1972. *Anat. Rec.* 173:333.
YELLIN, H. 1974. *Exp. Neurol.* 42:412.
YELLIN, H. 1975. *Anat. Rec.* 182:479.
YOSHIZATO, K., OBINATA, T., HUANG, H.-Y., MATSUDA, R., SHIOYA, N., and MIYATA, T. 1981. *Dev. Growth Differ.* 23:175.
YOUMANS, W. B. 1967. *Am. J. Phys. Med.* 46:173.
YOUN, B. W., and MALACINSKI, G. M. 1981. *J. Embryol. Exp. Morphol.* 66:1.
YOUNG, J. Z. 1950. The determination of the specific characteristics of nerve fibers. In P. Weiss (ed.), *Genetic Neurology*. University of Chicago Press, Chicago.
YOUNKIN, S. G., BRETT, R. S., DAVEY, B., and YOUNKIN, L. H. 1978. *Science* 200:1292.
ZALIN, R. J. 1977. *Dev. Biol.* 59:241.
ZALIN, R. 1979. *Dev. Biol.* 71:274.
ZALIN, R. J., and MONTAGUE, W. 1974. *Cell* 2:103.
ZELENA, J. 1962. The effect of denervation on muscle development. In E. Gutmann (ed.), *The Denervated Muscle*, Chap. 3. Publishing House of the Czechoslovak Academy of Sciences, Prague.
ZEMAN, R. J., and WOOD, D. S. 1980. *J. Histochem. Cytochem.* 28:714.
ZENKER, W., and ANZENBACHER, H. 1964. *J. Cell. Comp. Physiol.* 63:273.
ZENKER, W., and HOHBERG, E. 1973. *Z. Anat. Entwicklungsgesch.* 139:163.
ZEVIN-SONKIN, D., and YAFFE, D. 1980. *Dev. Biol.* 74:326.

7

The Cellular Basis of Postnatal Muscle Growth

INTRODUCTION

Animals are born with a certain number of fibers in each muscle. These fibers have already fallen under the influence of motor neurons, which will play a major role in the determination of the contraction speed, metabolism, and growth pattern of individual muscle fibers. During postnatal development, the fibers will grow radially and longitudinally. Radial growth is relatively easy to measure and it involves the longitudinal splitting and proliferation of myofibrils. Longitudinal growth in major carcass muscles is difficult to measure in individual fibers and it involves the formation of new sarcomeres at each end of the fiber. There is no guarantee that radial and longitudinal growth will be linearly related: in fact, it is rather unlikely. Different physiological fiber types might grow at different rates, depending on muscle activity.

Real fiber number and length are very difficult to estimate unless all fibers run from one end of a muscle to the other. This does not happen in major carcass muscles. The apparent number of fibers is the number seen in a cross section of a muscle. The apparent number of fibers depends on the fraction of the real fiber number that runs through the cross section. There is no guarantee that this fraction will remain constant during development. Muscle fibers are multinucleate at birth and new nuclei will be added to the fiber during its postnatal development. New nuclei are derived from the mitotic division of myoblasts which have been trapped on the muscle fiber surface as satellite cells.

Muscle fibers are the basic living units of muscles. Although muscle fibers are essentially cylindrical in shape, they are surprisingly difficult to measure or to count. In any serious attempt to understand the scientific literature on fiber numbers and dimensions, careful attention must be given to the details of microtechnique, muscle architecture, and simple geometry. In this chapter, these details are given in footnotes to the main text. They are presented in a concise form using a morphometric notation which is explained in an appendix to this chapter. Readers may question the need for the tedious translation of obvious parameters, such as the numbers of muscle fibers into a rather difficult morphometric notation, even if it is confined to footnotes. However, as Bertrand Russell (1963) said: "Obviousness is always the enemy to correctness. Hence we invent some new and difficult symbolism, in which nothing seems obvious. Then we set up certain rules for operating on the symbols, and the whole thing becomes mechanical. In this way we find out what must be taken as premiss and what can be demonstrated or defined."

The body of the text of this chapter is intended as a general review of the histology of muscle growth in meat animals, and the chapter starts with a review of muscle geometry. This is followed by some representative data on fiber numbers and dimensions in meat animals. The chapter concludes with a glance to the future—the measurement of fiber volume. This is the direction in which we must progress in order to integrate the study of muscle histology with the carcass dissection techniques which we use to evaluate meat yield in economic terms.

MUSCLE GEOMETRY AND MORPHOMETRY

Radial Dimensions

The ends of muscle fibers are rounded if they insert into tendons or they may be tapered if they insert *intrafascicularly* beneath the *endomysium* of another fiber (Schiefferdecker, 1891; Schafer, 1898; Bardeen, 1903; Huber, 1916; Le Gros Clark, 1971). Only very rarely are the fibers of carcass muscles branched at their ends, although branching might occur in the cutaneous muscles or in tongue muscles. The tight packing of fibers within a muscle compresses the fibers to an imperfect prism so that fibers appear somewhat polygonal in cross section (Figure 7-1). In the porcine longissimus dorsi, fiber polygons range from three to eight in number of sides. The most common shapes have five or six sides, depending on their position within the muscle (Swatland, 1975a). Variability in side number is related to variability in fiber cross-sectional area since small and large fibers fit together in a different pattern to that found when all the fibers have a similar size.

When fibers are isolated by maceration or microdissection, their diameters can

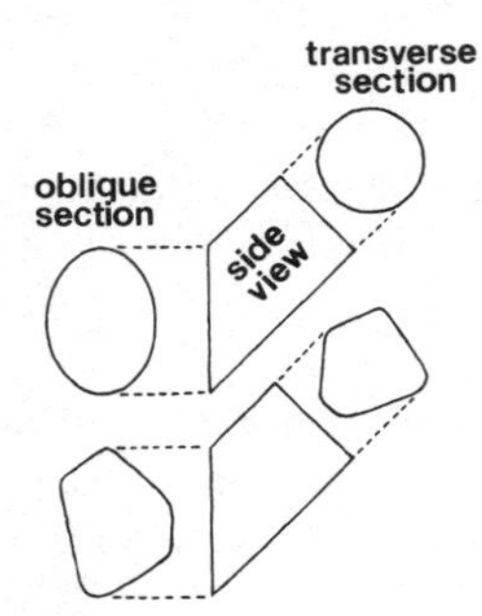

Figure 7-1 Oblique and transverse sections of rounded and prismatic muscle fibers. In the rounded fiber (top), the minimum diameter in an oblique section is the same as the fiber diameter. In an oblique section of a prismatic fiber (bottom), the minimum diameter is sometimes an inaccurate estimate of the mean fiber diameter.

be measured from a lateral view under the microscope. Care must be taken not to compress fibers between a microscope slide and its coverslip, nor to let the fibers dry out. Lateral view diameters are a meaningful measurement of polygonal mean diameters. This is because the muscle fibers are released from their compressed prismatic shape once they have been isolated from the muscle. In addition, as measurements are being taken with an eyepiece grid in a microscope, fiber diameters are unconsciously averaged through the depth of focus of the microscope. In regular polygons with an odd number of sides, a value close to the polygonal mean diameter is seen from a lateral view. In other shapes, the error is minimal and is canceled by sampling. Compared to muscle cross sections, one advantage of macerated muscle preparations is that tapered endings can be easily recognized, and can be excluded from a sample. With muscle cross sections, each small-diameter fiber must be checked through serial sections to determine whether it is really a small fiber or simply a tapered ending (Figure 7-2).

Muscle volume is almost, although not exactly, constant during contraction (Catton, 1957). Thus as fibers shorten in length, their radial dimensions become greater. When muscles become inextensible after death (*rigor mortis*), they can no longer contract by filament sliding. However, their radial dimensions are not stable. The occurrence of rigor mortis is accompanied by a decrease in fiber diameter as myofilaments are drawn closely together by the active heads of myosin molecules (Maughan and Godt, 1981). Fibers may also lose water as their pH declines toward the *isoelectric point* of myofibrillar proteins. If fibers are removed from a muscle prior to the onset of rigor mortis, they may contract in a manner that is difficult to control. Even if a muscle sample is clamped at a predetermined length, there is no guarantee that some parts of the sample will not contract strongly enough to stretch other parts. The measurement of radial fiber dimensions must be coupled with some method to reduce all data to a common muscle length. *Sarcomere length* is the obvious choice on which to base a correction factor for sample length when working with large animals. However, it is difficult to decide on a representative value for sarcomere length in a resting muscle.

Electron microscopy, given assiduous attention to tissue fixation and embedding, enables the accurate measurement of myofilament length. If related to a physiologically meaningful definition of the degree of *actin* and *myosin* overlap,

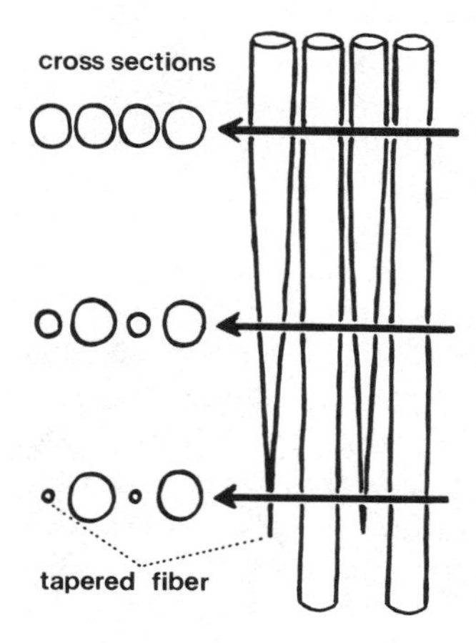

Figure 7-2 Effect of tapered intrafascicularly terminating fibers in the measurement of fiber cross sections.

resting sarcomere length can be calculated for each physiological type of muscle fiber, in each muscle, of each species, and at any particular age. It is difficult to justify this amount of extra work simply to correct other data for experimental error. The alternative is to relinquish the attempt to adjust data to an absolute basis. Radial dimensions might as well be adjusted to the standardized basis of the overall mean sarcomere length for the experiment. However, growth experiments that include fetal and neonatal animals may need special attention to correct for the growth-related elongation of sarcomere length which may occur in very young animals. Once the neonatal period has passed, however, the overall mean sarcomere length *for each muscle* provides a correction factor that is equal in usefulness to the true length of resting sarcomeres.

When muscles have been transversely sectioned to measure fiber areas, it is necessary to detach and remount the sample so as to obtain longitudinal sections for the measurement of sarcomere length. In macerated muscle preparations, however, both muscle fiber diameters and sarcomere length can be measured on the same sample by using phase-contrast or polarized light microscopy. Sarcomere lengths are best measured on myofibrillar fragments rather than on whole fibers. Measurements can be made on parts of fibers, however, if they are mounted in a fluid medium with a refractive index similar to that of the fiber itself (Huxley, 1972). The degree of fiber disruption is controlled by the duration of the maceration process. Another advantage of macerated muscle preparations is that *contracture knots* are easy to recognize and to reject from the sample, although if present in excessive numbers, the whole sample should be rejected from a growth study. Contracture knots are thickened lengths of muscle fiber which are usually caused by supercontracted sarcomeres (Schmalbruch, 1973). Whole fibers may be supercontracted if they have torn free from the connective tissue framework of their muscle. In living animals, such fibers probably degenerate. If detachment occurs during histological sampling, it may introduce a serious experimental error, particularly if supercontracted fibers are not randomly distributed. The major disadvantage of macerated muscle preparations is that they are difficult to handle in the histochemical procedures which are necessary to demonstrate the physiological differentiation of muscle fibers. Separation methods for fiber fragments (Mowafy et al., 1972) may be of some use. Macerated samples can also be handled in porcelain filters.

Histological fixation introduces an extra set of experimental variables. Fixation is necessary to preserve tissue structure, but some fixatives may cause tissue shrinkage while others may cause expansion. In conventional techniques for light microscopy, two or more fixatives may be combined in an attempt to cancel expansion against shrinkage, as well as to balance fast-diffusing and slow-diffusing reagents. In agricultural research, much of the classical work on the radial growth of muscle fibers in meat animals has been based on thick freehand sections of formaldehyde-fixed material (Hammond and Appleton, 1932; Meara, 1947; Joubert, 1956a). Joubert (1956a) showed that, after the initial fiber shrinkage associated with the development of rigor mortis, fiber diameters remained more or less constant while stored from 1 day to 6 months in 10 percent formalin (4 per cent formaldehyde). Variation in formaldehyde concentration, however, can change fiber diameters (Hegarty and Naude, 1973). Data may be biased by the technique with which they are collected, and this is difficult to distinguish from biological variation. This is a particular problem in the comparison of data from different studies. For example, the fiber diameter measurements reported by McMeekan (1940c) are considerably smaller than those reported in more recent studies. It is quite easy to make a mistake when a micrometer eyepiece is calibrated, and very difficult to detect this mistake in somebody else's research. If muscles are histologically fixed immediately after death, there may be problems if the slow penetration of the fixative allows deeper fibers to survive longer and to shrink to a greater extent than superficial fibers (Hegarty and Hooper, 1971). This problem can, however, be avoided (Goldspink et al., 1973).

As small a sample as 25 fibers in a macerated preparation can sometimes give accurate estimates of fiber diameter and sarcomere length (Hooper and Hanrahan, 1975). However, data may be affected by the degree of fiber mixing, and this needs to be checked constantly throughout an experiment. In another study, Khan et al. (1981) found that 25 fibers from a macerated sample did not give an accurate estimate of fiber diameters. Considerable biological variation between and within muscles can occur in muscle fiber diameters (Johnson and Beattie, 1973). The zonation of variance needs to be mapped before sampling sites are chosen for an experiment. In the bovine longissimus dorsi, for example, fiber diameters are at a minimum over rib 12 and along the lateral edge of the muscle (Swanson et al., 1965). In macerated muscle preparations the whole cross-sectional area of a muscle can be macerated in a blender to encompass intramuscular variation. Even in a rapidly rotating blender, however, there is a risk of nonrandom sampling if large numbers of fibers are dragged around together in their localized connective tissue framework. The answer to this problem is to use a large-capacity blender, and to use lots of chopped subsamples suspended in a large volume of liquid.

Although muscle tissues which have been hardened in formaldehyde can be sectioned by hand using a razor blade, histological cross sections require that the tissue should be hardened by impregnation with wax or by freezing. Wax embedding may cause shrinkage of muscle fibers (Naude and Hegarty, 1970). Freezing does not cause changes in muscle cross-sectional areas (Gunn, 1976), but there are other problems with frozen tissue. Frozen tissue melts just in front of the microtome blade as it cuts through the tissue (Thornberg and Mengers, 1957) and there is a risk of tissue deformation. Deformation is inevitable if frozen sections are not thawed symmetrically onto glass microscope slides or coverslips. Another potential problem with frozen sections is that unfixed myofibrils may shrink or expand in response to certain buffers used in histochemical reactions. Shrinkage leads to an apparent increase in the intercellular space. On the other hand, when fibers expand, the muscle section is forced upward, off the microscope slide, and it often adopts the appearance of a pie crust with a doubled-over edge. This occurs when dome-shaped individual muscle fibers are flattened back onto the microscope slide by a coverslip.

Prerigor (immediate postmortem) muscle samples which have been rapidly frozen in liquid nitrogen and sectioned while frozen will contract as soon as they are thawed onto a microscope slide. With longitudinal sections, this leads to the extensive disruption of the structure of the muscle fiber unless calcium ions are chelated as the section thaws out (Pearson and Sabarra, 1974). With transverse sections, the sectioned muscle fibers probably contract into the depth of the section. In a typical transverse section with a depth of 10 μm, there are about four sarcomeres in the depth of the section (assuming that the sample originated from a muscle at rest length). Very little thought has been given to the effect of the degree of overlap between thick and thin filaments in histochemical reactions for myofibrillar ATPase. Differences between muscle fibers in their degree of ATPase activity sometimes become more obvious at a short sarcomere length when there are more sarcomeres in the depth of the section (Swatland, 1975b).

Before the advent of relatively inexpensive microcomputers, the direct measurement of muscle fiber cross-sectional areas was extremely tedious. Using photomicrographs or tracings made with a projecting microscope, areas were measured with a *planimeter* or, even more tediously, by weighing cutout silhouettes, or by counting squares of graph paper. In the geometrical formula for the area of a circle, the diameter is squared. Thus simple substitution of an overall mean diameter into the formula gives an underestimate of mean fiber cross-sectional area (Table 7-1). With mechanical *particle size analyzers*, areas have to be calculated individually for each unit increment in diameter.

Simple modern methods use a *digitizer* to relay the X:Y coordinates of a movable pen or *stylus* to a microcomputer. The microcomputer is programmed to calculate the areas described by the manually operated stylus movements. The position of the tip of the stylus can be visualized directly in a microscope field by using a *camera lucida* and a *light-emitting diode* on the tip of the stylus. Alternatively, the human operator can work from a television screen in which are seen both the microscope field and a moving cursor which indicates the relative position of the stylus. In more sophisticated *image analysis systems*, the microscope field is scanned with a television camera and image patterns are recognized and measured by computer. A major problem with automated image analysis systems is the difficulty of resolving individual muscle fibers. Two separate circles that touch each other may be resolved by the computer as a figure of

TABLE 7-1. INADEQUACY OF MEAN FIBER CROSS-SECTIONAL AREA CALCULATED FROM MEAN FIBER DIAMETER[a]

Diameter (μm)	Area (sq μm)	Frequency	Mean diameter	Mean area
5	19.6	1		
6	28.3	1		
7	38.5	2		
8	50.3	2		
9	63.6	1		
10	78.5	3		
11	95.0	4	12.525	131.927
12	113.1	5		
13	132.7	5		
14	154.0	4		
15	176.7	4		
16	201.1	3		
17	227.0	2		
18	254.5	3		

[a]In this case the mean cross-sectional area calculated from the mean diameter would be 123.2 instead of 131.9 μm^2.

eight while two separate polygons that share a side become a single complex polygon. Even for the human observer, separation of fibers is often difficult if fibers are closely packed or if they exhibit identical histochemical reactions. Silver stains can be used to delineate muscle fibers by blackening the reticular fibers between the muscle fibers. However, silver stains often combine with any exposed aldehyde bonds originating from glycogen in the axis of a fiber. Although this is easily recognized by the human observer, it can present an additional problem for computer recognition. This problem can be avoided by pretreatment of sections with amylase to remove glycogen (Swatland, 1982a).

The radial growth of muscle fibers is associated with an increase in the apparent number of myofibrils seen in cross section (Heidenhain, 1913). The apparent number of myofibrils is a useful measure of radial growth, even though it is a rather artificial parameter. Myofibrils proliferate by longitudinal splits which are seldom continuous along the complete length of a muscle fiber. Thus what looks like one myofibril in a plane above the split may appear like two myofibrils in a lower plane. In stretch-induced growth of chicken muscles, the splitting of myofibrils starts in the center of an I band, and then spreads to the A band and to the periphery of the myofibril (Ashmore and Summers, 1981). Mean myofibrillar packing density[1] changes little during growth to market weight in pigs, so that mean muscle fiber cross-sectional area and packing density can be used to estimate the apparent numbers of myofibrils.[2] Neither light nor electron microscopy is ideal for this task since myofibrils are sometimes difficult to resolve individually with both methods. Myofibrils are difficult to count with a light microscope because of their small image size. With the electron microscope, residual bridges between myofibrils may be detected if the longitudinal separation of myofibrils is incomplete. The lateral interconnection between the Z lines of adjacent fibrils is probably due to the protein *desmin* (Granger and Lazarides, 1978).

As a muscle fiber accumulates contractile proteins during growth, it also increases its volume of sarcoplasm and the number of its mitochondria. Mitochondria are very abundant in the sarcoplasm of muscle fibers of young animals. The proliferation of mitochondria may lag behind the increase in sarcoplasmic volume that occurs with fiber growth. This is most evident in white fibers. Mitochondria appear to proliferate by fission after each mitochondrion has been internally subdivided by the formation of a septum (Duncan and Greenaway, 1981).

The cross-sectional areas of muscle fibers are increased if the fibers are cut obliquely rather than in exact cross section. Oblique sections may be due to poor alignment of the tissue on the microtome chuck, but even in correctly aligned tissue, they may occur if some fibers of the sample are contracted more than others. Fibers with a lesser degree of contraction may be thrown into sinuous folds which give rise to oblique sections at intervals along the fiber length. If fibers are cylindrical, this is not a problem since the minimum diameter gives the true diameter. However, if fibers tend

[1] Mean myofibrillar packing density:

MEANmflN/A:R,test

[2] Estimate of apparent number of myofibrils:

MEANmfrA:R × MEANmflN/A:R,test

to be polygonal in cross section, minimum diameters may underestimate true mean diameters (Figure 7-1).

Longitudinal Dimensions

The problem of muscle contraction and sarcomere length that was mentioned earlier, reappears in the measurement of muscle length, fascicular length, and fiber length. In complex muscles with fasciculi that diverge to follow an irregular muscle shape, muscle length cannot be resolved as a single parameter. In the masseter muscle of the pig's jaw, for example, Herring et al. (1979) found that there were complex patterns of variation in sarcomere length which followed muscle architecture. The masseter muscle is a particularly difficult example because its fasciculi are activated in a successive and partly overlapping sequence (Weijs, 1980), but many major carcass muscles are almost as complex.

In commercial carcasses, some muscles are stretched by the method of carcass suspension, whereas other muscles are free to contract. The extent to which unrestrained muscles contract is influenced by preslaughter factors (Chapter 9). One way to measure the longitudinal growth of a complex muscle without first undertaking a pilot study to define sarcomere length is to divide muscle length by sarcomere length. Thus the *sarcomere* becomes the unit of measurement for muscle length. Proper attention must be given to any possible growth-related changes of sarcomere length in neonatal animals. A broad sampling base along the muscle is also necessary. Samples are best taken soon after muscles have set in rigor mortis, so that both gross and microscopic measurements can be related to the same sarcomere length. In complex muscles, fascicular length may be related trigonometrically to muscle length, depending on the angle of fascicular divergence and on muscle fiber diameters (Figure 7-3). Sarcomere length can be measured by *laser diffraction* or by light microscopy; the former method does not require as many sample measurements (Cross et al., 1981).

In all except a few small pennate muscles whose fasciculi are composed of fibers that run from one end to the other, no routine method of measuring fiber length in meat animals has yet been devised. The factor that cannot easily be measured is the

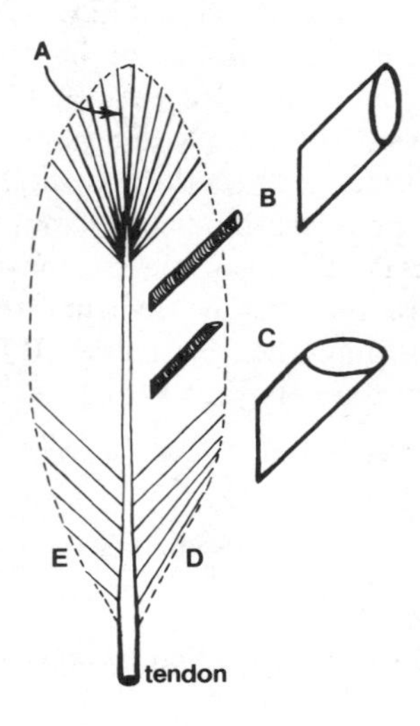

Figure 7-3 Patterns of fiber growth in a pennate muscle: (A) contribution to muscle length from the longitudinal growth of some muscle fibers; (B) contribution to muscle length from the radial growth of some muscle fibers; (C) contribution to muscle cross-sectional area from the radial growth of some muscle fibers. The pennate structure of the muscle may involve fibers of a similar length (D) or fibers with a similar angle of insertion (E).

fractional relationship that relates the length of intrafascicularly terminating muscle fibers to the length of their fasciculi.[3] There is no guarantee that this fraction is constant during development. Prenatally, fibers may grow in length faster than their fasciculi (Couteaux, 1941). Postnatally, if the initial difference between the length of intrafascicularly terminating muscle fibers and their fasciculi[4] remains constant, it becomes a smaller relative part of fascicular length.[5]

Fiber Number

In small muscles and in large muscles that have been longitudinally subdivided, muscle fibers can be counted directly. In the large muscles of the carcass, the numbers of muscle fibers can be estimated indirectly from representative samples. However, in the major carcass muscles of meat animals, it is unlikely that the *apparent number* of fibers in a cross section will include all the *real number* of fibers in the whole muscle. This discrepancy may be due to the fascicular arrangement within a muscle (Figure 7-4), or to the problem of fibers that do not extend the whole length of their fasciculus. In embryonic muscle, MacCallum (1898) first demonstrated this problem with reference to the determination of myoblast numbers. The same problem exists in adult animals with intrafascicularly terminating muscle fibers (Figure 7-5).

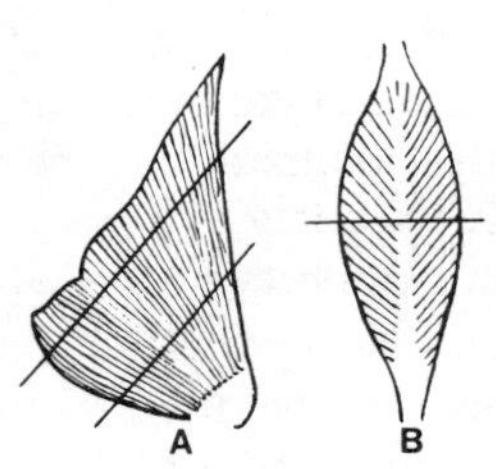

Figure 7-4 Apparent fiber numbers in complex muscles. In muscle A, a complex external shape causes the apparent fiber number to differ between planes of sectioning. In muscle B, fibers are sectioned obliquely and the maximum apparent number is always less than the real number.

Apparent fiber numbers are often estimated from a combination of muscle cross-sectional area and histological data which describe the packing density of fibers per unit area.[6] Another way to detect changes in apparent fiber numbers is to examine the regression of whole muscle area against fiber cross-sectional area. Slopes greater than, or less than 1 may indicate changes in apparent fiber numbers, provided that changes in the proportions of intramuscular adipose or connective tissue can be ruled out. There are a number of criteria that must be satisfied before estimates of apparent fiber

[3] Fiber length as a fraction of fascicular length:

$$\text{mfr:iftL} \;/\; \text{mfsL}$$

[4] Difference between fiber length and fascicular length:

$$\text{mfsL} - \text{mfr:iftL}$$

[5] If muscle fiber arrangement is constant, longitudinal growth may change the proportions of fiber length to fascicular length:

$$(\text{mfsL} - \text{mfr:iftL}) \;/\; \text{mfsL} \longrightarrow 1$$

[6] Apparent fiber number:

$$\text{EST:APPmfrN/A:R} = \text{mA:R} \times \text{MEANmfrN/A:R,test}$$

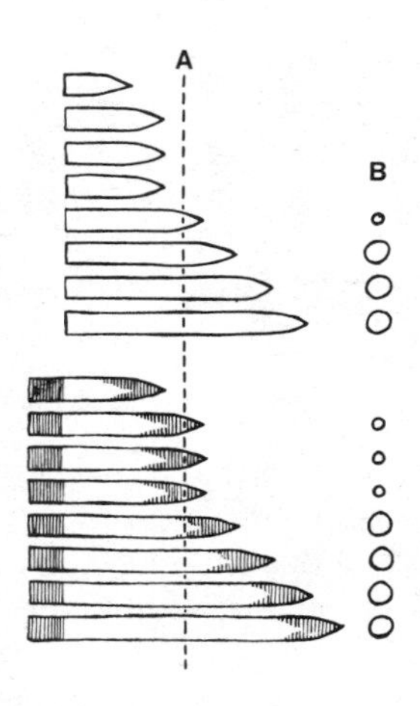

Figure 7-5 MacCallum's hypothesis. The arrangement and relative lengths of fibers may cause the number of fibers in a plane of sectioning (A) to vary, and the apparent number to change (B). All fibers have grown in length by the same amount (shaded area represents new sarcomeres) and both young (upper half of figure) and old (lower half of figure) muscles are sectioned at their midlength.

numbers acquire any validity. Unfortunately, it is not uncommon to find some or all of these criteria being violated in agricultural and biomedical research publications.[7]

1. Both gross and histological cross sections must be in the same or in nearby parallel planes. For example, if the cross-sectional area of the longissimus dorsi perpendicular to the vertebral column is measured as a gross muscle area, then histological sections must be made in a parallel plane. In this example, muscle fibers will be sectioned obliquely in both gross and histological planes since the fibers of the longissimus dorsi run at an angle to the vertebral column (Eisenhut et al., 1965).
2. No uncorrected tissue shrinkage or expansion can be allowed to occur between the measurement of muscle cross-sectional area and muscle fiber packing density. If large muscle areas are traced on a carcass, histological measurements must be taken without any subsequent change in tissue dimensions. This limits available histological techniques to frozen sections. With small muscles such as those found in poultry, whole or longitudinally subdivided muscles can be processed by a method that might entail tissue shrinkage. In this case, both whole muscle cross-sectional areas and fiber packing density per unit area can be determined on the same sections. Shrinkage then affects both measurements equally.
3. Fiber packing density[8] must be determined on a test area that is a true geometrical subunit of the whole muscle area. Having defined a unit area in the microscope field using a *micrometer grid* or the frame of a photomicrograph, fiber cross sections with a central axis within the defined area are counted. The central axis of a muscle fiber cross section is analogous to its center of gravity. If the central axis of a fiber is outside the defined area, this fiber is excluded, even if it partly protrudes into the defined area. If such fibers are included in any way,

[7]The violations listed here are used as the warnings (:WNG) for *class of measurement* in the morphometric notation described in the appendix.

[8]Muscle fiber packing density in a test area:

mfrN/A:R,test

it may lead to an overestimation of fiber numbers (Jiminez et al., 1975). An alternative method is to use a biologically defined test area, such as a muscle fasciculus. In this case, fascicular cross-sectional area is measured by planimetry and the fibers in the fasciculus are counted directly.

4. Sites for the measurement of fiber packing density must be selected using a predetermined plan that will encompass intramuscular inclusions (such as connective, adipose, neural, and vascular tissues) as well as intramuscular variation in packing density. The predetermined plan should be decided as a result of a preliminary study of the zonation of variation. The histological sampling base must include all the areas that are bounded by the whole muscle cross-sectional area. In cases where both gross and microscopic areas have been determined on shrunken tissue, this might include cracks and empty areas in the tissue block.
5. Incoming data on fiber packing density must be checked periodically to ensure that a sufficient number of observations is being made. A simple way to satisfy this requirement is to monitor the approach of the sample mean to the overall population mean as progressively more sites are sampled (Figure 7-6). Rhythmic peaks may appear on this curve as the line of sampling repeatedly scans through intramuscular zones of variation.
6. Great care is necessary in the histological and optical discrimination of individual muscle fibers. In fetal muscle, the separation of secondary fetal muscle fibers from their primary myotubes must be recognized. Whether secondary fibers are counted while they are still on their myotubes or after they have been detached is a point that must be standardized within the study. In postnatal muscle, intrafascicularly terminating fibers create a similar problem. The observed frequency of occurrence of intrafascicularly terminating fibers is affected by histological technique; their observed incidence is greater with techniques that make them clearly visible. In frozen sections, they are easily detected if the endomysial boundaries between fibers are stained with silver. In embedded tissue, they become more conspicuous if the fibers are shrunken and rounded in cross section.

Determination of muscle fiber packing density is a rather tedious operation, and automated methods have been developed. Thompson et al. (1979) developed a method in which relatively thick frozen muscle sections were broken up to free all their

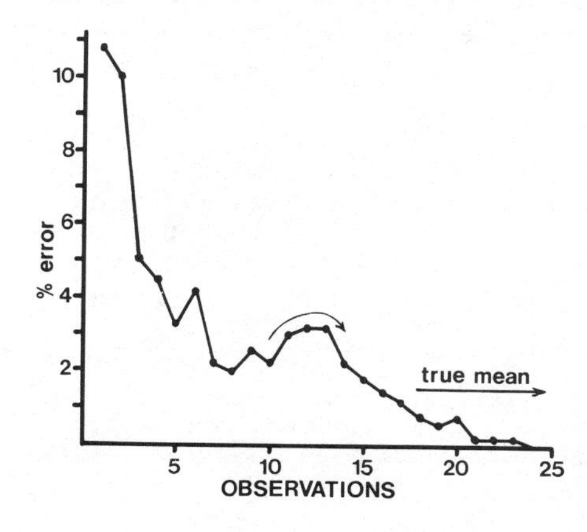

Figure 7-6 Typical incoming data on muscle fiber measurements. The mean value slowly approaches the true mean which would be known if all fibers were measured. As observations are collected across zones of variation, the error increases for a while (indicated by curved arrow). (From Swatland, 1979a.)

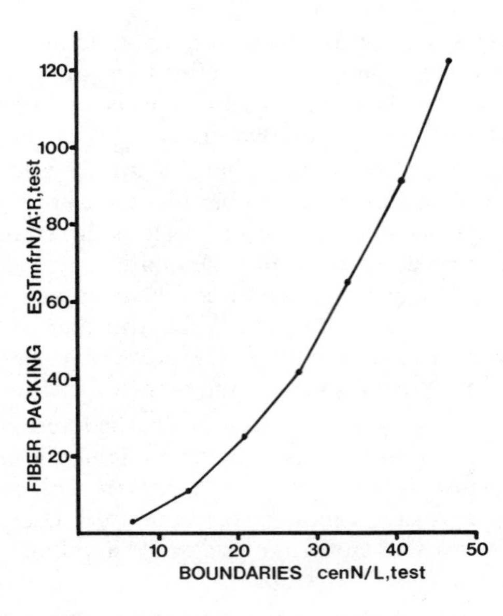

Figure 7-7 Relationship between muscle fiber packing density and endomysial boundary frequency. (From Swatland, 1979a.)

component muscle fibers; these were then counted with a *Coulter counter*. Alternatively, an automated *scanning stage* can be used to improve the yield of data in the estimation of fiber packing density. In this method, fiber boundaries rather than fiber axes are counted (Swatland, 1979a). Boundary frequency is related to packing density (Figure 7-7).[9] If boundaries are counted automatically by *scanning microphotometry*, a correction must be made for the incidence of nonfiber boundaries due to connective, adipose, vascular, and neural tissues within the muscle.[10]

Little progress has been made so far in the application of *automated image analysis* to the study of fiber numbers in meat animals. Most of our common statistical methods for the estimation of variability are for observations that are independent and identically distributed. These assumptions may no longer be valid in data that are produced by electronic pattern recognition (Nicholson, 1978). Statistical innovation, as well as histological, optical, and electronic development, is required if we are to use the power of modern computers for the study of muscle growth at the cellular level. The mathematical principles of stereology, morphometry, and scanning methods are reviewed by Weibel and Elias (1967) and DeHoff and Rhines (1968).

[9]Relationship between boundary frequency and muscle fiber packing density: if

L,test = sides of a square test area

then

$$\text{ESTmfrN/A:R,test} = (\text{MEANcenN/L,test} \ / \ 4)^2$$

[10]Correction for the presence of nonendomysial boundaries in automatic scanning:

$$((\text{MEANa\&c\&m\&n\&vN / L,test} \ / \ 4)^2 \times 0.57) + 27.1$$

This example is for longissimus dorsi muscles of pigs at market weight.

FIBER NUMBER

Comparative Survey

In much of the scientific literature on fiber numbers, it is difficult to decide whether certain authors are describing apparent or real fiber numbers. In relatively large creatures such as humans, fiber numbers are likely to be apparent fiber numbers. Thus Montgomery (1962) found that fiber numbers in the human sartorius muscle increased during prenatal development and that they reached a peak at about 4 months after birth. In smaller creatures such as mice and rats, apparent fiber numbers at a muscle midlength might well include all the fasciculi and intrafascicularly terminating fibers within a muscle so that apparent number is equal to or is a good measure of real number.

Luff and Goldspink (1967) and Aberle and Doolittle (1976) found that genetically large mice had more muscle fibers than did genetically small mice. However, different genetic strains that have not been selected for body size may also show appropriate differences in fiber numbers (Luff and Goldspink, 1970). Hanrahan et al. (1973) and Byrne et al. (1973) selected mice for high and low body weight and found that changes in fiber numbers were accompanied by changes in fiber diameters. The sex of the mice affected their fiber diameter but not their fiber number. Differences in fiber numbers between strains selected for body weight were examined on a constant-body-weight basis. Thus mice of the strain selected for large size were younger and had more, but smaller, muscle fibers (Hooper and McCarthy, 1976).

In rats, Morpurgo (1898) and Ott (1937) found that fiber numbers increased rapidly after birth. Ott (1937) found that neonates had about two-thirds of the adult fiber number. In the medial pterygoid muscle, Rayne and Crawford (1975) found that fiber numbers doubled between birth and 6 weeks, and then declined to the adult number. In a similar study on another rat muscle, Ontell and Dunn (1978) found that many of the apparently single fibers identified by light microscopy were revealed by electron microscopy as a cluster of small fibers. They suggested that neonatal increases in fiber numbers in rats might be an illusion caused by fiber separation rather than a real proliferation of fibers.

In guinea pigs, apparent fiber numbers decrease with age, but the rate of decrease may be reduced by exercise training (Faulkner et al., 1972). In this case, the real number of fibers remains constant (Maxwell et al., 1974). In cats, fiber numbers are also greater after exercise training, and this has been attributed to fiber splitting (Gonyea et al., 1977). It is not altogether clear how one can distinguish between fiber splitting and the growth of intrafascicularly terminating muscle fibers. In horses and dogs, apparent fiber numbers increase during postnatal growth, and animals selected for racing have greater apparent fiber numbers (Gunn, 1979). Whether or not superior racing breeds have an increase in real fiber number remains to be seen, although it seems quite reasonable that they might.

Fiber numbers in fish increase during growth after hatching (Mazanowska and Kordylewski, 1975; Weatherley et al., 1979, 1980). In the eel, only white fibers increase in number. New fibers appear on the surfaces of existing normal-diameter fibers (Willemse and Van Den Berg, 1978). The ATPase activity of new small fibers is initially acid-stable, but it gradually becomes acid-labile as the fibers grow in diameter

(Carpene and Veggetti, 1981). A similar process may occur in the muscles of amphibians (Watanabe et al., 1980). These observations add a new factor to the already difficult problem of understanding the nature of small diameter fibers in a growing muscle. Several possibilities have been listed so far:

1. tapered intrafascicular terminations;
2. fiber splitting;
3. newly formed small fibers;
4. degenerating fibers;
5. intrafusal fibers extending beyond spindle capsules.

In meat animals, the first category provides the most likely explanation of small-diameter fibers, but careful thought is needed before the other possibilities can be dismissed. Petrov (1976), for example, reached the conclusion that muscle growth in pigs follows the pattern found in fish, with the constant postnasal formation of new red fibers being coupled with the transition of red fibers to become white fibers. One hesitates to question how many authors have identified intrafascicularly terminating fibers as being new, split, or degenerating fibers. Myotubes formed during the regeneration of muscle tissue after injury or disease may also branch and give rise to the appearance of small-diameter muscle fibers in cross sections (Ontell and Feng, 1981).

Cattle

Bendall and Voyle (1967) estimated apparent fiber numbers in longissimus dorsi and semitendinosus muscles of Hereford and Friesian steers from 11 days to 2 years of age. The calculation of apparent fiber numbers was based on a complex series of propositions.[11] The longissimus dorsi started with 75 per cent of the peak postnatal number, which was reached at 5 months. After 5 months, numbers declined sharply and later reached 50 per cent of the peak number at 2 years. The semitendinosus started with its peak fiber number at 11 days, and then steadily lost fibers to drop below 60 per cent of the initial peak number at 2 years.

Fiber Hyperplasia in Cattle

For many years, cattle with greatly enlarged muscles and a scarcity of adipose tissue have caught the attention of beef producers (Plate 22). The large superficial muscles of the shoulder and proximal hind limb are often more enlarged than the distal limb

[11] Propositions used to calculate apparent fiber numbers:

MEAN:ASSUMmDEN = 1
ESTmA:R = mW / mL
standard msarL = 2.35E-3
A:R,test was set at approximately mfrA × 100
ESTmfrA:R = A:R,test / mfrN/A:R,test
EST:PROPmfr2R = $(\text{ESTmfrA:R})^{0.5} \times (\text{msarL} / 2.35\text{E-3})^{0.5}$
EST:APP:PROP:WNG1&2mfrN = ESTmA:R / $(\text{EST:PROPmfr2R})^2$

muscles or deep muscles of the carcass (Johnson, 1981). The condition has been given a number of colloquial names, such as *doppellender* (Wriedt, 1929), *double muscling* (Weber and Ibsen, 1934), *muscular hypertrophy* (Kidwell et al., 1952), *a groppa doppia* (Raimondi, 1962), and *culard* (Lauvergne et al., 1963). A detailed historical account of the condition is given by Oliver and Cartwright (1969). Carroll et al. (1978) examined differences in the palatability of cooked longissimus dorsi muscles between heterozygous double-muscled and normal animals. The effect of double muscling on meat flavor and overall acceptibility was minor. Although the greatly increased yield of lean meat in double-muscled animals is of considerable interest commercially, there is a long list of disadvantages and physiological abnormalities associated with the condition.

1. underdevelopment of male (Michaux et al., 1980) and female (Menissier, 1982b) reproductive tracts;
2. *macroglossia* (enlarged tongue), which may make it difficult for the calf to nurse (Oliver and Cartwright, 1969);
3. weak bones (Magliano, 1953).
4. erythrocyte fragility (King et al., 1976);
5. changes in *creatine* metabolism (Ansay and Gillet, 1966);
6. changes in pelvic dimensions associated with difficult calving (Vissac et al., 1973);
7. poor lactation (Lauvergne et al., 1963);
8. reduced adipose reserves when subjected to nutritional stress (Holmes and Robinson, 1970);
9. changes in growth hormone, luteinizing hormone, *testosterone*, and *insulin* (Michaux et al., 1982);
10. reduced weight of hide (MacKeller, 1960) and liver (Mason, 1963);
11. reduced collagen content, measured by reduced *hydroxyproline* content (Lawrie et al., 1964; Boccard, 1982) and histologically detectable as a reduction in the perimysium (Dumont and Schmitt, 1973);
12. longer gestation period (Raimondi, 1962);
13. lower levels of *thyroxine* (Novakofski and Kauffman, 1981);
14. susceptibility to heat stress (Halipre, 1973);
15. small feet and cannon bones (Mason, 1963);
16. growth in live weight may be less than normal (Menissier, 1982a).

The genetic basis of the condition is due to a single pair of autosomal genes. Heterozygous animals may range from normal to extremely double-muscled phenotypes (Hanset, 1967; Rollins et al., 1972; Oliver and Cartwright, 1969; Lauvergne et al., 1963). *Incomplete dominance, incomplete penetrance*, and *modifier genes* are the usual explanations that are proposed to explain the phenotypic range of heterozyous animals, but it is difficult to locate any proof that the range is solely due to genetic interactions at the DNA level. The gene for double muscling is also *pleiotropic*, and affects the numerous physiological and anatomical systems listed earlier. Vissac et al. (1973) reported that the expressivity of double muscling in heifers was affected by their plane of nutrition. It is equally plausible, therefore, that incomplete dominance and pleiotropy might be a result of *epigenetic* interactions. Epigenetics is the science concerned with the causal analysis of development (Waddington, 1952): it encompasses genetic interactions which are mediated by the interactions of gene products in the cells of the body tissues. The development of epigenetics as a distinct scientific discipline was, unfortunately, eclipsed by the dramatic discoveries made in the chemistry of nucleic acids. The literature relating to epigenetics is now widely scattered through a number of related disciplines, such as biochemistry, cell biology, and quantitative genetics. In animal agriculture, however, epigenetics provides the link between the genetic manipulation of genotypes and the pragmatic collection of data from slaughtered phenotypes.

Double muscling results primarily from an increase in the apparent number of muscle fibers. As far as is known at present, the increase in apparent fiber numbers is based on an increase in real fiber numbers. Together with the hyperplasia of extrafusal muscle fibers, double-muscled cattle may also have an increase in the number of the intrafusal fibers in their neuromuscular spindles (Swatland, 1982c). MacKeller (1968) estimated apparent fiber numbers[12] in extensor digitorum longus and extensor digitorum medialis muscles of a double-muscled animal. Two normal animals of the same breed (South Devon) were used as the basis of comparison. Results are given in Table 7-2.

TABLE 7-2. APPARENT FIBER NUMBERS IN DOUBLE-MUSCLED CATTLE

Parameter	Double-muscled	Normal	Normal
Live weight (kg)	498	559	504
Extensor digitorum longus	250,000	184,000	122,000
Extensor digitorum medialis	202,000	159,000	215,000

Source: Data from MacKeller (1968).

Although these data were inconclusive, MacKeller (1968) showed that muscle enlargement was not caused by radial enlargement of muscle fibers since, in extensor digitorum longus, extensor digitorum medialis, and cutaneus muscles, mean fiber diameter was less in the double-muscled animal. The latter finding was supported by the more extensive study of Ouhayoun and Beaumont (1969), who estimated fiber-packing density[13] by weighing fiber areas cut from photomicrographs. In sternohyoideus muscles (*not* enlarged in double-muscled animals) fibers were smaller

[12]Estimate of apparent fiber numbers:

EST:APPmfrN/A:R;0.5L,m

from

mA:R;0.5L × ESTmfrN/A:R;0.5L,test

[13]Fiber packing density was calculated from

MEANmfrN/A:R,test0.12

and

MEANmfrA:R

than in normal animals. In the long head of the triceps brachii (enlarged in double-muscled animals), fibers were the same size as in normal animals. This showed that double-muscled animals probably had extra fibers, even in carcass muscles which were not enlarged, as well as in those that were. Subsequent studies (MacKeller and Ouhayoun, 1973) showed that the double-muscled condition in South Devon and Charolais breeds was identical.

Swatland and Kieffer (1974) estimated apparent fiber numbers[14] in sartorius and extraocular anterior rectus muscles of double-muscled fetuses (Plate 23). Whole muscles were fixed, embedded, and sectioned. In the sartorius, apparent fiber numbers were doubled at each stage of development relative to normal fetuses (Figure 7-8). Fiber numbers were adjusted to a common basis relative to the number present at a crown-rump length of 26 cm, to find out if the extra fibers in double-muscled animals were formed before or after fetuses had reached a length of 26 cm. Evidence of extra fiber formation after 26 cm was minimal, and it was concluded that the hyperplastic condition had an earlier origin. The major origin of extra fibers was attributed to an increase in the real number of primary fetal fibers (myotubes) and to cellular changes prior to 26-cm crown-rump length. Fiber diameters were not significantly different between normal and double-muscled fetuses. In contrast to the situation in the sartorius muscle, fiber numbers were not doubled in the anterior rectus muscle. The embryological origin of extraocular muscles is from the *pro-otic somites* of the head (Goodrich, 1958). These somites are rather unusual in the way in which they develop and in their innervation, but whether this has any connection with their lack of response to the gene for double muscling is unknown. Dry weights of a range of carcass muscles were all increased in double-muscled fetuses. Muscle enlargement may also be widespread in the carcasses of adult cattle (Butterfield, 1966).

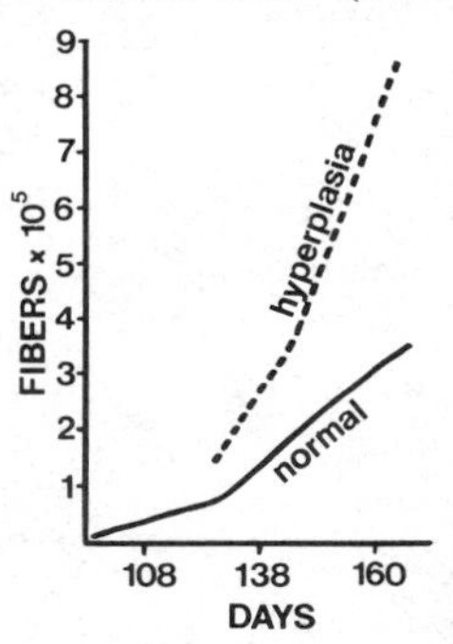

Figure 7-8 Development of apparent fiber numbers in sartorius muscles of normal and double-muscled bovine fetuses. Double-muscled fetuses have more muscle fibers (hyperplasia). The age scale is in days from conception, estimated from fetal length. (Data from Swatland and Kieffer, 1974.)

The extent to which increases in fiber diameter may contribute to muscle enlargement in double-muscled animals is unclear. Ashmore and Robinson (1969) examined semitendinosus and triceps brachii muscles, and concluded that double muscling was due to a disproportionate number of glycolytic-type fibers, and that the larger diameters of these fibers could explain muscle enlargement. Holmes and

[14]Apparent fiber numbers,

EST:APPmfr:fp&fsN/A:R;0.5L

were estimated from

mA:R;0.5L × MEANmfr:fp&fsN/A:R;0.5L,test2.88E-3

Ashmore (1972) examined triceps brachii, cutaneus, and semitendinosus muscles, and found that: (1) double-muscled cattle had fewer alpha red fibers; (2) conversion of alpha red to alpha white fibers was greater than in normal animals; and (3) alpha white fibers were much larger than red fibers relative to normal animals. Studies on biceps femoris, semitendinosus, semimembranosus, longissimus dorsi, and cutaneus muscles in double-muscled animals have confirmed these findings (Hendricks et al., 1973; West, 1974). Changes in fiber-type ratios have been traced back to fetal development in semitendinosus, radialis, and cutaneus muscles (Ashmore et al., 1974). Glycolytic metabolism is enhanced in the muscles of double-muscled animals, and this produces some degree of stress susceptibility and an increased incidence of pale meat (Holmes and Ashmore, 1972). In European double-muscled cattle, Ansay (1973) was unable to find any decrease in myoglobin concentration.

Pigs

Staun (1963, 1972) reviewed an extensive series of experiments undertaken on fiber numbers in Danish pigs. Fiber numbers in longissimus dorsi muscles[15] were estimated[16] using frozen sections which were cut from meat taken 24 hours post-mortem and fixed in formalin. Staun (1963, 1972) concluded that (1) fiber numbers differ between breeds (Table 7-3); (2) fiber numbers are unaffected by nutritional

TABLE 7-3. BREED DIFFERENCES IN APPARENT FIBER NUMBERS IN THE LONGISSIMUS DORSI MUSCLES OF PIGS

Animals	Fiber Number
Pietrain (P) gilts	1,078,000
Pietrain barrows	984,000
Danish Landrace (DL) gilts	1,124,000
Danish Landrace barrows	988,000
P × DL gilts	1,005,000
P × DL barrows	934,000
Duroc high-fat line	659,000
Duroc low-fat line	802,000
Yorkshire high-fat line	738,000
Yorkshire low-fat line	797,000
Veredeltes Landschwein gilts	869,000
Veredeltes Landschwein barrows	940,000
Deutsches weisses Edelschwein gilts	758,000
Deutsches weisses Edelschwein barrows	717,000

Source: Data from Staun (1963).

[15]Estimate of apparent fiber numbers:

EST:APP:WNG1&2&4mfrN/A:R;0.6L

[16]Basis and location of estimate:

mA:R;0.6L × MEANmfrN/A:X?L;0.6L,test0.17

TABLE 7-4. BREED AND LINE DIFFERENCES IN APPARENT FIBER NUMBERS IN THE LONGISSIMUS DORSI MUSCLES OF PIGS

Animals	Fiber Number
Hampshire, line 1	833,530
Hampshire, line 2	876,450
Yorkshire	749,910
Hampshire 1 × Yorkshire	860,730
Hampshire 2 × Yorkshire	931,670

Source: Data from Miller et al. (1975).

treatment; (3) fiber numbers do not differ between sexes; and (4) fiber numbers remain constant during growth to market weight. Heritability estimates for fiber numbers in gilts and barrows were 0.43 and 0.48, respectively.

Davies (1972) examined the relationship between the cross-sectional areas of whole longissimus dorsi muscles and the cross-sectional areas of their muscle fibers.[17] The slope did not differ significantly from 1. Davies (1972) concluded that, after 10 days of age, growth in muscle area was due to the growth of a constant population of muscle fibers.

Miller et al. (1975) also investigated apparent fiber numbers in the longissimus dorsi muscles of different genetic lines of pigs.[18] Genetic lines differed in their fiber numbers (Table 7-4). Fiber number was more closely related to measures of muscle mass than was the diameter of the fibers. However, the estimate of heritability for fiber number was zero. No differences in fiber numbers were detected between barrows and gilts.

Stickland and Goldspink (1973) developed the idea of using a small muscle from the foot of the pig to predict real fiber numbers in major carcass muscles. It was suggested that a small muscle could be surgically removed from potential breeding stock so as to select animals on the basis of a high fiber number. Real fiber numbers[19] were determined by direct counting in a number of small muscles:

Extensor digitalis communis

Rhomboideus capitis

Extensor digiti I longus

Flexor digiti V brevis (indicator muscle, fdvb)

[17]Comparison of muscle areas and fiber areas:

LOGmA:R versus LOGmfrA:R

[18]Apparent fiber numbers,

EST:APP:WNG1&2mfrN/A:R;0.4L

were calculated from

mA:R;0.4L × MEANmfrN/A;X?L;0.4L,mfs

[19]Real fiber numbers:

REALmfrN/A:R

Muscles were removed from several pigs which were obtained from local farms and which were usually less than 4 weeks old. Fiber numbers in the indicator muscle were correlated with fiber numbers in the other three muscles. It was concluded that the animals had differed genetically in their fiber numbers and that, in each animal, all muscles were similarly affected. Stickland et al. (1975) showed that the real number of fibers in the flexor digiti V brevis remained constant at different ages and that fiber numbers differed between litters. Miniature pigs were found to have reduced numbers of muscle fibers (Stickland and Goldspink, 1978).

Stickland and Goldspink (1973) also estimated apparent fiber numbers in longissimus dorsi (ld) and sartorius (sar) muscles during growth from 2 to 215 days. Apparent numbers showed no change with age.[20] Selection on the basis of an indicator muscle is a good idea, at least in theory, and deserves further consideration.[21] To make it worthwhile in practice, however, the correlation of real fiber number in the indicator muscle with apparent fiber number in major carcass muscles will have to be quite strong. In a later report, Stickland and Goldspink (1975) concluded that fiber numbers in the flexor digiti V brevis were inversely related to fat depth measurements. A deduction that might be drawn from this is that present selection methods which favor decreased fat depth are already acting on real fiber numbers.

A major problem with many of the studies that have been published on muscle fiber numbers in the longissimus dorsi muscles of pork chops is that they relate to apparent fiber numbers, not to real fiber numbers. If whole longissimus dorsi muscles are subdivided and the total apparent numbers of fibers are counted, it may be found that only a small and rather variable fraction (6% to 12%) of the total apparent number appears in the longissimus dorsi muscle when it is transected in a single pork chop (Swatland, 1975c). Thus any factors that influence the magnitude of this fraction are superimposed on what at first sight appears to be the number of fibers in the muscle. The factor that causes most concern, since it varies postnatally, is muscle fiber length. This factor is discussed later in this chapter.

At the midlength of sartorius muscles of pigs, apparent fiber numbers may show a steady increase throughout fetal development and onward during growth to market weight (Figure 7-9). Furthermore, if body growth is arrested by a low plane of nutrition, the increase in apparent fiber numbers is stopped or reversed. Apparent fiber

[20]Constancy of apparent fiber numbers:

ldEST:APPmfrN/A:R;0.6L = approximately 730,000

sarEST:APPmfrN/A:R;?L = approximately 65,000

fdvb:REALmfrN/A:R;0.5L = approximately 3,000

[21]To evaluate the potential of the indicator muscle as a basis for the prediction of fiber numbers in major carcass muscles, it is necessary to know the coefficient of correlation (r) between fiber numbers in the indicator muscle and fiber numbers in the major carcass muscles. Using data presented by Stickland and Goldspink (1973) in their figure 6, approximations can be calculated ($n = 5$) as

fdvb versus sar, $r = 0.9$, $P < 0.05$

fdvb versus ld, $r = 0.45$, NS

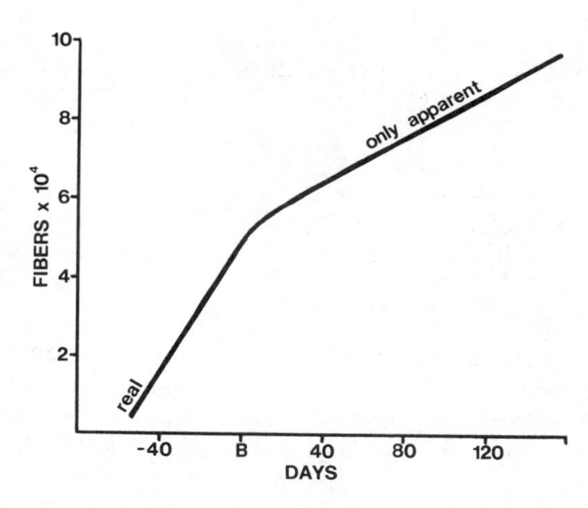

Figure 7-9 Increase in apparent fiber numbers at the midlength of the sartorius muscle in pigs. Time scale in days before and after birth (B). [Data from Swatland (1973, 1976a), but data may vary between breeds and with different counting methods.]

numbers[22] were estimated[23] using whole muscle sections from fixed and embedded tissue. Muscles from larger animals were subdivided longitudinally and the data were combined (Swatland, 1973, 1975b, 1976a). In pigs, therefore, there is a considerable difference of opinion on the interpretation of data on fiber numbers. There is one school of thought that considers that real and apparent fiber numbers are fixed after birth. Another school considers that real fiber numbers reach their maximum early in development and that increases in fiber number after the neonatal period are restricted to changes in apparent numbers. In a third school, it is thought that both real and apparent numbers of fibers continue to increase due to the formation of new fibers for several years after birth (Petrov, 1976). A consideration of longitudinal growth later in this chapter provides the answer to this dilemma.

Sheep

Hammond (1932) speculated that muscle fiber numbers in some breeds of sheep had been increased as a result of selective breeding for increased meat yield. Joubert (1956a) was unable to account for the greater muscularity of male lambs, relative to females, on the basis of differences in fiber diameter: he concluded that males had more muscle fibers. During growth to market weight, data (from table 23 of Joubert, 1956a) show that the radial growth of muscle fibers in the longissimus dorsi does not keep pace with the radial growth of the whole muscle. If the accumulation of intramuscular adipose tissue could be ruled out as a contributory factor, this might be regarded as evidence of an increase in apparent fiber numbers.

Everitt (1968) examined the effects of a high plane and a low plane of maternal nutrition on apparent fiber numbers in the semitendinosus muscles of fetal lambs.

[22] Apparent fiber numbers:

EST:APPmfrN/A:R;0.5L

[23] Basis of estimate of apparent fiber numbers:

mA:R;0.5L × MEANmfrN/A:R;0.5L,test

TABLE 7-5. EFFECT OF MATERNAL PLANE OF NUTRITION ON APPARENT FIBER NUMBERS IN SEMITENDINOSUS MUSCLES OF FETAL LAMBS

Fetal Age	High Plane	Low Plane
90 days gestation	63,810	55,270
140 days gestation	85,630	59,100

Source: Data from Everitt (1968).

Apparent fiber numbers[24] were estimated[25] using sections cut from fixed and embedded tissue (personal communication). The results are summarized in Table 7-5. Competition between twin and triplet fetuses may cause a similar effect on apparent fiber numbers[26] in sartorius muscles (Swatland and Cassens, 1973). In this case, however, data were somewhat equivocal because of a problem in muscle geometry.

Muscle fiber hyperplasia can occur as a rarity in sheep. Dennis (1972) described a single case of muscle fiber hyperplasia in the hindlimb muscles of a grossly deformed lamb. Naerland (1940) described fiber hyperplasia in the carcasses of three adult sheep. The increase in muscle weight, averaged over a large set of representative carcass muscles, was 122 per cent (relative to unaffected controls). The bulging shapes of muscles were separated by conspicuous creases, similar to those found in double-muscled cattle.

Poultry

Smith (1963) compared the radial dimensions of muscle fibers in White Gold and in White Leghorn chicks. Differences in the radial dimensions of muscle fibers could not account for breed differences in muscularity, and it was concluded that the more muscular breed had more muscle fibers. However, fiber size was found to be more important than fiber number in explaining the differences in muscle mass between chicks with large and with small bodies. Goldspink (1977) considered that selection on the basis of fiber numbers in an indicator muscle could be adapted for use in poultry.

Dwarf chicks with crooked necks, due to an autosomal recessive gene, exhibit *hypoplasia*, a decrease in their number of muscle fibers (Wick and Allenspach, 1978). The defect becomes apparent after primary myotubes have developed, and it may be a secondary consequence of a disorder in connective tissue development.

Fiber-Type Ratios

The biological basis of the physiological differentiation of muscle fibers was considered in Chapter 6. Fiber-type ratios describe the frequency of occurrence of different physiological fiber types. Fiber-type ratios may differ between muscles and

[24]Apparent fiber numbers:

EST:APP:WNG2&4&6mfrN/A:R;0.5L

[25]Basis of estimate of apparent fiber numbers:

mA:R;0.5L × MEAN:WNG2&4mfrN/A:R;0.5L,test

[26]Apparent fiber numbers:

EST:APP:WNG6mfrN/A:R;0.5L

between parts of muscles. They may also change during growth. Although the importance of fiber-type ratios may have been overestimated in earlier research, they are, nevertheless, an important parameter of muscle development in meat animals.[27]

In cattle, Melton et al. (1975) detected sire effects in fiber-type ratios and found a relationship between fiber-type ratios and the juiciness of cooked meat. In lambs, Moody et al. (1980) concluded that beta red fibers made the greatest contribution to palatability. Limousin × Angus crossbred steers have slightly more alpha white and slightly fewer alpha red fibers than Simmental × Angus or Hereford × Angus crosses (May et al., 1977). Miller et al. (1975) found that longissimus dorsi muscles of Yorkshire breed pigs had a greater proportion of fibers with a high content of lipid droplets.

RADIAL DIMENSIONS

Information concerning fiber numbers and longitudinal muscle growth in laboratory animals is important because these parameters are only poorly understood in meat animals. Our knowledge of the radial growth of muscle fibers in meat animals, however, is better established and is self-sufficient. Early work, mainly on sheep, showed that a combination of the radial and longitudinal growth of muscle fibers accounted for the increase in muscle mass that occurred between birth and the attainment of market weight. Longitudinal growth, however, was seldom measured and too much importance was attached to the more easily measured radial dimension. At one time it was even suggested that fiber diameters might be used to measure muscularlity in meat animals. Hopes in this direction faded rapidly once the complexity of the interactions between fiber number, length, and radius became known. Muscle fiber diameters may be negatively correlated ($r = -0.37$) with the tenderness of cooked meat (Ramsey et al., 1967). Possibly, the relationship between fiber diameter and meat tenderness is independent of secondary factors such as connective tissue strength and sarcomere length. The relationship between fiber diameter and tenderness is curvilinear in nature (Hiner et al., 1953).

Sheep

Although Hammond (1932, p. 483) knew of Huber's (1916) work on intrafascicularly terminating fibers, he concluded that the problems created by intrafascicularly terminating fibers could be ignored. Mean fiber diameters halfway along muscles were shown to be smaller than the mean diameters at each end.[28] Since the proportionality of these two mean values was approximately constant, it was decided that meaningful comparisons between muscles and between animals could be made solely on the basis of midlength samples. This decision simplified the study of muscle geometry and

[27]Following the nomenclature used for muscle fiber numbers, fiber-type ratios can be regarded as a component of muscle fiber packing density per unit area:

mfrN/A:R = mfr:rN/A:R + mfr:iN/A:R + mfr:wN/A:R

[28]Basis of fiber diameter measurement:

MEANmfr:ete&ift2R/A:R;0.5L = 40.1E-3

MEANmfr:ete2R/A:R;0&1L = 42.37E-3

enabled considerable progress to be made. Its unfortunate consequence, however, was that the analysis of muscle growth was reduced from three dimensions to two. Longitudinal growth and the involvement of intrafascicularly terminating fibers thus became invisible, and the vital parameter that links the radial geometry of a muscle to its weight was obscured.

Hammond (1932, p. 484) was also aware of the physiological differentiation of muscle fibers into red and white categories. It is now thought that red and white fibers may differ in diameter at certain stages of development. This possibility, however, was rejected by Hammond on the basis of a rather weak argument. He observed that frequency histograms of fiber diameters were unimodal. From this he argued that there was no distinct size difference between red and white fibers. This argument fails to take into account the possible coexistence of two subpopulations (red and white fibers) that might differ in their mean diameter but which might have a large overlap in their range of diameters. It is an irony, not uncommon in the history of science, that progress on one aspect of a problem should stifle curiosity concerning its other aspects. In this case, knowledge concerning fiber diameters was advanced at the expense of knowledge concerning fiber length and arrangement.

Hammond (1932) discovered that, in each region of the body, the muscles that showed the greatest percentage increase in weight after birth also developed the largest mean fiber diameters. This relationship failed in comparisons made between different body regions in which differences in fiber number and length outweighed the contributions accruing from differences in radial growth. Muscles with a dominant postural function were found to have smaller mean fiber diameters than the major propulsive muscles. Muscle fibers in rams had a larger mean diameter than those in ewes, while those of wethers were intermediate in diameter.

As lambs are grown to market weight, their muscle fibers increase in diameter. Thus if they are slaughtered at an incomplete stage of growth, fiber diameter is related to live weight and to muscle mass. Beyond a live weight of 45 kg, however, there is little further radial growth of fibers (Moody et al., 1970). An animal's plane of nutrition affects the radial growth of its muscle fibers. In ewes on a high plane of nutrition, Joubert (1956a) found that muscle fiber diameters increased in proportion to increases in the total muscle mass of the carcass. The increase in radial growth due to supermaintenance feeding was limited if the treatment was extended for a long time. Muscle fiber diameters decreased when the sheep were placed on a submaintenance plane of nutrition. Asghar and Yeates (1979) later showed that submaintenance feeding results in degeneration of actin and myosin filaments. In normal muscle fibers, the sarcoplasm between myofibrils makes an important contribution to the radial dimensions of muscle fibers. Asghar and Yeates (1979) found that submaintenance feeding decreased the number of mitochondria and glycogen granules, whereas the transverse tubular system and sarcoplasmic reticulum were unaffected.

The transformation of muscle fibers from one physiological type to another complicates studies on the radial growth of fibers. For example, fiber transformation from a type with a large diameter to a type with a small mean diameter might increase the mean diameter of the small-diameter type. This would be difficult to distinguish from radial growth. In a number of ovine muscles, White et al. (1978) and Moody et al. (1980) found an increase in the percentage of fibers with strong ATPase, and an increase in the percentage of fibers with weak aerobic enzyme activity. Differences in the radial growth of different fiber types may vary between muscles. In the ovine semimembranosus, all three major fiber types exhibit radial growth as live weight increases from 32 kg to 41 kg. However, in the longissimus dorsi, only aerobic fibers with weak ATPase exhibit radial growth during this period (Solomon et al., 1981). Fibers with weak ATPase activity and strong aerobic enzyme activity exhibit sustained radial growth and, in some situations, may grow to be larger than fibers with strong ATPase activity or fibers with weak aerobic enzyme activity. Submaintenance feeding has a greater effect on white fibers than on red fibers (Suzuki, 1973).

Pigs

Any hopes of using muscle fiber diameters to assess muscularity in pigs were quashed by Livingston et al. (1966), who measured fiber diameters[29] in longissimus dorsi samples taken by biopsy from live pigs and taken from chilled bacon carcasses. During growth from 40 kg to 125 kg live weight, mean diameters increased from 35.5 μm to 45.3 μm. In finished bacon carcasses, however, fiber diameters were only poorly correlated with indices of longissimus dorsi development.[30] Only 10 per cent of the variance in longissimus dorsi area could be attributed to mean fiber diameter in commercial carcasses. The zonation of variance in mean fiber diameter was examined at sites across the rib-eye area. Differences between sites were not statistically significant and the zonation varied from carcass to carcass. Within-pig variance was only slightly reduced by increasing the number of fibers measured from 50 to 200. The tapered ends of intrafascicularly terminating fibers were identified as the primary factor which invalidated mean fiber diameters as a measure of carcass muscularity. Miller et al. (1975) found that fiber diameters were only very weakly correlated with longissimus dorsi area (r = 0.12).

In meat animals, muscles in different parts of the body grow at different rates. This type of differential growth is called *allometric growth* (Chapter 8). Muscles located proximally within a limb have a greater relative growth than distal muscles. Muscles located posteriorly along the vertebral column have a greater relative growth than anterior muscles. McMeekan (1940a) showed that these differences in relative growth are *not* due to differences in the radial growth of muscle fibers. Longitudinal growth has so far been measured only in a few muscles of meat animals. However, the evidence which is available indicates that allometric growth of whole muscles might originate from differences in longitudinal growth (Swatland, 1982b).

Chrystall et al. (1969) found that fiber diameters in the longissimus dorsi reached 95 per cent maximum diameter by 150 days of age (Figure 7-10). Not all muscles, however, show a similar deceleration in radial growth. In the cutaneus muscle, radial growth continues at a more or less steady rate for at least 200 days (Judge, 1978).

[29]Fiber diameters:

MEANmfr2R/A:R;0.6L

[30]Relationship of fiber diameters to the size of the longissimus dorsi:

area, $r = 0.31$

width, $r = 0.23$

height, $r = 0.03$

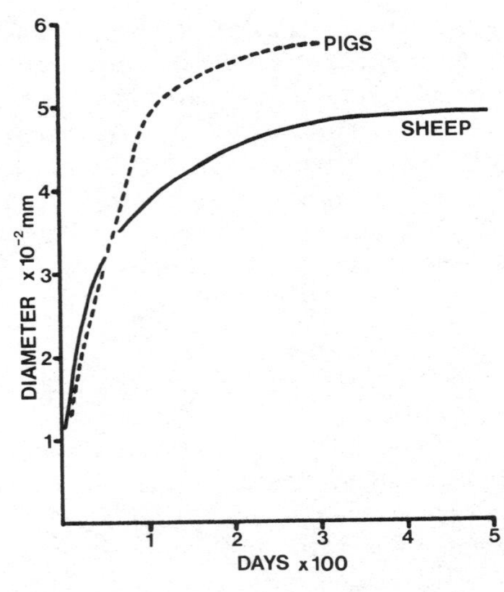

Figure 7-10 Radial growth of muscle fibers in the longissimus dorsi of pigs and in three muscles (averaged) from sheep. (Data from Chrystall et al., 1969, and from Joubert, 1956a.)

Carcasses from pigs on a low plane of nutrition have muscle fibers with a smaller diameter than those of well-fed pigs (McMeekan, 1940b). Fiber diameters are determined by an *interaction* between animal age and plane of nutrition. Thus, when compared on a constant-live-weight basis, pigs that are older and leaner may have larger-diameter fibers than pigs that are younger and fatter (McMeekan, 1940c). When pigs are fed a submaintenance ration, fiber diameters in the longissimus dorsi show a greater decrease in diameter than those of the trapezius muscle (Moody et al. 1969).

Just after farrowing, there is a dramatic increase in the radial growth of *extrafusal* muscle fibers but not of *intrafusal* fibers (Figure 7-11; Swatland, 1974). When the radial growth of different physiological types of extrafusal fibers is examined separately, fibers with weak ATPase (acid-stable ATPase) activity may show a delay of about 1 week after farrowing before radial growth can be detected. However, this delay occurs during a period when the percentage of fibers with weak ATPase activity is rapidly increasing. Consequently, the delay might be an illusion due to the recruitment of small-diameter fibers which previously had strong ATPase activity.

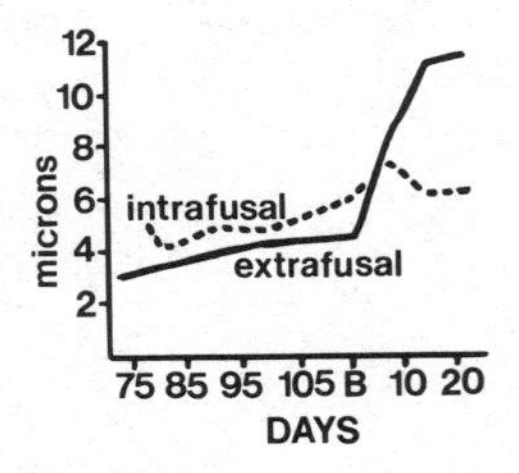

Figure 7-11 Radial growth of extrafusal and intrafusal muscle fibers in the sartorius of fetal and neonatal pigs. Age in days gestation before birth (B), and in days after birth. (Data from Swatland, 1974.)

Since the packing density of myofibrils in sartorius muscles shows little change once the neonatal period has passed,[31] the radial growth of muscle fibers is

[31]Packing density of myofibrils:

MEAN:APPmflN/A:R;0.5L,test1E-4 = approximately 40

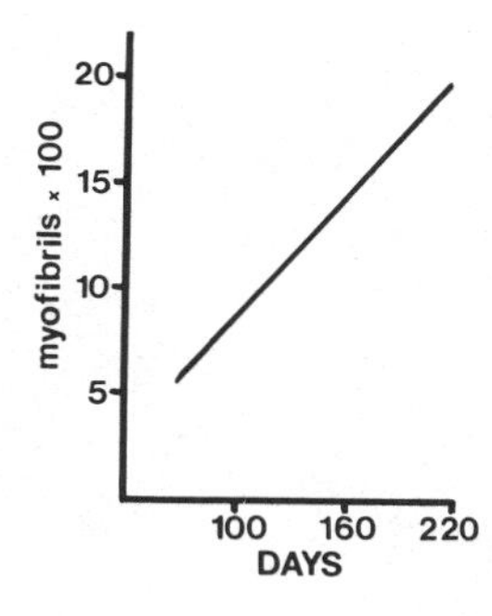

Figure 7-12 Rate of myofibrillar proliferation in red muscle fibers of longissimus dorsi muscles of pigs. (Data from Swatland, 1976b.)

approximately proportional to the rate of proliferation of myofibrils.[32] In red fibers, the apparent number of myofibrils[33] keeps pace with growth in animal live weight (Figure 7-12; Swatland, 1976b). In rats, Howells et al. (1978) found differences between fiber types in the packing density of myofibrils.[34] This might also occur in pork muscles.

Cattle

Muscle fiber growth in the longissimus dorsi muscles of cattle follows a pattern similar to that found in pigs. Radial growth is rapid at birth but levels off in older animals, and fiber diameters are only poorly related to rib-eye areas in animals of a similar age (Figure 7-13, Tuma et al., 1962). Differences in fiber diameters due to breed or sex may (Cornforth et al., 1980) or may not (LaFlamme et al., 1973) be detectable. There may be sire effects on fiber cross-sectional areas (Melton et al., 1974). Fiber diameters may be weakly correlated with meat tenderness (Lewis et al., 1977). Although radial growth of fibers slows down in older animals, it may persist in beta red fibers with a postural function. Johnston et al. (1975) found that the radial growth of beta red fibers was continued between 153 and 233 days of age in Angus and Hereford steers on a standard finishing ration.

The cross-sectional area of the longissimus dorsi is related to both muscle fiber diameter and muscle fiber length, since fibers pass through the rib-eye area at an angle. In cattle, the heritability of rib-eye area has been estimated as 68 per cent (Knapp and Clark, 1950), 69 per cent (Knapp and Norkskog, 1946), 72 per cent (Shelby et al., 1955), and 76 per cent (Christians et al., 1962). It is difficult to decide to what extent fiber diameters contribute to the heritability of rib-eye areas because of the as yet unmeasured contribution of muscle fiber length.

[32]Relationships of fibers and fibrils during radial growth:

$$\mathrm{mfrA{:}R} \propto \mathrm{APPmflN/A{:}R}$$

[33]Apparent number of myofibrils:

$$\mathrm{MEAN{:}APPmflN/A{:}R;0.5L,mfr}$$

[34]Packing density of myofibrils:

$$\mathrm{MEAN{:}APPmflN/A{:}R,mfr{:}r} > \mathrm{MEAN{:}APPmflN/A{:}R,mfr{:}i\&w}$$

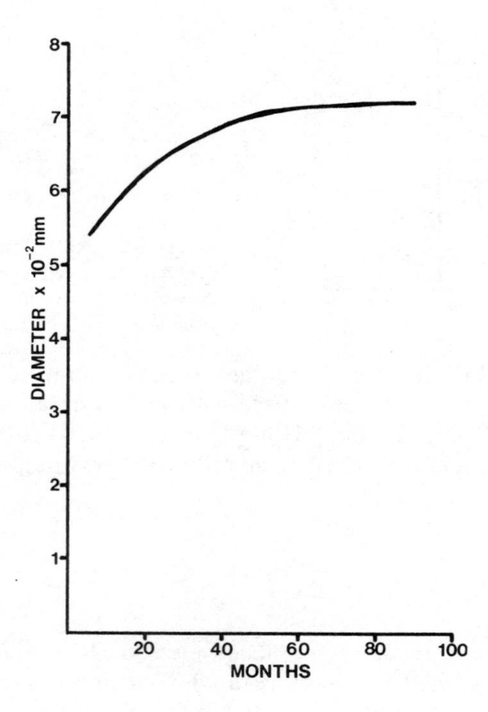

Figure 7-13 Relationship of longissimus dorsi fiber diameter to age in cattle. (Data from Tuma et al., 1962.)

Poultry

The radial growth of muscle fibers in poultry is slightly different from that found in cattle, sheep, and pigs. First, the velocity of radial growth is sometimes constant during most of the commercial growing period, so that the decelerating phase of the growth curve may not become evident. Second, differences between breeds in the radial growth of fibers are particularly well developed (Figure 7-14; Smith, 1963).

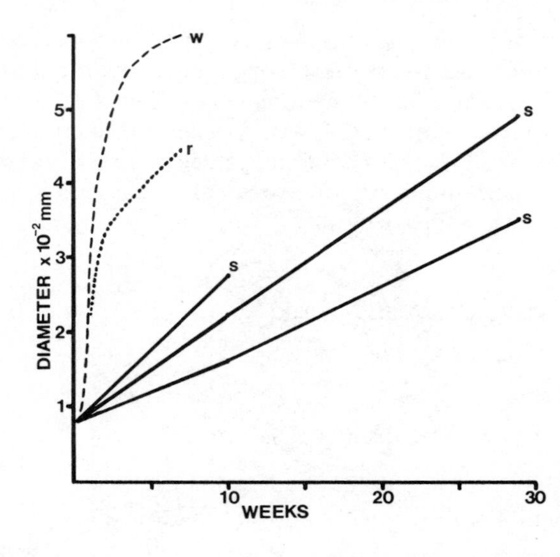

Figure 7-14 Radial growth of muscle fibers in chickens: (w) white fibers of pectoralis; (r) red fibers of pectoralis; (s) mixed fiber types of the sartorius in three breeds with different growth rates. (Data from Kiessling, 1977, and Smith, 1963.)

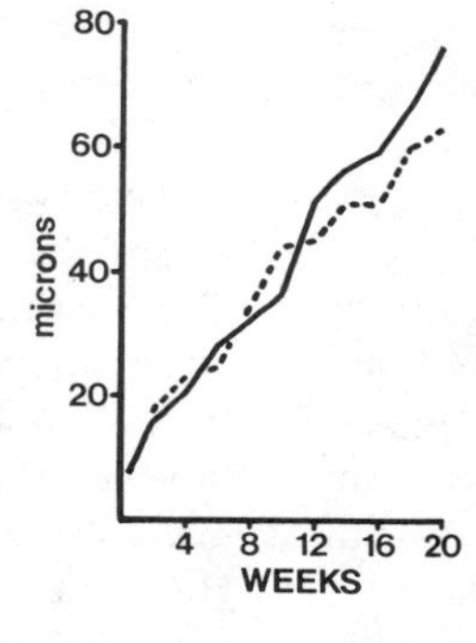

Figure 7-15 Radial growth (microns diameter) of pectoralis muscle fibers in two breeds of turkeys with the same growth rate but different meat yields. The breed with greater muscularity is indicated by the solid line. (Data from Swatland, 1980a.)

Breed differences in fiber diameters often disappear when farm mammals are compared on the basis of equal live weight. In turkeys, however, pectoralis muscles of birds from a genetic line with a high meat yield may have larger-diameter fibers than birds from a low-yielding line compared at the same live weight (Figure 7-15; Swatland, 1980a). In the sartorius (iliotibialis cranialis) muscles of turkeys, fibers with weak ATPase activity grow at a relatively faster rate than do fibers with strong ATPase activity (Figure 7-16; Swatland, 1979e). At first site, this might not be immediately obvious since growth curves for the two fiber types are more or less parallel. The difference in radius between the two fiber types, however, becomes a progressively smaller fraction of the total radius, so that, *per unit of initial radius*, fibers with weak ATPase activity may grow slightly faster, relative to fibers with strong ATPase activity. However, in the pectoralis of the chicken, the radial growth of white fibers is faster than that of red fibers (Figure 7-14; Kiessling, 1977). Thus four major factors are associated with differences in the radial growth of muscle fibers in a species of poultry: (1) whether the muscle is a fast-growing breast muscle or a leg muscle working against body weight; (2) the degree to which the birds have been selected for meat yield; (3) the physiological differentiation of muscle fibers; and (4) the sex of the bird if sexual dimorphism is well developed.

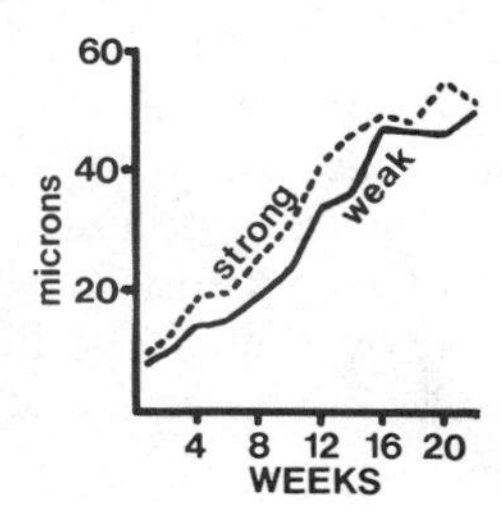

Figure 7-16 Radial growth (microns diameter) of fibers with strong ATPase and fibers with weak ATPase in sartorius muscles of turkeys. (Data from Swatland, 1979e.)

LONGITUDINAL GROWTH

The longitudinal growth of muscles is difficult to study in meat animals. Even when the longitudinal axes of muscle fibers and their fasciculi are parallel to the longitudinal axis of a whole muscle, it is extremely difficult to ascertain the length of individual fibers. In laboratory animals, longitudinal growth is thought to originate from the formation of new sarcomeres at the ends of the muscle fibers (Muir, 1961). Myofibrils may insert obliquely at their ends so that new filaments can be added without interfering with the mechanical continuity of the fibrils at their attachment to the cell membrane. Presumably, the longitudinal growth of muscle fibers in meat animals occurs by a similar mechanism.

Crawford (1954) marked rabbit muscles with ink and with implanted wires to find out where new muscle tissue was added during longitudinal growth. New tissue was added evenly along the muscle length rather than being restricted to the ends of the muscle. Crawford (1954) thought that this result might have been due to the formation of new sarcomeres at the ends of intrafascicuarly terminating muscle fibers. The problem with this type of experiment, however, is the possibility that ink marks or wire implants might follow the growth of intramuscular connective tissue rather than the growth of muscle fibers. Mackay and Harrop (1969) obtained results similar to those of Crawford (1954), even when they used a muscle whose fibers ran from one end to the other. Kitiyakara and Angevine (1963) investigated the source of new muscle fiber nuclei by injecting tritiated thymidine. Incorporation of new DNA was most active at the ends of the muscle, but there was evidence of minor growth zones associated with intrafascicularly terminating fibers in the belly of the muscle. The ratio of muscle belly length to free tendon length remains approximately constant during growth (Muhl and Grimm, 1974).

Meat Animals

Fiber number and fiber diameter both affect carcass muscularity in a straightforward manner—they both tend to increase muscle girth. The effects of muscle fiber length, however, are more subtle and are almost unknown in meat animals. Medial anatomical reference points on hindquarters of beef can be used to superimpose a triangle onto the shape of the hind limb (Figure 7-17). The shape of the posterior profile of the hind limb can then be measured as an area relative to the area of the triangle (Figure 7-18), or by linear measurements relative to other triangular bases (Bass et al., 1981). Simple linear measurements of muscle length may show a tendency to be greater in carcasses with a convex profile, but such data are seldom statistically significant (Colomer-Rocher et al., 1980). The problem is that overall muscle length in large animals conveys virtually no information about muscle fiber length, and the increased volume of muscle tissue due to the longitudinal growth of fibers usually contributes to muscle girth rather than to muscle length.

An increase in the girth of a deep muscle, perhaps as a result of the radial growth of its muscle fibers, causes any overlying superficial muscles to bulge outward. Thus the dominant pattern of cellular growth in the superficial muscle may be longitudinal

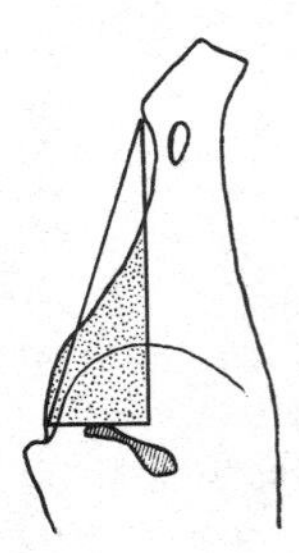

Figure 7-17 In a hanging side of beef, the base of the reference triangle is level with the posterior edge of the symphysis pubis and is parallel to the horizon. The apex of the triangle meets the projecting tip of the tuber calcis.

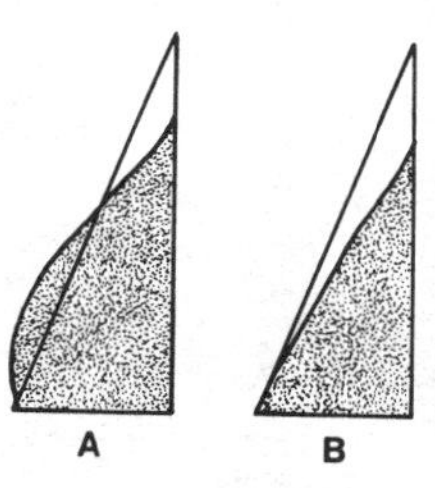

Figure 7-18 Well-muscled (A) and poorly muscled (B) carcasses measured from the profile area of the carcasses (stippled region) relative to the reference triangle.

growth, since its length from origin to insertion is increased in a curvilinear manner as the deep muscle bulges outward. The extra muscle mass gained by the longitudinal growth of a superficial muscle may be sufficient to compensate for a decrease in apparent fiber numbers and for a cessation of radial fiber growth. Even if this is true, however, it leaves unresolved the problem of whether the overall extra muscle mass is due to increases in fiber numbers, increases in fiber diameters, or to longer fibers. In summary, both longitudinal muscle growth and changes in the longitudinal distribution of muscle mass may be important for carcass conformation. When both the fatness and the length of a carcass are taken into account, a subjective appraisal of muscle conformation can be a useful guide to the anticipated lean yield of a carcass (Kempster and Harrington, 1980; Colomer-Rocher et al., 1980; Bass et al., 1981).

The longitudinal growth of muscles follows the longitudinal growth of bones. New sarcomeres are added in series, both during normal muscle growth and in experiments where adult muscle length is artificially increased by immobilization of limbs (Williams and Goldspink, 1976). Joubert (1956a) found that wethers had longer bones than ewes. On the basis of the assumption that muscle length and bone length are interrelated, he attributed part of the greater muscularity of wethers to their longer muscles. A similar assumption has been invoked to explain part of the greater muscularity of tom turkeys relative to hen turkeys (Mullen and Swatland, 1979).

In complex carcass muscles it may be possible to measure fascicular length, although fascicular length conveys no information about fiber length unless the distribution of intrafascicularly terminating muscle fibers is known. Fascicular length is one of the factors that contributes to the cross-sectional area of the longissimus dorsi muscle. The width and the depth of longissimus dorsi cross-sectional area have long been recognized as partially independent variables, with depth sometimes a better indicator than width for the prediction of muscularity (Palsson, 1939). For example, ewes fed at a supermaintenance level show proportionately greater gains in depth relative to width. Similarly, submaintenance feeding causes a greater reduction in depth than in width (Joubert, 1956a). In pork, however, the width of the longissimus dorsi cross-sectional area may be more strongly correlated with total carcass muscle than is depth ($r = 0.64$ versus $r = 0.50$, respectively; McMeekan, 1941). In pork carcasses (Livingston et al., 1966), width is weakly correlated with fiber diameter ($r = 0.23$, NS), whereas depth is not correlated with fiber diameter ($r = 0.03$).

The partial independence of depth and width in the longissimus dorsi is due to the fact that muscle fasciculi pass through the muscle cross-sectional area in a plane that is almost parallel to depth. Thus when fasciculi grow in length, they contribute to depth but not to width (Figure 7-19). If the longitudinal growth of fasciculi is proportionately faster than growth in muscle length, the increased dorsal overlapping of fasciculi

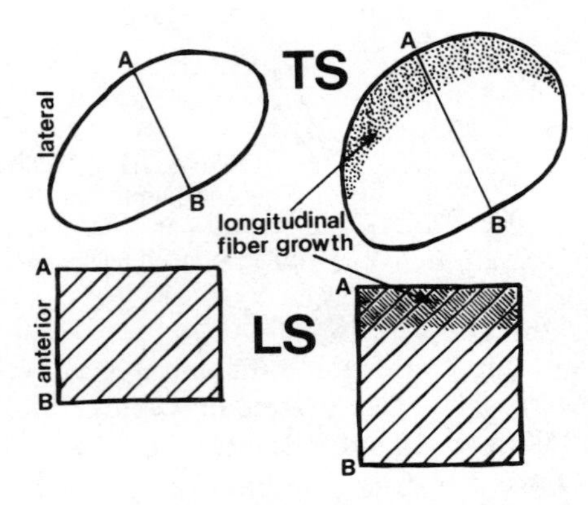

Figure 7-19 Transverse sections (TS) and longitudinal sections (LS) through longissimus dorsi muscles to show how the longitudinal growth of fibers adds to the depth of the muscle (line from A to B) and to the area of the muscle in cross section.

may cause an increase in apparent fiber numbers.[35] This deduction is supported geometrically by the relationship between fascicular length and the depth of the cross-sectional area of the longissimus dorsi muscle (Figure 7-20). Every millimeter of gain in fascicular length adds 0.77 mm to the depth of the longissimus dorsi. Thus if the longissimus dorsi muscle of a carcass has a large apparent number of muscle fibers, it may be because a greater fraction of the number of fibers in the whole muscle has been transected in the muscle cross-sectional area. In general, longissimus dorsi depth follows fascicular length and apparent fiber number, while width follows real fiber number and fiber diameter.

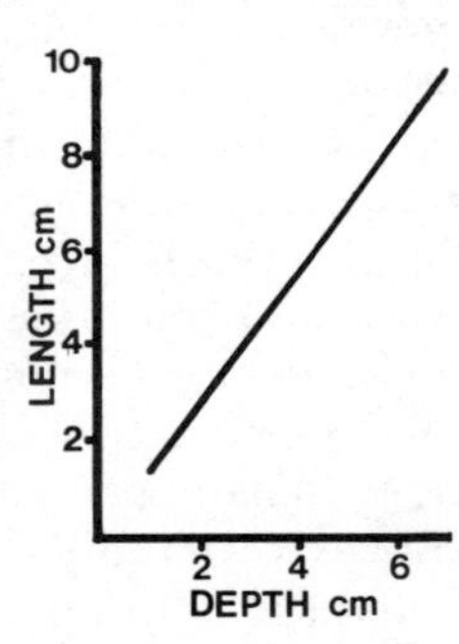

Figure 7-20 Relationship between fascicular length and the depth of the cross-sectional area of the porcine longissimus dorsi muscle. (Data from Swatland, 1976c.)

Intrafascicularly terminating muscle fibers are difficult to map in carcass muscles. At present, therefore, our knowledge of longitudinal fiber growth in meat animals is limited to the situation in small muscles without intrafascicularly terminating fibers. Only crude estimates of sarcomere formation times can be made

[35] Increase in apparent fiber numbers due to longitudinal growth: since

$$\text{mfsL is parallel to mX45L}$$

if

$$\text{D:PROPmfsL} > \text{D:PROPmL}$$

then

$$\text{mA:R} \propto \text{mfsL}$$

and

$$\text{APPmfrN/A:X45L,m} \propto \text{mfsL}$$

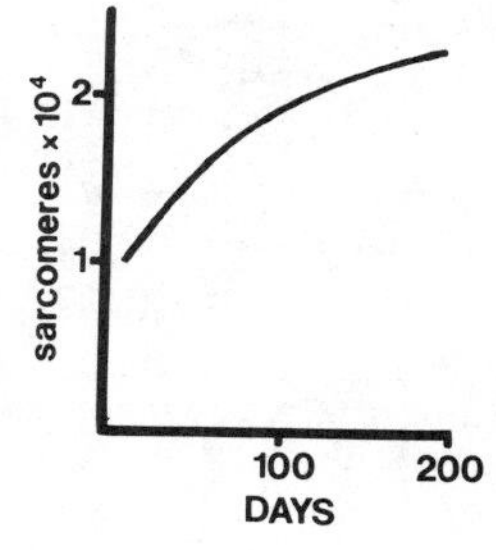

Figure 7-21 Approximate rate of sarcomere formation in the peroneus longus muscle of pigs. (Data from Swatland, 1978.)

over the commercial growing period of meat animals, but about every 20 minutes a new sarcomere is added to the fiber length of the peroneus longus muscles of pigs during growth to market weight (Figure 7-21). In muscles that have relays of intrafascicularly terminating fibers along their length, or relays of fasciculi along their length, the number of points at which longitudinal growth occurs may be increased. If two growth points, one at each end of a peroneus longus fasciculus, can add a sarcomere every 20 minutes, the formation of each sarcomere must take about 40 minutes. Thus a muscle with four relays of intrafascicularly terminating fibers (or four relays of fasciculi) might add a new sarcomere to its length in about 5 minutes because it has eight points for longitudinal growth. In pigs that are growing to market weight, muscles deep in the ham may add a new sarcomere to the length of the muscle every 5 minutes. In superficial muscles which are increasing in length at a faster rate, the time for sarcomere addition is faster, just over 2 minutes (Swatland, 1982b).

MUSCLE FIBER NUCLEI

Meat is composed of multinucleated striated muscle fibers. Muscle fiber nuclei contain DNA which is combined with *histones* and with other structural proteins to form *chromatin.* When DNA is being used to direct the synthesis of new or replacement proteins, the chromatin is dispersed and binds only weakly to histological stains; in this state, it may be called *euchromatin.* In nondividing cells, chromatin may form darkly stained irregular clumps called *chromatin particles.* Nuclei also contain RNA. Darkly stained clumps of RNA form *nucleoli.* The number of nucleoli may vary between animal species. Condensed regions of darkly stained *chromosomes* sometimes persist between cell divisions and are called *heterochromatin.* In the mononucleated cells of the body, such as those of the skin or liver, darkly stained chromosomes composed of inactive DNA are seen when cells divide. As explained below, the situation in multinucleated striated muscle fibers is more complex, and distinct chromosomes are not seen by light microscopy within a muscle fiber.

Several types of nuclei can be found among the several hundred nuclei which are located along each centimeter of fiber length (Figure 7-22). True muscle fiber nuclei are located within the *sarcoplasm* of a muscle fiber. Although the nuclei of satellite cells appear by light microscopy to be located within the sarcoplasm, electron microscopy shows that satellite cells are located in depressions in the fiber surface. Thus the nucleus of a satellite cell is separated from the sarcoplasm of its muscle fiber by a satellite cell membrane and by a muscle fiber membrane (Figure 7-23). The existence of satellite

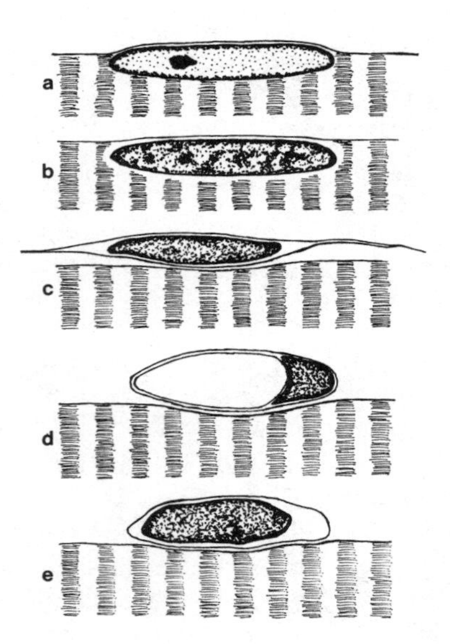

Figure 7-22 Nuclei associated with muscle fibers as they appear by light microscopy: (a) muscle fiber nucleus; (b) satellite cell nucleus; (c) fibroblast; (d) endothelial cell of capillary; (e) pericyte

cells was proved by Mauro (1961). In the older literature, histologists such as Baldwin (1913) had seen satellite cells by light microscopy, but their findings lacked credibility at the time and were soon forgotten. With thin (0.5 μm) sections of muscle embedded in plastic, criteria have now been established for the identification of satellite cells by light microscopy (Venable and Lorenz, 1970; Ontell, 1974) (Table 7-6).

Muscle nuclei increase in number during postnatal development. The relative magnitude of the increase varies from muscle to muscle. In rats, the biceps brachii muscle exhibits a twofold increase while the gastrocnemius exhibits a fourfold increase in nuclear numbers (Enesco and Puddy, 1964). Before the existence of satellite cells was proved by electron microscopy, it was rather difficult to explain how muscle nuclei were able to increase in number without any evidence, by light microscopy, of *mitosis* in muscle fiber nuclei. Muscle fiber nuclei are able to withstand longitudinal compression during muscle contraction by means of concertina-like wrinkles in their nuclear membrane. For many years, some histologists regarded wrinkled muscle nuclei as evidence of nuclear division. In the absence of the normal features of mitosis, the division of wrinkled nuclei was thought to be by *amitosis* (Carr, 1934). Amitosis was defined as cleavage of a nucleus without any threadlike formation of nuclear

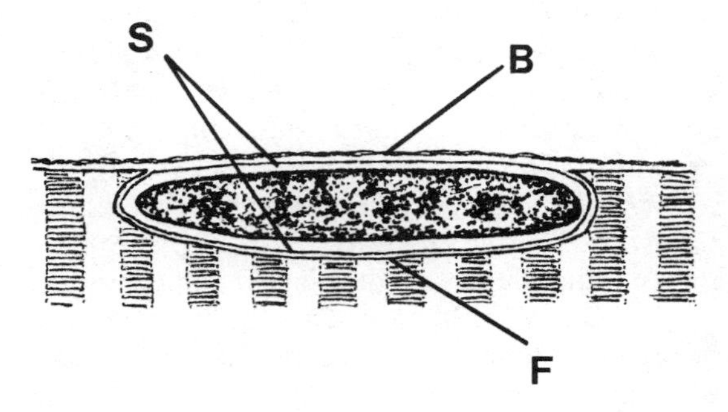

Figure 7-23 Membranes delineating the satellite cell as seen by electron microscopy. The satellite cell membrane (S) pushes against the muscle fiber membrane (F). The basement membrane (B) covers both the satellite cell and the muscle fiber surface.

TABLE 7-6. IDENTIFICATION BY LIGHT MICROSCOPY OF THE NUCLEI ASSOCIATED WITH MUSCLE FIBERS

Muscle fiber nuclei
- (1) No clear zone separates them from myofibrils.
- (2) They are more basophilic than are fibroblast nuclei.
- (3) Their nucleolus is usually well defined and large.
- (4) Their outer chromatin is tightly distributed along the nuclear membrane.
- (5) Their internal cromatin is evenly dispersed.

Satellite cells
- (1) They are indented into a fiber but sometimes they protrude from the profile of the fiber.
- (2) They have variable amounts of basophilic cytoplasm.
- (3) Their outer chromatin is heavily and unevenly deposited along the nuclear membrane.
- (4) Their internal chromatin is scattered in clumps.
- (5) Their nucleolus is small and is usually masked by internal chromatin.
- (6) Sometimes their nuclei are uniformly stained dark.
- (7) Eighty per cent have a clear space (0.5 to 0.2 μm) separating them from myofibrils.

Fibroblasts
- (1) Their light cytoplasm is located at the poles of the nucleus.
- (2) They are separated from myofibrils by at least 0.5 μm.
- (3) A nucleolus is usually present.
- (4) Chromatin clumping is less than in pericytes or satellite cells.

Endothelial cells
- (1) They have a basophilic nucleus.
- (2) They are associated with a capillary wall.

Pericytes
- (1) They have a basophilic nucleus.
- (2) The cells are smaller than satellite cells.
- (3) The separation from myofibrils is at least 0.5 μm.
- (4) Indentation into a fiber surface is only slight.

Sources: Compiled from Venable and Lorenz (1970) and Ontell (1974).

material, similar to the type of division which was then thought to occur in the meganuclei of ciliate protozoans.

The new nuclei which are added during the postnatal growth of muscle fibers are now thought to be derived from daughter cells generated by the mitosis of satellite cells. Mitosis in satellite cells can be identified by electron microscopy, and the synthesis of DNA by satellite cell nuclei can be detected by radioautography (Shafiq et al., 1968; Moss and Leblond, 1970, 1971; Allbrook et al., 1971; Cardasis and Cooper, 1975). Satellite cells are widely regarded as premyoblasts which have been trapped under the basement membrane that surrounds individual muscle fibers (Ishikawa, 1966). However, it is difficult to be sure of the morphological identification of premyoblasts in fetal muscle, and other types of cells, such as *mast cells*, may also be found beneath the basement membrane (Ontell, 1977).

In partially injured muscle fibers, large numbers of ribosomes may surround certain muscle fiber nuclei which may then be isolated from the sarcoplasm of the fiber by the coalescence of small vesicles (Reznik, 1976). Mononucleated cells formed in this way are involved in muscle fiber regeneration together with satellite cells. The nervous system exerts some control over the morphology and relative numbers of satellite cells in the muscles of fetal pigs. Destruction of the spinal cord in the cervical region causes a decrease in the relative numbers of satellite cells in the sartorius (Campion et al., 1978), but no effect is found after the decapitation of fetuses (Campion et al., 1981b).

Satellite cells with a myogenic potential can be released from the muscle fiber surface by enzymes that dissolve the basement membrane (Bischoff, 1974). When detached by freeze-fracturing, it appears that satellite cells have narrow projections radiating over the muscle fiber surface and it has been suggested that satellite cells are capable of movement (Schmalbruch, 1978). The plasma membranes of satellite cells which are apposed to muscle fiber membranes show varying numbers of caveolae and interdigitations (Reger and Craig, 1968). Satellite cells in older animals are metabolically less active than those of younger animals (Snow, 1977).

On a proportional basis, there are more muscle fiber nuclei in red muscles than in white muscles, and nuclei are more frequent at the ends of fibers than at their midlength (Schafer, 1898; Burleigh, 1977). Muscle fiber nuclei are often counted in histological cross sections of muscle. There may be a sampling problem if section thickness is small relative to nuclear length.[36] Although this problem can usually be ignored in the relatively thick paraffin sections normally used for light microscopy, appropriate corrections must be made for the very thin sections used in electron microscopy (Schmalbruch and Hellhammer, 1977).

In slow-contracting soleus muscles of rats, approximately 8 per cent of muscle nuclei are satellite cells, whereas in the fast-contracting tibialis anterior muscle, only 4 per cent are satellite cells (Schmalbruch and Hellhammer, 1977). This is probably due to the growth characteristics of the two muscles rather than to their differences in fiber-type composition (Gibson and Schultz, 1982). In soleus muscles and in the diaphragm, satellite cells have more cytoplasm and so are more easily detected by light microscopy. In the slow muscles of rats, DNA synthesis and cell division are more active than in fast muscles, and the correspondingly higher incidence of satellite cells is established before muscle fibers exhibit histochemical differentiation (Kelly, 1978a). DNA synthesis in satellite cells is initially decreased by muscle denervation (Kelly, 1978a), but later, satellite cell numbers are increased (Hanzlikova et al., 1975). In muscles of mature animals, satellite cells are more numerous in the vicinity of motor end plates (Kelly, 1978b). Both the relative incidence and the total number of satellite cells decrease with age in mouse muscles (Young et al., 1978), although the ability of surviving satellite cells to synthesize myofibrillar proteins is unimpaired (Young et al., 1979).

The amount of DNA in a nucleus varies greatly throughout the animal kingdom (Szarski, 1976). Birds have approximately 3.5 pg of DNA in each nucleus, whereas

[36]Since the number of nuclei per cross-sectioned muscle fiber is based on

$$mnN/V,mfr$$

where

$$V,mfr = A{:}R,mfr \times L,mfr$$

and

$$L,mfr = \text{section thickness}$$

a bias exists if

$$mnL > L,mfr$$

mammals have approximately 7 pg. In tissues with mononucleated cells, an estimate of the total DNA divided by the amount of DNA per nucleus gives an estimate of cell number. Estimates may be misleading, however, if large numbers of cells are about to divide by mitosis, and if there is a complex population of different cell types in the tissue. DNA content is difficult to use as a measure of cell numbers in skeletal muscle because skeletal muscle fibers are multinucleate. Also, the distribution of nuclei may vary between physiological fiber types and at different points along the length of individual fibers. The DNA content of skeletal muscles can be used to estimate the volume of muscle tissue or mass of protein which is serviced by each nucleus, but the value of this information is greatly decreased by the large and variable number of cells that contain very little cytoplasm (satellite cells, pericytes, and so on). The accurate histological data needed to interpret such biochemical data are difficult to locate for any particular species or muscle.

Pigs

The growth of pigs to market weight is accompanied by a steady increase in the DNA content of individual muscles. The DNA content may reach an asymptote at approximately 180 days in the semimembranosus, while DNA accumulation still continues in the biceps femoris (Durand et al., 1967). The semimembranosus may thus acquire a greater proportion of muscle mass per unit of DNA. Although differences due to breed and nutrition are to be expected in such data, this example serves as a general warning that data collected from a single muscle may not be representative of all carcass muscles.

The muscle mass supported by each unit of DNA generally increases with age. A sharp change occurs at the end of gestation (Farmer et al., 1980). This relationship may be expressed in a number of ways, based on the following relationships:

RNA to DNA
Protein to DNA
Wet weight to DNA

These relationships do not exactly correspond to histological relationships[37] since the histological ratio may exclude cells types that are included in the biochemical relationships. The approximate total of DNA in a muscle[38] is modified by the numbers of nuclei in G_2 of the mitotic cycle. Thus in young animals there may be fewer nuclei than predicted biochemically. Histological estimates are usually based on the numbers of nuclei associated with muscle fibers, but the cutoff point varies widely from author to author. Thick sections for light microscopy minimize the sampling problem due to

[37]Histological measurements of nuclear members, such as

mnN/V,mfr

[38]Biochemical estimates of DNA content are usually based on

SUMan&cn&mn&nn&vn × 7E-12

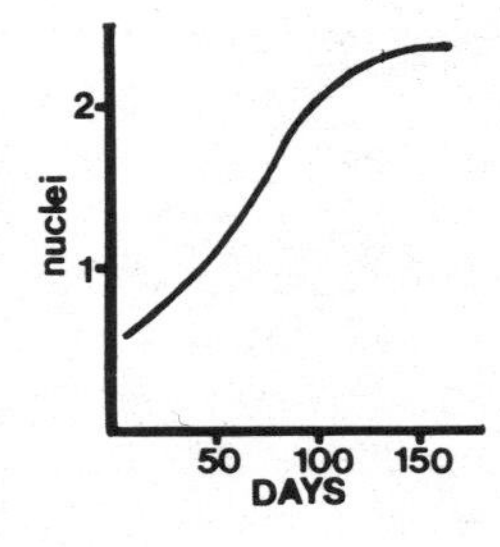

Figure 7-24 Increase in the mean number of nuclei in transverse sections 10μm thick from sartorius muscle fibers of pigs. (Data from Swatland, 1977.)

nuclear length but the discrimination of different types of nuclei may be unreliable.[39] Electron microscopy allows better discrimination of nuclei.[40]

Powell and Aberle (1975) examined biochemical DNA ratios in lightly muscled and in heavily muscled Durocs. The greater mass of biceps femoris muscles in heavily muscled pigs was associated with an increase in the biochemically estimated number of nuclei. Protein-to-DNA ratios were similar in both types of pigs. In genetically obese pigs, RNA-to-DNA ratios are decreased (Ezekwe and Martin, 1975; Harbison et al., 1976). The RNA-to-DNA ratio is elevated if pigs are fed a high-protein ration (Hogberg and Zimmerman, 1979).

Satellite cells were first identified in porcine muscle by Muir (1970). Campion et al. (1979) observed both absolute and relative decreases in numbers of satellite cells between 95 days gestation and 1 day after birth. Over the period from 1 to 64 weeks after birth, there is a decrease in the number of satellite cells relative to the number of muscle fiber nuclei. However, the absolute number of satellite cells increases (Campion et al., 1981a). Robinson (1969) found that inadequate maternal nutrition before farrowing and during lactation caused a long-term reduction of the DNA content of triceps and semitendinosus muscles in the progeny. The accumulation of nuclei in cross-sectioned sartorius muscle fibers appears to follow a sigmoid growth curve (Figure 7-24).[41] In pigs placed on a maintenance ration so as to arrest growth, the recruitment of new muscle fiber nuclei is abruptly halted.

[39]By light microscopy, the number of nuclei per cross-sectioned fiber

nN/V,mfr

includes

mn:sar&mn:sat&vn:peri

some

vn:cap&cn

but excludes

an

[40]Discrimination of nuclei by electron microscopy:

mnN/V,mfr = mn:sar&satN/V,mfr

[41]Relationship of radial fiber growth and nuclear numbers:

$$\text{mfrA:R} \propto \text{mnN/V,mfr}$$

Cattle

Herold and Nelms (1964) teased apart individual muscle fibers from biopsy samples of longissimus dorsi muscles. Samples were taken at 5 and at 67 days of age from Angus calves, and the numbers of nuclei per unit length of muscle fiber were counted microscopically (Table 7-7). The DNA content of longissimus dorsi muscles peaks at about 140 days (Trenkle et al., 1978), although this may be delayed in cattle on a low plane of nutrition. DNA concentration in cattle at market weight shows little or no effect due to castration, breed type, feed energy level, or feed protein level (La Flamme et al., 1973; Lipsey et al., 1978).

TABLE 7-7. COEFFICIENTS OF CORRELATIONS BETWEEN NUMBERS OF MUSCLE FIBER NUCLEI AND COMMERCIAL TRAITS IN LONGISSIMUS DORSI MUSCLES OF ANGUS CALVES

	Age at biopsy	
Trait	5 days	67 days
Body weight	0.74	0.87
Body weight at weaning	0.67	0.88
Daily gain until weaning	0.62	0.85
Daily gain on feed	0.70	0.67

Source: Data from Herold and Nelms (1964).

Poultry

Moss et al. (1964) examined the DNA content of chicken breast muscles at hatching and at 1, 2, and 4 weeks after hatching. Biochemical estimates of nuclear numbers (DNA/2.5E-9) increased with growth and were slightly higher for males. The amount of cytoplasm per nucleus increased (Figure 7-25). It was concluded that the number of

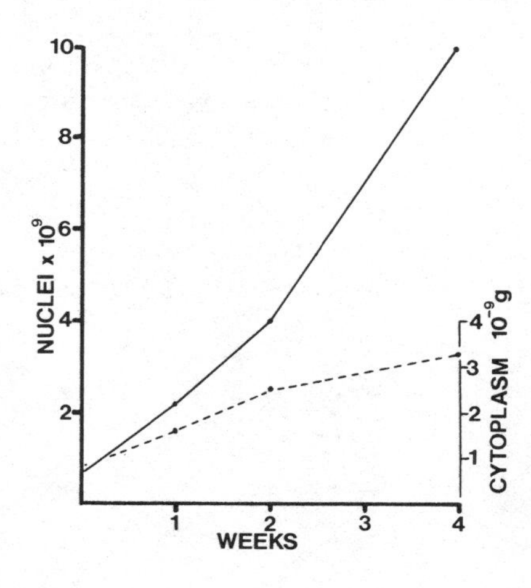

Figure 7-25 Increase in number of muscle nuclei estimated biochemically, and the mass of cytoplasm associated with each nucleus. (Data on White Leghorn chickens from Moss et al., 1964.)

nuclei increased in proportion to area dimensions,[42] while an estimate of the amount of cytoplasm per nucleus was proportional to linear dimensions.[43] Moss (1968) also found that muscle fiber cross-sectional areas were proportional to biochemical estimates of the total number of nuclei in the muscle.

VOLUMETRIC ANALYSIS OF MUSCLE GROWTH

Much of the early research on muscle growth in meat animals was undertaken in an attempt to find a simple biochemical or histological method for the measurement of muscle mass in meat animals or their carcasses. The search for single parameters that are closely related to muscle mass appears to have discouraged a volumetric approach to muscle growth. In an attempt to describe the muscle mass of meat animals in volumetric terms, it is necessary to examine all the parameters that are related to fiber number, volume, and arrangement. Implicit in this approach is the assumption that the muscle fiber is the basic unit of muscle tissue through which sensory feedback and control information is mediated during muscle growth and development. However, there are other opinions on this subject. Cheek (1975) and Landing et al. (1974) suggest that each nucleus within a muscle fiber has jurisdiction over a certain volume of cytoplasm called a DNA unit. They suggest that the DNA unit, rather than the multinucleated muscle fiber, must be the cellular unit through which growth processes are mediated. Cheek (1975) even suggests that fiber diameter has little significance with respect to cell growth.

The Multinucleated State of Muscle Fibers

Multicellular animals are presumed to have arisen from unicellular animals during the early evolution of life on earth. Many possible evolutionary routes can be postulated (Kerkut, 1960), and the subject will always be amenable to scientific debate. The evolutionary origin of the multinucleated state of skeletal muscle fibers is a less profound but equally difficult problem. During ontogeny, the multinucleated state is created by the fusion of mononucleated cells, not by a failure of cytoplasmic separation in a line of dividing cells. Thus the phylogenetic origin of the multinucleated state in skeletal muscle fibers may have followed a similar route. Cardiac muscle contains striated myofibrils, and shares many other features in common with skeletal muscle, but most cardiac muscle cells are mononucleate. Thus there appear to be some distinct advantages which result from the multinucleated state in skeletal muscle fibers. Even in anomalous locations, such as in the *pineal gland*, striated muscle fibers are multinucleated (Diehl, 1978).

In skeletal muscles, there are a number of features that are not well developed in cardiac muscle, which might relate to the possible advantages of the multinucleated state.

[42]Relationship of radial fiber growth and nuclear numbers:

$$mnN \propto mW^{0.5}$$

[43]Relationship of amount of cytoplasm per nucleus to linear dimensions:

$$mW \ / \ mn \propto mW^{0.33}$$

1. Individual motor neurons directly control the contraction of skeletal muscle fibers and probably play a major role in the physiological differentiation of fibers.
2. Skeletal muscle fibers are usually unbranched.
3. In most skeletal muscles, fibers contract to draw two distant points closer together. This greatly simplifies the working geometry of individual fibers.

A lack of transverse cellular boundaries along the length of a fiber enables a relatively direct system of activation and mechanical continuity either side of the neuromuscular junction. In cardiac muscle, on the other hand, fibers contract from one end to the other in a precisely regulated sequence which directs the flow of blood within the heart. The answers to many basic questions concerning the multinucleated state are unknown.

1. Do all sarcoplasmic nuclei perform a similar function, or is there some degree of regional differentiation along the fiber? For example, nuclei located near the neuromuscular junction might specialize in the regulation of proteins concerned with neuromuscular transmission.
2. When the synthesis of myofibrillar proteins is modified during ontogeny or as a result of crossed reinnervation, do all nuclei change their activity simultaneously or are intermediate stages in the conversion of myofibrillar proteins due to the activity of a heterogeneous population of nuclei?
3. Does each nucleus have its own territory of sarcoplasm, or is messenger RNA soon dispersed over a wide volume of sarcoplasm by changes in fiber shape during contraction?
4. Does sarcoplasmic volume regulate the number of sarcoplasmic nuclei, or vice versa?
5. Does a sarcoplasmic nucleus have a limited life span with regular replacement by daughters of satellite cells, or does it survive indefinitely like the nucleus of a neuron?

Muscle Dimensions in Chickens

In agricultural research, we are only just beginning to escape from an obsession with fiber diameters, and to develop a volumetric comprehension of fiber growth in meat animals. Meara (1947) first attempted this with rabbit muscle, but much of the information that is available at present relates to poultry. Moss (1968) concluded that fiber length and diameter maintain a constant ratio during the growth of chicken pectoralis muscles from hatching to 266 days. In gastocnemius muscles, however, this proportionality is lost after 2 months because of the completion at this time of longitudinal, but not radial growth. Halvorson and Jacobson (1970) examined muscle growth in well-muscled, fast-growing crossbred chickens and in less well endowed Leghorns. The meatiness of crossbreds was due mainly to increases in the length and width of the pectoralis and to increases in the length, but not in the width of the supracoracoideus. Two thigh muscles were also examined: both the length and width of the iliotibialis were increased, whereas the semitendinosus was only increased in length. In general, however, thigh muscles contribute little to breed differences in muscularity.

Variability in muscle geometry is a major problem in understanding the volumetric growth of muscles. Helmi and Cracraft (1977) found that, in biceps femoris and adductor superficialis muscles of chickens, the lengths of the longest and shortest fasciculi grew at equal rates, whereas in adductor profundus, the shortest fasciculi grew faster than the longest fasciculi.

Turkeys

Carcass conformation is visible to the consumer who purchases a whole bird. In turkeys, the depth of the keel of the sternum does not grow as fast, on a relative basis, as the length of the keel. Thus the desirable convexity in breast shape lateral to the keel is directly related to the depth of meat over the sternum, but is inversely related to keel depth (Swatland, 1979b). Genetic lines with a higher meat yield relative to live weight have larger-diameter fibers in the pectoralis muscle (Swatland, 1980a). Male turkeys usually have larger muscles than females. In males, the supracoracoideus muscle of the breast has a greater cross-sectional area and length than in females. However, because of the angular insertion of fibers in this pennate muscle, the greater volume in males is due to radial growth and not to longitudinal growth of muscle fibers (Swatland, 1979c). In the sartorius (iliotibialis cranialis) muscle of the thigh, the opposite situation prevails at the cellular level, and the greater muscle volume in males is due to longitudinal growth rather than to the radial growth of fibers (Swatland, 1979d). In the sartorius, fibers with strong ATPase and fibers with weak ATPase grow radially at different relative rates (Swatland, 1979e), but the situation for longitudinal growth is unknown.

Ducks

Differences in the volumetric growth patterns of two duck muscles due to breed and sex are shown in Table 7-8 (Plates 18 and 19). This table gives some idea of the great complexity of muscle growth in meat animals and poultry. Each vertical column in the table is incomplete, since major parameters such as fiber number are missing, and the table includes just two of a large number of carcass muscles.

General Principles of Volumetric Growth

Although many muscles are irregular in shape, their muscle fibers are basically cylindrical. Figure 7-26 illustrates the radial and longitudinal growth of a hypothetical cylinder; the volume of the cylinder is called a *projected volume* (PV). The projected volume is the product of length and maximum cross-sectional area.[44] Muscle volume

[44]Projected volume:

$$PV = L \times AMAX{:}R$$

In a true cylinder

$$PV = V$$

but in a structure such as a whole muscle,

$$PV > V$$

TABLE 7-8. SUMMARY OF GROWTH PATTERNS IN TWO DUCK MUSCLES

Parameter	Sartorius	Supracoracoideus
Function	Walking, standing, or swimming	Flying
Fiber arrangement	Parallel	Pennate
Maximum volumetric growth velocity	In first 5 weeks	After 5 weeks
Major source of increments in first 5 weeks	Longitudinal and radial fiber growth	Longitudinal and radial fiber growth
Major source of increments after 5 weeks	Not detected	Radial fiber growth
Volumetric basis of sexual dimorphism	Fiber diameter, muscle length, perhaps relative fiber length	Not detected
Volumetric basis of breed difference in early growth	Muscle length and fiber diameter	Not detected
Largest-diameter fibers at 10 weeks	Strong ATPase but only after fast growth	Weak ATPase
Allometry between ATPase fiber types	Weak ATPase fibers grow faster	Not detected
Difference in SDH activity between ATPase fiber types	Mean values similar	Stronger SDH in strong ATPase fibers
Correlation of SDH with mean fiber cross-sectional area	Inverse in strong ATPase fibers	Inverse

Source: Swatland (1981).

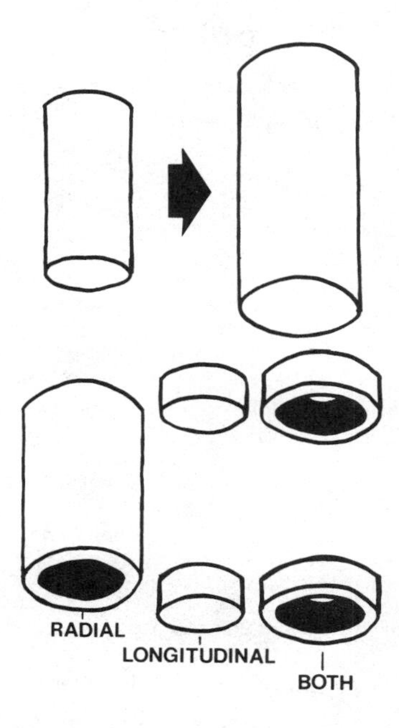

Figure 7-26 Diagram to illustrate the components of growth in a cylindrical system. The growth of the cylinder at the top of the figure is composed of the three sets of components shown at the bottom of the figure.

can be measured directly or calculated from dissection weight and density. Projected volume is one of two links through which muscle weight can be related to cellular growth patterns. The second link is the arrangement of muscle fibers within the projected volume of a whole muscle.

In many muscles, the long axes of fibers are parallel to the long axis of the muscle. In the case of a muscle with parallel fibers, the proportionality between the growth of cylindrical muscle fibers and the growth of the whole muscle can be simplified to a *shape factor*.[45] The muscle shape factor can be used to estimate the contribution made by volumetric growth increments of muscle fibers to growth increments of muscle volume or weight. The projected volume of muscle fibers is estimated from histological measurements[46] and is an overestimate of true volume (because tendinous insertions are conical and intrafascicular terminations are tapered). In carcass muscles of meat animals, fiber arrangement is so complex that it defies routine measurement. If the shape factor of a whole muscle with parallel fibers is known *at a particular point in time*, it can be used as a correction factor with which to estimate the contribution of fiber volume to muscle volume.[47] Figure 7-27 shows the increments to projected volume which are derived from radial and longitudinal growth.[48]

The contributions of the radial and longitudinal growth of muscle fibers to the growth of a whole muscle sometimes change during development. Changes in the source of volumetric increments can be detected only by collecting data at a series of points in time. Data are collected from a group of animals which are slaughtered in a series so as to form a *pseudo-ontogenetic* sequence. A true ontogenetic sequence is impossible to obtain since measurements are destructive. In a pseudo-ontogenetic

[45]Definition of shape factor:

$$\text{mPV} \times \text{SHAPE FACTOR} = \text{mV}$$

[46]Projected volume of muscle fibers:

$$\text{mfrA:R} \times \text{mfrL} = \text{mfrPV}$$

[47]The

$$\text{SHAPE FACTOR} \times \text{mfrPV} \propto \text{mV}$$

and

$$\text{mV} \times \text{DEN} = \text{W at dissection}$$

[48]At a point in time t, the total increment added since the last measured point in time $t - 1$ is

$$\text{mPV(t)} - \text{mPV(t} - 1) = \text{DmPV}$$

Time dimensions are incorporated into the morphometric notation as parenthetical statements immediately following the *parameter of measurement*. The conventional notation for an increment, delta or δ, is written in uppercase "D," since it is a *class of measurement*. The volumetric increment derived from radial growth alone is a hollow cylinder:

$$\text{DmPV/A:R} = [\text{mA(t):R} \times \text{mL(t}-1)] - [\text{mA(t}-1)\text{:R} \times \text{mL(t}-1)]$$

The increment from longitudinal growth alone is composed of a cylinder at each end:

$$\text{DmPV/L} = [\text{mA(t}-1)\text{:R} \times \text{mL(t)}] - [\text{mA(t}-1)\text{:R} \times \text{mL(t}-1)]$$

There is also an increment obtained from an interaction of radial and longitudinal growth:

$$\text{DmPV/L\&A:R} = \text{DmPV} - \text{DmPV/L} - \text{DmPV/A:R}$$

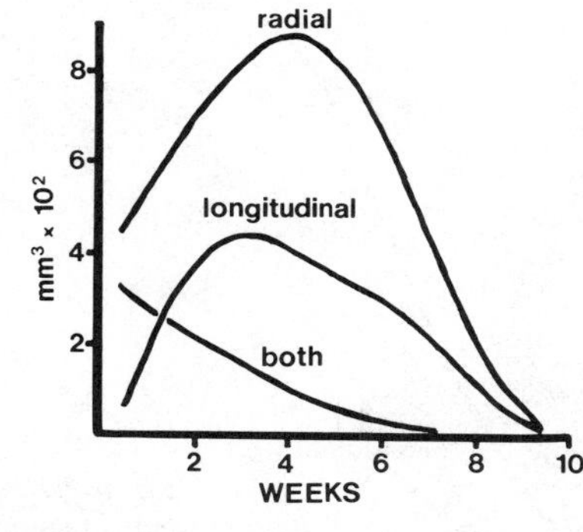

Figure 7-27 Volumetric growth in a whole muscle, with contributions from radial, longitudinal, and mixed sources. This is a muscle with parallel fibers, but pennate muscles are similar. (From Swatland, 1980c.)

sequence, differences in rate of growth between animals are compounded with the experimental error of measurement, and the resulting growth curves are seldom smooth when plotted. Irregular curves make the analysis of rate of growth rather difficult since an animal that has grown slowly may create a zero or negative growth increment. If a curve is fitted mathematically to the data, it is rather difficult to understand the biological significance of the terms in a polynomial expression. A simple alternative is to fit a smooth curve whose coordinates are calculated from the sequential averages of pairs of data points. The subjective decisions that must be made are as follows: (1) whether to allow the first and last points to be changed; and (2) the number of times to repeat the process of sequential averaging.

These decisions should be reasoned from the existing knowledge of the biological basis of the system described by the curve. For example, in curves that describe muscle growth, it is not uncommon to find a dip at the final end of the curve. For example, if the experiment terminates with young adult males under constant environmental conditions and with no change in animal behavior, the final dip is most likely to be due to differences in growth attainment of the animals in the pseudo-ontogenetic sequence. In this case, it is reasonable to use a smoothing technique that progressively elevates the data points in the final dip. On the other hand, if the experiment terminates with animals in which a decrease in muscle mass is anticipated as a result of reduced feed intake, or senescence, or a change in animal behavior, it is reasonable to use a smoothing technique which preserves the final dip. Figure 7-27 shows the origin of volumetric increments to muscle mass in a muscle with parallel fibers. The volumetric growth of individual fibers follows a similar pattern because the fibers are parallel with the long axis of the muscle.

In muscles with a complex fiber arrangement, volumetric analysis is far more difficult (Figure 7-3). If fibers are inserted at an angle to the longitudinal axis of a muscle,[49] growth in fiber diameters may contribute to muscle length[50] and to muscle cross-sectional area.[51] The contribution of fiber length to muscle dimensions is more

[49]With fibers inserted at an unknown angle to the muscle long axis:

mX?L

[50]Contribution of radial fiber growth to muscle length:

mfr2R / SINEmX?L → mL

[51]Contribution of radial fiber growth to muscle cross-sectional area:

PI × mfrR × (mfrR / COSINEmX?L) → mA:R

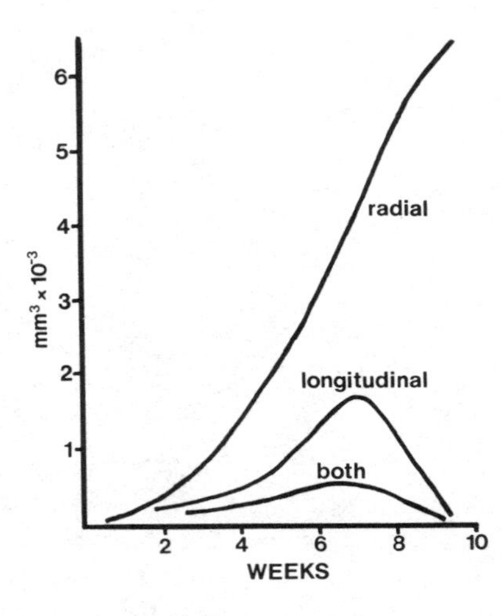

Figure 7-28 Volumetric growth of individual muscle fibers in a pennate muscle. In this case, the contributions from radial fiber growth outweigh those due to longitudinal growth. (From Swatland, 1980b.)

complex. At the end of the muscle, where fibers radiate from the intramuscular tendon, increments to fiber length may be added directly to muscle length. Halfway along the muscle, increments to fiber length may contribute more to cross-sectional area than to the length of the muscle. The magnitude of these contributions may depend on whether muscle fasciculi have a standard length with a variable angle of insertion, or a variable length with a standard angle of insertion. Figure 7-28 shows the origin of increments to muscle fiber volume in a pennate muscle. In contrast to the parallel-fibered muscle, radial growth is the dominant process.

Fiber Volume and Postmortem Metabolism

In the preceding paragraphs, the volumetric analysis of muscle fiber growth has been used in an attempt to unite histological data and gross dissection data. Volumetric analysis may also enable histological data to be united with gross biochemical data. Anaerobic glycolysis and the production of lactate from glycogen during the conversion of muscles to meat determine the rate and extent of pH decline postmortem. In turn, pH has a profound effect on the appearance of meat and on its water-holding capacity. Muscle glycogen is partitioned between individual muscle fibers which may differ in volume, in glycogen concentration, and in metabolic activity during the conversion of muscles to meat. Glycogen concentration in individual fibers is affected by cellular patterns of glycogen storage and depletion prior to slaughter.

Postmortem glycogenolysis in individual muscle fibers can be monitored postmortem in frozen cross sections of muscle that have been stained by the *periodic acid–Schiff* reaction for glycogen. The total glycogen content of a muscle at slaughter is determined by the product of total fiber volume and mean glycogen concentration. Different physiological fiber types may differ in their volume, in their glycogen concentration at slaughter, and in their rate of glycogenolysis postmortem. For example, most of the lactate that is formed soon after the slaughter of market weight pigs originates from white fibers. This topic is pursued further in Chapter 9.

REFERENCES

ABERLE, E. D., and DOOLITTLE, D. P. 1976. *Growth* 40:133.

ALLBROOK, D. B., HAN, M. F., and HELLMUTH, A. E. 1971. *Pathology* 3:233.

ANSAY, M. 1973. *Ann. Med. Vet.* 117:459.

ANSAY, M., and GILLET, A. 1966. *Ann. Med. Vet.* 7:512.

ASGHAR, A., and YEATES, M. F. 1979. *Agric. Biol. Chem.* 43:445.

ASHMORE, C. R., and ROBINSON, D. W. 1969. *Proc. Soc. Exp. Biol. Med.* 132:548.

ASHMORE, C. R., and SUMMERS, P. J. 1981. *Am. J. Physiol.* 241:C93.

ASHMORE, C. R., PARKER, W., STOKES, H., and DOERR, L. 1974. *Growth* 38:501.

BALDWIN, W. M. 1913. *Z. Allg. Physiol.* 14:146.

BARDEEN, C. R. 1903. *Anat. Anz.* 23:241.

BASS, J. J., JOHNSON, D. L., COLOMER-ROCHER, F., and BINKS, G. 1981. *J. Agric. Sci., Camb.* 97:37.

BENDALL, J. R., and VOYLE, C. A. 1967. *J. Food Technol.* 2:259.

BISCHOFF, R. 1974. *Anat. Rec.* 180:645.

BOCCARD, R. 1982. *Curr. Top. Vet. Med. Anim. Sci.* 16:148.

BURLEIGH, I. G. 1977. *J. Cell. Sci.* 23:269.

BUTTERFIELD, R. M. 1966. *Aust. Vet. J.* 42:37.

BYRNE, I., HOOPER, J. C., and MCCARTHY, J. C. 1973. *Anim. Prod.* 17:187.

CAMPION, D. R., RICHARDSON, R. L., KRAELING, R. R., and REAGAN, J. O. 1978. *Growth* 42:189.

CAMPION, D. R., RICHARDSON, R. L., KRAELING, R. R., and REAGAN, J. O. 1979. *J. Anim. Sci.* 48:1109.

CAMPION, D. R., RICHARDSON, R. L., REAGAN, J. O., and KRAELING, R. R. 1981a. *J. Anim. Sci.* 52:1014.

CAMPION, D. R., HAUSMAN, G. J., and RICHARDSON, R. L. 1981b. *Biol. Neonate* 39:253.

CARDASIS, C. A., and COOPER, G. W. 1975. *J. Exp. Zool.* 191:347.

CARPENE, E., and VEGGETTI, A. 1981. *Experientia* 37:191.

CARR, R. W. 1934. *Am. J. Anat.* 49:1.

CARROLL, F. D., THIESSEN, R. B., ROLLINS, W. C., and POWERS, N. C. 1978. *J. Anim. Sci.* 46:1201.

CATTON, W. T. 1957. *Physical Methods in Physiology*, p. 228. Sir Isaac Pitman & Sons, London.

CHEEK, D. B. 1975. *Fetal and Postnatal Cellular Growth*, p. 489. John Wiley & Sons, New York.

CHRISTIANS, C. J., CHAMBERS, D., WALTER, L. E., WHITEMAN, J. V., and STEPHENS, D. F. 1962. *J. Anim. Sci.* 21:387.

CHRYSTALL, B. B., ZOBRISKY, S. E., and BAILEY, M. E. 1969. *Growth* 33:361.

COLOMER-ROCHER, F., BASS, J. J., and JOHNSON, D. L. 1980. *J. Agric. Sci., Camb.* 94:697.

CORNFORTH, D. P., HECKER, A. L., CRAMER, D. A., SPINDLER, A. A., and MATHIAS, M. M. 1980. *J. Anim. Sci.* 50:75.

COUTEAUX, R. 1941. *Bull. Biol. Fr. Belg.* 75:101.

CRAWFORD, G. N. C. 1954. *J. Bone Joint Surg.* 36B:294.

CROSS, H. R., WEST, R. L., and DUTSON, T. R. 1981. *Meat Sci.* 5:261.

DAVIES, A. S. 1972. *J. Anat.* 113:213.

DEHOFF, R. T., and RHINES, F. N. 1968. *Quantitative Microscopy.* McGraw-Hill Book Company, New York.

DENNIS, S. M. 1972. *Cornell Vet.* 62:263.

DIEHL, B. J. M. 1978. *Cell Tissue Res.* 190:349.

DUMONT, B.-L., and SCHMITT, O. 1973. *Ann. Genet. Sel. Anim.* 5:499.

Duncan, C. J., and Greenaway, H. C. 1981. *Comp. Biochem. Physiol.* 69A:329.
Durand, G., Fauconneau, G., and Penot, E. 1967. *C. R. Acad. Sci. Paris* 264D:476.
Eisenhut, R. C., Cassens, R. G., Bray, R. W., and Briskey, E. J. 1965. *J. Food Sci.* 30:955.
Enesco, M., and Puddy, D. 1964. *Am. J. Anat.* 114:235.
Everitt, G. C. 1968. Prenatal development of uniparous animals with particular reference to the influence of maternal nutrition in sheep. In G. A. Lodge, and G. E. Lamming (eds.), *Growth and Development of Mammals.* Butterworth & Company (Publishers), London.
Ezekwe, M. O., and Martin, R. J. 1975. *Growth* 39:95.
Farmer, L. J., Mackie, W. S., and Ritchie, P. J. 1980. *J. Agric. Sci., Camb.* 95:563.
Faulkner, J. A., Maxwell, L. C., and Lieberman, D. A. 1972. *Am. J. Physiol.* 222:836.
Gibson, M. C., and Schultz, E. 1982. *Anat. Rec.* 202:329.
Goldspink, G. 1977. The growth of muscles. In K. N. Boorman and B. J. Wilson (eds.), *Growth and Poultry Meat Production*, pp. 13–28. British Poultry Science, Edinburgh.
Goldspink, G., Gelder, S., Clapison, L., and Overfield, P. 1973. *J. Anat.* 114:1.
Gonyea, W., Ericson, G. C., and Bonde-Petersen, F. 1977. *Acta Physiol. Scand.* 99:105.
Goodrich, E. S. 1958. *Studies on the Structure and Development of Vertebrates*, Vol. 1, Chap. 5. Reprint by Dover Publications, New York.
Granger, B. L., and Lazarides, E. 1978. *Cell* 15:1253.
Gunn, H. M. 1976. *Histochem. J.* 8:651.
Gunn, H. M. 1979. *J. Anat.* 128:821.
Halipre, A. 1973. *Ann. Genet. Sel. Anim.* 5:441.
Halvorson, D. B., and Jacobson, M. 1970. *Poult. Sci.* 49:132.
Hammond, J. 1932. Growth and Development of Mutton Qualities in the Sheep. Oliver & Boyd, London.
Hammond, J., and Appleton, A. B. 1932. Study of the leg of mutton. In J. Hammond, *Growth and Development of Mutton Qualities in the Sheep.* Oliver & Boyd, London.
Hanrahan, J. P., Hooper, A. C., and McCarthy, J. C. 1973. *Anim. Prod.* 16:7.
Hanset, R. 1967. *Ann. Med. Vet.* 111:140.
Hanzlikova, V., Mackova, E. V., and Hnik, P. 1975. *Cell Tissue Res.* 160:411.
Harbison, S. A., Goll, D. E., Parrish, F. C., Wang, V., and Kline, E. A. 1976. *Growth* 40:253.
Hegarty, P. V. J., and Hooper, A. C. 1971. *J. Anat.* 110:249.
Hegarty, P. V. J., and Naude, R. T. 1973. *Proc. R. Ir. Acad.* 73:87.
Heidenhain, M. 1913. *Arch. Mikrosk. Anat.* 83:427.
Helmi, C., and Cracraft, J. 1977. *J. Anat.* 123:615.
Hendricks, H. B., Aberle, E. D., Jones, D. J., and Martin, T. G. 1973. *J. Anim. Sci.* 37:1305.
Herold, A., and Nelms, G. E. 1964. *J. Anim. Sci.* 23:592.
Herring, S. W., Grimm, A. F., and Grimm, B. R. 1979. *Am. J. Anat.* 154:563.
Hiner, R. L., Hankins, O. G., Sloane, H. S., Fellers, C. R., and Anderson, E. E. 1953. *Food Technol.* 18:364.
Hogberg, M. G., and Zimmerman, D. R. 1979. *J. Anim. Sci.* 49:472.
Holmes, J. H. G., and Ashmore, C. R. 1972. *Growth* 36:351.
Holmes, J. H. G., and Robinson, D. W. 1970. *J. Anim. Sci.* 31:776.
Hooper, A. C., and Hanrahan, J. P. 1975 *Life Sci.* 17:775.
Hooper, A. C. B., and McCarthy, J. C. 1976. *Anim. Prod.* 22:131.
Howells, K. F., Jordan, T. C., and Howells, J. D. 1978. *Acta Histochem.* 63:177.
Huber, G. C. 1916. *Anat. Rec.* 11:149.
Huxley, H. E. 1972. Molecular basis of contraction in cross-striated muscles. In G. H. Bourne (ed.), *The Structure and Function of Muscle*, Vol. 1, Pt. 1, pp. 301–387. Academic press, New York.
Ishikawa, H. 1966. *Z. Anat. Entwicklungsgesch.* 125:43.

JIMINEZ, A. S., CARDINET, G. H., SMITH, J. E., and FEDDE, M. R. 1975. *Am. J. Vet. Res.* 36:375.
JOHNSON, E. R. 1981. *Anim. Prod.* 33:31.
JOHNSON, E. R., and BEATTIE, A. W. 1973. *J. Agric. Sci., Camb.* 81:9.
JOHNSTON, D. M., STEWART, D. F., MOODY, W. G., BOLING, J., and KEMP, J. D. 1975. *J. Anim. Sci.* 40:613.
JOUBERT, D. M. 1956a. *J. Agric. Sci., Camb.* 47:59.
JOUBERT, D. M. 1956b. *J. Agric. Sci., Camb.* 47:449.
JUDGE, M. D. 1978. *J. Anim. Sci.* 47:156.
KELLY, A. M. 1978a. *Dev. Biol.* 65:1.
KELLY, A. M. 1978b. *Anat. Rec.* 190:891.
KEMPSTER, A. J., and HARRINGTON, G. 1980. *Livestock Prod. Sci.* 7:361.
KERKUT, G. A. 1960. *Implications of Evolution.* Pergamon Press, Oxford.
KHAN, H. A., HARRISON, D. L., and DAYTON, A. D. 1981. *J. Food Sci.* 46:294.
KIDWELL, J. F., VERNON, E. H., CROWN, R. M., and SINGLETARY, C. B. 1952. *J. Hered.* 43:63.
KIESSLING, K.-H. 1977. *Swed. J. Agric. Res.* 7:115.
KING, W. A., BASRUR, P. K., BROWN, R. G., and BERG, R. T. 1976. *Ann. Genet. Sel. Anim.* 8:41.
KITIYAKARA, A., and ANGEVINE, D. M. 1963. *Dev. Biol.* 8:322.
KNAPP, B., and CLARK, R. T. 1950. *J. Anim. Sci.* 9:582.
KNAPP, B., and NORDSKOG, A. W. 1946. *J. Anim. Sci.* 5:194.
LAFLAMME, L. F., TRENKLE, A., and TOPEL, D. G. 1973. *Growth* 37:249.
LANDING, B. H., DIXON, L. G., and WELLS, T. R. 1974. *Hum. Pathol.* 5:441.
LAUVERGNE, J. J., VISSAC, B., and PERRAMON, A. 1963. *Ann. Zootech.* 12:133.
LAWRIE, R. A., POMEROY, R. W., and WILLIAMS, D. R. 1964. *J. Agric. Sci., Camb.* 62:89.
LE GROS CLARK, W. E. 1971. *The Tissues of the Body.* Clarendon Press, Oxford.
LEWIS, P. K., BROWN, C. J., and HECK, M. C. 1977. *J. Anim. Sci.* 45:254.
LIPSEY, R. J., DIKEMAN, M. E., KOHLMEIER, R. H., and SCOTT, R. A. 1978. *J. Anim. Sci.* 47:1095.
LIVINGSTON, D. M. S., BLAIR, R., and ENGLISH, P. R. 1966. *Anim. Prod.* 8:267.
LUFF, A. R., and GOLDSPINK, G. 1967. *Life Sci.* 6:1821.
LUFF, A. R., and GOLDSPINK, G. 1970. *J. Anim. Sci.* 30:891.
MACCALLUM, J. B. 1898. *Johns Hopkins Hosp. Bull.* 9:208.
MACKAY, B., and HARROP, T. J. 1969. *Acta Anat.* 72:38.
MACKELLER, J. C. 1960. *Vet. Rec.* 72:507.
MACKELLER, J. C. 1968. Muscular hypertrophy in South Devon cattle. Fellowship thesis, Royal College of Veterinary Surgeons, London.
MACKELLER, J. C., and OUHAYOUN, J. 1973. *Ann. Genet. Sel. Anim.* 5:163.
MAGLIANO, A. 1953. *15th Int. Vet. Congr.*, Pt. 1, pp. 578–580.
MASON, I. L. 1963. *Anim. Prod.* 5:57.
MAUGHAN, D. W., and GODT, R. E. 1981. *J. Gen. Physiol.* 77:49.
MAURO, A. 1961. *J. Biophys. Biochem. Cytol.* 9:493.
MAXWELL, L. C., FAULKNER, J. A., and HYATT, G. J. 1974. *Am. J. Appl. Physiol.* 37:259.
MAY, M. L., DIKEMAN, M. E., and SCHALLES, R. 1977. *J. Anim. Sci.* 44:571.
MAZANOWSKA, J., and KORDYLEWSKI, L. 1975. *Acta Biol. Cracov* 18:221.
MCMEEKAN, C. P. 1940a. *J. Agric. Sci., Camb.* 30:276.
MCMEEKAN, C. P. 1940b. *J. Agric. Sci., Camb.* 30:387.
MCMEEKAN, C. P. 1940c. *J. Agric. Sci., Camb.* 30:511.
MCMEEKAN, C. P. 1941. *J. Agric. Sci., Camb.* 31:1.
MEARA, P. J. 1947. *Onderstepoort J. Vet. Sci. Anim. Ind.* 21:329.
MELTON, C., DIKEMAN, M., TUMA, H. J., and SCHALLES, R. R. 1974. *J. Anim. Sci.* 38:24.
MELTON, C. C., DIKEMAN, M. E., TUMA, H. J., and KROPF, D. H. 1975. *J. Anim. Sci.* 40:451.

MENISSIER, F. 1982a. *Curr. Top. Vet. Med. Anim. Sci.* 16:23.
MENISSIER, F. 1982b. *Curr. Top. Vet. Med. Anim. Sci.* 16:350.
MICHAUX, C., VAN SICHEM, R., BECKERS, J.-F., DE FONSECA, M., and HANSET, R. 1982. *Curr. Top. Vet. Med. Anim. Sci.* 16:350.
MILLER, L. R., GARWOOD, V. A., and JUDGE, M. D. 1975. *J. Anim. Sci.* 41:66.
MONTGOMERY, R. D. 1962. *Nature (Lond.)* 195:194.
MOODY, W. G., KAUFFMAN, R. G., and CASSENS, R. G. 1969. *J. Anim. Sci.* 28:746.
MOODY, W. G., TICHENOR, D. A., KEMP, J. D., and FOX, J. D. 1970. *J. Anim. Sci.* 31:676.
MOODY, W. G., KEMP, J. D., MAHYUDDIN, M., JOHNSTON, D. M., and ELY, D. G. 1980. *J. Anim. Sci.* 50:249.
MORPURGO, B. 1898. *Anat. Anz.* 15:200.
MOSS, F. P. 1968. *Am. J. Anat.* 122:555.
MOSS, F. P., and LEBLOND, C. P. 1970. *J. Cell Biol.* 44:459.
MOSS, F. P., and LEBLOND, C. P. 1971. *Anat. Rec.* 170:421.
MOSS, F. P., SIMMONDS, R. A., and MCNARY, H. W. 1964. *Poult. Sci.* 43:1086.
MOWAFY, M., KASTENSCHMIDT, L. L., and CASSENS, R. G. 1972. *J. Anim. Sci.* 34:563.
MUHL, Z. F., and GRIMM, A. F. 1974. *Growth* 38:389.
MUIR, A. R. 1961. Observations on the attachment of myofibrils to the sarcolemma at the muscle-tendon junction. In J. D. Boyd, F. R. Johnson, and J. D. Lever (eds.), *Electron Microscopy in Anatomy*, Chap. 19. Edward Arnold (Publishers), London.
MUIR, A. R. 1970. *J. Comp. Pathol.* 80:137.
MULLEN, K., and SWATLAND, H. J. 1979. *Growth* 43:151.
NAERLAND, G. 1940. *Skand. Vet. Tidskr.* 30:811.
NAUDE, R. T., and HEGARTY, P. V. J. 1970., *Proc. S. Afr. Soc. Anim. Prod.* 9:217.
NICHOLSON, W. L. 1978. *J. Microsc.* 113:223.
NOVAKOFSKI, J. E., and KAUFFMAN, R. G. 1981. *J. Anim. Sci.* 52:1437.
OLIVER, W. M., and CARTWRIGHT, T. C. 1969. *Tex. Agr. Exp. Stn. Tech. Rep. 12.*
ONTELL, M. 1974. *Anat. Rec.* 178:211.
ONTELL, M. 1977. *Anat. Rec.* 189:669.
ONTELL, M., and DUNN, R. F. 1978. *Am. J. Anat.* 152:539.
ONTELL, M., and FENG, K. C. 1981. *Anat. Rec.* 200:11.
OTT, E. 1937. *Biol. Gen.* 12:510.
OUHAYOUN, J., and BEAUMONT, A. 1969. *Ann. Zootech.* 17:213.
PALSSON, H. 1939. *J. Agric. Sci., Camb.* 29:544.
PEARSON, J., and SABARRA, A. 1974. *Stain Technol.* 49:143.
PETROV, J. VON. 1976. *Zentralbl. Veterinaermed. C* 5:224.
POWELL, S. E., and ABERLE, E. D. 1975. *J. Anim. Sci.* 40:476.
RAIMONDI, R. 1962. *I bovini Piemontesi a groppa doppia.* G. B. Pavaria, Turin.
RAIMONDI, R. 1965. *Ann. Sper. Agric., N.S.* 17:471.
RAMSEY, C. B., COLE, J. W., and TENDICK, E. W. 1967. *Univ. Tenn. Agric. Exp. Stn. Bull. 424.*
RAYNE, J., and CRAWFORD, G. N. C. 1975. *J. Anat.* 119:347.
REGER, J., and CRAIG, A. S. 1968. *Anat. Rec.* 162:483.
REZNIK, M. 1976. *Differentiation* 7:65.
ROBINSON, D. W. 1969. *Growth* 33:231.
ROLLINS, W. C., TANAKA, M., NOTT, C. F. G., and THIESSEN, R. B. 1972. *Hilgardia* 41:433.
RUSSELL, B. 1963. *Mysticism and Logic.* Barnes & Noble, New York.
SCHAFER, E. A. 1898. Fibres; their figure and measurement. In E. A. Schafer, and G. D. Thane (eds.), *Quain's Elements of Anatomy*, 12th ed. Vol. 1, Pt. 2., pp. 286–287, Longmans, London.
SCHIEFFERDECKER, P. 1891. Morphologie des Muskelgewbes; Bauder Muskeln. In W. Behrens, A. Kossel, and P. Schiefferdecker (eds.), Die Gewebe des menschlichen Kǒrpers und ihre mikroskopische Untersuchung, *Zweiter Band.* Harald Bruhn, Braunschweig.
SCHMALBRUCH, H. 1973. *Brain* 96:637.

SCHMALBRUCH, H. 1978. *Anat. Rec.* 191:371.
SCHMALBRUCH, H., and HELLHAMMER, U. 1977. *Anat. Rec.* 189:169.
SHAFIQ, S. A., GORYCKI, M. A., and MAURO, A. 1968. *J. Anat.* 103:135.
SHELBY, C. E., CLARK, R. T., and WOODWARD, R. R. 1955. *J. Anim. Sci.* 14:372.
SMITH, J. H. 1963. *Poult. Sci.* 42:283.
SNOW, M. H. 1977. *Cell Tissue Res.* 185:399.
SOLOMON, M. B., MOODY, W. G., KEMP, J. D., and ELY, D. G. 1981. *J. Anim. Sci.* 52:1019.
STAUN, H. 1963. *Acta Agric. Scand.* 13:293.
STAUN, H. 1972. *World Rev. Anim. Prod.* 8(3):18.
STICKLAND, N. C., and GOLDSPINK, G. 1973. *Anim. Prod.* 16:135.
STICKLAND, N. C., and GOLDSPINK, G. 1975. *Anim. Prod.* 21:93.
STICKLAND, N. C., and GOLDSPINK, G. 1978. *J. Agric. Sci., Camb.* 91:255.
STICKLAND, N. C., WIDDOWSON, E. M., and GOLDSPINK, G. 1975. *Br. J. Nutr.* 34:421.
SUZUKI, A. 1973. *Jap. J. Zootech. Sci.* 44:50.
SWANSON, L. A., KLINE, E. A., and GOLL, D. E. 1965. *J. Anim. Sci.* 24:97.
SWATLAND, H. J. 1973. *J. Anim. Sci.* 37:536.
SWATLAND, H. J. 1974. *J. Anim. Sci.* 39:42.
SWATLAND, H. J. 1975a. *J. Anim. Sci.* 41:78.
SWATLAND, H. J. 1975b. *Zentralbl. Vet. Med. A* 22:756.
SWATLAND, H. J. 1975c. *J. Anim. Sci.* 41:794.
SWATLAND, H. J. 1976a. *Growth* 40:285.
SWATLAND, H. J. 1976b. *J. Anim. Sci.* 42:1434.
SWATLAND, H. J. 1976c. *J. Anim. Sci.* 42:63.
SWATLAND, H. J. 1977. *J. Anim. Sci.* 44:759.
SWATLAND, H. J. 1978. *J. Anim. Sci.* 46:118.
SWATLAND, H. J. 1979a. *Mikroskopie* 35:280.
SWATLAND, H. J. 1979b. *J. Agric. Sci., Camb.* 93:1.
SWATLAND, H. J. 1979c. *Zentralbl. Veterinaermed. C* 8:227.
SWATLAND, H. J. 1979d. *Zentralbl. Veterinaermed. A* 26:159.
SWATLAND, H. J. 1979e. *J. Anat.* 129:591.
SWATLAND, H. J. 1980a. *J. Agric. Sci., Camb.* 94:383.
SWATLAND, H. J. 1980b. *Growth* 44:139.
SWATLAND, H. J. 1980c. *Growth* 44:355.
SWATLAND, H. J. 1981. *Growth* 45:58.
SWATLAND, H. J. 1982a. *Mikroskopie* 39:317.
SWATLAND, H. J. 1982b. *J. Agric. Sci., Camb.* 98:629.
SWATLAND, H. J. 1982c. *Experientia* 38:855.
SWATLAND, H. J., and CASSENS, R. G. 1972. *J. Anim. Sci.* 35:336.
SWATLAND, H. J., and CASSENS, R. G. 1973. *J. Agric. Sci., Camb.* 80:503.
SWATLAND, H. J., and KIEFFER, N. M. 1974. *J. Anim. Sci.* 38:752.
SZARSKI, H. 1976. *Int. Rev. Cytol.* 44:93.
THOMPSON, E. H., LEVINE, A. S., HEGARTY, P. V. J., and ALLEN, C. E. 1979. *J. Anim. Sci.* 48:328.
THORNBURG, W., and MENGERS, P. E. 1957. *J. Histochem. Cytochem.* 5:47.
TRENKLE, A., DEWITT, D. L., and TOPEL, D. G. 1978. *J. Anim. Sci.* 46:1597.
TUMA, H. J., VENABLE, J. H., WUTHIER, P. R., and HENRICKSON, R. L. 1962. *J. Anim. Sci.* 21:33.
VENABLE, J. H., and LORENZ, M. D. 1970. Trial analysis of the cytokinetics of a rapidly growing skeletal muscle. In A. Mauro, S. A. Shafiq, and A. T. Milhorat (eds.), *Regeneration of Striated Muscle, and Myogenesis*, pp. 271–278. Exerpta Medica, Amsterdam.
VISSAC, B., MENISSIER, F., and PERREAU, B. 1973. *Ann. Genet. Sel. Anim.* 5:23.
WADDINGTON, C. H. 1952. *The Epigenetics of Birds.* At the University Press, Cambridge.
WATANABE, K., SASAKI, F., TAKAHAMA, H., and ISEKI, H. 1980. *J. Anat.* 130:83.

WEATHERLEY, A. H., GILL, H. S., and ROGERS, S. C. 1979. *Can. J. Zool.* 57:2385.
WEATHERLEY, A. H., GILL, H. S., and ROGERS, S.C. 1980. *Can. J. Zool.* 58:1535.
WEBER, A. D., and IBSEN, H. L. 1934. *Proc. Am. Soc. Anim. Prod.* 27:228.
WEIBEL, E. R., and ELIAS, H. 1967. *Quantitative Methods in Morphology*. Proc. Symp. Quantitative Methods in Morphology, 8th Int. Congr. Anatomists, Weisbaden, Germany. Springer-Verlag, Berlin.
WEIJS, W. A. 1980. *Am. Zool.* 20:707.
WEST, R. L. 1974. *J. Anim. Sci.* 38:1165.
WHITE, N. A., MCGAVIN, M. D., and SMITH, J. E. 1978. *Am. J. Vet. Res.* 39:1297.
WICK, R. A., and ALLENSPACH, A. A. 1978. *J. Morphol.* 158:21.
WILLEMSE, J. J., and VAN DEN BERG, P. G. 1978. *J. Anat.* 125:447.
WILLIAMS, P. E., and GOLDSPINK, G. 1976. *J. Anat.* 122:455.
WRIEDT, C. 1929. *Z. Indukt. Abstamm- Vererbungsl.* 51:482.
YOUNG, R. B., MILLER, T. R., and MERKEL, R. A. 1978. *J. Anim. Sci.* 46:1241.
YOUNG, R. B., MILLER, T. R., and MERKEL, R. A. 1979. *J. Anim. Sci.* 48:54.

APPENDIX

Morphometric Notation

Morphometry is a technique that uses numerical data to describe the shape, size, or frequency of occurrence of objects. *Stereology* is a technique that uses morphometric data to make predictions concerning dimensions which are not directly accessible in the sample data. Morphometry and stereology are widely used in metallurgy, and they are of growing importance in anatomy and histology. Notations for the communication and identification of morphometric parameters already exist but they are difficult to use with skeletal muscles. Skeletal muscle is an unusual tissue because its two dominant axes, length and radius, relate to different dimensions of cellular structure with unique properties. Muscle fibers are not randomly distributed in a matrix but are arranged longitudinally in fasciculi, and fascicular arrangement is the basis of muscle architecture. Notations for stereological parameters have been established in several fields, but they often use subscripts and superscripts, and Greek symbols for statistical operations. These characters are rather difficult to enter on most computer terminals. Some type of concise notation is essential when data are stored in a computer. The following notation is based on three essential units of information:

CLASS OF MEASUREMENT, SUBJECT, and PARAMETER

These units are condensed to a single statement by omitting spaces and by using upper- and lowercase characters:

CLASSsubjectPARAMETER

When lowercase characters are not available on the keyboard, any spare symbol may be used to separate the string:

CLASS|SUBJECT|PARAMETER

In everyday use, the data from individual animals and muscles must be labeled. This is done by preceding the statement with an animal number and a muscle abbreviation (in lowercase characters). For example,

99ldCLASSsubjectPARAMETER

Having constructed statements without internal spaces, it is now possible to use spaces to separate statements in mathematical operations:

CLASSsubjectPARAMETER × CLASSsubjectPARAMETER

The CLASSsubjectPARAMETER system is not unusual—its equivalent may be found in almost any research report on muscle stereology. The extra feature of the notation presented here is the way in which it is expanded to encompass the details, which become very important when morphometric data on muscle are used stereologically.

1. *Expansion of "class"* The class of measurement includes such familiar items as

SUM arithmetic sum
MEAN mean
EST estimate
D increment per unit time, dx/dt
LOG logarithm

Other classes may be abbreviated when needed but they are always written in uppercase characters. Following the class of measurement, and separated from it by a colon, may come any warning concerning an inference that has been made about the system under study, or about any technical problem that occurred relative to the data in the statement.

:ASSUM assumed value
:PROP proportional to, rather than equal to, a value
:REAL supposed to include all elements of a system
:APP apparent, only the elements that appear
:WNG1 warning of a type 1 experimental error (types of error are discussed above footnote 7)

2. *Expansion of "subject of measurement"* The subject of measurement is written in lowercase characters. The first letter identifies the tissue types found within a muscle:

a adipose tissue
c connective tissue
m muscle tissue
n neural tissue
v vascular tissue

The subsequent letters of the code identify a tissue unit:

ac adipose cell
an adipose cell nucleus
cen connective tissue, endomysium
cep connective tissue, epimysium
cpe connective tissue, perimysium
cn connective tissue, fibroblast nucleus
mfs muscle fasciculus
mfr muscle fiber
mfl muscle fibril
mft muscle filament
mn muscle fiber nucleus
msar muscle sarcomere
nn nucleus associated with nerve tissue
na nerve axon
nta terminal axon
va artery
vc capillary
vv vein
vn nucleus of cell associated with blood vessel

There are certain subtypes of tissue unit that may need to be identified. These are appended following a colon:

:ift a fiber that terminates intrafascicularly
:ete a fiber that runs from end to end of a muscle
:sat satellite cell of muscle fiber
:sar located in sarcoplasm
:peri perivascular cell
:cap capillary
:r red fiber (beta red or Type I red)
:i intermediate fiber (alpha red or Type II red)
:w white fiber (alpha white or Type II white)
:fp fetal primary fiber
:fs fetal secondary fiber

When several different units are considered as one group, this is indicated with an ampersand,

mfr:r&i aerobic muscle fibers
cpe&cenN number of peri- and endomysial boundaries

3. *Expansion of "PARAMETER OF MEASUREMENT"* Some parameters of measurement are as follows:

A area
AMAX maximum area
L length
N number
R radius
RMAX maximum radius
RMIN minimum radius
2R diameter
PI 3.14159
V volume
PV projected volume (AMAX × L)
W weight
X angle (relative to a specified orientation which appears next; for example, an angle of 45° to the longitudinal muscle axis = mX45L)
DEN density = W/V

If it is necessary to specify the time at which a measurement is taken, time in days is stated parenthetically after the parameter of measurement. It is necessary to define zero time, which can be relative to conception (C), birth (B), hatching (H), or the start of a trial (T). For example, the volume of a muscle 150 days after birth is written as

mV(150B)

The parameter of measurement usually requires some information on the basis of sampling. This is appended following a diagonal "/":

/A per area
/H homogenized or macerated sample
/L per length
/R per radius
/V per volume

The sample basis is usually oriented in a particular manner. Orientation is appended following a colon:

/A:R area in a radial plane (cross-sectional area)

Information on orientation may be supplemented by information on position which is appended after a semicolon. Position is specified relative to the longitudinal axis (L) and to the radial plane (R) of the muscle. Relative position is specified by a fraction, following these rules:

Muscle of axial skeleton
;0.0L is anterior
;1.0L is posterior
;0.0R is nearest to the middorsal line
;1.0R is furthest from the middorsal line

Muscle of appendicular skeleton

;0.0L is proximal

;1.0L is distal

;0.0R is nearest to limb or girdle bone

;1.0R is furthest from limb or girdle bone

This system is far from perfect but works quite well in meat research. For example, the quadrants of the longissimus dorsi muscle can be described by the 0.0R to 1.0R notation. The length of the longissimus dorsi muscle is from the last cervical to the last sacral vertebra. In limb muscles, the major zonation of sample variance is coaxial with the limb because meat nearer the bone is often composed of redder and often thinner fibers.

In some cases it may be necessary to define sample position by two coordinates. These are both appended following a semicolon, as in the following example for the mean minimum diameter of muscle fibers measured two-thirds the way along the longissimus dorsi muscle in the quadrant of the muscle nearest the vertebral column:

MEANmfr2RMIN/A:R;0.6L;0.25R

Sometimes it is necessary to give a third muscle coordinate. Fractional length (as defined earlier) has located a plane somewhere along the length of the muscle, while fractional radius (as defined earlier) has located sample position along one axis of this plane. Exact location needs two coordinates in a plane; the second is given by PFR, the perpendicular to the fractional radius. Note that R by itself is a plane but fractional R is a single linear dimension. A common example involving both sample orientation and position is the estimated apparent number of muscle fibers in a cross section of a whole muscle halfway along its length:

EST:APPmfrN/A:R;0.5L,m

Appended after the final comma is a lowercase code describing the subject of the sample, in this case, a whole muscle. Thus the subject of the measurement precedes the parameter of measurement, while the description of the sample base follows the sample base. This avoids confusion when these elements are isolated as column headings for a matrix:

mA muscle area as a primary measurement on a carcass

A,m area of muscle in which something else was measured

In some cases, where a parameter is not followed by a sample basis, it may be necessary to use the sample basis notation directly appended to the parameter. For example.

mA:R muscle cross-sectional area

mA:R;0.5L muscle cross-sectional area at the muscle midlength

Data are not always collected using a biologically defined sample area such as the number of myofibrils appearing in a cross section of a muscle fiber:

APPmflN/A:R,mfr

More often, data are collected within an arbitrary test dimension such as a micrometer grid in a microscope eyepiece. This information is appended to the sample basis following a comma and using "test" in lowercase characters to follow the rule set for tissue units:

mflN/A:R,test0.001

This conveys the information that the number of myofibrils was counted in 0.001 square millimeter of a cross-sectional area. The dimension of the sample basis (per length, per area, per volume) automatically sets the power of the unit of measurement. Millimeters are used throughout the notation since they are at the interface of macroscopic and microscopic measurements. With very large or very small numbers an exponential notation indicated by E can be used.

A major source of confusion in the literature on muscle fiber numbers in meat animals concerns the plane of measurement in which "test" exists. In simple muscles, the longitudinal axis of the muscle is parallel to the longitudinal axes of both muscle fasciculi and muscle fibers. The number of muscle fibers in a cross-sectional test area

mfrN/A:R,test

can be taken in only one plane. In complex muscles, however, the longitudinal axes of fasciculi and their fibers may diverge from the longitudinal axis of the whole muscle. Thus there may be some confusion as to whether the test area is a cross section relative to the whole muscle or relative to the muscle fibers. In the notation used here, "test" is in geometrical agreement with the planes and axes that relate to the "subject of measurement." Fiber number in a cross section relative to the muscle fibers is

mfrN/A:R,test

Even if the cross section is made relative to the whole muscle, it is described relative to the muscle fibers, the subject of measurement. For example, if the cross-sectional area of the whole muscle is at an angle of 45° to the axes of the muscle fibers, it is written as

A:X45L

If fibers are counted in a test area in that plane, it is written as

mfrN/A:X45L,test

If fibers are counted in the whole cross section of the muscle so that the muscle defines the sample area, it is written as

mfrN/A:X45L,m

In the case of a complex muscle, therefore, it is necessary to measure the angle of fiber insertion relative to the longitudinal axis of the muscle. If the angle is unknown, it can be written as a question mark. If the angle is unknown, the information content of the data is greatly reduced. In actual computations involving "test," it is necessary to standardize the units of "test." For example,

EST:APPmfrN/A:R,m = MEANmfrN/A:R,test × (mA / A:R,test)

8

Animal Growth and Development

INTRODUCTION

Four or five thousand years ago, the ancient Egyptians invented the feedlot as a system for methodically fattening the *iwz* or sacrificial bull. Other animals, such as the *ng*, were lean and rangy in conformation because they were put out to forage after they were captured from the wild (Smith, 1969). At a practical level, some of the fundamentals of both meat animal development and meat quality had obviously become apparent to the high priests or temple gourmets. Despite this promising early start, however, there are still many aspects of meat animal growth and development which we can see happening but which we cannot explain.

It is quite difficult to provide a universal definition of meat animal growth since some of the changes that occur in meat animals before they reach a marketable condition are reversible. This raises the problem of whether or not growth itself is reversible, or whether reversible increments to body weight should be separated from irreversible growth. If an animal increases its live weight by drinking, few people would maintain that the resulting increment to live weight is due to growth. If an animal increases its body weight by accumulating adipose tissue between and within its muscles, most people might agree that these are true growth increments. Yet these adipose increments might readily be lost if the animal is placed on reduced feed. Similarly, even the myofibrillar proteins of lean meat can be used as an energy reserve in fasting animals. However, the growth of the vital organs and nervous system is practically irreversible.

In commonsense terms, growth can be appreciated as an increase in height, length, girth, and weight which occurs when a healthy young animal is given adequate food, water, and shelter. Live weight is the most important and most commonly measured of these parameters and, if recorded at regular intervals, it yields a simple growth curve. For the practical purpose of determining feed conversion efficiency, the same data can be expressed as a curve for rate of gain. When live weight data are scrutinized for the small differences in growth that may be important when large numbers of animals are being raised, even the measurement of live weight can become a problem because of weight changes due to defecation, urination, and excess weight on the skin (Koch et al., 1958). Thus, in experimental conditions it is desirable to use measurements of body weight that have been mathematically reassembled from the weights of emptied viscera and clean hides obtained after slaughter.

Absolute measurements of body composition can be made on animals only after they have been killed. Further growth is then impossible. There are several ways around this problem, but none of them is entirely satisfactory. First, the animals may be kept alive and allowed to grow, and the meat yield may be measured with a nondestructive method of limited accuracy. If nondestructive measurements of growth are made so that animals can be kept alive and remeasured at intervals, the data require careful statistical analysis since each measurement on the same animal is partially related to preceding and succeeding measurements (Kowalski and Guire, 1974). As a second way around the problem, a homogeneous group of animals can be individually slaughtered in a sequence that enables the growth attainment of living animals to be visualized. The problem with this method is that any differences between the animals are superimposed on the visualized growth sequence. For example, a slow-growing animal that is slaughtered in sequence after a fast-growing animal may lead to a dip in the reconstituted growth curve. A third solution to the problem is used in the comparison of genotypes or diets. The animals are grown to a set point and are slaughtered, and then the data from different types of animals are compared. The problem here is to identify an equivalent slaughter point at which comparisons can reasonably be made. Animals can be slaughtered when they reach a set age, when they reach a set degree of fat deposition, or when they reach a set live weight. Different slaughter end points often yield different results.

Although studies of carcass composition have an important place in agricultural research, they are not the best means of investigating the process of growth itself. Linear measurements are largely a reflection of skeletal development rather than muscular development, and it is difficult to make accurate skeletal measurements on large animals when they are still alive. Curvilinear measurements made around the girth of the hind limb are a reasonable index of muscle development, but these are difficult to standardize. A method for the multiple recording of linear measurements has been developed which is based on the animal's shadow falling on a series of photocells (Clark et al., 1976). With the rapid developments that have been made in automated image analysis, there may be a revival of interest in linear measurements of growth, although the geometrical analysis of biological shapes is quite difficult. Some interesting possibilities are discussed by Oxnard (1980).

If a simple geometrical structure such as a sphere shows increases in its height, its surface area will increase in proportion to the square of its height, while its volume will increase in proportion to the cube of its height. If isolated tumor cells are grown in suspension in a constantly replenished medium, they form spheroidal clumps with a limited maximum diameter (10^6 cells with a spheroid diameter from 3 to 4 mm; Folkman and Hochberg, 1973). These spheroids of cells reach the limit to which radial diffusion and surface-to-volume ratio can supply nutrients and oxygen, and the limit at which waste products can still be removed from the center of the spheroid. If these cells were primitive life forms early in geological time, little further evolutionary increase in size would be possible until they had solved these problems. One solution is *gastrulation*, to increase the surface-to-volume ratio by becoming cup-shaped. Another solution is to develop a vascular system to enhance the transport of nutrients, oxygen, and waste materials. If the tumor cells mentioned earlier were in their original host, they would have to await *angiogenesis* (invasion by the vascular system) before they could start their lethal phase of exponential growth. Even at the spheroid level of geometrical development, however, there are often more subtle factors that limit or regulate growth. Biophysical studies with microelectrodes have revealed mechanisms of intercellular communication between normal cells. Communication between tumor cells may be reduced relative to normal cells (Loewenstein and Kanno, 1966). In healthy organisms there is evidence of a genetic limit to growth which involves cellular communication of some sort (Bonner, 1963). Regulation may involve inhibition by a factor such as cyclic AMP, as in the density-dependent inhibition of cellular proliferation (Froehlich and Rachmeler, 1972), or it may be part of a more extensive regulatory system involving an intercellular feedback control. In some cellular systems, genetic factors cause a programmed sequence of cell death (Lockshin and Beaulaton, 1974). Programmed cell death is often responsible for the development of the basic shape of an animal, as in the development of the chick's wing (Saunders et al., 1962).

As meat animals grow from birth to slaughter weight, they do not maintain a constant shape, and they deviate from a simple pattern of geometric growth. In general, however, even a small increase in linear dimensions causes a proportionately greater increase in body weight since body weight is a function of volume. In newborn farm animals, the rate of heat loss from the body is a serious problem. In small animals, the surface area for heat loss is relatively high, while the volume of muscle and brown fat that is capable of generating heat is relatively small. Since animals maintain an approximately constant body temperature during growth, there are probably subtle interactions between homeostatic mechanisms and body size. Basal metabolism is generally proportional to body weight raised to the power of 3/4 (Kleiber, 1961), rather than to the power of 2/3, as would be expected on simple geometrical grounds (Gray, 1981).

Among meat animals, it is difficult to find a situation in which an animal can grow to any marked extent without also exhibiting *development* of one type or another. Whereas growth can be considered as an increase in height, length, girth, or weight, development can be considered as a change in composition, structure, or ability, although neither definition is completely satisfactory. Seebeck (1968a) considered development to be the sum of growth plus differentiation, where differentiation was isolated from its histological connotations and was defined as the process in which parts of an animal change in their relative proportions. The concept is valid, but unfortunately, it conflicts with the widely accepted histological meaning of differentiation.

In functional systems such as the locomotory system, development may occur in direct response to growth in live weight. For example, as an animal grows heavier, it may need to use more of its muscle mass to oppose gravity, and there may be changes in the physiological properties of the diverted muscle mass. Similar functional changes occur in the digestive system, in bones, and in other body systems. After growth has ceased in the adult animal, developmental changes continue to occur as the animal passes through maturity to senescence. Among biologists, there seems to be an implicit understanding that developmental changes are directed toward the attainment of a mature composition, structure, or ability. Thus the retrogressive changes that occur later in life which are associated with a decline in composition, structure, and ability are regarded as senescent changes rather than developmental changes. Development and senescence are merely the early and late stages of *ontogeny*, the progress of an individual animal through its life cycle.

Although some aspects of growth, such as fat deposition, appear to be reversible, this is rarely true of developmental processes. As an animal is growing, a vast number of developmental changes are taking place concurrently. These changes are not usually undone if an animal simply loses weight. For example, a reversible accumulation of triglyceride may be accompanied by an increase in the number and size of fat cells, but the loss of a moderate amount of triglyceride may occur by all cells releasing a share of their fatty acids. Thus after "degrowth" appears to have occurred, the animal has not reverted to the developmental state (adipose cell number) it had before the period of growth started.

PRENATAL GROWTH

If the start of growth is measured from the time of conception rather than from the time of birth or of hatching, it is found that a considerable proportion of the time taken to reach market weight is spent in the uterus or in the egg (Table 8-1).

The data in Table 8-1 are approximate, because of variability in the duration of gestation between breeds and the even greater flexibility of animal age at slaughter. However, the orders of magnitude seen in the bottom line of the table hold true for many commercial operations. The control that can be exercised over the prenatal growth of meat animals is minimal since the environment which is created for the fetus by the uterus or *cleidoic* egg is buffered against the vicissitudes of the farm environment. A cleidoic egg is one which, except for temperature and the exchange of gases, is isolated from its environment by membranes and a shell, as in poultry.

The selective breeding of meat animals should be based on the selection of

TABLE 8-1. APPROXIMATE TIME TAKEN TO GROW TO A MARKET WEIGHT, AS MEASURED IN DAYS FROM THE TIME OF CONCEPTION

Development	Beef	Lamb	Pork	Chicken
Forelimb bud	24	20	16–17	2.2
Hindlimb bud	26	21	17–18	2.2
Fetus	45	21	20	5
Birth or hatching	285	150	112	22
At slaughter	850	350	270	70
Percentage time prenatal	33	43	42	31

animals bearing phenotypic characters which are related to the economic yield of meat. There is a biological relationship between the genes, which are manipulated by selective breeding, and the phenotype, upon which breeding decisions are made. The biological linkages between genotypes and phenotypes pass through an *epigenetic* space in which gene products react among themselves and with environmental factors. The major epigenetic interactions involved in the formation of a new animal take place in the prenatal period. Thus prenatal development may be the most important but least accessible phase of meat animal development. While the newly formed animal is developing its major histological types of tissues, it is called an *embryo*. After these tissues are acquired and up until birth or hatching, it is called a *fetus*.

Synopsis of Prenatal Development

A mammalian egg contains very little yolk (*oligolecithal egg*), whereas an avian egg contains a lot of yolk (*telolecithal egg*). In a telolecithal egg, cell division in the relatively small amount of protoplasm does not spread very far into the viscous yolk mass. However, mammals appear to have evolved from reptiles which had telolecithal eggs, and they retain features which suggest that their early embryos once had to cope with the problem of a large inert yolk mass.

Cleavage is the process by which a *zygote* (fertilized ovum) is subdivided into smaller cells called *blastomeres*. When a new embryo has become subdivided into a ball of eight or more blastomeres, but has not yet formed layers of blastomeres, it is called a *morula*. The morula starts to form a new animal from an *inner cell mass* (Figure 8-1). The morula also forms a fluid-filled sphere of cells called the *trophoblast*. The trophoblast contributes to the *placenta* and so is lost at birth. The inner cell mass together with the trophoblast form a *blastocyst*. The embryo that will develop from the inner cell mass is roofed over by *amniotic folds* which later fuse. *Endoderm* from the inner cell mass spreads over the inner surface of the trophoblast; at this stage, the blastocyst is said to be *bilaminar*. The blastocyst becomes *trilaminar* when *mesoderm* from the inner cell mass spreads *between* the outer trophoblast layer and the inner

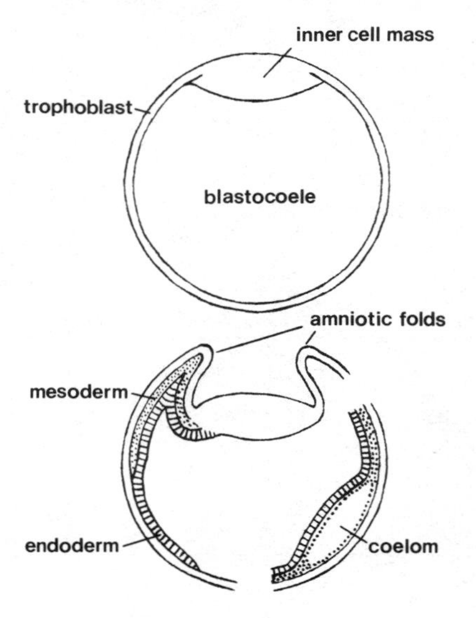

Figure 8-1 Blastocyst. The top diagram shows the initial structure of the blastocyst. The bottom composite diagram shows the expansion of the endoderm and mesoderm, the formation of amniotic folds, and the formation of extraembryonic coelom within the mesoderm.

endoderm layer (Figure 8-1). The mesoderm layer then splits within itself to form a cavity within the mesoderm, the *extraembryonic coelom*.

Expansion of the extraembryonic coelom allows the *yolk sac* and *allantois* to expand into the coelom from the gut of the embryo (Figure 8-2). In chicks, the yolk sac contains the yolk. In cattle, sheep, and pigs, the yolk sac is rudimentary. In chicks, the allantois is used for respiration and for storage of waste products. In cattle, sheep, and pigs, the allantois enables the vascular communication of the developing animal and its mother. The double layer of cells formed by the mesoderm and the ectodermal trophoblast is called the *chorion*. The allantoic membrane is pushed against the chorion to form the *allantochorion* (Figure 8-2). The interface between the allantochorion and the mammalian uterus is increased in surface area by the formation of chorionic *villi*. In pigs, villi are scattered over the chorion (*diffuse placenta*). In cattle and sheep, villi are grouped into about 100 patches called *cotyledons*.

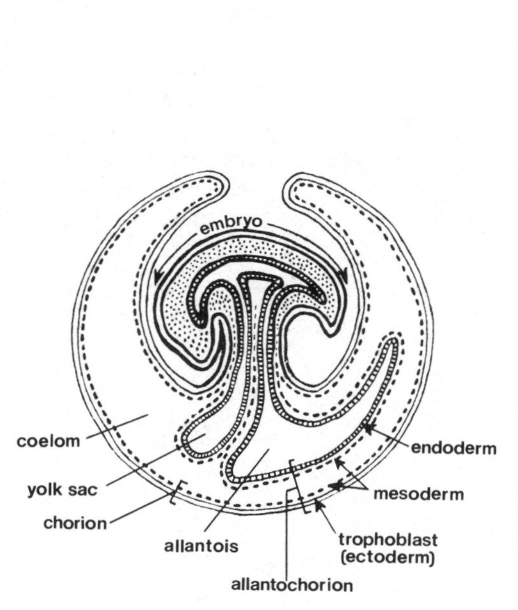

Figure 8-2 Expansion of the yolk sac and allantois into the extraembryonic coelom, and the formation of the allantochorion.

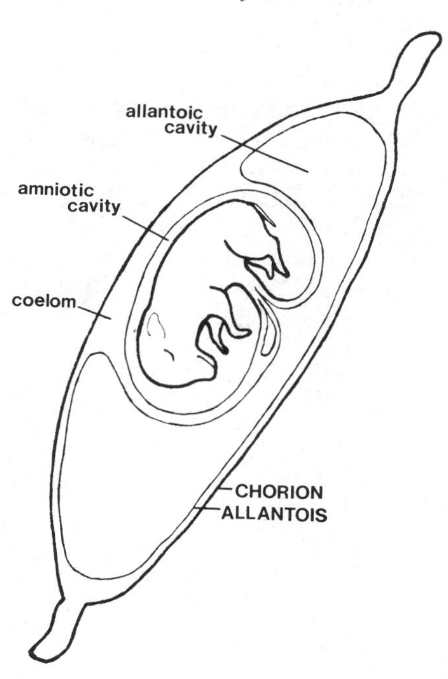

Figure 8-3 Fetal pig and its membranes.

The space within an avian egg is limited, and the growth of the fetus and allantois is balanced by shrinkage of the yolk sac. The mammalian uterus, however, is capable of considerable expansion, and the allantoic cavity grows quite large (Figure 8-3).

Cattle

The most widely used measure of embryonic and fetal growth is *crown-rump length* (Figure 8-4). This is a straight line measured from the crown of the head to the base of the tail. The crown may be taken as a point midway between the eye orbits on a line traversing the approximate position of the frontal eminence (Committee on Bovine

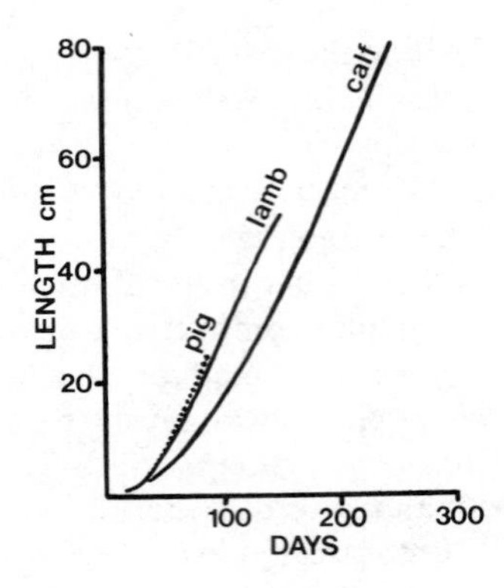

Figure 8-4 Prenatal growth of crown-rump length in cattle, sheep, and pigs. Time in days of gestation. (Data from Evans and Sack, 1973.)

Reproductive Nomenclature, 1972). Growth in weight of the calf is usually more variable than crown-rump length. The weight of the fetus does not usually overtake the weight of the fetal membranes until approximately 3 months after conception (Figure 8-5). Data on visceral growth are given by Hubbert et al. (1972). The relative water content of muscles decreases during fetal development. The concentration of DNA and RNA remains high until approximately 6 to 7 months gestation; after this time, the relative DNA content sharply declines and there is an increase in dry matter and total nitrogen (Ansay, 1974). Extensor carpi radialis and extensor digitorum communis muscles have a high prenatal growth rate which decreases postnatally. Rectus abdominis and longissimus dorsi muscles grow slowly until birth and then their growth accelerates after birth (Johnson, 1974).

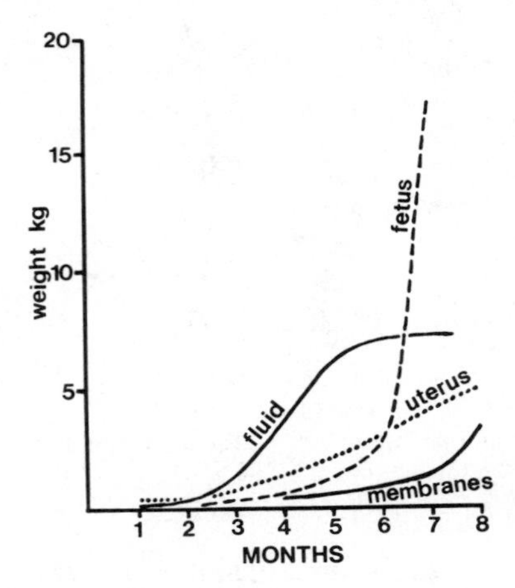

Figure 8-5 Uterine composition during gestation in cows. (Data from Hammond, 1927.)

Sheep

In fetuses of unknown age, crown-rump length may be used to determine the age of the fetus from 50 to 100 days gestation. After this time, estimates of fetal age can be made from brain weight, long bone length, and the number of appendicular ossification centers (Richardson et al., 1976). The morphology of visceral development is described by Bryden et al. (1972a, 1972b, 1973). The age of fetal lambs can also be estimated from the state of development of the hair follicles in the skin (Galpin, 1935). Undernutrition of the ewe during the last 6 weeks of gestation causes a severe decrease in the birth weight of lambs (Wallace, 1948a).

Joubert (1956) undertook an extensive study of fetal development in single male lambs. Embryonic growth and development proceed in an anterior to posterior sequence so that the head reaches a relatively large size early in development. During fetal development, the remainder of the body catches up to, and overtakes the growth of the head. The head-to-body ratio, however, is rather variable between animals. The feet and the tail develop last. Inadequate maternal nutrition and competition between twin and triplet fetuses may inhibit skeletomuscular growth (Everitt, 1968; Swatland and Cassens, 1973). In competitive situations, males sustain less inhibition of growth than females (Burfening, 1972). Differences in body weight between twins and triplets are related to differences in placental development and position in the uterus (McDonald et al., 1981). In prolific ewes, high fetal losses early in development are associated with the suboptimal growth of surviving fetuses (Robinson, 1981).

Pigs

The morphology and crown-rump growth of prenatal pigs are described by Altman and Dittmer (1962), Ullrey et al. (1965), Pomeroy (1960), Patten (1948), and Marrable and Ashdown (1967). Sows are *multiparous* and usually have 10 or more offspring in each litter. Wild pigs have a longer gestation period (130 days) and fewer offspring per litter than do modern commercial pigs, so that it appears that selection for larger litters may have decreased the length of gestation (Salmon-Legagneur, 1968). Fetuses compete with each other for the nutrients which are made available by the uterus, and the success of a fetus is related to its position within the uterus.

Longitudinal growth of bones depends more on nutritional status, and hence fetal body weight, than on the age of the fetus. Maturation of the epiphyses, however, depends more on age than on nutritional status (Adams, 1971). The muscles of runt pigs are smaller than those of their more successful littermates. Although the DNA content of muscles is decreased in runts, the decrease in protein is even greater, so that protein-to-DNA ratios are decreased (Widdowson, 1971). Runt pigs appear to have fewer muscle fibers. This effect is most pronounced in Type II white fibers (Powell and Aberle, 1981), most of which are formed in the late fetal phase of fiber mass production (Chapter 6). The smaller pigs of a litter, if they are able to compete successfully for milk after birth, may be able to catch up in their growth by the time of weaning. Below a birth weight of about 1 kg, however, runt piglets are doomed to produce carcasses with less muscle and extra fat (Powell and Aberle, 1980).

During fetal development, there are extreme changes in body shape as the limbs compensate for their late start relative to the head and trunk (Figure 8-6). At the time of birth, some muscles have already reached their maximum growth velocity, while others are still accelerating (Figure 8-7). The relative water content of muscles generally decreases during prenatal development, and some care is necessary in the choice of a suitable base parameter on which to measure growth or biochemical composition. In the 24 hours following birth, there is a marked increase in the relative water content of muscles (Pownall and Dalton, 1976).

Sows sometimes drop on their unwary offspring and kill them, and any degree of

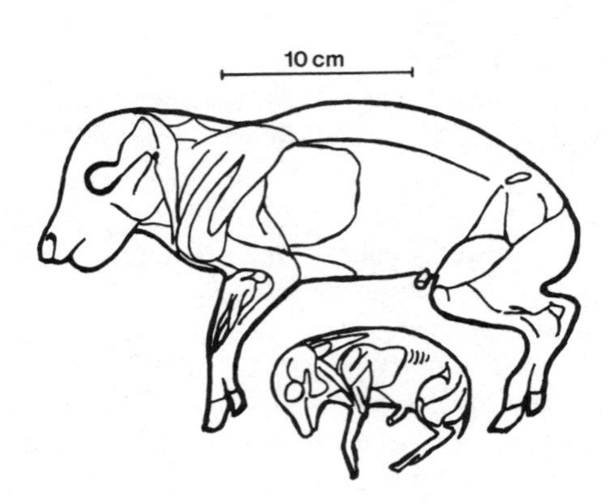

Figure 8-6 Partial muscle dissections of fetal pigs at 55 days (bottom) and 105 days gestation (top).

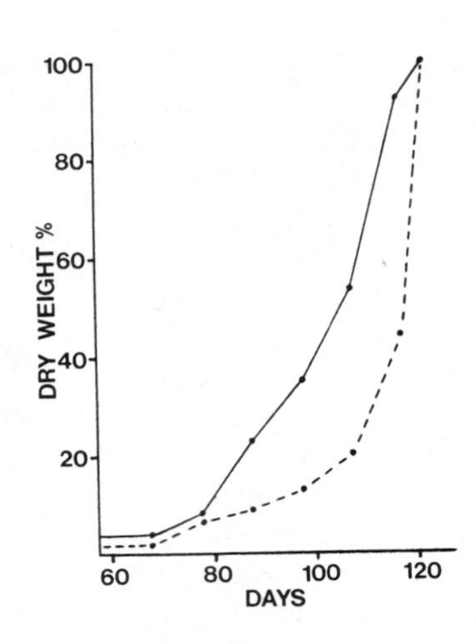

Figure 8-7 Allometric growth of peroneus longus (solid line) and longissimus dorsi (broken line) in fetal pigs. Time scale in days of gestation.

locomotor impairment in baby pigs increases this risk. *Splayleg* is a condition in which the hind limbs, and sometimes the forelimbs, of a newborn pig are temporarily unable to support the body. The condition has been reported in most countries with a well-developed swine industry (Thurley et al., 1967; Cunha, 1968; Dobson, 1968; Olson and Prange, 1968; Melhorn et al., 1970; Lax, 1971; Smidt, 1972; Bergstrom, 1972). If affected pigs are nursed through the period of high risk just after farrowing, they may compensate by enhanced synthesis of muscle proteins, and they may then grow quite normally (Hajek, 1980). Slippery floors make the splayleg problem worse (Kohler et al., 1969). Affected pigs can be helped with a loose coupling between the hind limbs to prevent the limbs splaying outward (Bollwahn and Pfeiffer, 1969; Krudewig, 1971). Splayleg may be due to *choline* deficiency in the sow's diet (Cunha, 1968), but choline and methionine supplements do not necessarily reduce the incidence of splayleg (Dobson, 1971). Miller et al. (1973) identified *zearalenone* toxin in moldy feed as another possible cause. Electromyography of splayleg piglets may reveal a decreased muscle response to neural activation (Hnik and Vejsada, 1979a), but neuromuscular conduction is normal (Hnik and Vejsada, 1979b).

Splayleg piglets sometimes exhibit *myofibrillar hypoplasia*, a reduction in the apparent number of myofibrils (Thurley et al., 1967). Muscle fibers with fewer myofibrils may contain more glycogen granules (Deutsch and Done, 1971) and a higher polysaccharide content (Patterson and Allen, 1972), but this is difficult to isolate from the large variation in biochemical composition which is normally found among litters of normal pigs (Patterson et al., 1969). The rate of incorporation of tritiated thymidine into DNA shows that mitotic rates are reduced in the muscles of splayleg piglets (Farmer et al., 1981). Whether this is a primary cause or a secondary result of the condition is unknown. The splayleg condition might well be a result of physiological immaturity due to selection for larger litters. The topic is reviewed in detail by Ward (1978a, 1978b).

Poultry

Somites develop in an anterior to posterior sequence so that posterior somites are younger. However, the initial size of posterior somites is larger, and they grow more rapidly (Herrmann, 1952). Thus, regardless of age, they are similar in size and can be studied collectively. Volumetric growth and DNA accumulation in chick somites are shown in Figure 8-8.

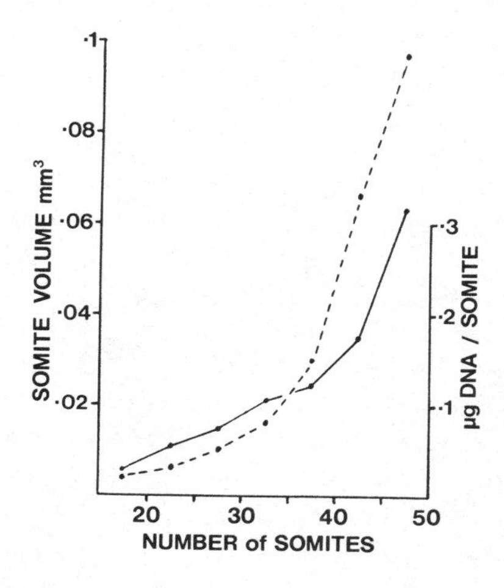

Figure 8-8 Relationships of somite volume (broken line) and amount of DNA per somite (solid line) to the number of somites in chick embryos. (Data from Herrmann, 1952.)

Abnormal Development

Developmental mechanisms sometimes go wrong. In view of their tremendous complexity, this is hardly surprising—the more remarkable fact is that they do not go wrong more often. Much of the research effort in animal agriculture has been directed at maximizing the productivity of animals whose developmental mechanisms function properly and in a natural way. Although many developmental defects in farm animals have been categorized (Lerner, 1944), little is known about their causes at the cellular and biochemical level.

The general principles of developmental defects in meat animals are reviewed by Binns et al. (1969). Defects may be genetic (Hutt, 1934a, 1934b; Roberts, 1929) or may be due to environmental factors, such as a deficit or excess of nutritional factors (Palludan, 1961; Dyer et al., 1964; Dyer and Rojas, 1965), high temperature (Hartley et al., 1974; Wildt et al., 1975), plant toxins (James, 1972; Crow and Pike, 1973; Leipold et al., 1973; Edmonds et al., 1972; Selby et al., 1971), diseases (Kennedy, 1971; Sever, 1971), drugs (Vente et al., 1972), or pollution (Mulvihill, 1972). The results of defective development are most conspicuous in newborn animals, and most malformed animals are culled at this time. However, many animals may have already been lost by this time, either by embryonic loss or by fetal mummification. Animals with inconspicuous defects or which are physiologically immature may die during the neonatal period.

A number of general types of defects are recognized.

1. Development may be arrested so that certain tissues are either not formed

(*aplasia*) or else suffer a reduction in number of cells or components (*hypoplasia*).

2. Organs such as the central nervous system and heart have parts which should become closed off or fused, and the mechanisms involved may fail.
3. Embryonic structures such as the allantois may persist instead of disappearing.
4. Tissues may occur in the wrong location.
5. Body parts such as limbs and visceral organs may be duplicated. Visceral duplication is sometimes harmless and may remain undiscovered until animals reach the abattoir as adults.

The experimental manipulation of the prenatal development of farm animals has resulted in commercial methods such as *embryo transfer techniques* (Seidel, 1981). This early stage of animal development might also be the most amenable stage for genetic engineering. In the DNA molecule, the purines *adenine* and *guanine*, and the pyrimidines *thymine* and *cytosine*, all project inwardly from a double helix of sugar molecules which are connected by their phosphate bonds. Each adenine is paired with a thymine on the opposite strand of the helix. Each guanine is opposite a cytosine. The exact sequence or permutation of several tens of thousands of these pairs of bases forms a word written in a four-letter alphabet—that word is a *gene*. The gene can be copied for the transmission of genetic information, or it can be used to dictate the sequence of amino acids during the assembly of a protein.

Genes were first transferred experimentally into new locations by techniques such as chromosome assimilation and cell fusion. In order to show that the transfer of genes had actually taken place, the recipient cell had to be placed in an environment in which the transferred genes would reveal themselves. For example, the recipient cells might suddenly acquire the ability to survive in a medium that lacked a hitherto essential nutrient. Next, three new techniques were developed for the transfer of isolated DNA: (1) direct injection into the nucleus of a recipient cell using a micropipette; (2) the use of a virus to carry the DNA into the recipient cell; and (3) precipitation of the DNA with calcium phosphate to make it attractive for uptake by the recipient cell.

The genetic engineering of mammalian development by such techniques has only just started and, at the present time, it depends heavily on the use of relatively simple and extensively studied genes. Apart from important exceptions such as muscle fiber hyperplasia in cattle, most of the commercially important traits of meat animals are regulated by large numbers of genes, and they might be very difficult to manipulate by genetic engineering. In laboratory animals, growth rates can be increased dramatically by introducing genes that lead to increased levels of growth hormone (Palmiter et al., 1982), but whether or not genetic engineering at the molecular level will enable us to develop superior types of meat animals remains to be seen. The challenge is to make meat animals more efficient, not merely larger.

A critical link that is needed for the genetic engineering of meat animals is a cell line that can be both asexually propagated in vitro and then reincorporated back into the cycle of normal sexual reproduction in meat animals. An in vitro phase is necessary for genetic manipulation, for testing the outcome of the manipulation, and for cloning successful products. It must be possible, however, to return the modified cells to an early phase of embryogenesis in which they can usefully contribute to the formation of a new animal. If the genetic modification does not get incorporated into the gonads, it cannot be propagated without repeating the reintroduction of cloned cells. This whole sequence of events, from genetic manipulation to the production of fertile progeny, has already been performed in mice (Stewart and Mintz, 1981).

THE MEASUREMENT OF GROWTH

The measurement of growth and development in live animals by indirect and nondestructive methods is extremely important, since it allows growth to be studied in ongoing experiments or in commercial agriculture. Except in the case of the most successful indirect method, *ultrasonic* probing for back fat thickness, indirect methods are not very accurate. The basis of the ultrasonic method is that high-frequency sound waves are generated by a transducer on the animal's back. Since the velocity of the waves through the tissue is known, the time taken for echoes to return from tissue boundary layers can be used to determine their depth. Strong echoes are returned from the muscle-fat boundary over the longissimus dorsi muscle, but other echoes are also returned from connective tissue septa in the subcutaneous adipose tissue and from the axial skeleton ventral to the longissimus dorsi. By moving the transducer in an arc over the animal's back, depth measurements can be assembled into a map of the carcass section. Ultrasonic scanning and other indirect methods of carcass assessment are reviewed in detail by Stouffer (1969). An alternative ultrasonic method is to measure changes in the velocity of the signal as it passes through the tissue, since the proportion of adipose tissue is correlated with the reciprocal of velocity (Miles and Woods, 1979). Sophisticated equipment is not always necessary to measure fat depth. In live broiler chickens, for example, the amount of abdominal fat can be measured with calipers (Pym and Thompson, 1980).

The relative degree of development of commercially valuable carcass components such as the loin and hindquarter is difficult to assess subjectively in the live animal. The perception of volume or mass can be simulated by a photographic method known as *photogrammetry*. This method utilizes the parallax effect of a stereo photograph which is compounded from two photographs taken simultaneously by two cameras separated by a fixed and accurately known distance. The volume of the animal body or anatomically defined parts of the body can be calculated from the degree of the stereo effect. Although this method is considered to be quite accurate, its cost and complexity have so far limited its use.

Animal volume has also been investigated as a means of calculating density or *specific gravity*. Instead of being immersed in water by the classical method of Archimedes, live animals are handled by the air displacement technique, which uses negative air pressures around an animal in an experimental chamber (Liuzzo et al., 1958). The accumulation of fat (low density) decreases the overall density of the body so that density then provides an index of fatness. In practice, however, it is difficult to obtain accurate volumetric measurements which allow for the offsetting effects of bones. In cattle, bone density increases with age, and it varies between sexes and between breed types (Jones et al., 1978a).

An alternative method for adipose tissue volume is based on the fact that adipose tissue has a very low water content. Chemicals such as the drug *antipyrin* or one of its derivatives are injected into the vascular system from whence they diffuse into the total volume of aqueous body fluids. Provided that the injected substance is only slowly metabolized or excreted by the body, and provided that it becomes adequately dispersed, the extent to which the injected substance becomes diluted indicates the total diluting volume. In general, fat animals have a proportionately smaller aqueous volume than do lean animals. Density methods can also be used for dressed carcasses (Jones et al., 1978b).

Another indirect method of estimating body composition is based on the fact that nearly all the *potassium* in the body is intracellular. Since adipose cells contain only a trace of potassium in their scanty amount of cytoplasm, the total amount of potassium in the body is approximately proportional to the lean body mass. A constant proportion (0.012 per cent) of potassium atoms are radioactive (the isotope ^{40}K), and they emit gamma rays which can be converted into small flashes of light and counted electronically. If an animal is shielded from the gamma rays coming from the environment, its total gamma emission gives an index of its total muscle mass. A background count of the extrinsic radiation that penetrates the shield is subtracted from the count which is made during the 3 minutes when the animal is being measured. Animals are taken off their feed and water for 24 hours prior to being measured so as to avoid complications from potassium in the rumen. Different types of animals require different calibration factors from which to predict their muscle mass (Keck, 1970). The individual muscles of the carcass may differ considerably in their concentration of potassium (Lawrie and Pomeroy, 1963). Radioactivity can also be employed in another method, using *tritium*, an isotope of hydrogen, to estimate the water volume and, indirectly, the fat content of live pigs (Sheng et al., 1977). In the EMME system, the volume of electrolytes in the animal body is estimated from the disturbance that their electrical conductivity creates in an externally applied electromagnetic field.

After an animal has been slaughtered, its composition can be determined with an accuracy which is limited only by the available research resources. When a research project is properly planned, the numbers of experimental animals required to satisfy the research objectives can often be estimated statistically. A point that should receive equal attention but which is often ignored is the extent to which the analysis of each carcass should be pursued. At one extreme, all the available research resources can be used to collect simple and often misleading data from large numbers of animals. At the other extreme, just a few animals can be analyzed in great detail. In general, agricultural researchers have tended toward the first extreme.

In examining the published research on animal growth and carcass composition, individual reports should be examined critically to see if they describe growth and development in a more or less homogeneous group of animals such as might be found in commercial agriculture. Sometimes, hopefully always for scientific reasons, a report will describe the conditions found in a heterogeneous group of animals with a range of types not normally found in commercial agriculture. Results and conclusions derived from the latter type of experiment, while they may be vital in contributing to an understanding of the physiology of growth, are often extremely misleading when applied in the context of commercial agriculture. For example, when muscle growth in meat animals was first analyzed microscopically, it was found that muscle fibers grew in diameter as the amount of muscle on the carcass increased. This led to the suggestion that muscle fiber diameters might be used as an index of muscularity in meat animals.

For reasons that are presented in detail in Chapter 7, this method has little or no potential use in commercial agriculture. If one were to compare tail length with muscle mass over a complete span from embryo to adult, quite a high coefficient of correlation would be anticipated. Attempts to predict the lean meat content of commercial carcasses from the tail length of market-weight animals would probably be very disappointing.

Many scientific disciplines are launched by the publication of a series of concepts and experiments which linger as guiding lights long after they are superseded in theory and technique. The scientific study of carcass composition was most conspicuously launched in Victorian times by Lawes (1814–1900) and Gilbert (1817–1901) at Rothamsted in England. Following Liebig's initially controversial idea that animals could produce fat from nonfat components in their diet, Lawes and Gilbert set out to prove that much of the fat in meat carcasses originated by *de novo* synthesis. The outcome of their research was a series of carcass dissection studies, the like of which continues to the present day (Hall, 1905).

Lawes and Gilbert examined the body composition of cattle, sheep, and pigs at different stages of fattening. In Victorian times, fattening was taken to extremes that would strike contemporary butchers with horror—the carcass fat content of one very fat sheep was well over 50 per cent. The experimental data, however, were analyzed in the same way that they are today, with body compartments being divided into mineral ash, nitrogenous substances (mainly protein), fat, and water. Appropriate weight corrections were made for the gut contents. Lawes and Gilbert soon became aware that the accumulation of nitrogenous substances was accompanied by accumulation of large amounts of water, whereas the accumulation of fat was not. They found the relevance of this observation in the diminishing increments to body weight which accompany the change from protein accumulation to fat deposition in older animals.

Although the first stone in the foundation of the study of meat animal growth was set in place by Lawes and Gilbert, the remaining basic elements were contributed by Hammond and his students at Cambridge. The greatest tribute to their work is that we now accept their original discoveries as common knowledge. An elementary account of this founding knowledge is presented in an indispensable volume titled *Hammond's Farm Animals* (Hammond et al., 1971): its lasting influence on present research is concisely surveyed by Elsley (1976).

Growth Curves

If the dimensions of an animal are measured from conception to senescence, the data usually follow a sigmoid curve (Figure 8-9). The growth curves of meat animals raised under commercial conditions appear as relatively flat slopes, and their sigmoid shape becomes apparent only if animals are kept beyond a typical market weight. Thus the maximum growth velocity occurs within the commercial growing period (Figure 8-10).

At first, the fertilized ovum divides mitotically with little or no growth in mass. As soon as it develops the means of assimilating energy, however, the growth of the embryo begins to accelerate. At birth and at weaning there may be a temporary deceleration of growth as an animal switches from one source of nutrients to another. Except for a slight acceleration at puberty, subsequent growth usually maintains a steady velocity until the terminal deceleration, which occurs as animals reach their

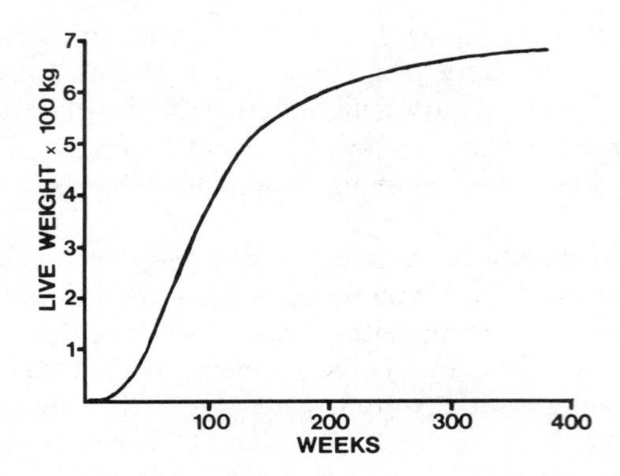

Figure 8-9 Sigmoid growth curve for Holstein cows. Time scale starts at conception. (Data from the Committee on Bovine Reproduction Nomenclature, 1972, and from Altman and Dittmer, 1962, subjected to sequential averaging, Swatland, 1980a.)

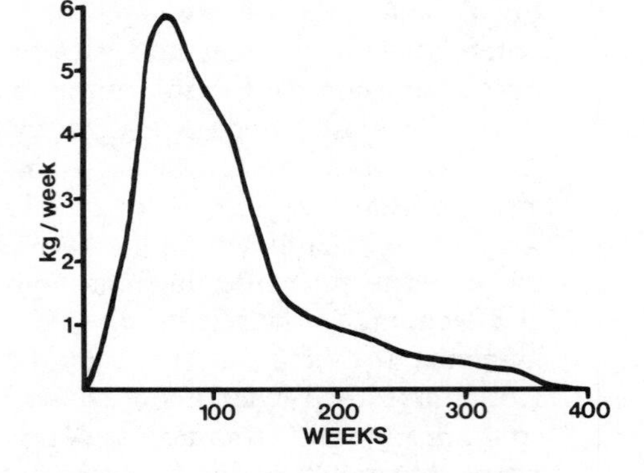

Figure 8-10 Velocity of growth or rate of gain, calculated from the data in Figure 8-9. Time scale starts at conception.

mature size. Even in an apparently smooth growth curve, however, the possibility of an underlying *circadian* periodicity in growth rate (Barr, 1973) should not be forgotten.

Growth curves have been extensively investigated from a mathematical viewpoint, but as yet, the mathematical approach has contributed little to our understanding of the biology of growth. Complex curvilinear functions do not help to explain the nature of growth unless some biological meaning can be attached to their terms. If all that is needed is a good empirical fit for a growth curve, a polynomial function is as useful as any (Donald, 1940; Needham, 1950, 1957), or a systematic method of selecting an appropriate transformation may be used (Box and Cox, 1964).

If protein turnover rates are measured in rapidly growing young pigs, the half-life for protein synthesis is shorter than the half-life for protein catabolism (Table 8-2). In the muscle tissue of growing animals, *insulin* is the most important factor that inhibits the degradation of myofibrillar proteins when anabolism exceeds catabolism (Tischler, 1981). In fasting animals, where catabolism exceeds anabolism, proteolysis is promoted by *glucocorticoids. Epinephrine* and *serotonin* have a minor effect.

The *growth equation of von Bertalanffy* (1957) is based on the concept that each increment to body weight (W) represents the triumph of anabolism over catabolism (Figure 8-11),

TABLE 8-2. PROTEIN TURNOVER TIMES IN YOUNG PIGS

	Half-life (days)	
Protein type	Synthesis	Catabolism
Sarcoplasmic	4.8	9.4
Myofibrillar	5.7	16.4
Connective tissue	8	20

Source: Data from Perry (1974).

$$\frac{dW}{dt} = aW^b - cW^d$$

$$\text{body weight increment} = \text{anabolic increment} - \text{catabolic loss}$$

This yields values for parameters b and d which are between 2/3 and 1, much like the power relationships of body mass or volume to body surface area. Heat loss tends to be proportionate to skin area, while gut areas may be rate limiting for nutrient uptake. Both heat generation and anabolic systems tend to be related to body volume.

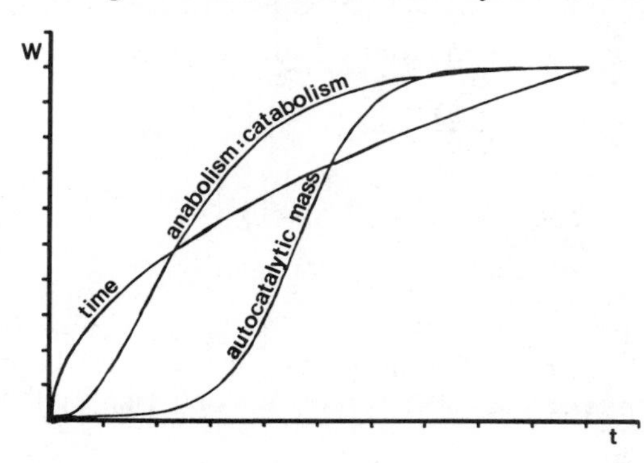

Figure 8-11 Examples of mathematically calculated growth curves controlled by time, autocatalytic mass, and the balance of anabolism to catabolism: (W) weight; (t) time. (Courtesy of J-P. Schoch.)

Body weight is partitioned between several commercially important compartments. Less energy is required to form a kilogram of lean meat than to form a kilogram of dissectable fat. Thus a progressive diversion of feed energy away from protein accumulation and toward fat deposition may contribute to the decelerating curve of body weight growth in meat animals, although in consistently fat animals such as pigs this effect may be only slight. Robison (1976) concluded that the decrease in feed efficiency that occurs with increased live weight in pigs is the result of increased maintenance cost and not increased fat deposition. The time at which the diversion to fat deposition takes place may depend on the breed of animal, and it occurs at a relatively later time in animals with a large adult size. The relative amount of water in the lean tissues of the body decreases with age. The rate of decrease in water content may be altered at weaning and at the onset of sexual maturity (Setiabudi et al., 1975).

In attempting to give a biological meaning to the terms b and d in von Bertalanffy's equation, one cannot simply assume that the proportional growth of the surface areas of the body keeps pace with the growth of the whole body. In pigs, for example, the small intestine grows rapidly after birth but reaches its mature length between 5 and 6 months of age (McCance, 1974). A similar early growth attainment of final size is found in the alimentary tract of cattle (Carnegie et al., 1969). Similarly, with heat loss from the body surface, the increased insulation provided by subcutaneous fat must be taken into account. Various regions of the body may differ in their depth of subcutaneous adipose insulation, and there are considerable differences between breeds in the contribution of each body region to the total body area (Deighton, 1932). The properties and uses of the von Bertalanffy equation are reviewed by Fabens (1965) and Richards (1959).

Robertson (1923) found a resemblence between sigmoid growth curves and those of *autocatalytic* or self-accelerating chemical reactions in which the product of a simple reaction acts as a catalyst to accelerate further reaction. In a first-order chemical reaction, the reaction rate is proportional to only one concentration. For example, in a decomposition reaction the disappearance of the decomposing substance is given by

$$-\frac{dc}{dt} = k \times c$$

$$\text{rate of disappearance} = \text{reaction constant} \times \text{concentration of substance}$$

The initial concentration of the substance a is reduced by the amount lost, x, so that its concentration becomes $a - x$ and the equation soon changes to

$$-\frac{d(a-x)}{dt} = k(a-x)$$

Since da/dt, the derivative of a constant, is zero, then

$$\frac{dx}{dt} = k(a-x)$$

Robertson (1923) took the case of an autocatalytic reaction where the reaction constant (k) was multiplied by the mass or concentration of the catalyst, which was the reaction product itself (x):

$$\frac{dx}{dt} = kx(a-x)$$

The curve of this reaction was sigmoid (Figure 8-11) and resembled growth curves for body weight (W):

$$\frac{dW}{dt} = kW(a-W)$$

Its initial acceleration results from the increase of the catalytic reaction product, and its later deceleration results from a depletion of the raw material or starting substance. Robertson was aware of the fact that overall body growth was the sum of a very large number of individual reactions rather than a single reaction, but he forced himself to choose between two alternatives: either the similarity of autocatalytic reaction curves and animal growth curves was a mere superficial resemblence or, alternatively, it was a meaningful indication of an underlying "master reaction" working on autocatalytic principles. Robertson chose the latter alternative on an intuitive basis. It is usually considered that there is little or no evidence in support of Robertson's choice, and that his *logistic* equation survives today only as a means of fitting a curve to growth data. It might be wise, however, to refrain from passing judgment on Robertson's intuition until we actually find out how growth is regulated.

Robertson's initial chemical reaction catalyzed itself; animals initially add new cells that form further new cells, and for a short while the system may behave like a population of microorganisms proliferating in an abundant-nutrient medium. Even under these conditions, however, mitotic rates usually show some deceleration. For example, in chick embryos at the 17-somite stage, the number of cells is doubled in 4 hours, but to double the numbers again then takes about 13 hours (Rosenquist, 1982).

As young farm animals grow larger, they are able to consume more feed and they begin to store increasing amounts of chemical energy. After a certain point, however, the capacity for energy storage becomes limited by the increasing demands of cellular maintenance and repair, and by the diversion of energy toward reproduction. These self-accelerating and limiting factors work in the same way as those in von Bertalanffy's equation (1957). If the postnatal body tissues of birds and mammals were composed solely of undifferentiated cells such as those of the embryo, self-acceleration by cell proliferation might be taken seriously. In meat animals, however, this is not the case.

In the skeletal muscles, massive amounts of protein are assembled into myofibrils. Myofibrils grow in length by forming new sarcomeres at their ends, and they grow radially before they subdivide by longitudinal splitting. New proteins for the myofibril are formed by the activity of ribosomes. Thus myofibrillar growth per se is not self-accelerating, except in an indirect manner—stronger animals might get more to eat. The growth of adipose tissue is dominated by triglyceride accumulation. Triglyceride accumulation is not self-accelerating, and it provides an even smaller indirect advantage to the animal's growth. Thus, the accelerating phase of growth in meat animals becomes all the more remarkable when we realize that much of the newly added biological material is *not* self-replicating.

In the equations considered so far, the final deceleration necessary to create a sigmoid curve has originated from, or has been proportionate to, the enlarged body mass itself. Another way of thinking about the control of deceleration is to envisage it a function of time, so that the growth rate decreases in simple inverse proportion to animal age:

$$\frac{1}{W}\frac{dW}{dt} = \frac{k}{t}$$

$$\text{specific growth rate} = \frac{\text{species constant}}{\text{age}}$$

Growth curves constructed in this manner, with animal age as the controlling factor in deceleration, never reach a point of zero growth in older animals (Figure 8-11). Although perpetual growth may occur in lower vertebrates, until cut short by accidental death, there is a finite upper limit to the size of mammals and birds. In agriculture, age-based growth curves may provide a reasonable fit to growth data only because the commercial growing period is in the first half of the sigmoid curve.

Instead of growth rate declining in simple inverse proportion to age, it may decline in logarithmic proportion, as in the Gompertz equation (1825). Thus in a Gompertz relationship, the logarithm of specific growth rate plotted against time yields a negative linear slope (Laird et al., 1965). The Gompertz equation is compared to other common growth functions in Table 8-3.

During the commercial growth of meat animals to market weight, growth may appear almost linear with a constant velocity. During this time, myofibrils, triglyceride, and bone matrix are passively accumulating (*accretionary growth*), and the constraints to growth have no measurable impact that can be isolated from sampling error and experimental error. This biologically distinct period of linear growth can be isolated from its sigmoid context to improve the fit of predicted curves to actual data (Laird, 1966).

TABLE 8-3. COMMON GROWTH FUNCTIONS WITH TIME AS THE INDEPENDENT VARIABLE[a]

Exponential:	$W = be^{kt}$
Monomolecular:	$W = a(1 - be^{-kt})$
Logistic:	$W = \dfrac{a}{1 + be^{-kt}}$
Gompertz:	$W = ae^{-be^{-kt}}$
Parabola:	$W = bt^k$

[a]Logarithms are expressed to the natural base *e*, *W* stands for weight, and *t* for time. *Source:* Medawar (1945).

In the comparison of breeds that differ in their rate of physiological development, it is difficult to justify the use of *chronological* age rather than *physiological* age to predict the decline in the logarithm of the specific growth rate. Another problem with time as the regulatory parameter of growth is that it makes no allowance for periods when an animal may cease growing because of restricted nutrient intake or stress. This problem does not arise with body weight as the regulatory parameter, and it could also be avoided by using physiological age rather than chronological age. There are, however, no simple units with which to measure physiological age. The concept of physiological age expounded by Brody (1945) has been used to create a physiological time scale called *metabolic age* (Taylor, 1965). This scale is based on the relationship between mature size and the chronological time taken to reach it. The chronological time taken to reach mature weight is made proportionate to the mature weight raised to the 0.27th power, so that the units of metabolic age are those of chronological age divided by the 0.27th power of mature weight. Another approach to the problem of measuring physiological age is to evaluate animals on the basis of a set of criteria which are known to change as animals grow older (Hofecker et al., 1980; Ludwig and Smoke, 1980). Some of the suggested criteria are:

1. running capacity;
2. spontaneous motor activity;
3. reactive motor capacity;
4. physical and chemical tests of the degree of cross-linking in collagen;
5. elasticity of the aorta;
6. accumulation of brain cell pigment (lipofuscin);
7. levels of plasma triglyceride and total cholesterol;
8. mineral and trace element levels in the heart and kidney;
9. thickness of the glomerular basement membrane;
10. numbers of neurons counted in macerated brain tissue;
11. frequency of abnormal chromosomes observed in experimental liver regeneration;
12. weight of the thymus gland;

13. rate of thyroxine degradation;
14. mitotic potential of cultured cells.

Genetic selection for leaneness in pigs may have delayed their physiological maturation (Doornenbal, 1975). Thus genetic selection might be more effective if animals are given a high plane of nutrition and suitable environmental conditions for optimal growth. In model systems, however, the responses to selection appear to be more complex, and genetic advances made under optimal conditions may not automatically be expressed if the animals are switched from optimal to suboptimal environmental growth conditions (Park et al., 1966). One aspect of this problem is that physiological maturation may be delayed in a population whose growth is environmentally restrained.

Intrinsic *seasonal rhythms* in muscle growth are difficult to identify in farm animals that reach market weight in less than a year since intrinsic rhythms are difficult to isolate from seasonal reproductive cycles and seasonal cycles in the nutrient quality of animal feed. In laboratory animal experiments, intrinsic seasonal changes in rates of growth and cellular regeneration have been found to persist even in animals kept under constant conditions of temperature and day length (Pennycuik, 1972; Schauble, 1972). In human populations, seasonal variation in growth shows some complex interactions with subject age and sex (Gindhart, 1972). Seasonal rhythms in milk and egg production are described by Brody (1945). In Scottish free-grazing sheep, the average empty body weight of ewes peaks in November and reaches a minimum at the end of April (Russel et al., 1968). Subcutaneous fat provides the primary energy reserve, which is depleted during gestation.

In poultry, light has a profound effect on aspects of animal behavior such as feeding and roosting (Siegel and Guhl, 1956), which then secondarily affect growth rate. For example, poultry may be reared at a low intensity of light so that they expend less energy on muscular activity. However, poultry may then start to use other clues, such as temperature or sounds of activity, to adjust or entrain their circadian rhythm of activity (Cain and Wilson, 1974). Chickens exhibit maximum growth if they can feed just before night-time, since during the night they can then work through the feed that they have stored in their crop. However, to do this, the chickens have to learn how to anticipate the start of their period of darkness. They can do this more readily if the light intensity is slowly decreased to simulate dusk (Savory, 1976a). The more rapid growth of chickens under these conditions is due to increased feed intake and to increased feed efficiency (Savory, 1976b). Chickens also show enhanced growth if kept at a continuously low light intensity (Beane et al., 1965) but in this case there is no increase in feed efficiency. The effects of illumination can be found in the levels of thyroxine production in baby chicks (Kleinpeter and Mixner, 1947). Supplemental illumination also stimulates growth in cattle (Peters et al., 1978) and may act via the gonads, since stimulation of growth may occur in heifers but not in steers (Tucker and Ringer, 1982).

Proportional or Allometric Growth

In the graphical analysis of animal shape, D'Arcy Thompson (1917) used rectangular coordinates to analyze and manipulate anatomical plan views. The coordinates were transformed mathematically and the anatomical plan was stretched to its new shape,

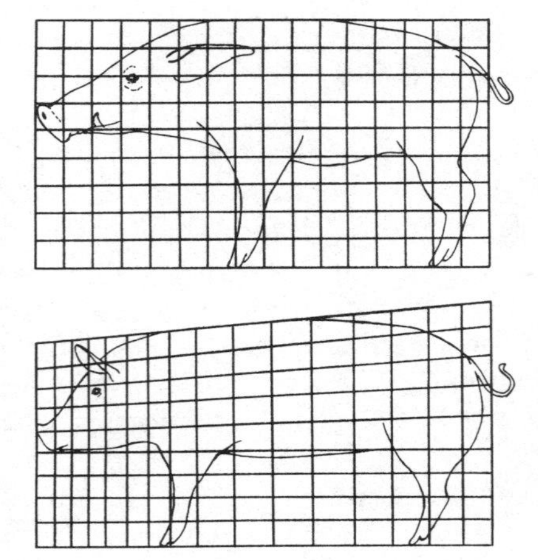

Figure 8-12 Cartesian transformation to illustrate the changes in body shape that may have occurred in the development of modern domestic pigs (bottom) from wild pigs, *Sus scrofa* (top).

thus facilitating the perception of ontogenetic and phylogenetic changes in shape (Figure 8-12). Before the advent of computer graphics, this technique was too difficult for the routine analysis of body shape. Without a computer, the technique is largely subjective, since in most cases it is easier to fit a transformed Cartesian grid to a drawing than vice versa. Thus there is a risk of merely using the Cartesian transformation to illustrate a preconceived idea of a shape change, as in Figure 8-12. Hammond (1950) simplified the application of this method of analysis by making photographic prints of animals at magnifications that resulted in a constant skull length. Lateral views of animals at constant skull length were superimposed onto graph paper grids to demonstrate the major changes in animal shape resulting from the domestication and selective breeding of farm animals. Thus when skull length was held constant photographically, the loins and hams of meat-type pigs appeared to balloon out phylogenetically from the diminutive rump of the wild boar. The same technique revealed a comparable inflation of the rear end as pigs grew from birth to slaughter weight.

At birth, meat animals tend to have large heads and long slender limbs. Subsequent growth is marked by an increase in body length and depth until finally, the hams fill out convexly. Hammond's (1932) technique showed the animal shape that would be most likely to yield the greatest mass of edible meat. Unfortunately, shape then became a more important criterion than yield, despite the fact that many generations of blocky meat animals failed to yield their promised return of lean meat on the butcher's block. The culprit, of course, was subcutaneous and intermuscular fat. To estimate meat yield in live animals it is necessary to visualize the muscle mass beneath the animal's outward shape. Body regions where subcutaneous fat is scarce, and where muscle mass can be judged by the stance between the limbs are, therefore, particularly important in the judgment of live meat animals.

Hammond's (1932) concept of the ontogeny of shape in sheep was that a wave of growth starts from the head and passes back along the vertebral column. Secondary waves in the limbs pass from distal to proximal. The union of the anterior to posterior wave in the vertebral column with the distal to proximal wave in the hind limb results in the late, but extensive growth of the pelvic region. Thus the difference in shape

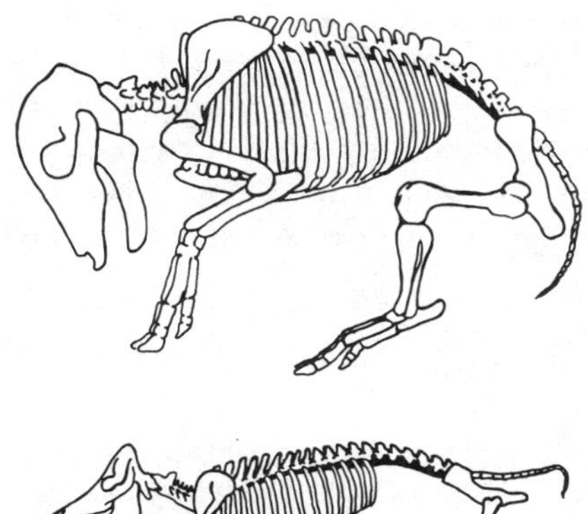

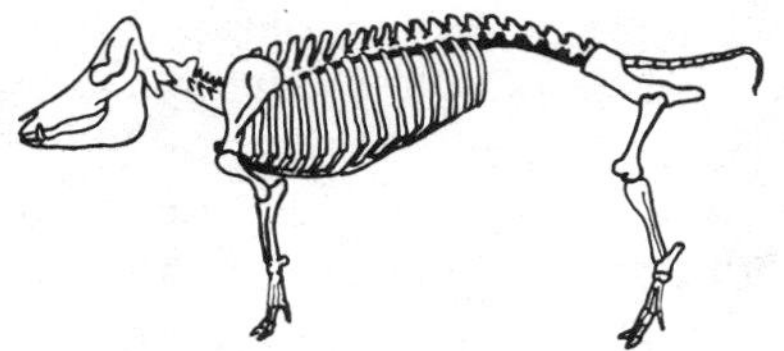

Figure 8-13 Allometric growth of the skeleton. The diagram of the fetal pig skeleton at the top of the figure has been adjusted to the same length as the diagram of the adult pig skeleton at the bottom of the figure.

between the fetal and adult skeleton (Figure 8-13) resembles the shape change that occurred in the transformation of wild to domestic farm mammals. The major changes in both situations are a reduction in the head relative to the body, and a reduction in the limbs relative to the trunk of the body. In pigs, however, the anterior to posterior wave may (Davies and Kallweit, 1979) or may not (Richmond et al., 1979) be detectable. The allometric growth ratios of individual vertebrae are negative (about 0.95) in the anterior cervical and midthoracic vertebrae, and are positive (about 1.05) in the lumbar region (Davies, 1979). In the posterior cervical and anterior thoracic regions the allometric ratios are close to 1. These allometric ratios are derived as follows.

Huxley (1932) described a simple mathematical method for the detection and measurement of the *allometric growth* patterns revealed by D'Arcy Thompson's graphical methods. Allometry is the study of relative growth, of changes in proportion with increase in size (Henderson et al., 1966). In order to compare the relative growth of two components (one of which may be the whole body), they are plotted logarithmically on X and Y axes:

$$y = bx^k$$

$$\log y = \log b + k \log x$$

The slope of the resulting regression is called the *allometric growth ratio*, often designated as k. With $k = 1$, both components are growing at the same rate. With $k < 1$, the component represented on the Y axis is growing more slowly than the component on the X axis. With $k > 1$, the Y axis component is growing faster than the X-axis component. Smith (1980) suggested that the logarithmic transformation in the allometric growth equation has become popular because it *appears* to work well; scattered data points are pulled into a neat line with a high coefficient of correlation. However, these features do not automatically prove that a logarithmic transformation is necessarily the best or the only way in which to transform a set of data.

Allometric growth is normally regarded in a positive light. Thus a unit of the body with a positive allometric growth ratio ($k > 1$) is thought of as exhibiting rapid relative growth. When we attempt to fit the phenomenon of allometric growth into the general framework of our knowledge of sigmoid growth curves, the concept of allometric growth as a positive phenomenon is inappropriate. As pointed out by Wallace (1948b), during the period of growth after birth (and even somewhat before birth) the velocity of growth tends to decline. Thus those parts of the body with high postnatal allometric growth ratios can also be regarded as retarded parts, relative to those parts which have already decelerated their growth.

Allometric Growth of Carcass Muscles

This topic is the subject of a comprehensive monograph by Berg and Butterfield (1976) in which much of the published literature is surveyed. Berg and Butterfield (1976) analyzed beef muscle growth patterns using Huxley's (1932) allometric growth ratio to categorize muscles into one of three monophasic categories (high, average, and low impetus) depending on whether their allometric growth ratio was greater than, equal to, or less than a value of 1 (Figure 8-14).

Berg and Butterfield (1976) found that many of the postnatal allometric growth ratios of muscles changed during development so that a diphasic or dual categorization was needed. The dominant diphasic types were high followed by average, average followed by high, and low followed by average. The growth patterns of certain muscles in the beef carcass may also show an abrupt change in growth rate at birth (Johnson, 1974). To simplify the analysis of the large number of muscles that are present in the carcass, muscles were considered in regional groups. Although this is almost an essential step in arriving at generalizations about muscle growth patterns, it is very difficult to identify regional muscle groups that are also functional muscle groups. Thus within a regional anatomical group, there are usually a few muscles that pull against distant parts of the skeleton and which belong functionally to a more distant group.

In Chapter 4, functional muscle groups are arranged dichotamously by their anatomical origins and insertions. Unfortunately, translation from functional groups defined by origin and insertion to regional anatomical groups as used by Berg and Butterfield (1976) is rather difficult. To simplify the following overview, the functional groups of Chapter 4 are identified by their numerical designation in Chapter 4, while the regional anatomical muscle groups given by Berg and Butterfield (1976) are identified by their regional name.

Berg and Butterfield (1976) found that proximal hindlimb and spinal muscle groups (most of group 5 in Chapter 4) were nearly all either monophasic low, monophasic average, or biphasic high-average impetus muscles. The only group 5

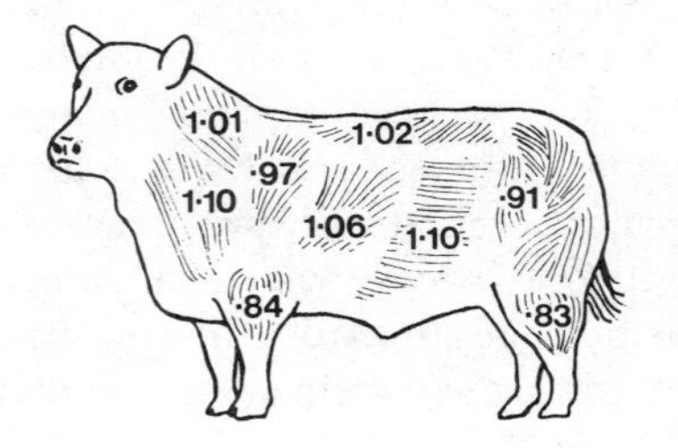

Figure 8-14 Allometric growth ratios of regional muscle groups in steers, calculated from muscle groups relative to total muscle weight. (Data from Berg and Butterfield, 1976.)

muscles of Chapter 4 that reached a monophasic high level of growth were the tensor fascia lata and the obliquus abdominis internus. A possible explanation is that the abdominal support provided by these two muscles might continue to expand in response to continued visceral growth.

Muscles identified by Berg and Butterfield (1976) as distal hindlimb muscles (mostly group 3 muscles in Chapter 4) and distal forelimb muscles (mostly group 2 in Chapter 4) had either a monophasic low or a biphasic low-average impetus. Berg and Butterfield (1976) suggested that the immediate requirement for neonatal locomotion was responsible for an early prenatal acceleration of growth in these muscles, so that subsequent postnatal growth was slower than in the remaining more proximal muscles. Since many of the distal limb muscles have a complex pennate structure, constraints resulting from muscle structure might also be involved, as outlined in Chapter 7. Pennate muscles are restricted in their potential for growth, whereas overlying muscles that are composed of relays of intrafascicularly terminating fibers are able to grow radially and longitudinally for a longer period.

The proximal forelimb muscles (mostly group 4 muscles in Chapter 4) were found to be mostly either monophasic low or average impetus muscles. Berg and Butterfield (1976) found monophasic high and diphasic average-high impetus muscles in the thorax-forelimb and neck-forelimb groups. They suggested that this might be a relic of the heavy head and neck muscles which wild bulls once used for fighting. Rams also show considerable growth of their neck muscles (Lohse, 1973). Since many of the ventral neck muscles act as antagonists to the ligamentum nuchae, it is possible that this is an ontogenetic rather than a phylogenetic feature. In Zebu cattle (*Bos indicus*) the shoulder hump is composed primarily of the rhomboideus cervicis muscle (Heath, 1979). The growth of the skull and the ligamentum nuchae are probably interrelated, so that the enlargement of ventral neck muscles is required to counterbalance a stronger ligamentum nuchae. The distribution of the carcass muscle mass is changed when steers lose weight on a low plane of nutrition (Seebeck and Tulloh, 1968b). Growth changes in muscle weight distribution in pigs are shown in Figure 8-15.

Berg and Butterfield (1976) and Broadbent et al. (1976) compared muscle growth in conventional early-maturing beef breeds, some dairy breeds, and crossbreds from large draft breeds, but they were unable to find any economically important differences in anatomical muscle weight distribution. Such differences, however, may occur between breeds of pigs and sheep. Davies (1974b) found that Pietrains had greater muscularity in the loin and hind limb, while Large Whites had relatively greater development of neck, shoulder, and forelimb muscles. These two breeds do not differ in bone distribution (Davies, 1975). Breed differences in muscle weight distribution in sheep were reported by Taylor et al. (1980).

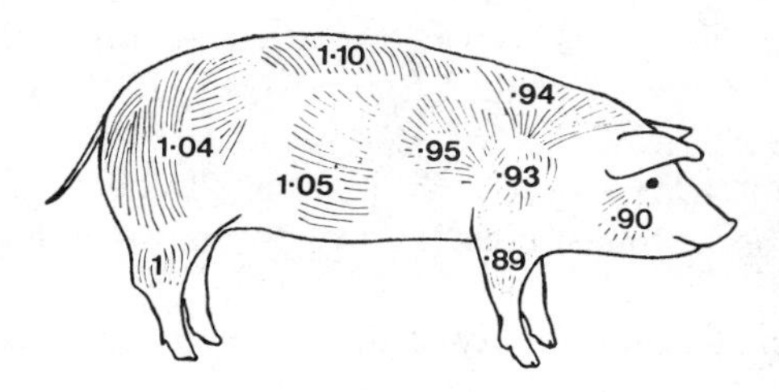

Figure 8-15 Allometric growth ratios of regional muscle groups in gilts, calculated from muscle groups relative to total muscle weight. (Data from Davies, 1974b with breeds averaged.)

In cattle, muscle distribution is influenced by sex. Proximal hindlimb and abdominal muscles are heavier in heifers than in steers, and heavier in steers than in bulls. The order is reversed for muscles of the neck and thorax (Mukhoty and Berg, 1973). In cattle, castration causes a marked decrease (up to 55 per cent) in the growth of shoulder muscles (Brannang, 1971). This effect is centered on the splenius muscle at the cervical-thoracic junction. Ewes may have a lower percentage of shoulder, shank, and neck than that in wethers (Kemp et al., 1981). In sheep, however, sex differences in muscle weight distribution may be absent in situations where breed differences do exist (Seebeck, 1968a). In pigs, sex and level of feeding have only a small effect on the proportion and distribution of muscle and bone (Davies et al., 1980). Differences in muscle weight distribution which are under hormonal control might be caused by differences in skeletal growth. Allometric growth of the skeletons of female rats is more stable than that of males when subjected to experimental endocrine manipulation (Reisenfeld, 1976).

Many carcass muscles may acquire appreciable amounts of intramuscular fat in older animals. This cannot be removed by dissection and is, therefore, included in the muscle weight. Fortunately for carcass dissectors, growth gradients for intramuscular fat in different muscle groups are similar to those for the muscles (Johnson et al., 1973; Davies and Pryor, 1977). When expressed on the same basis (weight of muscle plus bone), heifers are generally fatter than steers, and steers are fatter than bulls. These differences are related to the time of onset of fat deposition (Mukhoty and Berg, 1971). Since the anatomical distribution of muscle mass in different breeds of cattle is fairly constant, the genetic reduction of fat content probably provides the best means of selecting for an increased proportion of lean meat (Koch et al., 1981).

Allometric growth is quite conspicuous in the muscles and bones of poultry (Figure 8-16), although not along the vertebral column as in mammals (Harrison, 1970). In ducks, the leg muscles are well developed at an early age, while the growth of the breast muscles is quite late (Stadelman and Meinert, 1977; Swatland, 1980b). The same overall pattern of growth occurs in chickens (Halvorson and Jacobson, 1970) and in turkeys (Harshaw and Rector, 1940; Hartung and Froning, 1968; Swatland, 1979), but to a less extreme degree. This pattern of muscle development is an obvious advantage to an animal which in the wild state would depend on its legs for early locomotion and only later would learn to fly. Poultry exhibit considerable variability in their body proportions between breeds. Broad Breasted Bronze turkeys, for example, may have shorter legs than other smaller strains (Asmundson and Lerner, 1942). At present, it is doubtful whether differences in conformation have any relation to the relative distribution of the muscle mass. Differences in nutrition do not appear to affect the distribution of the muscle mass in poultry (Wilson, 1954).

In chickens and turkeys, the early deposition of subcutaneous fat is a desirable trait, but in ducks and geese the reverse is true. Genetic selection for live weight gain is difficult to uncouple from increased fatness (Clayton et al., 1974). The heritabilities of the weight and the proportion of the breast muscles are relatively high in ducks (Clayton and Powell, 1979). But because of the marked allometry of muscle growth in ducks, selection for gain in the breast muscles may not improve the meat yield of the leg.

In summary, the genetic inflexibility of the allometric growth patterns of beef

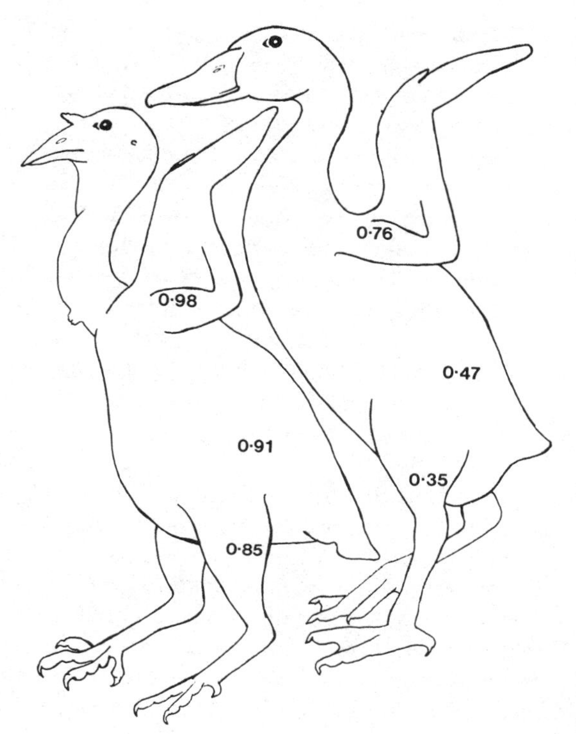

Figure 8-16 Allometric growth ratios of representative muscles of the wing, thigh, and leg in turkeys and ducks, calculated relative to breast muscle weight. (Data from Swatland, 1979, 1980b.)

muscles discourages any attempt to change the distribution of muscles in the carcass. In cattle, the obverse tactic is more attractive, namely, selection against fatness throughout the carcass. In poultry, large allometric growth gradients extend along the breast-leg axis, but we are not yet sure how they will respond to genetic selection. Perhaps leg muscle weights have a lower heritability than breast muscle weights (Johnson and Asmundson, 1957) precisely because they are at the low end of an allometric growth gradient.

Allometry and Domestication

The domestication of farm animals several thousand years ago was accompanied by allometric changes in body shape and by an overall decrease in body size (Zeuner, 1963). The archeological record of animal domestication is difficult to interpret since traces of different species may overlap. In cattle, for example, the bones of the ancestral wild auroch (*Bos primigenius*) may give way to those of the extinct shorthorn of the Iron Age (*Bos longifrons*) and finally to the ancestors of our present species (*Bos taurus*). The bones of *B. longifrons*, however, might simply have been those of *B. primigenius* cows. The frequency of occurrence of bones may reflect the balance of hunting, containment, and progressive breeding at any point in time. The identification of individual bones is confounded by the nonrandom nature of specimens left by hunters, and by technological practices such as castration and the development of herds of cows. Selection may have encouraged the development of *paedomorphic* forms with persistent juvenile characteristics, such as pliable behavior, as well as *neotenic* forms with precocious reproductive development. Superimposed on these genetic changes there may have been changes in body structure which were associated with the new environment created by animal husbandry.

Domestication was associated with considerable reductions in the weight of the brain; a 24 per cent reduction in sheep (Ebinger, 1974), a 33 per cent reduction in pigs (Kruska and Rohrs, 1974), and a 16 per cent reduction in horses (Kruska, 1973). The diminution of brain size relative to body weight was accompanied by allometric changes between parts of the brain. In sheep, the cortical reduction of white matter exceeded that of the gray matter, and the parts of the brain that were involved with the senses of smell, sight, and hearing showed the greatest reduction, perhaps in relation to a safer environment and controlled feeding (Ebinger, 1975). Pigs that have reverted to a wild state do not restore their original brain size, although allometric changes may occur; feral pigs have a smaller cerebellum and a larger medulla oblongata (Kruska and Rohrs, 1974).

Domestication is accompanied by allometric changes in the endocrine glands. The relatively recent derivation of laboratory rats from wild Norway rats (*Mus norvegicus*) provides an interesting model of events that may have occurred during the domestication of farm animals. Richter (1954) concluded that the dominant changes in rats were a reduction in the adrenal cortex and an increase in the activity of the gonads. In wild and domesticated ducks, appropriate differences occur in the pituitary-adrenal system, but the nature of the relationship between domestication and changes in behavior and adrenocortical activity is still open to question (Martin, 1978).

Sheep provide a striking example of phenotypic changes which can be produced by what Charles Darwin called unconscious selection. These changes include (1) the ratio of wool to hair; (2) an increase in tail length or adiposity; (3) development of the lop-eared condition; (4) increased convexity of the nose due to a decrease in jaw length; and (5) a reduction in the number and complexity of horns (Zeuner, 1963).

ENERGY FOR GROWTH

Energy is the capacity of doing work. A suitable unit of energy and work is the *joule*, named after James Prescott Joule (1818–1889) and abbreviated J. To raise the temperature of one gram of water by one degree Celsius requires 4.184 J (1 cal). In mechanical terms, if applied to a free body with a mass of one kilogram so as to cause an acceleration of one meter per second per second, one joule causes a displacement of one meter in the direction of the force. Heat production and muscular work are essential for the animal to stay alive, and growing tissues compete with each other and with these essential services to obtain chemical energy from nutrients circulating in the vascular system. Lean meat with a water content of approximately 80 per cent contains approximately 4.7 kJ/g, while the energy content of fat is approximately eight times greater (Webster, 1976). Mammalian basal metabolic rate is approximately 12.5 $kJ/kg^{0.75}$ per hour (Kleiber, 1961).

In the nomenclature used by the International Biological Programme (Petrusewicz and Macfadyen, 1970), the growth equation of von Bertalanffy (1957) becomes

$$P = A - R$$

$$\text{production in defined time} = \text{assimilation} - \text{maintenance cost}$$

When extended to an ecological perspective of *biomass* (growth of a population of animals), production can be considered as the sum of production due to animal reproduction (P_r) and production due to body growth (P_g):

$$P = P_r + P_g$$

When applied at the other end of the scale, to individual animal cells, energy requirements for biosynthesis and maintenance can be expressed in terms of chemical energy as moles of adenosine triphosphate (ATP). For example, in cultured mammalian cells, Kilburn et al. (1969) found that 1.6×10^{-11} mole of ATP per cell was required for biosynthesis, while a greater amount, 2.9×10^{-11} mole of ATP per cell, was required for maintenance.

Feed energy which is digested and absorbed, and which ends up circulating in the vascular system, is only a fraction of that which passes along the gut. From the total feed intake or *consumption* (C), a certain portion of *rejecta* (FU) is lost to the animal. The rejecta may be composed of feces or regurgitated feed (*egesta*, F) and substances which are excreted by the kidneys or lost through the skin (*excreta*, U). *Metabolizable energy*, or *assimilation* (A) in the ecological nomenclature, is the portion of the feed that can be utilized by an animal or by a population of animals for *respiration* (R):

$$A = R + P = C - FU$$

The term "respiration" is used here in a broad sense which is equivalent to oxidative metabolism, or to the liberation of energy for all aspects of body maintenance, such as heat production and muscular work. Quite large differences (15 per cent) in heat production may exist between breeds of cattle (Webster, 1977), and cattle with a low skin temperature under normal conditions may tend to exhibit a high growth rate (Wood and Hill, 1914). However, in practice, it is difficult to use the relationship between heat loss and rate of growth to assess productivity. Bulls, for example, exhibit a high heat loss coupled with a rapid growth rate. In this case, the energy lost as heat is more than offset by the greater efficiency of lean meat deposition relative to fat deposition (Webster, 1977).

Animals have considerable control over the amount and the nature of the things they eat. In many practical and experimental situations this behavior can be a primary factor in the regulation of growth. A number of possible control systems for the regulation of energy balance by consumption are reviewed by Hervey (1971). Control may be passive, as in the increased energy cost of locomotion as an animal grows heavier. Active controlling elements may involve hypothalamic feedback circuits, although the lines of communication are difficult to identify. The medium of communication may be represented by the levels of circulating factors, such as glucose, amino acids, fatty acids, and steroids (which are partitioned between aqueous and lipid compartments of the body).

Ruminants consume a considerable volume of roughage, and the rumen may be filled before energy requirements are satisfied. Normally, however, feeding behavior changes to accommodate differences in the energy content of the feed. Control signals may be transmitted by acetate and propionate sensors in or near the rumen (Bines, 1971). Animal behavior is also involved in the effect of long-term stress on animal growth. Stress diminishes live weight gains, mainly because of a decrease in fat deposition (Judge and Stob, 1963).

Digested energy (D), or *total digestible nutrients*, is of fundamental importance in the formulation of rations for the commercial growth of meat animals. For example, in ruminants the digested energy varies considerably between different types of animal feeds, ranging from 40 per cent for wheat straw to 80 per cent for corn (Byerly, 1967).

$$D = P + R + U = A + U$$

Many subtleties are needed to ammend these equations in laboratory research. Ruminants, for example, may lose about 8 per cent of digestible energy as methane (Byerly, 1967), while in all animals heat may be unavoidably liberated during the assimilation of nutrients by the body (*specific dynamic action*).

Although the growth equation of von Bertalanffy (1957) can be translated into energy terms, the other growth equations in which deceleration is a function of body weight or of age present more of a problem. One approach to the problem is given by Hammond's (1950) idea that body tissues compete for circulating nutrients (Figure 8-17). There are a number of astute points about this model; (1) it concerns itself only with assimilation; (2) it is based on a priority system where vital organs make the greatest claim; and (3) it encompasses the division of energy between animal growth and animal reproduction. In the pregnant female, the uterus attains a priority for food distribution which is between A and B. The model also corresponds fairly well with our present understanding of the endocrine control of the flow of energy. In ruminants, for

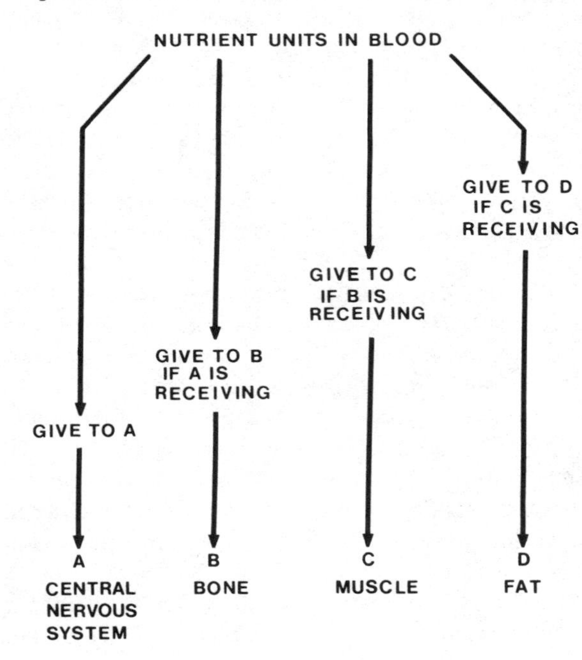

Figure 8-17 Distribution of nutrients according to the priority system proposed by Hammond. The absolute amounts available to each type of tissue are not specified, only the priority in which they are available. The type of nutrient is not specified. The physical basis of the distribution system might be the competitive ability or relative numbers of nutrient receptors in each type of tissue. The model implies that a low plane of nutrition affects the tissue types in a fixed sequence starting with the fat.

example, the energy flow to skeletal muscle is increased by STH, while the flow to adipose tissue is increased by insulin (Trenkle, 1981). In other words, circulating factors such as hormones may govern the energy distribution system by switching on and off the energy assimilation systems of each type of tissue.

Hammond's model was also meant to explain the effect of reduced energy intake on muscle development. When Pomeroy (1941) placed pigs with a live weight of 150 kg on a submaintenance diet, he found that the tissues of the carcass were affected in reverse order to their order of anatomical development. Fat was reduced most, bone was reduced least, and muscle was intermediate in reduction. However, this aspect of Hammond's model has been criticized by Berg and Butterfield (1976), who point out that recent research shows that muscle-to-bone ratios may not change in animals on a low plane of nutrition. The responses of animals to a low plane of nutrition appear to be rather variable. For example, Seebeck and Tulloh (1968a) placed Angus steers on a restricted diet so that live weights were reduced. Bone and connective tissue were relatively unaffected, muscle mass was reduced, but there were only relatively slight reductions in carcass fat content. Chemical analyses showed that the loss of muscle weight was due to protein, not to dehydration or to mobilization of intramuscular fat (Seebeck and Tulloh, 1969).

Within a breed, muscle-to-bone ratios may not be a fixed function of carcass weight if animals have been placed on a low plane of nutrition which causes weight loss or inhibition of further growth (Murray et al., 1972). Muscle-to-bone ratios may also remain constant despite intensive genetic selection for leaness and growth rate; Davies (1974a) found no change in ratio over a 30-year period of genetic improvement in the Large White breed. Dairy and beef breeds may differ in their muscle-to-bone ratios as a consequence of selection for milk yield versus meat yield (Broadbent et al., 1976). Differences in muscle-to-bone ratios between breeds of beef cattle appear to be a relict of an earlier period of breed improvement when certain breeds were selected for their muscular strength as draft animals (Berg and Butterfield, 1976).

Berg and Butterfield (1976) proposed an alternative model of nutrient distribution during growth (Figure 8-18). Energy distribution between tissues was envisaged as a combination of nutrient availability and tissue capacity. Hammond's (1950) model resembles growth equations in which specific growth rate is a function of age, essentially of physiological age. As animals become older they become capable of reproduction and extensive fat deposition, and their priorities change accordingly. Conversely, the model of Berg and Butterfield resembles growth equations in which specific growth rate is a function of body size, because of the implied limits to tissue volume in all except adipose tissue. Thus according to these two models, the animal "knows" either its physiological age or what size it should be.

Weiss and Kavanau (1957) postulated that overall body dimensions might be dependent on each tissue of the body regulating its own growth. They proposed that each tissue type produced both *templates*, with a catalytic action on growth, and *antitemplates*. Templates were thought to be confined to the cell that produced them, whereas antitemplates diffused out of the cell to become uniformly diluted through the remainder of the body. The concentration of antitemplates in the body at any particular time was thought to be a balance between their continuous formation and degradation. Thus definitive size was reached when the concentration of antitemplates

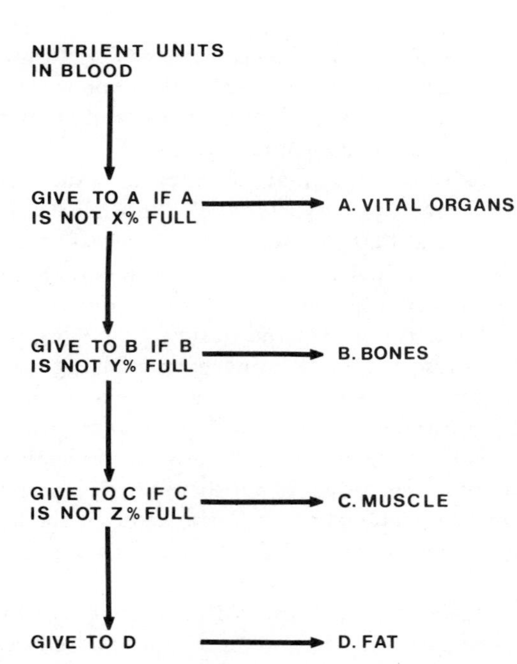

Figure 8-18 Distribution of nutrients on a combined priority and volumetric basis, as proposed by Berg and Butterfield (1976). The values of X, Y, and Z are variable. They depend on the plane of nutrition, and their interaction determines the compound ratio of muscle to fat to bone at any point in time. The effects of a low plane of nutrition are balanced among the tissue types, depending on the severity of starvation.

in the whole body finally reached a level where it blocked all intracellular templates. Compensatory growth due to experimental removal of tissue would, in this proposed mechanism, be detected by remaining cells as a reduction in antitemplate concentration due to removal of a source of antitemplates. Thus remaining cells would be freed to grow until the antitemplate level was restored.

A radically different mechanism was proposed by Tanner (1963), who suggested that certain brain cells were able to keep track of animal age. These cells were thought to operate a *time tally mechanism* which matched animal age against the body mass anticipated at any particular time. Information on animal mass was brought to the time tally by substances resembling antitemplates which were produced by one or a small number of body tissues or organs. Evidence from experimental animals suggests that animals do have a "set point" for body weight, and that this can be decreased by experimental lesions in the lateral hypothalamus (Keesey, 1976).

The rates of many reactions within cells, including the synthesis of DNA for cell division, are regulated by calcium ions which are bound to a protein, *calmodulin*, to form a biocatalyst. Reactions are accelerated when the cell membrane alters its permeability and allows a brief influx of calcium ions from outside the cell. The calcium-calmodulin complex remains active until calcium ions are ejected from the cell by a vigorous calcium ion pump in the cell membrane (Cheung, 1980).

It might be possible to combine all these proposed growth-controlling mechanisms into a single system, with each tissue compartment being given (1) a priority for obtaining energy; (2) an essential respiratory requirement; and (3) a preferred size calculated from an interaction between age and body weight. This could be incorporated into an algorithm with (4) a biological clock; (5) an intrinsic system for monitoring body mass; and logical questions concerning (6) pregnancy; (7) required muscular work; (8) environmental heat loss; (9) nutrient intake; and (10) stress. It would, however, be extremely difficult to place meaningful limits on parameters 1 to 3 and to program metabolic interactions resulting from inputs 4 to 10. The reason is that, although the vascular system may resemble a branching pipeline system, it carries a number of different fuels at once. Metabolically, it is compartmentalized (glucose, glycerol, amino acids, fatty acids, ketone bodies, lactate, pyruvate, high-density lipoproteins, low-density lipoproteins, very low-density lipoproteins, etc.) with a comparable number of interacting hormonal and biochemical regulators. However, in cattle, blood glucose concentration can in fact be correlated with the rate of growth in live weight (Rowlands et al., 1973). Required muscular work (7) is a particularly difficult parameter to evaluate since, as well as diverting energy to a mechanical output, work also stimulates muscle growth so that muscle mass and physical activity might be correlated (Millward et al., 1976). Cunningham et al. (1963) showed that the rate of muscle growth in pigs could be increased by electrical stimulation for 20 minutes per day. Feed efficiency was not impaired, and the experimental treatment may simply have increased the uptake of amino acids by stimulated muscles. Moderate differences in the degree of voluntary muscle activity usually have no detectable effect on the growth of pigs (Skjervold et al., 1963).

Whittemore and Fawcett (1976) and Whittemore (1976) have defined a number of these parameters and logical questions in a model system for protein and lipid accumulation in pigs. Proteins have only a limited life and are continuously being broken down. Their resynthesis, however, is not at first complete, so that losses must be made up by the synthesis of new protein. Initial losses during resynthesis are probably due to the fact that hydroxyproline and the methylated amino acids, methyl histidine and methyl lysine, are difficult to reincorporate into protein and are excreted (Lewis et al., 1974).

Production (P) can be defined as the net balance of food transferred to the tissue of a population during a defined time period, whereas *total biomass growth* (G) is the total amount of body tissue generated by a population in a defined time period. *Weight loss* (L) is the biomass used for metabolic needs within a defined time period when respiration exceeds assimilation also:

$$L = G - P$$

Because of the continuous turnover of proteins, Whittemore and Fawcett (1976) demonstrate the necessity of some extra terms at the cellular level:

- pb protein breakdown (mass/time)
- pn total new protein (mass/time)
- ϕ proportion of pb not resynthesized and thus lost from the system: made good by new synthesis
- pa protein accretion by new synthesis (mass/time)
- ps total protein synthesis

Unfortunately, the terminology used by Whittemore and Fawcett (1976) used capital P for protein and this conflicts with the ecological nomenclature (Petrusewicz and Macfadyen, 1970). However, lowercase p is free, and is used here to indicate protein:

$$pn = \phi pb + pa$$

$$ps = pb + pa$$

Whittemore and Fawcett (1976) proposed that the ratio pn/ps is a function of protein mass. There might also be an interaction with use-induced hypertrophy in the case of muscle proteins since both protein synthesis and degradation are increased during hypertrophy (Laurent and Millward, 1980). Murray et al. (1974) subjected pigs to strenuous treadmill exercise. There was no effect up to 7 weeks of age but, from 7 to 9 weeks, exercised animals grew faster and consumed less feed per kilogram of live weight.

Using a number of published values for parameters, Whittemore and Fawcett (1976) calculated that the energy cost for the synthesis of 1 kg of protein was 7.3 MJ of assimilation or metabolizable energy. When considering total body protein, it is worth remembering that protein synthesis for the maintenance and activity of the intestine, liver, and vascular system accounts for a major part of total activity. The alimentary tract also provides protein and energy for maintenance when animals are placed on a low plane of nutrition (Carnegie et al., 1969).

In agriculture, feed costs are of paramount importance in both experimental research on farm animals and in the analysis of productivity in animal systems. This view of animal growth as an input-output system has led to the development of an interesting group of growth equations that encompass both feed inputs and animal growth. Titus (1927) used the "law of diminishing returns" postulated by Spillman and Lang (1924) to analyze the growth and feed consumption of domestic ducks:

$$W = W_{max} - dW \left(\frac{P}{C}\right)^{x}$$

where W = live weight for any given number of feed units consumed
W_{max} = theoretical maximum live weight attainable
dW = gain in live weight above the theoretical initial weight
P/C = ratio between gains in live weight for successive units of feed
x = number of feed units consumed

As well as giving an acceptable fit to the live weight and feed consumption data examined by Titus (1927), this relationship was also compatible with the live weight-to-age relationships described by Robertson's (1923) logistic equation. Together with a further equation relating feed intake to animal age, the "law of diminishing returns" has been fitted to feed consumption and live weight growth in pigs (Parks, 1970).

Students of farm animal nutrition may disapprove of the use of ecological terms in place of names more familiar to animal nutritionists. However, at a time in history when orbiting satellites are routinely scanning photosynthetic molecules across the face of the planet, there is a pressing need for a unifying concept of energy flow which can integrate molecular aspects in an ecological context. This broad vista extends from cell biology, through the economic study of agricultural systems, to the global ecology which encircles and permeates our agricultural lands. For those whose primary interest lies in one small part of the overall energy system, it is wiser to elaborate on small parts of an ecological model rather than attempting to create an ecological model by lumping together the terminologies of a large number of independent subdisciplines.

CATCH-UP OR COMPENSATORY GROWTH

If animals are deprived of food and their growth slows down, they are often able to catch up by growing faster if their food supply is restored (Wilson and Osbourn, 1960). In agriculture, this is often called compensatory growth: The animal is compensating for lost growth time. In experimental zoology and pathology, however, there are many other types of compensatory growth. For example, the experimental removal of one kidney often stimulates the growth of the remaining one. The loss of muscle fibers by disease may be compensated by extra radial growth of the survivors.

In many animal production systems, catch-up growth is rarely required, while in other systems it may be an important aspect of overall growth to market weight. If growth is arrested very early or very late in development, catch-up growth requires cautious scrutiny. If a prenatal reduction of growth affects cell division and causes *hypoplasia* (less than normal numbers of cells), it may be too late after birth to acquire a normal number of cells if the food supply is restored. In this situation, if catch-up growth does occur, it may be achieved by cellular *hypertrophy* (greater than normal cell size). Catch-up growth may fail if the cause of the growth arrest has damaged cell populations, as in the case of cortisone-induced growth arrest in rats (Mosier, 1972). Catch-up growth does not necessarily occur at the same rate in all tissues. For example, in turkeys the linear growth of bones is more readily restored than the growth of fat (Auckland, 1972). However, it is important to bear in mind the different dimensions of these two systems (the problem of x, x^2, and x^3). Starvation may cause a reduction in the amount of DNA in a muscle, and this may then limit the extent to which catch-up growth can occur (Allen et al., 1979).

Catch-up growth in the live weight of older animals is sometimes due to fat deposition rather than to a restoration of anticipated body composition. In starved pigs, Lister and McCance (1967) found that restoration of adequate food for growth at an age of 1 year resulted in animals with normal muscle-to-bone ratios but with a lower live weight and with extra fat. In turkeys, on the other hand, the body protein content is more readily restored than the body fat content (Auckland and Morris, 1971). Tissue wet weight, however, can be misleading. For example, undernourished pigs have an increase in the water content of their skin and muscles (Dickerson and McCance, 1964).

Interference with growth just before puberty may change the rate of physiological or metabolic aging so that puberty is delayed, and the animal has an extended growth period to catch up (Kerr, 1975). Another subtle point in the interpretation of experiments on catch-up growth is that appetite may be increased following the restoration of an adequate food supply. Thus catch-up growth in cattle is achieved by a combination of both increased feed intake and increased feed utilization (Meyer et al., 1965).

Catch-up growth in sheep and lambs can be accomplished without detectable deleterious effects on meat quality (Winter, 1970; Asghar and Yeates, 1979a, 1979b). Muscle fiber diameters are reduced during starvation but they recover during catch-up growth; fiber numbers remain constant (Yeates, 1964). Catch-up growth may not occur to the same extent in all muscles. Lohse et al. (1973) found it to be greater in high-impetus muscles. As starved sheep lose weight, their ratio of circulating STH per unit of body size increases and, during catch-up growth, the adenohypophysis becomes hypertrophied (McManus et al., 1972). Thus the adenohypophysis appears to be involved, either directly or indirectly, in the phenomenon of catch-up growth. Pigs that differ genetically in their degree of fatness also differ in their potential for catch-up growth. Genetically fat pigs can compensate for low crude protein levels in their starter ration, whereas genetically lean pigs suffer a reduction in the cross-sectional area of their longissimus dorsi muscles (Hogberg and Zimmerman, 1978). Catch-up growth may be more rapid in gilts than in barrows (Robinson, 1964).

REGULATORY MECHANISMS IN MEAT ANIMAL GROWTH

In comparison with other body cells, skeletal muscle fibers are unusual because they are multinucleated, and because they derive new nuclei from the mitosis of satellite cells. Not long ago, STH was thought to be the main factor responsible for stimulating mitosis (Jenkin, 1970). Much of the information on STH levels in meat animals supported this belief. Injection of STH into meat animals may lead to increases in live weight gain, feed conversion, or leaneness (Henricson and Ullberg, 1960). In pigs, plasma STH declines with age, and may, perhaps, be correlated with average daily gain ($r = 0.46$; Chappel and Dunkin, 1975). In the pituitary glands of sheep, peak STH activity is found at birth and then it declines postnatally (Charrier, 1973). There is also some evidence that anabolic steroids such as *diethylstilbestrol* may act in ruminants by increasing STH secretion (Preston and Gee, 1957; Davis et al., 1977). Breeds of cattle with rapid early growth and a large mature size may secrete more STH than small, slow-growing breeds (Ohlson et al., 1981). All these findings, however, may be due to the indirect action of STH in maximizing lipid utilization for energy requirements and in maintaining essential protein synthesis under unfavorable circumstances (Turner and Munday, 1976). *Somatostatin* is a peptide that inhibits STH release (McQuillan, 1977). In skeletal muscle, it may have a role in the regulation of alanine and glutamine metabolism (Magno-Sumbilla et al., 1980).

Whether or not STH regulates mitosis in normal healthy animals by acting as a stimulus to mitosis is a difficult question to answer. The restoration of STH to muscles from hypophysectomized rats or hypopituitary human dwarfs does stimulate nuclear replication (Cheek and Wyllie, 1975), and the restoration of testosterone after its absence produces a similar response (Galavazi and Szirmai, 1971). These restoration effects are difficult to interpret, however, since STH and androgens both stimulate protein synthesis in muscle (Figure 8-19). Enhanced protein synthesis may lead to muscle fiber hypertrophy, which, in turn, may trigger satellite cell division. Even exercised muscles may develop both a higher protein content and a greater concentration of DNA (Bailey et al., 1973), and the ratio of satellite cells to true fiber nuclei may increase when muscle hypertrophy is induced by tenotomy of synergists (Hanzlikova et al., 1975).

Under optimal conditions for muscle growth, does the rate of fiber hypertrophy set the pace of satellite cell division, or vice versa? In cultured myoblasts, testosterone does produce a modest stimulation of proliferation (Powers and Florini, 1975) and the time spent by myoblasts in G_1 is reduced. The well-known anabolic effect of androgens on muscles in the living animal is mediated by androgen receptors in the muscle fiber cytosol (Michel and Baulieu, 1974; Snochowski et al., 1981). The glucocorticoid

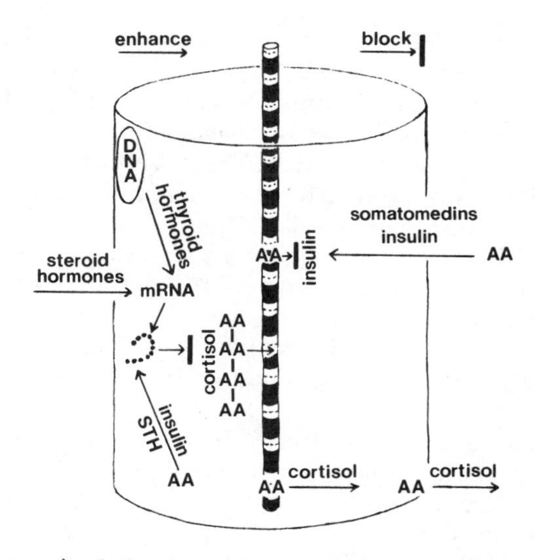

Figure 8-19 Hormonal control of amino acid (AA) metabolism in a muscle fiber. A polyribosome is represented as a string of beads. Several of the regulatory factors involved (insulin, STH, and somatomedins) may be bound at the cell membrane so that they achieve their effect by way of intermediary factors within the fiber.

receptors in skeletal muscle are also located in the cytosol (Mayer et al., 1974; Ho-Kim et al., 1981).

Since the work of Salmon and Daughaday in 1957, the importance of STH as a primary growth stimulant has been eclipsed by *somatomedins*. With in vivo experiments, it became established that the restoration of STH to hypophysectomized animals led to a growth spurt in cartilagenous growth zones. The growth of cartilage was monitored by the rate of sulfate incorporation into the matrix. In vitro, however, even high levels of STH produced only slight increases in the rate of sulfate incorporation. The action of STH in vivo was then found to be mediated by a *sulfation factor* or somatomedin which was released from the liver in response to STH. The generic term "somatomedin" now designates a whole family of STH-dependent polypeptides with insulinlike properties (Uthne, 1973). High levels of one of the somatomedins have been shown to exert a negative feedback on STH production by enhancing the negative effects of somatostatin on STH production (Berelowitz et al., 1981).

Insulin stimulates protein synthesis by acting at the level of the peptide chain (Jefferson et al., 1977) and thus it stimulates mitosis in cultured myoblasts (Florini and Roberts, 1979). The effects of somatomedins are widespread; they stimulate (1) the transport of glucose; (2) the uptake and incorporation of amino acids in the muscles of hypophysectomized animals; (3) DNA synthesis in glialike cells; and (4) collagen synthesis (Merimee and Rabin, 1973). Their effect, like *neurotrophic factor* (Ozawa and Kohama, 1978), can even be detected during myogenesis (Merrill et al., 1977). Somatomedins are distinguished from insulin by not being blocked by anti-insulin serum, and by being present in hypophysectomized diabetic rats treated with STH. A very high concentration of insulin (relative to that normally found in the blood) stimulates differentiation in cultured myoblasts. In this case, the insulin appears to act as an analog of somatomedins. In contrast to this, somatomedins produce an equivalent stimulation of myoblast growth and differentiation while at their normal low physiological concentration (Ewton and Florini, 1981).

Salmon and DuVall (1970) examined the effects of somatomedins on rat costal cartilage and diaphragm muscle. Muscle was less sensitive than cartilage to STH in vivo, and to somatomedins in vitro. As well as their generally agreed upon origin from the liver, somatomedins might also be produced in muscle (Hall, 1971). In an agricultural context, this hints at a possible two-way regulatory interrelationship between muscle growth and growth of the whole body. Animals with a greater mass of lean muscle might be able to produce more somatomedins and thus acquire longer bones as a result of extra growth at their epiphyseal plates.

Lambs with a rapid live weight growth have higher serum somatomedin levels than do those with slower growth (Wangsness et al., 1981). Serum somatomedin levels in lambs decline between 2 and 6 weeks postnatally, and this parallels a decline in specific growth rate (Olsen et al., 1981). In chick embryos, somatomedins become measurable at 13 days, they peak at 15 days, and they may not be detectable in the serum after hatching (Gaspard et al., 1981).

Several somatomedins have been identified (Charrier, 1978). Somatomedin-A stimulates sulfate uptake by chick cartilage, somatomedin-B stimulates DNA synthesis in human glialike cells, somatomedin-C stimulates sulfate uptake by rat cartilage, and somatomedin-P stimulates sulfate uptake by rib cartilage in pigs. Charrier (1980) and Charrier and Vezinhet (1980) found evidence of two somatomedins acting separately on thymidine uptake in DNA synthesis and on sulfate uptake in the growth of cartilage matrix. In the first case, the somatomedin was thought to be stimulating mitosis during tissue growth by cellular proliferation (hyperplasia). In the second case, the somatomedin was thought to be stimulating cellular hypertrophy.

The proliferation of cultured cells usually requires some encouragement from certain substances in the culture medium. *Multiplication-stimulating activity* (MSA) has been found in a number of substances that occur in the serum of various culture media (Temin et al., 1972). However, cells such as rat liver cells are able to proliferate in the absence of serum MSA. Dulak and Temin (1973) grew cultures of these cells and found that the cells had produced their own MSA. The newly produced MSA was a polypeptide, it had nonsupressible insulinlike activity, and it had sulfation factor activity—just like a somatomedin. Merrill et al. (1977) found that MSA also increased the uptake of amino acids by myoblasts just before it caused an increase in cell numbers and protein content. Thus natural somatomedins, like the experimentally produced analog MSA, are currently regarded as a dominant factor in the control of muscle growth (Florini et al., 1977). MSA is the analog of somatomedin A, and both of these substances act on both myoblasts and myotubes (Ewton and Florini, 1980).

If early cellular hyperplasia and later cellular hypertrophy are regulated by separate somatomedins, as Charrier (1980) suggests, what regulates the decelerating phase of the sigmoid growth curve? The name *chalone*, originally proposed as a general term for inhibitory chemical messengers, is now used for chemical messengers that inhibit mitosis (Bullough, 1973). Different types of body tissue have their own regulatory chalones, but they are identical or very similar in many different species of animals. Although not all attempts to identify chalones have been successful, a growing volume of scientific literature supports their existance. In lower vertebrates, chalones participate in the regulation of limb regeneration (Coomber and Scadding, 1983).

The foremost technical problem in the isolation of chalones is to ensure that cytotoxic substances are not confused with chalone activity (Houck, 1973). For example, if cells are treated with a tissue extract which inadvertently contains a cytotoxic substance, their normal metabolism may be disturbed so that any reduction in mitotic rate is secondary. Another problem is to avoid the short-term inhibition of mitosis by epinephrine (Jenkin, 1970). Most of the research on chalones has been undertaken by oncologists. It has been suggested that tumor cells lose their sensitivity to mitotic regulation by chalones (Iverson, 1978). Therapeutically, it might be possible to synchronize mitosis in tumors so that the impact of chemotherapy can be increased (Houck, 1973).

From a mathematical model of rat liver regeneration, Bard (1978) estimated that liver chalone has a half-life of 3 hours and a negative exponential dose-response curve. Thus after surgical removal of part of the liver, the chalone concentration in the blood would soon drop, and liver cell mitosis would be partially freed from inhibition. As chalone levels were restored by the formation of new liver cells, they would inhibit further mitosis. Chalones might also directly or indirectly switch cells from mitosis to differentiation, since De Paermentier et al. (1979) found that arrested cells formed trabeculae separated by ducts resembling biliary canaliculi. Hepatic chalone activity in cattle was reported by Sekas et al. (1979).

There is evidence of two types of skin chalone (Okulov et al., 1978). G_1 chalone prevents entry into the S phase of mitosis; it has a molecular weight of 21,000, and an isoelectric pH of 5.55. It is interesting to note that calmodulin is also thought to act at the G_1-S boundary (Chafouleas et al., 1982). G_2 chalone prevents entry into the M phase of mitosis; it has a molecular weight of 34,000 and an isoelectric pH of 5.85. Both chalones are glycoproteins and antibodies to G_2 do not react with G_1. Houck et al. (1977) also concluded that fibroblast chalones were glycopeptides, but with a molecular weight of about 5000. An epidermal chalone from pig skin was reported by Laurence et al. (1979). Bertsch and Marks (1979) found that epidermal G_1 chalone acts primarily on stem cells in an advanced phase of differentiation, while G_2 chalone acts on more primitive cell types. Estrogens may stimulate mitosis in their target tissues by lowering their levels of G_2 chalone (Okulov et al., 1979). Testicular G_1 and G_2 chalones have also been reported (Irons and Clermont, 1979). Chalones do not appear to change qualitatively with age since Leith (1978) successfully exchanged them between young and old mice.

Despite many years of intensive investigation, no unified theory has yet emerged to explain how animal growth is regulated. Are somatomedins and trophic factors (Chapter 6) responsible for the autocatalytic acceleration of animal growth? Are chalones responsible for the deceleration of animal growth, like the theoretically predicted antitemplates? The pieces of the jigsaw puzzle may be in front of us at this very moment.

REFERENCES

ADAMS, P. H. 1971. *Biol. Neonate* 19:341.

ALLEN, R. E., MERKEL, R. A., and YOUNG, R. B. 1979. *J. Anim. Sci.* 49:115.

ALTMAN, P. L., and DITTMER, D. S. 1962. *Growth.* Federation of American Societies for Experimental Biology, Washington, D.C.

ANSAY, M. 1974. *Zentralbl. Veterinaermed. A* 21:603.

ASGHAR, A., and YEATES, N. T. M. 1979a. *Agric. Biol. Chem.* 43:437.

ASGHAR, A., and YEATES, N. T. M. 1979b. *Agric. Biol. Chem.* 43:455.

ASMUNDSON, V. S., and LERNER, I. M. 1942. *Poult. Sci.* 21:505.
AUCKLAND, J. N. 1972. *Br. Poult. Sci.* 13:251.
AUCKLAND, J. N., and MORRIS, T. R. 1971. *Br. Poult. Sci.* 12:137.
BAILEY, D. A., BELL, R. D., and HOWARTH, R. E. 1973. *Growth* 37:323.
BARD, J. B. L. 1978. *J. Theor. Biol.* 73:509.
BARR, M. 1973. *Teratology* 7:283.
BEANE, W. L., SIEGEL, P. B., and SIEGEL, H. S. 1965. *Poult. Sci.* 44:1009.
BERELOWITZ, M., SZABO, M., FROHMAN, L. A., FIRESTONE, S., CHU, L., and HINTZ, R. L. 1981. *Science* 212:1279.
BERG, R. T., and BUTTERFIELD, R. M. 1976. *New Concepts of Cattle Growth.* John Wiley & Sons, New York.
BERGSTROM, J. G. 1972. *Sven. VetTidn.* 24:219.
BERTSCH, S., and MARKS, F. 1979. *Cancer Res.* 39:239.
BINES, J. A. 1971. *Proc. Nutr. Soc.* 30:116.
BINNS, W., JAMES, L. F., KEELER, R. F., and VAN KAMPEN, K. R. 1969. Developmental abnormalities. In E. S. E. Hafez and I. A. Dyer (eds.), *Animal Growth and Nutrition*, Chap. 8. Lea & Febiger, Philadelphia.
BOLLWAHN, W. VON, and PFEIFFER, A. 1969. *Dtsch. Tierarztl. Wochenschr.* 76:239.
BONNER, J. T. 1963. *Morphogenesis. An Essay on Development*, pp. 142–143. Atheneum Publishers, New York.
BOX, G. E. P., and COX, D. R. 1964. *J. R. Stat. Soc.* 26:21.
BRANNANG, E. 1971. *Swed. J. Agric. Res.* 1:69.
BROADBENT, P. J., BALL, C., and DODSWORTH, T. L. 1976. *Anim. Prod.* 23:341.
BRODY, S. 1945. *Bioenergetics and Growth.* Reinhold Publishing Corporation, New York.
BRYDEN, M. M., EVANS, H. E., and BINNS, W. 1972a. *J. Morphol.* 138:169.
BRYDEN, M. M., EVANS, H. E., and BINNS, W. 1972b. *J. Morphol.* 138:187.
BRYDEN, M. M., EVANS, H., and BINNS, W. 1973. *Anat. Rec.* 175:725.
BULLOUGH, W. S. 1973. The chalones: a review. In B. K. Forscher and J. C. Houck (eds.), *Chalones: Concepts and Current Researches.* Natl. Cancer Inst. Monogr. 38, p. 5.
BURFENING, P. J. 1972. *Anim. Prod.* 15:61.
BYERLY, T. C. 1967. *Science* 157:890.
CAIN, J. R., and WILSON, W. O. 1974. *Poult. Sci.* 53:1438.
CARNEGIE, A. B., TULLOH, N. M., and SEEBECK, R. M. 1969. *Aust. J. Agric. Res.* 20:405.
CHAFOULEAS, J. G., BOLTON, W. E., HIDAKA, H., BOYD, A. E., and MEANS, A. R. 1982. *Cell* 28:41.
CHAPPEL, R. J., and DUNKIN, A. C. 1975. *Anim. Prod.* 20:51.
CHARRIER, J. 1973. *Ann. Biol. Anim. Biochem. Biophys.* 13:155.
CHARRIER, J. 1978. *Ann. Biol. Anim. Biochem. Biophys.* 18:33
CHARRIER, J. 1980. *Reprod. Nutr. Dev.* 20:301.
CHARRIER, J., and VEZINHET, A. 1980. *Reprod. Nutr. Dev.* 20:93.
CHEEK, D. B., and WYLLIE, R. G. 1975. Postnatal cellular growth: hormonal consideration. In D. B. Cheek (ed.), *Fetal and Postnatal Cellular Growth*, pp. 415–435. John Wiley & Sons, New York.
CHEUNG, W. Y. 1980. *Science* 207:19.
CLARK, J. L., EAKINS, R. L., HEDRICK, H. B., and KRAUSE, G. F. 1976. *J. Anim. Sci.* 42:569.
CLAYTON, G. A., and POWELL, J. C. 1979. *Br. Poult. Sci.* 20:121.
CLAYTON, G. A., FOXTON, R. N., NOTT, H., and POWELL, J. C. 1974. *Br. Poult. Sci.* 15:153.
CLAYTON, G. A., NIXEY, C., and MONAGHAN, G. 1978. *Br. Poult. Sci.* 19:755.
Committee on Bovine Reproductive Nomenclature. 1972. *Cornell Vet.* 62:216.
COOMBER, B. L., and SCADDING, S. R. 1983. *Cell Tissue Kinet.* 16:77.
CROW, M. W., and PIKE, H. T. 1973. *J. Am. Vet. Med. Assoc.* 162:453.
CUNHA, T. J. 1968. *Feedstuffs* 40(10):25.

CUNNINGHAM, H. M., FRIEND, D. W., and NICHOLSON, J. W. G. 1963. *J. Anim. Sci.* 22:226.
DAVIES, A. S. 1974a. *Anim. Prod.* 19:367.
DAVIES, A. S. 1974b. *Anim. Prod.* 19:377.
DAVIES, A. S. 1975. *Anim. Prod.* 20:45.
DAVIES, A. S. 1979. *Zentralb. Veterinaermed.* C 8:164.
DAVIES, A. S., and KALLWEIT, E. 1979. *Z. Tierz. Zuechtungsbiol.* 96:6.
DAVIES, A. S., and PRYOR, W. J. 1977. *J. Agric. Sci., Camb.* 89:257.
DAVIES, A. S., PEARSON, G., and CARR, J. R. 1980. *J. Agric. Sci., Camb.* 95:251.
DAVIS, S. L., OHLSON, D. L., KLINDT, J., and ANFINSON, M. S. 1977. *Am. J. Physiol.* 233:E519.
DEIGHTON, T. 1932. *J. Agric. Sci., Camb.* 22:418.
DE PAERMENTIER, F., BARBASON, H., and BASSLEER, R. 1979. *Biol. Cell.* 34:205.
DEUTSCH, K., and DONE, J. T. 1971. *Res. Vet. Sci.* 12:176.
DICKERSON, J. W., and MCCANCE, R. A. 1964. *Clin. Sci.* 27:123.
DOBSON, K. J. 1968. *Aust. Vet. J.* 44:26.
DOBSON, K. J. 1971. *Aust. Vet. J.* 47:587.
DONALD, H. P. 1940. *J. Agric. Sci., Camb.* 30:582.
DOORNENBAL, H. 1975. *Growth* 39:427.
DULAK, N. C., and TEMIN, H. M. 1973. *J. Cell. Physiol.* 81:153.
DYER, I. A., and ROJAS, M. A. 1965. *J. Am. Vet. Med. Assoc.* 147:1393.
DYER, I. A., CASSETT, W. A., and RAO, R. R. 1964. *Bioscience* 14:31.
EBINGER, P. 1974. *Z. Anat. Entwicklungsgesch.* 114:267.
EBINGER, P. 1975. *Anat. Embryol.* 144:267.
EDMONDS, L. D., SELBY, L. A., and CASE, A. A. 1972. *J. Am. Vet. Med. Assoc.* 160:1319.
ELSLEY, F. W. H. 1976. *Proc. Nutr. Soc.* 35:323.
EVANS, H. E., and SACK, W. O. 1973. *Zentralbl. Veterinaermed. C* 2:11.
EVERITT, G. C. 1968. Prenatal development of uniparous animals with particular reference to the influence of maternal nutrition in sheep. In G. A. Lodge, and G. E. Lamming (eds.), *Growth and Development of Mammals*, pp. 131–151. Butterworth & Company (Publishers), London.
EWTON, D. Z., and FLORINI, J. R. 1980. *Endocrinology* 106:577.
EWTON, D. Z., and FLORINI, J. R. 1981. *Dev. Biol.* 86:31.
FABENS, A. J. 1965. *Growth* 29:265.
FARMER, L. J., MACKIE, W. S., and RITCHIE, P. J. 1981. *J. Agric. Sci., Camb.* 97:569.
FLORINI, J. R., and ROBERTS, S. B. 1979. *In Vitro* 15:983.
FLORINI, J. R., NICHOLSON, M. L., and DULAK, N. C. 1977. *Endocrinology* 101:32.
FOLKMAN, J., and HOCHBERG, M. 1973. *J. Exp. Med.* 138:745.
FROEHLICH, J. E., and RACHMELER, M. 1972. *J. Cell Biol.* 55:19.
GALAVAZI, G., and SZIRMAI, J. A. 1971. *Z. Zellforsch.* 121:548.
GALPIN, N. 1935. *J. Agric. Sci., Camb.* 25:344.
GASPARD, K. J., KLITGAARD, H. M., and WONDERGEM, R. 1981. *Proc. Soc. Exp. Biol. Med.* 166:24.
GINDHART, P. S. 1972. *Hum. Biol.* 44:335.
GOMPERTZ, B. 1825. *R. Soc. Philos. Trans.* pp. 513–585.
GRAY, B. F. 1981. *J. Theor. Biol.* 93:757.
HAJEK, I. 1980. *Physiol. Bohemoslov.* 29:145.
HALL, A. D. 1905. *The Book of the Rothamsted Experiments.* John Murray, London.
HALL, K. 1971. *Acta Endocrinol.* 66:491.
HALVORSON, D. B., and JACOBSON, M. 1970. *Poult. Sci.* 49:132.
HAMMOND, J. 1927. *The Physiology of Reproduction in the Cow.* At the University Press, Cambridge.

HAMMOND, J. 1932. *Growth and Development of Mutton Qualities in the Sheep*. Oliver & Boyd, Edinburgh.
HAMMOND, J. 1950. *Proc. R. Soc. Lond. B* 137:452.
HAMMOND, J., MASON, I. L., and ROBINSON, T. J. 1971. *Hammond's Farm Animals*. Edward Arnold (Publishers), London.
HANZLIKOVA, V., MACKOVA, E. V., and HNIK, P. 1975. *Cell Tissue Res*. 160:411.
HARRISON, T. J. 1970. *J. Anat*. 106:165.
HARSHAW, H. M., and RECTOR, R. R. 1940. *Poult. Sci.* 19:404.
HARTLEY, W. J., ALEXANDER, G., and EDWARDS, M. J. 1974. *Teratology* 9:299.
HARTUNG, T. E., and FRONING, G. W. 1968. *Poult. Sci*. 47:1348.
HEATH, E. 1979. *Acta Anat*. 105:56.
HENDERSON, I. F., HENDERSON, W. D., and KENNETH, J. H. 1966. *A Dictionary of Biological Terms*. Oliver & Boyd, Edinburgh.
HENRICSON, B., and ULLBERG, S. 1960. *J. Anim. Sci*. 19:1002.
HERRMANN, H. 1952. *Ann. N.Y. Acad. Sci*. 55:99.
HERVEY, G. R. 1971. *Proc. Nutr. Soc*. 30:109.
HNIK, P., and VEJSADA, R. 1979a. *Physiol. Bohemoslov*. 28:251.
HNIK, P., and VEJSADA, R. 1979b. *Physiol. Bohemoslov*. 28:385.
HOFECKER, G., SKALICKY, M., KMENT, A., and NIEDERMULLER, H. 1980. *Mech. Ageing Dev*. 14:345.
HOGBERG, M. G., and ZIMMERMAN, D. R. 1978. *J. Anim. Sci.* 47:893.
HO-KIM, M. A., TREMBLAY, R. R., and DUBE, J. Y. 1981. *Endocrinology* 109:1418.
HOUCK, J. C. 1973. General introduction to the chalone concept. In B. K. Forscher and J. C. Houck (eds.), *Chalones: Concepts and Current Researches*. Natl. Cancer Inst. Monogr. 38, p. 1.
HOUCK, J. C., KANAGALINGAM, K., HUNT, C., ATTALLAH, A., and CHUNG, A. 1977. *Science* 196:896.
HUBBERT, W. T., STALHEIM, O. H. V., and BOOTH, G. D. 1972. *Growth* 36:217.
HUTT, F. B. 1934a. *J. Hered.* 25:41.
HUTT, F. B. 1934b. *Cornell Vet*. 24:1.
HUXLEY, J. S. 1932. *Problems of Relative Growth*. Methuen & Co., London.
IRONS, M. J., and CLERMONT, Y. 1979. *Cell Tissue Kinet*. 12:425.
IVERSON, O. H. 1978. *Virchows Arch. B Cell. Pathol*. 28:271.
JAMES, L. F. 1972. *Am. J. Vet. Res.* 33:835.
JEFFERSON, L. S., LI, J. B., and RANNELS, S. R. 1977. *J. Biol. Chem*. 252:1476.
JENKIN, P. M. 1970. *Control of Growth and Metamorphosis*. Pergamon Press, Oxford.
JOHNSON, A. S., and ASMUNDSON, V. S. 1957. *Poult. Sci.* 36:959.
JOHNSON, E. R. 1974. *Res. Vet. Sci.* 17:273.
JOHNSON, E. R., PRYOR, W. J., and BUTTERFIELD, R. M. 1973. *Aust. J. Agric. Res*. 24:287.
JONES, S. D. M., PRICE, M. A., and BERG, R. T. 1978a. *Can. J. Anim. Sci*. 58:105.
JONES, S. D. M., PRICE, M. A., and BERG, R. T. 1978b. *J. Anim. Sci*. 46:1151.
JOUBERT, D. M. 1956. *J. Agric. Sci., Camb*. 47:382.
JUDGE, M. D., and STOB, M. 1963. *J. Anim. Sci*. 22:1059.
KECK, C. R. 1970. *Am. Hereford J*. July 1.
KEESEY, R. E. 1976. *Proc. Reciprocal Meat Conf*. 29:105.
KEMP, J. D., ELY, D. G., FOX, J. D., and MOODY, W. G. 1981. *J. Anim. Sci*. 52:1026.
KENNEDY, P. C. 1971. *Fed. Proc*. 30:110.
KERR, G. R. 1975. Skeletal growth in the fetal Macaque. In D. B. Cheek (ed.), *Fetal and Postnatal Cellular Growth*, pp. 289–298. John Wiley & Sons, New York.
KILBURN, D. G., LILLY, M. D., and WEBB, F. C. 1969. *J. Cell Sci.* 4:645.

Kleiber, M. 1961. *The Fire of Life*. John Wiley & Sons, New York.
Kleinpeter, M. E., and Mixner, J. P. 1947. *Poult. Sci*. 26:494.
Koch, R. M., Schleicher, E. W., and Arthaud, V. H. 1958. *J. Anim. Sci*. 17:604.
Koch, R. M., Dikeman, M. E., and Cundiff, L. V. 1981. *J. Anim. Sci*. 53:992.
Kohler, E. M., Cross, R. F., and Ferguson, L. C. 1969. *J. Am. Vet. Med. Assoc*. 155:139.
Kowalski, C. J., and Guire, K. E. 1974. *Growth* 38:131.
Krudewig, B. 1971. Die Behandlung der Gratschellung neugeborener Ferkel. Inaug. Diss. Tierarztl. Hochschule, Hanover.
Kruska, D. von. 1973. *Sonderdruk Z. Zool. Systematik Evolutionsforsch*. 11:81.
Kruska, D., and Rohrs, M. 1974. *Z. Anat. Entwicklungsgesch*. 144:61.
Laird, A. K. 1966. *Growth* 30:349.
Laird, A. K., Tyler, S. A., and Barton, A. D. 1965. *Growth* 29:233.
Laurence, E. B., Spargo, D. J., and Thronley, A. L. 1979. *Cell Tissue Kinet*. 12:615.
Laurent, G. J., and Millward, D. J. 1980. *Fed. Proc*. 39:42.
Lawrie, R. A., and Pomeroy, R. W. 1963. *J. Agric. Sci., Camb*. 61:409.
Lax, T. 1971. *J. Hered*. 62:250.
Leipold, H. W., Oehme, F. W., and Cook, J. E. 1973. *J. Am. Vet. Med. Assoc*. 162:1059.
Leith, J. T. 1978. *Cell Tissue Kinet*. 11:433.
Lerner, I. M. 1944. *J. Hered*. 35:219.
Lewis, D., Boorman, K. N., and Buttery, P. J. 1974. Efficiency of protein utilization. In D. Lister, D. N. Rhodes, V. R. Fowler, and M. F. Fuller (eds.), *Meat Animals. Growth and Productivity*, pp. 103–113. Plenum Press, New York.
Lister, D., and McCance, R. A. 1967. *Br. J. Nutr*. 21:787.
Liuzzo, J. A., Reineke, E. P., and Pearson, A. M. 1958. *J. Anim. Sci*. 17:513.
Lockshin, R. A., and Beaulaton, J. 1974. *Life Sci*. 15:1549.
Loewenstein, W. R., and Kanno, Y. 1966. *Nature (Lond.)* 209:1248.
Lohse, C. L. 1973. *Growth* 37:177.
Lohse, C. L., Pryor, W. J., and Butterfield, R. M. 1973. *Aust. J. Agric. Res*. 24:279.
Ludwig, F. C., and Smoke, M. E. 1980. *Exp. Aging Res*. 6:497.
Magno-Sumbilla, C., Collins, R. M., and Ozand, P. T. 1980. *Horm. Metab. Res*. 12:439.
Marrable, A. W., and Ashdown, R. R. 1967. *J. Agric. Sci., Camb*. 69:443.
Martin, J. T. 1978. *Am. Zool*. 18:489.
Mayer, M., Kaiser, N., Milholland, R. J., and Rosen, F. 1974. *J. Biol. Chem*. 249:5236.
McCance, R. A. 1974. *J. Anat*. 117:475.
McDonald, I., Robinson, J. J., and Fraser, C. 1981. *J. Agric. Sci., Camb*. 96:187.
McManus, W. R., Reid, J. T., and Donaldson, L. E. 1972. *J. Agric. Sci., Camb*. 79:1.
McQuillan, M. T. 1977. *Somatostatin*, Vol. 1. Eden Press, Churchill Livingstone, Edinburgh.
Medawar, P. B. 1945. Size, shape and age. In W. E. Le Gros Clark and P. B. Medawar (eds.), *Essays on Growth and Form Presented to D'Arcy Wentworth Thompson*, pp. 157–186. Clarendon Press, Oxford.
Melhorn, I., Rittenbach, P., and Seffner, W. 1970. *Mh. VetMed*. 25:781.
Merimee, T. J., and Rabin, D. 1973. *Metabolism* 22:1235.
Merrill, G. F., Florini, J. R., and Dulak, N. C. 1977. *J. Cell. Physiol*. 93:173.
Meyer, J. H., Hull, J. L., Weitkamp, W. H., and Bonilla, S. 1965. *J. Anim. Sci*. 24:29.
Michel, G., and Baulieu, E.-E. 1974. *C. R. Acad. Sci. (D), Paris*. 279:421.
Miles, C. A., and Woods, M. O. 1979. A study of the transmission of ultrasound and the measurement of fatness. *Index of Research*, pp. 65–66. Meat and Livestock Commission, Bletchley, England.
Miller, J. K., Hacking, A., Harrison, J., and Gross, V. J. 1973. *Vet. Rec*. 93:555.
Millward, D. J., Garlick, P. J., and Reeds, P. J. 1976. *Proc. Nutr. Soc*. 35:339.

MOSIER, H. D. 1972. *Growth* 36:123.
MUKHOTY, H., and BERG, R. T. 1971. *Anim. Prod.* 13:219.
MUKHOTY, H., and BERG, R. T. 1973. *J. Agric. Sci., Camb.* 81:317.
MULVIHILL, J. J. 1972. *Science* 176:132.
MURRAY, D. M., TULLOH, N. M., and WINTER, W. H. 1972. *Proc. Aust. Soc. Anim. Prod.* 9:360.
MURRAY, D. M., BOWLAND, J. P., BERG, R. T., and YOUNG, B. A. 1974. *Can. J. Anim. Sci.* 54:91.
NEEDHAM, A. E. 1950. *Proc. R. Soc. B* 137:115.
NEEDHAM, A. E. 1957. *Nature* (*Lond.*) 180:1293.
OHLSON, D. L., DAVIS, S. L., FERRELL, C. L., and JENKINS, T. G. 1981. *J. Anim. Sci.* 53:371.
OKULOV, V. B., KETLINSKII, S. A., RATOVITSKII, E. A., and KALINOVSKII, V. P. 1978. *Biokhimiya* 43:770.
OKULOV, V. B., IVANOV, M. N., and ANISIMOV, V. N. 1979. *Endokrinologie* 74:20.
OLSEN, R. F., WANGSNESS, P. J., PATTON, W. H., and MARTIN, R. J. 1981. *J. Anim. Sci.* 52:63.
OLSON, L. D., and PRANGE, J. F. 1968. *Vet. Med. Small Anim. Clin.* 63:714.
OXNARD, C. E. 1980. *Am. Zool.* 20:695.
OZAWA, E., and KOHAMA, K. 1978. *Muscle Nerve* 1:230.
PALLUDAN, B. 1961. *Acta Vet. Scand.* 2:32.
PALMITER, R. D., BRINSTER, R. L., HAMMER, R. E., TRUMBAUER, M. E., ROSENFELD, M. G., BIRNBERG, N. C., and EVANS, R. M. 1982. *Nature* 300:611.
PARK, Y. I., HANSEN, C. T., CHUNG, C. S., and CHAPMAN, A. B. 1966. *Genetics* 54:1315.
PARKS, J. R. 1970. *Am. J. Physiol.* 219:833.
PATTEN, B. M. 1948. *Embryology of the Pig*, 3rd ed. McGraw-Hill Book Company, New York.
PATTERSON, D. S. P., and ALLEN, W. M. 1972. *Br. Vet. J.* 128:101.
PATTERSON, D. S. P., SWEASEY, D., ALLEN, W. M., BERRETT, S., and THURLEY, D. C. 1969. *Zentralbl. Veterinaermed. A* 16:741.
PENNYCUIK, P. R. 1972. *Aust. J. Biol. Sci.* 25: 627.
PERRY, B. N. 1974. *Br. J. Nutr.* 31:35.
PETERS, R. R., CHAPIN, L. T., LEINING, K. B., and TUCKER, H. A. 1978. *Science* 199:911.
PETRUSEWICZ, K., and MACFADYEN, A. 1970. *Productivity of Terrestrial Animals. Principles and Methods*. Blackwell Scientific Publications, Oxford.
POMEROY, R. W. 1941. *J. Agric. Sci., Camb.* 31:50.
POMEROY, R. W. 1960. *J. Agric. Sci., Camb.* 54:31.
POWELL, S. E., and ABERLE, E. D. 1980. *J. Anim. Sci.* 50:860.
POWELL, S. E., and ABERLE, E. D. 1981. *J. Anim. Sci.* 52:748.
POWERS, M. L., and FLORINI, J. R. 1975. *Endocrinology* 97:1043.
POWNALL, R., and DALTON, R. G. 1976. *Br. Vet. J.* 132:259.
PRESTON, T. R., and GEE, I. 1957. *Nature* (*Lond.*) 179:247.
PYM, R. A. E., and THOMPSON, J. M. 1980. *Br. Poult. Sci.* 21:281.
REISENFELD, A. 1976. *Acta Anat.* 94:321.
RICHARDS, F. J. 1959. *J. Exp. Bot.* 10:290.
RICHARDSON, C., HEBERT, C. N., and TERLECKI, S. 1976. *Vet. Rec.* 99:22.
RICHMOND, R. J., JONES, S. D. M., PRICE, M. A., and BERG, R. T. 1979. *Can. J. Anim. Sci.* 59:471.
RICHTER, C. P. 1954. *J. Natl. Cancer Inst.* 15:727.
ROBERTS, J. A. F. 1929. *J. Genet.* 21:110.
ROBERTSON, T. B. 1923. *The Chemical Basis of Growth and Senescence*. J. B. Lippincott Company, Philadelphia.
ROBINSON, D. W. 1964. *Anim. Prod.* 6:227.
ROBINSON, J. J. 1981. *Livestock Prod. Sci.* 8:273.

ROBISON, O. W. 1976. *J. Anim. Sci.* 42:1024.
ROSENQUIST, G. C. 1982. *Anat. Rec.* 202:95.
ROWLANDS, G. J., PAYNE, J. M., DEW, S. M., and MANSTON, R. 1973. *Vet. Rec.* 93:48.
RUSSEL, A. J. F., GUNN, R. G., and DONEY, J. M. 1968. *Anim. Prod.* 10:43.
SALMON, W. D., Jr., and DAUGHADAY, W. H. 1957. *J. Lab. Clin. Med.* 49:825.
SALMON, W. D., and DUVALL, M. R. 1970. *Endocrinology* 87:1168.
SALMON-LEGAGNEUR, E. 1968. Prenatal development in the pig and some other multiparous animals. In G. A. Lodge and G. E. Lamming (eds.), *Growth and Development of Mammals*, pp. 158–194. Butterworth & Company (Publishers), London.
SAUNDERS, J. W., GASSELING, M. T., and SAUNDERS, L. C. 1962. *Dev. Biol.* 5:147.
SAVORY, C. J. 1976a. *Br. Poult. Sci.* 17:341.
SAVORY, C. J. 1976b. *Br. Poult. Sci.* 17:557.
SCHAUBLE, M. K. 1972. *J. Exp. Zool.* 181:281.
SEEBECK, R. M. 1968a. *Proc. Aust. Soc. Anim. Prod.* 7:297.
SEEBECK, R. M. 1968b. *Anim. Breed. Abstr.* 36:167.
SEEBECK. R. M., and TULLOH, N. M. 1968a. *Aust. J. Agric. Res.* 19:477.
SEEBECK, R. M., and TULLOH, N. M. 1968b. *Aust. J. Agric. Res.* 19:673.
SEEBECK, R. M., and TULLOH, N. M. 1969. *Aust. J. Agric. Res.* 20:199.
SEIDEL, G. E. 1981. *Science* 211:351.
SEKAS, G., OWEN, W. G., and COOK, R. T. 1979. *Exp. Cell Res.* 122:47.
SELBY, L. A., MENGES, R. W., HOUSER, E. C., FLATT, R. E., and CASE, A. A. 1971. *Arch. Environ. Health* 22:496.
SETIABUDI, M., KAMONSAKPITHAK, S., SHENG, H.-P., and HUGGINS, R. A. 1975. *Growth* 39:405.
SEVER, J. L. 1971. *Fed. Proc.* 30:114.
SHENG, H.-P., KAMONSAKPITHAK, S., NAIBORHU, A., CHUNTANANUKOON, S., and HUGGINS, R. A. 1977. *Growth* 41:139.
SIEGEL, P. B., and GUHL, A. M. 1956. *Poult. Sci.* 35:1340.
SKJERVOLD, H., STANDAL, N., and BRUFLOT, R. 1963. *J. Anim. Sci.* 22:458.
SMIDT, W. J. 1972. Congenital defects in pigs. *VIIth Int. Congr. Anim. Reprod. A. I. Munchen*, Vol. II, pp. 1145–1148.
SMITH, H. S. 1969. Animal domestication and animal cult in dynastic Egypt. In Ucko, P. J. and Dimbleby, G. W. (eds.), *The Domestication and Exploitation of Plants and Animals*. Aldine-Atherton, Chicago.
SMITH, R. J. 1980. *J. Theor. Biol.* 87:97.
SNOCHOWSKI, M., LUNDSTROM, K., DAHLBERG, E., PETERSSON, H., and EDQVIST, L.-E. 1981. *J. Anim. Sci.* 53:80.
SPILLMAN, W. J., and LANG, E. 1924. *The Law of Diminishing Returns*. World Book Company, Chicago.
STADELMAN, W. J., and MEINERT, C. F. 1977. *Poult. Sci.* 56:1145.
STEWART, T. A., and MINTZ, B. 1981. *Proc. Natl. Acad. Sci. USA* 78:6314.
STOUFFER, J. R. 1969. Techniques for the estimation of the composition of meat animals. In *Techniques and Procedures in Animal Science Research*. American Society of Animal Science, Albany, N.Y.
SWATLAND, H. J. 1979. *J. Agric. Sci., Camb.* 93:1.
SWATLAND, H. J. 1980a. *Growth* 44:139.
SWATLAND, H. J. 1980b. *Poult. Sci.* 59:1773.
SWATLAND, H. J., and CASSENS, R. G. 1973. *J. Agric. Sci., Camb.* 80:503.
TANNER, J. M. 1963. *Child Dev.* 34:817.
TAYLOR, St. C. S. 1965. *Anim. Prod.* 7:203.

TAYLOR, St. C. S., MASON, M. A., and MCCLELLAND, T. H. 1980. *Anim. Prod.* 30:125.

TEMIN, H. M., PIERSON, R. W., and DULAK, N. C. 1972. The role of serum in the control of multiplication of avian and mammalian cells in culture. In G. H. Rothblat and V. C. Cristofalo (eds.), *Growth, Nutrition, and Metabolism of Cells in Culture*, Vol. 1, pp. 49–81. Academic Press, New York.

THOMPSON, D'A. W. 1917. *On Growth and Form*. At the University Press, Cambridge.

THURLEY, D. C., GILBERT, F. R., and DONE, J. T. 1967. *Vet. Rec.* 80:302.

TISCHLER, M. E. 1981. *Life Sci.* 28:2569.

TITUS, H. W. 1927. *Poult. Sci.* 7:254.

TRENKLE, A. 1981. *Fed. Proc.* 40:2536.

TUCKER, H. A., and RINGER, R. K. 1982. *Science* 216:1381.

TURNER, M. R. 1972. *Proc. Nutr. Soc.* 31:205.

TURNER, M. R., and MUNDAY, K. A. 1976. Hormonal control of muscle growth. In D. Lister, D. N. Rhodes, V. R. Fowler and M. F. Fuller (eds.), *Meat Animals. Growth and Productivity*. Plenum Press, New York.

ULLREY, D. E., SPRAGUE, J. I., BECKER, D. E., and MILLER, E. R. 1965. *J. Anim. Sci.* 24:711.

UTHNE, K. 1973. *Acta Endocrinol. Suppl.* 175.

VENTE, J. P., WRATHALL, A. E., HEBERT, N., and HOSKIN, B. D. 1972. *Res. Vet. Sci.* 13:169.

VON BERTALANFFY, L. 1957. *Q. Rev. Biol.* 32:217.

WALLACE, L. R. 1948a. *J. Agric. Sci., Camb.* 38:93.

WALLACE, L. R. 1948b. *J. Agric. Sci., Camb.* 38:243.

WANGSNESS, P. J., OLSEN, R. F., and MARTIN, R. J. 1981. *J. Anim. Sci.* 52:57.

WARD, P. S. 1978a. *Vet. Bull.* 48:279.

WARD, P. S. 1978b. *Vet. Bull.* 48:381.

WEBSTER, A. J. F. 1976. Efficiencies of energy utilization during growth. In D. Lister, D. N. Rhodes, V. R. Fowler, and M. F. Fuller (eds.), *Meat Animals. Growth and Productivity*, pp. 89–102. Plenum Press, New York.

WEBSTER, A. J. F. 1977. *Proc. Nutr. Soc.* 36:53.

WEISS, P., and KAVANAU, J. L. 1957. *J. Gen. Physiol.* 41:1.

WHITTEMORE, C. T. 1976. *Proc. Nutr. Soc.* 35:383.

WHITTEMORE, C. T., and FAWCETT, R. H. 1976. *Anim. Prod.* 22:87.

WIDDOWSON, E. M. 1971. *Biol. Neonate* 19:329.

WILDT, D. E., RIEGLE, G. D., and DUKELOW, W. R. 1975. *Am. J. Physiol.* 229:1471.

WILSON, P. N. 1954. *J. Agric. Sci., Camb.* 44:67.

WILSON, P. N., and OSBOURN, D. F. 1960. *Biol. Rev.* 35:324.

WINTER, W. H. 1970. *Proc. Aust. Soc. Anim. Prod.* 8:283.

WOOD, T. B., and HILL, A. V. 1914. *J. Agric. Sci., Camb.* 6:252.

YEATES, N. T. M. 1964. *J. Agric. Sci., Camb.* 62:267.

ZEUNER, F. E. 1963. *A History of Domesticated Animals*. Harper & Row, Publishers, New York.

9

The Conversion of Muscles to Meat

INTRODUCTION

The main purpose of this chapter is to integrate many of the topics that have been introduced in earlier chapters. Although the structure and development of meat animals may be an intrinsically fascinating subject to those interested in animal production, we cannot afford to lose sight of the fact that the eventual financial reward for animal production comes from the sale of meat and meat products, not from the sale of living muscles. Too many published studies in animal agriculture have dealt at length with meat quantity but have neglected meat quality. Meat is one of the more expensive ways to produce proteins for human consumption and in order to maintain our existing animal industries we must pay particular attention to meat quality.

The conversion of muscles to meat occurs after animals are slaughtered. Muscle is a living tissue whose characteristic contractile activity is normally regulated in a purposeful manner by the nervous system. When muscles are completely converted to meat they are no longer capable of contraction by filament sliding. At no single instant in time does a commercially slaughtered meat animal pass from a living to a dead state and, similarly, the commercial conversion of muscles to meat is seldom an instantaneous event. When a live animal is brought from the farm to the abattoir it is aware of its new surroundings and is capable of locomotion. These are manifestations of the dominance of the central nervous system over a hierarchy of subservient, but equally complex physiological, biophysical, and biochemical systems. However, even the high state of cellular coordination in the live animal is only an equilibrium between a majority of living cells and a minority of moribund cells and damaged organelles. In many body tissues, individual cells have a relatively short life span, and their own organelles wear out and are replaced many times in the life of the cell. Nerve cells are unusual since they are not usually replaced at the end of their life span.

Except for animals which are ritually slaughtered for religious reasons, commercial meat animals are first rendered unconscious, either by concussion, by electrical stunning, or by a high concentration of carbon dioxide gas. Thus the highest level of regulation by the brain is lost, leaving only lower levels of reflex activity. In an unconscious state, animals are exsanguinated (bled out), and body tissues become *hypoxic* (with reduced oxygen level) and, finally, *anoxic* (without oxygen). In more cases than is generally realized, it is possible to stun animals humanely without violating religious laws (Mann, 1960).

After the exsanguination of an animal, there is a variable length of time during which the animal is neither completely alive nor completely dead. Higher levels of regulated coordination by the brain are gone but many cells, particularly muscle fibers, continue to pursue an independent existence. The cellular activity that persists after the death of a whole animal is called *necrobiosis*. The term "necrobiosis" also includes the activity of parts of a cell which survive after a whole cell has become disorganized. For example, some biochemical pathways in meat may continue to function long after muscle fibers have lost their regulatory membrane systems and their cellular integrity.

Many of the cells in the body are prepared for working without oxygen for a short time, since animal tissues often use oxygen at a faster rate than it can be obtained from the bloodstream. For example, during strenuous locomotion, muscles use *anaerobic glycolysis* to release energy from stored glycogen without oxygen. Body cells that are poorly adapted for anaerobic glycolysis run out of energy quite rapidly if they are deprived of oxygen. In this state, it may not be long before they begin to digest themselves by *autolysis*. Cells that produce digestive enzymes undergo particularly rapid autolysis after death. Skeletal muscle tissue, however, is relatively resistant to autolysis, and may be preserved for as long as 20,000 years under suitable conditions (Zimmerman and Tedford, 1976). If autolytic mechanisms were readily available inside a healthy muscle fiber, they might be too easily activated by strenuous muscle activity. The slowness of autolysis in healthy muscle tissue is an essential feature of meat which makes it an easily handled food product in temperate climates.

The delectable taste of pheasant meat calls for intact birds to be hung for about 10 days postmortem in a cool cellar at about 10°C. Provided that the viscera have not been perforated, there is virtually no bacterial invasion of the skeletal muscles (Mead, et al., 1974; Barnes et al., 1973), yet their tenderness is greatly increased. Thus an important point to bear in mind is that the enzyme systems of meat, including the autolytic enzymes, are mostly adapted for maximum efficiency at body temperature. When meat is stored at temperatures much below the original body temperature, the activity of degradative enzymes is relatively slow. In cold-blooded animals such as fish and marine invertebrates, however, storage temperatures just above 0°C may allow normal high rates of enzyme activity.

After an animal is exsanguinated, muscle fibers may survive for some time by anaerobic glycolysis. Sooner or later, however, they run out of energy. Either their primary store of carbohydrate, *glycogen*, is depleted, or else the end product of anaerobic glycolysis, *lactate*, deactivates biochemical systems by its acidity (*lactic acid*). When energy is no longer available, muscle fibers begin to lose their cellular integrity. However, some of the enzymes in biochemical pathways are still able to function for a prolonged time.

The events that should occur for the optimum conversion of muscles to meat are quite complex. The pH should be lowered by the formation of lactate from anaerobic glycolysis. Inadequate lactate formation may leave the meat dark, firm, and dry (DFD), while too much lactate, formed too quickly while muscles are still warm, may leave the meat pale, soft, and exudative (PSE; Briskey, 1964). Muscles should become firm as a result of their thick and thin filaments becoming cross-linked. This condition is known as *rigor mortis*. However, the formation of too many cross-links because of short sarcomeres may cause meat toughness. Beef that has just set in rigor mortis is tougher and has less taste than beef that has set in rigor mortis, and which has been kept cool but not frozen for a week or more. This extended period in the conversion of muscles to meat is called *aging, conditioning*, or *hanging*.

The chemical and microstructural changes that occur during the conversion of muscles to meat have mostly been investigated by the biochemical analysis of homogenized muscle and meat samples, and by random microscopic examination of muscle fibers within a sample. This approach to the subject has generated most of the widely accepted knowledge which forms the foundation for understanding the conversion of muscle to meat. Excellent accounts of this foundation knowledge may be found in the standard meat science text books (Lawrie, 1974; Price and Schweigert, 1971; Forrest et al., 1975) and in scientific reviews (Briskey et al., 1966, 1970; Bendall, 1973). Almost all commercial meat, however, is derived from muscles with muscle fibers which are specialized for different types of activity in the live animal. Fibers are basically either fast-contracting or slow-contracting, and they may obtain their energy aerobically, anaerobically, or by a combination of both metabolic adaptations. There is histochemical evidence that the conversion of muscles to meat occurs at different rates in different types of muscle fibers. This chapter emphasizes cellular heterogeneity in the conversion of muscles to meat. In Chapter 7 it was shown how muscle fiber volume was a parameter that could unite our knowledge of the histology of muscle growth with our knowledge of meat yields and productivity. In this chapter, fiber volumes and histochemical fiber types will provide a link between muscle growth and meat quality.

ANAEROBIC POTENTIAL IN LIVING MUSCLE

Skeletal muscles can use fatty acids, acetoacetate, and amino acids to obtain energy. In the conversion of muscles to meat, however, only the anaerobic pathway by which glycogen is converted to lactate can continue, since most of the muscle mass is anaerobic. Events on the meat surface, where oxygen is available to a limited depth, are a little different.

Controlled Release of Energy

The chemical process of *oxidation* was originally defined as the addition of oxygen in a reaction. The reverse process, *reduction*, was defined as the addition of hydrogen. These two processes later came to be defined in the same terms, oxidation as an increase in positive charge or a decrease in negative charge, and reduction as a decrease in positive charge or an increase in negative charge. In ionic terms, oxidation is equivalent to a loss of negatively charged electrons, while reduction is equivalent to a gain of electrons. Thus oxygen is not immediately needed within the muscle fiber for an oxidation reaction to take place. An appropriate example of oxidation and reduction in muscle metabolism is the reaction

$$NADH \longrightarrow NAD^+ \quad \text{(oxidation)}$$

$$NAD^+ \longrightarrow NADH \quad \text{(reduction)}$$

where NADH and NAD^+ are abbreviations for reduced and oxidized forms of the coenzyme nicotinamide-adenine dinucleotide. A coenzyme is a substance that activates or helps an enzyme, in this case by facilitating the transfer of reactive chemical groups.

In living muscle, the complete oxidation of carbohydrate to carbon dioxide and water requires oxygen, and it releases a lot of energy. Much of this energy is captured by adding a phosphate group to another molecule which already has two phosphate groups. Thus *adenosine diphosphate* (ADP) is converted to *adenosine triphosphate* (ATP). The ATP molecule carries this energy within the muscle fiber, and it can be released to another biochemical system by cleaving off the added phosphate (ATP → ADP + P). Muscle contraction is a primary user of ATP in the living animal, but substantial amounts of ATP are used by the membranes around and within the fiber for maintaining ionic concentration gradients. One molecule of a carbohydrate ($C_6H_{12}O_6$) may be combined with oxygen ($6O_2$) to add a phosphate group to each of 36 molecules of ADP. In addition to 36 molecules of ATP, this produces carbon dioxide ($6CO_2$) and water ($42H_2O$). However, if the muscle is anoxic, the most it can normally gain by oxidizing each carbohydrate molecule is two molecules of ATP.

Stored Energy in Muscle

Glycogen is the primary storage carbohydrate in muscle fibers. In electron micrographs, glycogen appears as single granules or clumps of granules which are located in the sarcoplasm between myofibrils and under the cell membrane. In longitudinal sections of skeletal muscle, glycogen and its associated enzymes appear to be concentrated at the I bands (Pette, 1975; Dolken et al., 1975). Glycogen is a polysaccharide formed by the linking together of large numbers of D-*glucose* units. The "D-" prefix simply indicates that the glucose is a member of a family of carbohydrates which are theoretically related to one of the mirror-image forms of a parent molecule. The structure of D-glucose and the numbering system for its carbon atoms are shown in Figure 9-1. Straight chains are formed by linkages between carbon atoms 1 and 4 (1-4 link-

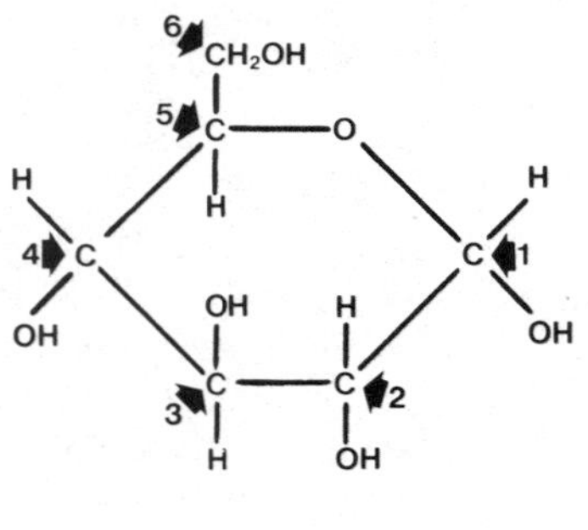

Figure 9-1 α-D-glucose with its carbon atoms numbered (arrows).

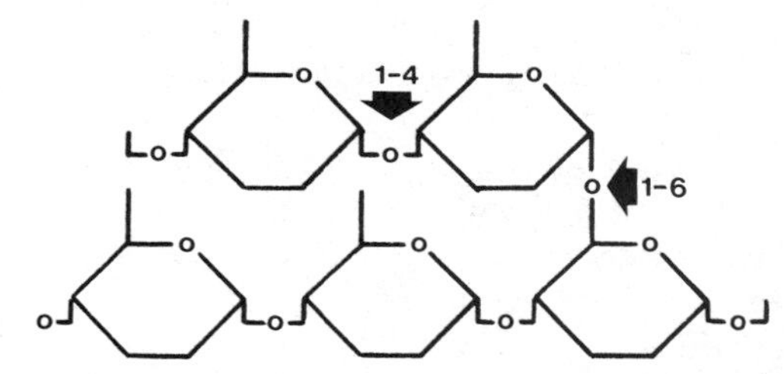

Figure 9-2 Formation of straight chains (1-4 linkage) and branch points (1-6 linkage) in glycogen.

age), while branch points are formed by linkages between carbon atoms 1 and 6 (1-6 linkage), as shown in Figure 9-2.

The concentration of glycogen in skeletal muscle rises steadily toward the end of gestation, and it may peak at about the time of birth. Pigs have particularly high muscle glycogen levels at birth (Shelley, 1961); the levels decline rapidly after birth to reach future adult levels within a week (Dalrymple et al., 1973). The accumulation of glycogen is a result of *glycogen synthetase* activity. In chick muscles, this enzyme becomes apparent once myoblasts have fused to form myotubes (de la Haba et al., 1968), and its activity is independent of insulin at this time (Grillo and Ozone, 1962). Different body organs may exhibit different patterns of change in glycogen levels; for example, the levels in cardiac muscle decline continuously both before and after birth (Shelly, 1961).

In humans there are a number of *glycogen storage diseases* which are caused by the genetic absence of glycogenolytic enzymes (Bethlem, 1970). In one of these diseases, *Pompe's disease*, glycogen accumulates in muscles due to the absence of a lysosomal enzyme, *acid maltase* (amylo-1,4-glucosidase). This enzyme can remove glucose residues from the external branches of glycogen (Banks et al., 1976). It is difficult to decide whether or not this enzyme has any role in the conversion of muscles to meat since lysosomes are difficult to locate in healthy muscle. Pompe's disease has been reported in Australian Shorthorn cattle (Richards et al., 1977) and possibly, in Corriedale sheep from New Zealand (Manktelow and Hartley, 1975).

Energy Release Without Oxygen

Glycogenolysis is the enzymatic degradation of glycogen, and it is the first step in the release of energy by the oxidation of glucose units (*glycolysis*). The enzyme which erodes straight chains at their nonreducing ends (ends exposed at carbon position 4) is usually called *phosphorylase*. As its name suggests, phosphorylase attaches a phosphate group to the carbon atom at position 1 as it removes a glucose unit from the straight chain. Thus the molecule released is α-D-glucose-l-phosphate, where the prefix "α" simply indicates one of two positions that the —H and —OH groups can adopt on the carbon atom in position 1 relative to the plane of the remainder of the molecule. Phosphorylase erodes straight chains until it comes to the fourth glucose unit preceding a branch point. The three glucose units before the fourth one which carries the branch are removed together, and are added to an adjacent free straight chain so that the 1-6 linkage thus exposed at the branch point can be severed. A second enzyme called *debranching enzyme* performs this task. Instead of being released as glucose-l-phosphate, the glucose unit released from a branch point remains as free glucose. Thus total glycogenolysis liberates glucose-l-phosphate and glucose in a ratio which indicates the ratio between the mean length of straight chains and the number of branch points.

Whereas the structure of glycogen and the mechanism of glycogenolysis are important in understanding the conversion of muscles to meat, the remainder of the pathway by which glycogen is anaerobically oxidized to lactate need only be covered in general principle. The most important question to be answered is why lactate is produced under anaerobic conditions, yet hardly at all under aerobic conditions. After a series of steps in the glycolytic pathway, molecules with six carbon atoms derived from the glucose units of glycogen are split to produce two molecules of *pyruvate*, each with three carbon atoms. If aerobic conditions prevail, the pyruvate which is formed in the cytosol of the muscle fiber now enters a *mitochondrion*. The cytosol is the aqueous phase of the sarcoplasm, which is located between membrane-bounded organelles such as mitochondria. After entering a mitochondrion, pyruvate is converted to a substance called *acetyl-CoA*. Acetyl-CoA becomes fused to *oxaloacetate* to form *citrate*. The citrate is then oxidized in a circular pathway, called the *Krebs cycle* or the *tricarboxylic acid cycle*. The cycle is completed by the regeneration of oxaloacetate. Continuous activity of the Krebs cycle is fueled by a range of carbohydrates, fatty acids, and amino acids, and it is the primary system for the aerobic generation of energy. Large numbers of molecules of ATP are produced from ADP by a series of reactions, *oxidative phosphorylation*, which occur in the mitochondrial membrane.

The reason for this brief excursion into aerobic metabolism is to explain why lactate is produced under anaerobic conditions. Under aerobic conditions, the production of two pyruvate molecules from a glucose-l-phosphate molecule results in the reduction of $2NAD^+$. Thus somewhere else in the muscle fiber, NADH must be reoxidized for glycolysis to continue. Aerobically, this occurs as a consequence of mitochondrial Krebs cycle activity, although NADH and NAD^+ do not actually cross the mitochondrial membrane. In anaerobic living muscles and in meat, the Krebs cycle is halted, and NADH is reoxidized in the cytosol by *lactate dehydrogenase* during the conversion of pyruvate to lactate. Pyruvate is of no immediate use anaerobically since mitochondrial oxidation has ceased, but its conversion to lactate ensures a continued supply of NAD^+ for the continuation of glycogenolysis and anaerobic glycolysis in the cytosol. However, since these events form only the initial stages of complete carbohydrate oxidation, they do not regenerate much ATP. The net gain of ATP is reduced to only two molecules of ATP per molecule of glucose-l-phosphate. Molecules of glucose which are released from glycogen branch points generate a total net gain of 3ATP.

Lactate dehydrogenase (LDH) is the enzyme that adds hydrogen atoms to pyruvate to produce lactate. LDH exists in a number of molecular forms (*isoenzymes*) which can be separated electrophoretically on the basis of their net electrical charge. If LDH-1 is prevalent, it facilitates aerobic metabolism, where possible, since it is inhibited by pyruvate and lactate. LDH-5 is not inhibited by high levels of lactate and pyruvate, and this isoenzyme facilitates anaerobic metabolism. LDH-1 is typical of cardiac muscle, while LDH-5 is typical of skeletal muscles, particularly those adapted for anaerobic conditions during contraction. In skeletal muscles, the ratio of LDH-1 to LDH-5 corresponds to the dominant activity pattern of a muscle. For example, muscles that are capable of sustained activity, and which can only be supported by aerobic metabolism, have more LDH-1 (Gutierrez et al., 1974). LDH-5 is the dominant isoenzyme in skeletal muscles from slaughter-weight pigs (te Brake, 1971).

Regulation of Glycolysis

The regulation of glycolysis in a muscle fiber of a live animal is integrated with the metabolic state of the fiber and its immediate energy needs. The metabolic state of the fiber is profoundly affected by hormones, particularly *epinephrine* (*adrenalin*), and by the extent of recent contractile activity of the fiber. Phosphorylase is particularly important in the conversion of muscle to meat since it is probably a primary control site for postmortem glycolysis (Kastenschmidt et al., 1968). Phosphorylase in muscle is most active when it is, itself, phosphorylated (in the *a form*). When dephosphorylated, it is less active (in the *b form*). In general terms, therefore, phosphorylase is switched on and off by the addition or removal of its phosphate, with on and off states being relative rather than absolute. The activity of phosphorylase b is dependent on the presence of AMP, but the activity of phosphorylase a is not. There are two conflicting requirements which make the mechanism for the activation of phosphorylase rather complex. First, since phosphorylase initiates the release of considerable amounts of chemical energy, there must be safeguards to prevent its uncontrolled activity. In stress-susceptible pigs, for example, the uncontrolled activity of anaerobic glycolysis may lead to excessive heat production and to a level of acidity that may soon prove fatal to the animal. The conflicting requirement is that the vast amounts of phosphorylase which are spread throughout the muscle mass must be rapidly activated by relatively small amounts of epinephrine. The epinephrine activation of severely frightened animals is often called the "fight or flight" response: neither of these responses is likely to be of much survival value if the anaerobic energy supply to body muscles is delayed.

The conflicting demands for fail-safe but rapid activation are satisfied by two particular features of the activation system. First, the conversion of phosphorylase b to phosphorylase a is inhibited locally in each muscle fiber by high concentrations of ATP and glucose-6-phosphate (an intermediate in the conversion of glycogen to lactate). Thus, if the energy released by phosphorylase activity is not rapidly consumed, the energy release system shuts down. If the energy is used, however, AMP and phosphate (from ATP $\rightarrow$ ADP + P and from ADP $\rightarrow$ AMP + P) facilitate the activation of phosphorylase. Second, to enable the rapid activation of phosphorylase throughout the musculature, the relatively small amounts of epinephrine that arrive at the muscle initiate a series of biochemical changes which function as an amplifier. A small input leads to a large output. Epinephrine causes *adenyl cyclase* to increase its formation, from ATP, of a compound called *cyclic AMP*. Cyclic AMP then activates another enzyme, *protein kinase*. With ATP and magnesium ions present, protein kinase then phosphorylates another enzyme, *phosphorylase b kinase b*. The active form, *phosphorylase b kinase a*, in the presence of magnesium ions, finally activates phosphorylase b to phosphorylase a (Gross and Mayer, 1974).

When animals require energy anaerobically during normal activity, phosphorylase b kinase is activated by calcium ions which have been released from the *sarcoplasmic reticulum* as the trigger for muscle contraction (Wanson and Drochmans, 1972). This activation system which links muscle contraction and glycogenolysis has a brief duration, so as to avoid the continuous use and depletion of glycogen reserves (Conlee et al., 1979). Glycogen is a rapidly available source of energy for either brief muscle activity or for the early stages of sustained muscle activity. The system to shut down phosphorylase activity when muscular activity ceases, or when an animal recovers from fright, involves *phosphatase* enzymes which progressively dephosphorylate phosphorylase a and phosphorylase b kinase a. Many features of the system for the activation of phosphorylase are shared with another system which activates glycogen synthesis. However, the shared features are integrated so as to be opposite in their effect. Thus the muscle fiber does not attempt to synthesize new glycogen at the same time that it is breaking it down, and vice versa.

Neonatal pigs often have difficulty maintaining their blood glucose levels, and they may die of *hypoglycemia* after even short periods of starvation (Dvorak, 1973). Given that they have high glycogen levels at birth (Shelley, 1961; Dalrymple et al., 1973), this is not easy to explain, although stored glycogen may simply be inadequate in amount if milk is not available. Despite an adult predisposition to the excessive accumulation of adipose tissue, newborn pigs have very little adipose tissue (<1 per cent), so that liberation of free fatty acids is severely limited. Dalrymple et al. (1973) found that most phosphorylase was in the relatively inactive b form just after birth. In adult pigs, *hyperglycemia* (high blood glucose level) readily occurs in response to exercise and epinephrine secretions (Muylle et al., 1968).

The Fate of Lactate in Living Animals

Under different agricultural systems, meat animals may exhibit a considerable range in their general muscular and cardiovascular fitness. At one extreme, sheep and cattle may roam in search of feed and water for part of the year, and may be expected to cope with adverse conditions much like wild animals. At the other extreme, animals may be reared in close confinement, partly to prevent them wasting feed energy by converting it to muscular work. Pigs which are infrequently exercised may have a low oxygen transport capacity (Steinhardt et al., 1974). Like horses, however, pigs can be put through a muscle training program which will reduce lactate production in response to a standard exercise test (Lindberg et al., 1973b).

In living animals, the lactate that is produced by contracting muscles is removed by the circulatory system. However, it must first travel out of the muscle fiber and into the interstitial space, before it can enter the bloodstream. The release of lactate by a muscle fiber may be retarded in situations such as exhaustive exercise (Wardle, 1978). A fraction of the lactate leaving a fiber may be in the form of undissociated lactic acid, and this fraction may increase with a rise in external pH (Mainwood and Worsley-Brown, 1975). In living animals, blood flow is increased in active muscles (*hyperemia*). When hyperemia occurs as a secondary response to hypoxia, it may be mediated by the release of adenosine (from AMP → A + P) into the interstitial fluid between muscle fibers (Phair and Sparks, 1979).

On arriving in the liver, lactate may be converted to *glucose-6-phosphate*, and then stored as glycogen or released as glucose back into the blood. Circulating glucose is available to muscles to be stored as glycogen or to be used directly as energy for contraction. This cycle of events is called the *Cori cycle*. After exhaustive exercise, an animal may continue to consume oxygen at an increased rate for some time (*oxygen debt*). During this recovery period, some lactate may be completely oxidized to release enough energy for the remaining lactate to be converted to glucose (*gluconeogenesis*) in the liver. When an animal is exsanguinated, the blood can no longer perform its transport function in the Cori cycle, and lactate is left to accumulate in the musculature. There are, however, a number of other possible fates for lactate in living animals.

The heart muscle receives its own blood supply from the *coronary artery*, straight off the aorta, and cardiac muscle fibers may use circulating lactate as an energy source. The high oxygen concentration in the aorta usually allows complete oxidation of lactate by cardiac muscle. Any lactate that remains in the aorta does not all get pumped to the liver or kidneys to participate in the Cori cycle, since the aorta supplies other major arteries in addition to the hepatic and renal arteries. Much of the circulating lactate may pass through noncontracting muscles. These also have the ability to oxidize lactate as an energy substrate, particularly if circulating fatty acids are unavailable (Dunn and Critz, 1975). Resting muscles may also take up circulating lactate and reconvert it directly to glycogen (Bendall and Taylor, 1970), probably via a pathway which is independent of the mitochondrial Krebs cycle (Connett, 1979). Finally, it is also possible for all the lactate consumed during an oxygen-debt period to be oxidized to CO_2 (Brooks et al., 1973).

Contraction causes an initial drop in the pH within muscle fibers, due to the hydrolysis of ATP. This is followed by an increase in pH due to the breakdown of *creatine phosphate* (CP) (MacDonald and Jobsis, 1976). CP acts as a short-term store of energy since it has a phosphate which can be transferred to ADP (CP + ADP → ATP + C) by the enzyme *creatine phosphokinase* (CPK). CP is now recognized as the dominant carrier of energy from mitochondria to myofibrils. CPK and creatine are normally contained within muscle fibers, but they may leak into the blood from damaged or diseased fibers. In healthy muscle, creatine is slowly but continuously converted to *creatinine*. Creatinine is lost in the urine in approximate proportion to the muscle mass of the body. The production of lactate may later cause another drop in pH. The interstitial pH between muscle fibers (Steinhagen et al., 1976) and the pH at the muscle surface (Smith et al., 1969) are profoundly affected by overall pH levels in the blood.

BIOLOGICAL VARIABLES IN THE CONVERSION OF MUSCLES TO MEAT

The breeds within a species may exhibit considerable variation in their muscle metabolism. For example, at a particular time and place, one breed of pig may provide PSE pork for experimental analysis, while another breed may be used to provide experimental control animals with normal meat. However, after a few years of remedial work by the producers of the first breed, the between breed range may be canceled. Similarly, at two widely separated geographical locations, variation within a breed may equal or exceed variation between breeds. Bendall (1979) found that the degree of muscle activity during slaughter had a major effect on metabolic patterns in the conversion of muscles to meat, even in the context of interspecific comparisons. In other words, the differences between quiescent and kicking carcasses may exceed those found between quiescent carcasses of different species.

Muscle Color

The postural muscles of a pork carcass may appear darker than the muscles used for rapid locomotion because the postural muscles contain more fibers with a high *myoglobin* concentration (Cassens and Cooper, 1971). Myoglobin is a soluble protein which is formed from a single polypeptide chain twisted around an oxygen-carrying *heme* group. The heme group is composed of an atom of iron and a *porphyrin* ring. The oxygen storage capacity of myoglobin in terrestrial animals is very limited (James, 1972), so that the primary function of myoglobin is probably to transport oxygen within the muscle fiber. The myoglobin concentration of skeletal muscles, unlike the hemoglobin content of the blood, is not increased by physiological adaptation to a low atmospheric pressure (Poel, 1949). Similarly, exercise training (Mitchell and Hamilton, 1933; Fitts et al., 1974) and level of nutrition (Lawrie, 1950) have little or no effect on the myoglobin concentration of skeletal muscles. Cattle fed on grass produce lean meat with a normal color for their age and type (Black et al., 1940). However, pigs that have been immobilized in a small pen may produce less myoglobin than normal (Lawrie, 1950), and milk-fed calves deprived of dietary iron produce pale veal. The muscles of pigs reared under constant temperature conditions contain more myoglobin than those of pigs reared in fluctuating temperature conditions (Thomas and Judge, 1970). Heritability estimates for pork muscle color may range from 0.17 to 0.55 (Pease and Smith, 1965).

Muscle fibers with a high myoglobin concentration may be almost completely aerobic, or they may have a dual aerobic-anaerobic capability. Conversely, within a carcass, pale muscles usually contain a greater proportion of anaerobic fibers with high levels of stored glycogen (unless depleted by struggling) and appropriate pathways for the rapid conversion of pyruvate to lactate. In living muscles, the *lactate threshold*, or work load at which blood lactate starts to accumulate, increases in proportion to the percentage of aerobic fibers (Ivy et al., 1980). Fibers with a slow speed of contraction are predominantly aerobic. Thus muscle color is often related to ATPase activity, so that a dark red color often indicates slow ATPase activity. The porcine longissimus dorsi is an exception to the rule (Bendall, 1975). In pork, color differences within muscles can be related to differences in muscle metabolism postmortem (Beecher et al., 1965b). Differences in myoglobin concentration between fiber types are less well developed in species with uniformly dark meat. Thus problems with variegated cuts of meat are conspicuous only in pork. At the same site within certain muscles, there may be differences in muscle color between breeds of pigs (Kastenschmidt et al., 1968).

In pork muscles, the fibers with the greatest aerobic activity are grouped toward the center of each fasciculus. Morita et al. (1969) showed that these central aerobic fibers contain most of the myoglobin that is present in pork. With histochemical methods, myoglobin does not become apparent in porcine muscle fibers until about 1 week after birth. Three weeks after birth, differences in myoglobin content between aerobic and anaerobic fibers can be detected (Morita et al., 1970a). Biochemical determinations of the myoglobin content in pork muscles show that it increases steadily until about 1 year and then the rate of accumulation slows down (Lawrie, 1950). Myoglobin can be located histochemically by a fluorescent antibody method (Kagen and Gurevich, 1967) or by taking advantage of its *peroxidase* activity to form insoluble pigments from benzidine or diaminobenzidine (Goldfischer, 1967; James, 1968). Benzidine is a carcinogen and is unsuitable for routine laboratory use. With the peroxidase method, glutaraldehyde is required to fix the myoglobin and to inhibit peroxidase activity from other sources. Beef muscles have a higher myoglobin content than pork muscles, and nearly all bovine muscle fibers contain at least some myoglobin (Morita et al., 1970b). In beef, the concentration of myoglobin in aerobic fibers with strong ATPase activity is equal to or almost as strong as that in aerobic fibers with weak ATPase activity (Swatland, 1979b).

In the absence of oxygen, myoglobin has a dark red color which is almost purple. With oxygen present, *oxymyoglobin* is bright red. The transformation to oxymyoglobin is seen a few minutes after cutting into the anaerobic center of a piece of meat. After prolonged exposure to the atmosphere, however, the iron atom of myoglobin may be converted to the ferric form, and brown *metmyoglobin* is formed. This occurs most rapidly just under the surface of the meat. In meat, the levels of myoglobin and *cytochromes* are closely related, both biochemically (Lawrie, 1953) and histochemically (Swatland, 1978b).

Cytochromes have an important indirect effect on meat color. The rate at which cytochromes utilize oxygen determines the depth to which oxygen is found in the meat. In the absence of oxygen, cytochromes may reduce metmyoglobin back to myoglobin. Thus the effect of histochemical fiber types on meat color may be rather more complicated than the simple relationship between myoglobin distribution and fiber types. For example, *peroxisomes* (or microbodies) are membrane-bounded organelles which exhibit *catalase* and *peroxidase* activity. Hand (1974) identified peroxisomes in skeletal muscle fibers and suggested that they might provide alternate pathways between pyruvate and lactate. Jenkins and Tengi (1981) found that red muscles had about twice the catalase activity of white muscles. The glutathione peroxidase activity of poultry muscles and its relationship to the oxidation of myoglobin are described by Lin and Hultin (1978). Postmortem changes in myoglobin have important effects on the packaging and sale of meat. These are discussed by Huffman (1980), Kropf (1980), and Ernst (1980).

Age and Sex of Animals

The sex of an animal may affect postmortem muscle metabolism, and differences between sexes may involve a differential response to preslaughter stress (Martin and Fredeen, 1974; Tarrant et al., 1979). Animal age affects a number of muscle properties which are important in the conversion of muscles to meat. In pigs, muscle contraction speeds become faster during postnatal development (Campion et al., 1973) and, immediately after exsanguination, heavy pigs often have lower levels of stored glycogen (Sink and Judge, 1971). In white muscles, a sharp drop in stored muscle glycogen occurs about 1 week after birth, together with increased phosphorylase activity (Dalrymple et al., 1973). Mean chain length of glycogen shows little change during growth to market weight (Kjolberg et al., 1963). Both Sink and Judge (1971) and Dalrymple et al. (1973) found breed differences in the glycogen levels of pork.

Animal Response to Preslaughter Stress

The *porcine stress syndrome* (PSS) described by Topel et al. (1968) is the best known example of a biological factor (hereditary susceptibility to stress) which interacts with an environmental factor (preslaughter stress) to alter meat quality. Early studies on the heritability of PSS were reviewed by Christian (1972). Although an exact mode of inheritance was not decided, an autosomal recessive gene with variable penetrance was suspected. In Landrace pigs of the Netherlands, inheritance of PSS is simple and results from an autosomal recessive gene with complete penetrance (Eikelenboom et al., 1977). In a Missouri herd, inheritance was found to be a complex dominant with a strongly modified single dominant gene or two dominant genes acting together to generate a range of phenotypes (Williams et al., 1977). In Poland China pigs, Britt et al. (1977) also found that inheritance of PSS involved more than one abnormal gene. These possible differences in apparent mode of inheritance serve as a caution that PSS might not be due to exactly the same physiological malfunction in all breeds of pigs in all countries. This makes the definition of the syndrome rather important, and here, too, the situation is far from simple. In a veterinary context, the term "stress" has a variety of meanings in nutrition and pathology (Fraser et al., 1975). Relative to PSS pigs in commercial agriculture, however, the stress which initiates events leading to transport death or poor-quality meat is simply the movement of pigs from farm to abattoir.

If a PSS pig is unable to cope with the stress of being moved to an abattoir: (1) it may become lethargic; (2) its body temperature may increase (*hyperthermia* or *hyperpyrexia*); (3) it may start to take rapid, shallow breaths with its mouth open (*dyspnea*); (4) its skin may appear very pale with a slight blue venous coloration (*cyanosis*); and (5) its body muscles may tremble, or become rigid or weak (Ludvigsen, 1954). If a PSS pig survives as far as an abattoir where it can be legally slaughtered for its meat, some or all of its carcass muscles may appear unusually wet and pale, and the muscle fasciculi may separate very readily (Ludvigsen, 1954). The paleness of PSE pork takes about 1.5 hours after slaughter to develop fully (Swatland, 1981b), so that it is often difficult to detect moderate PSE visually before carcasses are placed in the meat cooler (usually within 45 minutes of slaughter).

Transport death due to PSS and the formation of PSE pork after slaughter are both due to the uncontrolled and very rapid production of lactate from glycogen. Lactate production may cause an unusually low blood pH while animals are alive, and an unusually low pH in the meat soon after slaughter. The early postmortem rate of decline in pH is particularly important since early acidity strikes at proteins which are still at body temperature and which are more readily denatured. Paradoxically, however, some PSS pigs produce meat which is dark, firm, and dry (DFD) instead of being PSE. Pigs that produce DFD meat may have been stressed but they have survived so as to enter an abattoir with their muscles almost entirely depleted of glycogen. In this state, lactate production and the postmortem decline in pH will be minimal, and a condition opposite to PSE will result. In commercial crossbred pigs, the incidence of DFD is higher after overnight lairage, particularly if animals from the same farm have been kept together (Moss, 1980).

Dark-cutting beef is another result of preslaughter stress. Climatic stress or aggressive behavior between animals may cause cattle to deplete their muscle glycogen. This greatly limits the amount of lactate that can be formed postmortem, and the ultimate pH of their meat stays high (typically above pH 5.9). Climatic stress often creates a seasonal rhythm in the incidence of dark-cutting beef, and aggressive behavior is more common between males. Elevated pH values are most severe in longissimus dorsi, semitendinosus, semimembranosus, adductor, and gluteus medius. In other muscles, pH values may be near to normal (Tarrant and Sherington, 1980). Thus the overall changes produced in beef carcasses by preslaughter stress are quite variable. Some muscles may increase their juiciness, whereas others become dry (Lewis et al., 1962a). This pattern of pH change between muscles differs from the generalized elevation of pH throughout the musculature which can be created by the injection of epinephrine. Glycogen depletion due to aggressive behavior between young bulls is most rapid in fast-contracting muscle fibers with strong ATPase activity. Depletion in slow-contracting fibers with weak ATPase activity is far slower but it lasts for a longer time after the period of stress. Glycogen levels do not return to normal until approximately 3 days after a stress period of 5 hours (Lacourt and Tarrant, 1980).

DFD meat is commonly found in the carcasses of stressed or exhausted pigs. Lewis et al. (1961, 1962b, 1967) investigated the effects of stress prior to slaughter by subjecting pigs to periodic electric shocks (18 per hour) for 5 or more hours prior to slaughter. This treatment caused glycogen depletion in the live animals, so that meat with a high pH was produced postmortem. This treatment caused favorable changes in a number of the characteristics of cooked pork such as (1) aroma; (2) flavor; (3) texture; (4) tenderness; and (5) juiciness. Thus, although DFD is a problem in beef, some taste panels may consider it to be a desirable condition in pork. The duration of the experimental treatment is important, since shorter treatments prior to slaughter might have the opposite effect, leading to PSE instead of DFD. With PSS pigs, shock treatment might easily cause premature death. Muscle glycogen levels are depleted in pigs that have been taken from warm conditions and immersed in very cold water (0.5° C) prior to slaughter (Sayre et al., 1961). When pigs from different producers are mixed while awaiting slaughter, aggressive activity lasts about 30 minutes; then the pigs begin to rest (Moss, 1978).

Carcass pH

Although techniques are available for the continuous monitoring of carcass pH (Jeacocke, 1977a), commercial data on meat pH are based on abattoir surveys of pH at one or two times postmortem. One measurement is usually taken soon after slaughter (within 45 minutes) to find the initial decline in the hot carcass, while a later measurement (usually 24 hours) indicates the ultimate pH of the carcass. In the measurement of meat pH, continuous attention is needed with glass pH electrodes since their sensitivity may change after a number of samples have been measured. When glass electrodes are pushed into meat, it may be difficult to obtain an accurate temperature correction. In the laboratory, maceration of meat samples in several times

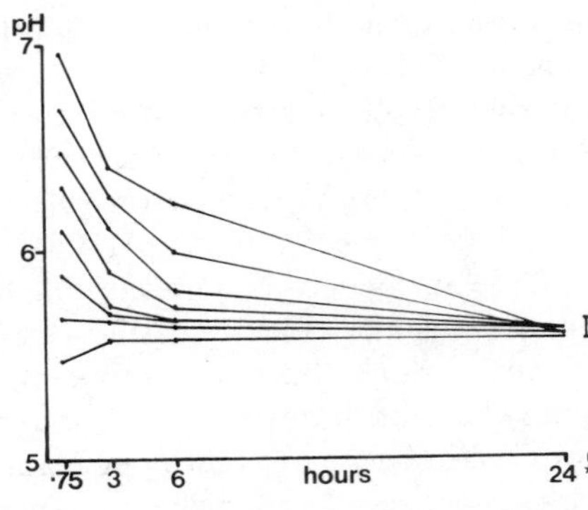

Figure 9-3 Postmortem pH decline in pork. Carcasses are grouped by their initial (45-minute) pH. (*) Range at 5 days postmortem. (Data from Scheper, 1971.)

their own weight of 5 mM sodium iodoacetate (to arrest further glycolysis) and 150 mM potassium chloride (to avoid changing the pK values of muscle buffers) is preferred (Bendall, 1973). The pK of a weak acid (HA) gives the pH of the point of half ionic dissociation, $HA \rightleftharpoons H^+ + A^-$. Other problems encountered in surveys of pork pH are variations between muscles and between seasons (Gallwey and Tarrant, 1979).

The data reported by Scheper (1971) provide a useful guide to the patterns of postmortem pH decline in pork carcasses. These are shown in Figures 9-3 and 9-4. Figure 9-3 shows how the pH decline in carcasses with different pH values at 45 minutes is likely to proceed. Figure 9-4 shows the different patterns of early postmortem change which can lead to different pH values at 24 hours. Scheper (1971) considered that pH at 45 minutes was a good predictor of PSE, while pH at 24 hours was better for predicting DFD. Sometimes (for example, the lowest line in Figure 9-3) pH is seen to increase postmortem. Bendall and Wismer-Pedersen (1962) explained similar occurrences as a result of cooling (from 37° C to 10° C) on the pK values of ionizable groups of proteins and buffering substances. Carcass traits, such as muscling and leaness, have been investigated in relation to meat quality. It has been suggested that meat yield is inversely related to meat quality. Weak relationships between fatness and early postmortem pH sometimes show up (Wood et al., 1979) and sometimes they do not (Eikelenboom et al., 1974).

Transmission values or turbidity determinations of acidified muscle extracts give another measure of pork quality. Proteins that are affectd by a low pH while they are still in a warm carcass precipitate more readily and leave a clearer solution when tested.

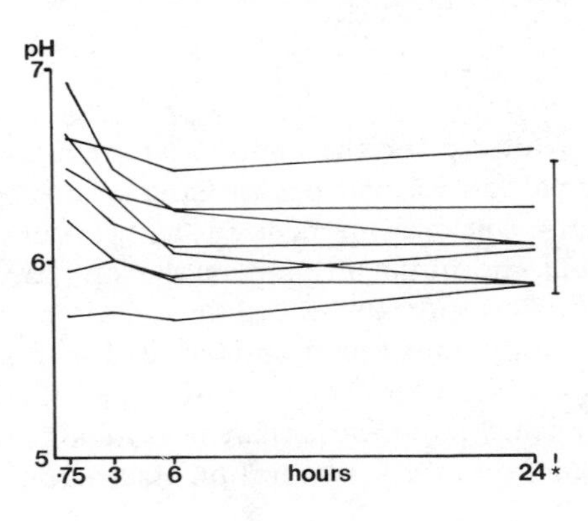

Figure 9-4 Postmortem pH decline in pork. Carcasses are grouped by their pH at 24 hours. (*) Range at 5 days postmortem.

Transmission values have been related to subjective scores of carcass muscling (Eikelenboom et al., 1974), and can be predicted from meat reflectance at 633 and 627 nm (Davis et al., 1978a) or from the scatter coefficient of a laser beam in intact pork chops (Birth et al., 1978).

In beef, pork, and lamb, our interest in carcass pH is focused on the appearance of the lean meat and its water-binding capacity. In poultry the focal point is different. Every treatment that results in an acceleration of the depletion of ATP or that leads to a faster decline in pH tends to make poultry meat tough (deFremery and Pool, 1960).

Evolution of Stress Problems

The origin of the one or more genes that cause PSS is unknown. They may have originated from relatively recent mutations, or they may be an ancient part of the genome which has been revealed by modern selection for lean muscle growth. In the nineteenth and early twentieth centuries, pork at the low end of the pH spectrum might have been viewed quite differently than at present. Low pH pork has a greater resistance to bacterial spoilage (Rey et al., 1976), and it allows faster penetration of curing ingredients in the traditional brine tank or dry salt methods of curing (Banfield, 1935). With limited refrigeration, if any, and the risk of botulism developing at a high pH in cured meat, PSE pork might not have been reported as a problem. In Great Britain in the 1930s, DFD pork caused by increased preslaughter transport was soon reported once it was identified as a problem (Callow, 1936). In the past, joints of meat were probably larger and fatter than they are now. The drip and evaporation losses from pork which bother the modern meat industry might have been much less. Old-fashioned sausage formulations often contain bread crumbs and rusk, which can absorb meat exudate.

Apart from muscle fiber hyperplasia in cattle, most heritable aspects of muscle development appear to be quantitative polygenic traits. It is possible, however, that in breeds such as the Pietrain, exceptional muscle enlargement and stress susceptibility might be due to a single *pleiotropic gene* (Ollivier, 1980b). Exceptional muscle enlargement in Pietrains is most conspicuous in the hind limbs, and Dumont and Schmitt (1970) have shown that this is associated with increased muscle fiber diameters in the semimembranosus. In the triceps brachii of the forelimb, fiber diameters in Pietrains are the same as those of breeds with a normal conformation, such as the Large White.

A definitive proof of the popular belief that fast live weight growth and muscularity cause poor-quality meat is difficult to find. Murray (1970) was unable to find differences in cooked meat quality in steers with different nutritionally regulated growth rates (0.8 kg/day versus 0.4 kg/day). In a comprehensive review of British breed improvement and pork quality over a 10-year period, Kempster (1979) found no evidence of a deterioration in pork quality. In France, however, Ollivier's (1977, 1980a) study of the responses of Large Whites to 10 years of genetic improvement has revealed some evidence of deleterious changes in meat color.

The mechanism by which selection for increased lean growth may cause a deterioration of meat quality (hypothesis 1) may someday be known. At present, however, it is difficult to reject an alternative hypothesis (hypothesis 2) that certain breeds tend to produce poor-quality pork, and that pigs of these breeds have become more common in some localities. These two hypotheses differ in a rather subtle way. The first suggests that selection for rapid growth and muscularity will place all breeds at risk and, if there has been a decrease in pork quality, that it is probably due to a change involving nearly the whole population of pigs. The second hypothesis suggests that certain segments of the whole population, namely breeds that tend to produce poor quality pork, have increased in proportion so as to increase the overall incidence of poor-quality pork. These two hypotheses are difficult to separate since the incidence of PSE due to PSS is superimposed on the incidence of PSE due to industrial factors. Another aspect of this problem is that, although PSE pork is undesirable, low-pH pork which is not PSE may be preferred in some situations. For example, bacon made from low-pH pork may be more attractive when evaluated visually (Warriss and Akers, 1980). A final complication is that pigs are now slaughtered at a much younger age than they used to be. PSE is visually less conspicuous in the muscles of older pigs which have a high myoglobin concentration in their muscles.

Water Content of Meat

An extensive monograph on the colloidal chemistry of meat by Hamm (1972) provides access to the extensive literature on this topic. If dried meat protein is rehydrated by exposure to increasingly damp air, three water compartments can be detected by the way in which water is taken up; 4 per cent of the water becomes firmly bound as a monolayer around muscle proteins, another 4 per cent is taken up as a looser second layer, and 10% of the water accumulates loosely between protein molecules.

Water-binding capacity can be defined as the ability of meat to bind its own water or, under the influence of external forces such as pressure and heat, to bind added water. *Water absorption* or gelling capacity can be defined as the ability of meat to absorb water spontaneously from a suitable aqueous environment. Water-binding capacity and water absorption are closely related. Although water absorption can be measured by increases in the weight and volume of meat samples when placed in various aqueous solutions, and although water-binding capacity can be determined by centrifugation, Hamm (1972) preferred the method of pressing 0.3-g meat samples on a filter paper at 35 kg/sq cm between two plates. On removal after 5 minutes, the areas covered by the flattened meat sample and the stain from the meat sample are marked and measured. After subtracting the meat-covered area from the total stained area to obtain the wetted area, the water content may be calculated as

$$\text{mg } H_2O = \frac{\text{wetted area (sq cm)}}{0.0948} - 8.0$$

Water-binding capacity is modified by pH and drops from a high around pH 10 to a low at the *isoelectric point* of meat proteins between pH 5.0 and 5.1. At its isoelectric point a protein bears no net charge and its solubility is minimal. Below pH 5, a value attained only if the pH of a processed meat product is deliberately lowered, water-binding capacity starts to increase again. Water absorption follows water-binding capacity in this regard. Thus as the pH of pork declines postmortem, its water-binding capacity decreases, and much of the water associated with the living muscle proteins before slaughter is free to leave the muscle fiber. At their isoelectric point, thick and thin filaments in myofibrils move closer together and so reduce the water space between them. By this time, muscle fibers are depleted of their ATP, and their membranes no longer confine the cell water (Heffron and Hegarty, 1974).

Water escapes from the spaces between muscle fiber bundles when they are cut, and the drip loss from PSE pork is increased to about 1.70 per cent from a normal value of about 0.77 per cent of trimmed carcass weight (Smith and Lesser, 1979). In sliced pork, drip losses increase with storage, from 9 per cent after 1 day to 12.3 per cent after 6 days (Lundstrom et al., 1977). This creates a serious weight loss from the carcass. In the United States in 1972, Hall (1972) combined the estimated incidence of PSE carcasses (18 per cent) and their typical extra shrink loss (5 to 6 per cent) to estimate total national losses as 94 to 95 million dollars. Obviously, this is only a rough guess, but it serves to emphasize the commercial importance of pH in pork. In the United States, Kauffman et al. (1978) estimated that the excess weight loss from PSE pork during transport, storage, and processing could exceed 1 million kilograms per annum. Exudate-filled spaces between muscle fiber bundles probably account for the soft texture and easily separated fiber bundles in PSE pork. The amount of water that is bound within muscle fibers may have an effect on meat tenderness (Currie and Wolfe, 1980).

As well as affecting water distribution, pH also has a direct effect on many muscle proteins. Under the microscope, dark-stained deposits are seen in muscle fibers from PSE pork (Bendall, 1973). These deposits are due to the precipitation of sarcoplasmic proteins (Bendall and Wismer-Pedersen, 1962) such as phosphorylase (Fischer et al., 1979). Myofibrillar proteins are also affected: (1) ATPase activity may be reduced (Sung et al., 1976); (2) interactions between actin and myosin may be changed (Park et al., 1975); and (3) myofibrils are less easily fragmented because Z-line disintegration is reduced (Kang et al., 1978).

Myoglobin is also changed in PSE pork, probably as a result of a rapid postmortem decline in pH (Bembers and Satterlee, 1975). The pH range at which myoglobin deterioration occurs is increased from 4.3 (normal pork) up to 5.1, although this point is seldom reached even in PSE pork (Figure 9-3). The conversion of pink myoglobin to pale brown metmyoglobin is accelerated in PSE pork, and this may contribute to the paleness of the meat. The paleness of PSE pork is due primarily to the increased scattering of light within the meat.

Histochemical Fiber Types

The involvement of histochemical fiber types (Chapter 6) in PSS pigs and in PSE pork has attracted considerable attention. A working hypothesis that appears to have been under examination in most studies is that muscle fiber histochemistry is changed in PSS pigs or in pigs that produce PSE pork. Increases in the relative numbers or diameters of anaerobic fibers may, perhaps, be related to PSE and PSS. Histochemical and analytical methods for typing muscle fibers, however, are constantly being improved. Thus one of the problems in evaluating early research on the topic is that the accuracy of fiber typing may be considered to be rather poor by present standards. This is a natural consequence of scientific progress, which carries no denigration of the efforts of early investigators.

A reasonable first approach to the topic was to liken the contrast between normal and PSE pork to that between dark and pale muscle areas within an individual carcass. Beecher et al. (1965a) found that pale areas of the semitendinosus muscle had a faster rate of postmortem glycolysis than did dark areas. However, early postmortem pH values for dark and pale muscles were quite similar, and were not meaningfully related to the frequency of red fibers identified by staining with sudan black (Beecher et al., 1965b). Sair et al. (1970) improved their fiber-typing technique by adding succinate dehydrogenase (SDH) and cytochrome oxidase reactions as markers for aerobic fibers: they compared longissimus dorsi muscles in PSS and non-PSS pigs. Contrary to the working hypothesis outlined earlier, both types of pigs had similar myoglobin levels and PSS pigs had higher SDH activity. The unexpected finding of higher SDH activity in PSS pigs was explained once the red fiber category was subdivided by incorporating serial reactions for ATPase (Cooper et al., 1969) and phosphorylase (Sair et al., 1972). Thus in the earlier research, some of the intermediate fibers with strong ATPase and phosphorylase reactions had accidentally been included in the red fiber category. These researchers modified their working hypothesis and attributed PSS and PSE to increased numbers of intermediate fibers (fibers with strong ATPase activity and a dual aerobic-anaerobic type of metabolism). However, increases in the percentage of red fibers in well muscled pigs (both PSS and non-PSS) cannot always be attributed to increased numbers of intermediate fibers. Andersen et al. (1975) typed red fibers using alkali-stable ATPase as a marker (this would exclude intermediate fibers), and they found increased numbers of red fibers in muscular pigs.

Dildey et al. (1970) typed fibers with sudan black, and they included intermediate fibers with moderate numbers of stained droplets in the red fiber category together with fibers with large numbers of stained droplets. Both PSE carcasses and well-muscled normal carcasses had a high ratio of white fibers with few stained droplets. Research by Merkel (1971) and Nelson et al. (1974) has provided further evidence of increased numbers of white fibers in PSE pork. Thus the experimental evidence is rather confusing. Sometimes PSE pork has more white fibers, sometimes it has more intermediate fibers, and sometimes it has more red fibers. In a heterogeneous group of PSS pigs, animals with large numbers of either white or intermediate fibers may be found (Swatland and Cassens, 1973). It is rather difficult to evaluate these conflicting studies on muscle fiber histochemistry in PSS and PSE pigs. Present standards for histochemical fiber typing require more extensive reference to serial reactions and stricter regulation of the subjective categorization of each fiber type. Since fiber-type ratios are now known to change during development, closer attention to the matching of animal groups for weight and age is now required. In evaluating any new research, a critical approach to the magnitude of the proportional increase in anaerobic muscle mass is required since, in many of the early studies, the extra anaerobic muscle mass was not large enough to explain the marked differences in pH decline between PSE and normal pork (Swatland, 1974). At present, there is now even some support for a null hypothesis since Gallant (1980) found no differences in fiber-type ratios between PSS and normal pigs.

Another working hypothesis which has generated considerable research activity is that large-diameter fibers are forced to rely on anaerobic metabolism because of a long diffusion pathway for oxygen from the muscle fiber surface to the fiber axis. The implication here is that pigs which are well muscled because of their large-diameter fibers are more likely to become anaerobic. In pigs, evidence both for (Merkel, 1971) and against (Sair et al., 1972) this working hypothesis has been gathered.

Endocrinology

Endocrine changes have been suspected since the time that PSS and PSE were first reported. Ludvigsen (1954) considered that PSE was caused by *hypothyroidism* (decreased thyroid gland function) since the incidence of PSS and PSE was worse in summer months when thyroid glands would have been least active. It was also thought that the condition could be created by experimentally blocking thyroid activity and that it could be alleviated by feeding iodinated casein. Pastea et al. (1972) were able to create muscle degeneration associated with thyroidal hyperplasia and adrenal atrophy by administering methylthiouracyl.

It is difficult to follow all the speculative working hypotheses that have been proposed since 1954 to link breed improvement to PSS and PSE. Thyroid hormones normally stimulate body heat production by increasing aerobic metabolism. Many researchers considered that PSS and PSE were related to an increase in anaerobic metabolism at the expense of aerobic metabolism since PSS pigs are susceptible to even mild heat stress. Thyroid hormones increase the uptake of triglyceride by aerobic fibers (Kaciuba-Uscilko et al., 1980) and, over a longer period of time, they may increase the activity and number of mitochondria (Wooten and Cascarano, 1980).

A simplistic analogy for thyroid activity may explain why it took so long to test and, subsequently, to reverse the original working hypothesis of hypothyroidism. Consider the case of a stream filling a mill pond which is used to drive a water wheel. If adequate functioning of the system is judged by revolutions of the water wheel, inadequate functioning of the system may occur in a number of ways: (1) the stream may dry up; (2) the water may arrive in the mill pond but may fail to turn the water wheel (if the water freezes); or (3) the water wheel may get jammed by a log and be unable to turn. In cases (2) and (3), the water level in the mill pond may rise until it overflows from the pond. By analogy, one could still claim support for hypothyroidism even if serum protein-bound iodine (PBI, a measure of thyroid hormones circulating in the bloodstream) was higher in PSS pigs. Thus when Judge et al. (1966) reported elevated serum PBI, decreased iodine uptake by the thyroid, and deficiencies in adrenocortical hormone production in pigs prone to developing PSE meat, it showed that thyroid physiology was upset, but not in which direction. Following similar results in an experiment on the effects of heat stress on PSS pigs, it was thought that circulating thyroid hormone might have failed to stimulate oxidative metabolism in skeletal muscles (Judge et al., 1966). Although there was an early report of increased levels of circulatory thyroid hormones in PSS pigs (Eikelenboom and Wiess, 1972), it was not until Marple et al. (1975) showed that pigs without thyroids had a retarded rate of postmortem glycolysis, whereas *hyperthyroid* (thyroxine supplemented) pigs had a faster rate of pH decline in their carcass muscles, that the working hypothesis of hypothyroidism was rejected. The evidence now favors hyperthyroidism (Eighmy et al., 1978), but the subject is still open to debate.

Adrenal Cortex

Changes in *adrenal* physiology have also been considered since Ludvigsen's (1954) report in which a deficiency in the *adrenal cortex* was proposed. Each adrenal gland is composed of an inner part, the *adrenal medulla*, surrounded by the adrenal cortex. The subdivisions of the adrenal gland were shown in Table 1-3. Rantsios (1972) found that the *zona glomerulosa* (*multiformis*) of Pietrain pigs with a low early postmortem muscle pH was decreased in depth. Passbach et al. (1970) found that intravenous injections of *aldosterone* increased pork paleness, and that an appropriate blocking agent improved muscle color. Quite how these two reports should be reconciled is not immediately obvious.

In the *zona reticularis of PSS* pigs, there may be lipid accumulations which suggest degeneration (Cassens et al., 1965), but ultrastructural and histochemical features may suggest the opposite condition—increased function (Ball et al., 1973). In 1966, Judge et al. concluded that there was still some support for a degree of adrenal insufficiency in PSS pigs. Marple et al. (1969) induced adrenal atrophy by injecting *prednisolone*, an exogenous glucocorticoid, up to 48 hours before slaughter; pigs with adrenal insufficiency exhibited signs of stress susceptibility, but their final meat quality was not affected. Next came two findings: (1) that PSS might or might not be associated with higher levels of plasma adrenal corticoids, depending on the degree of stress; and (2) that ACTH regulation appeared to be adequate in PSS pigs since they had higher ACTH levels that those of controls (Marple et al., 1972). Sebranek et al. (1973) then found that the adrenal response to ACTH in PSS pigs was impaired. Marple and Cassens (1973b) examined the metabolic clearance and turnover rates of cortisol and found both rates to be higher in PSS pigs. Thus by 1974, further evidence of increased plasma cortisol in PSS pigs (Marple et al., 1974) and the accumulation of other new information cast doubt on adrenal hypofunction as the cause of PSS and PSE (Ball et al., 1974; Lister, 1976). The reversal of the original working hypothesis (from adrenal insufficiency to hyperactivity) was used by Marple and Cassens (1973a) to explain the increased propensity for glycogenolysis in PSS pigs: higher cortisol levels were thought to stimulate gluconeogenesis so that muscles carried increased amounts of glycogen.

Pituitary

Following the early and apparently correct report of hyperthyroidism in PSS pigs (Eikelenboom and Wiess, 1972), Kraeling and Gerrits (1973) examined the regulation of thyroid activity by the *anterior pituitary* in PSS pigs (the structure and function of the pituitary were briefly summarized in Table 1-2). *Hypophysectomy* (removal of the pituitary) was found to slow down postmortem glycolysis. Further studies led to a working hypothesis that PSE pork might be derived from animals that have hypersecretion of one or more pituitary hormones (Kraeling et al., 1975). The induction of PSS and PSE by prolonged injection of porcine pituitary extract supports this hypothesis, as well as being in agreement with studies on the thyroid and adrenals

(Kraeling and Rampacek, 1977). Lean pigs tend to have increased plasma levels of growth hormone (Wangsness et al., 1977). Injection of porcine growth hormone may cause deterioration in pH-dependent aspects of pork quality (Wassmuth and Reuter, 1973).

Central Nervous System

The gray matter of the central nervous system is formed by the cell bodies of neurons. However, groups of neurons are also found in the central white matter of the brain where they are called *nuclei* (not to be confused with the term nuclei used in a cytoplasmic context). In lower vertebrates with a poorly developed cerebrum, the *caudate nucleus* is one of the nuclei that help to regulate muscular coordination during locomotion. In farm animals, the caudate nucleus still contributes to this activity, although it is subservient to a well-developed cerebrum. *Dopamine* is a neurotransmitter which is involved in the transmission of nerve impulses between neurons. Dopamine levels are very high in the caudate nucleus, and dopamine is thought to be involved in the indirect regulation of muscle activity. Altrogge et al. (1980) found that dopamine levels are lower than normal in the caudate nuclei of stress-susceptible pigs, and they have suggested that the normal inhibitory role of dopamine is reduced so as to cause excessive muscle stimulation during stress.

Adrenal Medulla

Arguments supporting the involvement of *catecholamines* (*epinephrine* and *norepinephrine*) in PSS and PSE were assembled by Topel (1971). Epinephrine and some norepinephrine are produced by the adrenal medulla. Norepinephrine is also released by *adrenergic* (postganglionic) fibers of the *sympathetic* division of the *autonomic nervous system*. The activity of postganglionic sympathetic fibers causes increased cardiac output and constriction of many blood vessels, but dilation of the vessels that supply muscles. Postganglionic fibers release norepinephrine, and they may also stimulate the adrenal medulla to release epinephrine. Plasma epinephrine levels in PSS pigs rise quickly in response to stress, but later they may drop to lower levels than in non-PSS pigs (Topel, 1971).

Although epinephrine and norepinephrine tend to act together to achieve similar end results in some body systems, differential responses may occur because of the distribution of *alpha* and *beta receptors*. Stimulation of alpha receptors is associated with an increase in membrane excitability and sodium ion permeability of vascular smooth muscle cells which then cause *vasoconstriction* (decreased blood vessel diameter). Stimulation of beta receptors is associated with decreased membrane excitability and enhanced activity of the sodium pump; this causes relaxation of vascular smooth muscle cells and *vasodilation* (increased blood vessel diameter).

Arterial and venous systems are dominated by alpha receptors, except in skeletal and cardiac muscle, where both alpha and beta receptors occur. Norepinephrine stimulates alpha receptors and causes vasoconstriction, whereas epinephrine stimulates both alpha and beta receptors, so that either vasoconstriction or vasodilation may occur. Thus the two key points which determine the vascular response of a particular organ to catecholamines are (1) the relative circulating levels of epinephrine and norepinephrine, and (2) the ratio of alpha to beta receptors in the organ.

In skeletal muscle (Youmans, 1967), the effect of very high levels of norepinephrine or epinephrine on alpha receptors (vasoconstriction) overrides their effect on beta-receptors (vasodilation). Topel (1971), however, supported an application of the "beta theory" of shock to the porcine stress syndrome. The dominant feature that caused anoxia, excessive anaerobic glycolysis and death, was thought to be excessive beta-receptor stimulation. Vasodilation due to beta-receptor stimulation was thought to lead to the pooling of blood in venules and to a partial bypass of the alveolary capillaries of the lungs. In research by Althen et al. (1977), circulating plasma catecholamines were not found to be an important factor in initiating the rapid glycogenolysis associated with PSE. The injection of extra epinephrine, however, does cause PSE (Althen et al., 1979). Norepinephrine and thyroxine may act synergistically to increase metabolic rates (Williams et al., 1977).

Alpha-receptor blocking agents (*phenoxybenzamine*) and beta-receptor blocking agents (*propranolol*) have been used to study the physiology of alpha receptors and beta receptors in PSS pigs. *Reserpine* reduces the amount of catecholamines available for release. Weiss et al. (1974) found that alpha blockade in PSS pigs caused skin blotching, tremors, gasping, and collapse. After stress, treated PSS pigs had high plasma glucose and lactate levels. Pigs with beta blockade had lower plasma lactate levels. Neither treatment, however, caused any change in meat quality. Lister et al. (1976) found that neither reserpine nor alpha blockade protected PSS pigs against the development of stress syndrome triggered chemically by suxamethonium. Althen et al. (1979) found that desensitization of beta receptors reduced the severity of epinephrine-induced PSE. *Carazolol*, a more powerful beta-blocker than propranolol, has now been found to reduce the incidence of PSE in PSS pigs (Warriss and Lister, 1982).

The involvement of the sympathetic nervous system in PSS is currently regarded as a secondary result of stress rather than a primary cause of PSS (Gronert et al., 1978; Gronert et al., 1980). Although the adrenal medulla, epinephrine, and beta receptors are known to be involved in experimentally induced PSS and PSE, the extent to which they are responsible for PSS and PSE in commercial agriculture is not clear.

Myogenic Factors

There is some evidence of a myogenic or intrinsic muscle factor in the porcine stress syndrome (Campion and Topel, 1975). The evidence includes reports of histological abnormalities in muscle, changes in contractile properties, and changes in the biochemistry of mitochondria and sarcoplasmic reticulum. With any of these factors, it is rather difficult to show that differences are a cause rather than a result of stress or low pH postmortem.

Mitochondrial abnormalities may occur in PSE meat (Bergmann and Wesemeier, 1970), but ultrastructural changes in mitochondria also occur in rigor mortis (Hegarty et al., 1978). Thus it is difficult to dissociate these changes from the earlier rigor mortis that occurs with PSE. The hypothesis that PSS is due to the uncoupling of oxidative phosphorylation now appears unlikely (Cheah and Cheah, 1976; Williams et al., 1977). However, mitochondria also have an essential role in regulating calcium ion concentration in the cytosol (Bygrave, 1978). An abnormal release of mitochondrial calcium which facilitates muscle contraction and anaerobic glycolysis has been proposed as the cause of PSS and PSE (Cheah and Cheah, 1976).

Changes in the contractile properties of muscle from PSS pigs have been interpreted as evidence that the sarcoplasmic reticulum is also implicated (Campion et al., 1974). Greaser et al. (1969) found that the calcium-accumulating ability of the sarcoplasmic reticulum was impaired in samples from PSS pigs. Gronert et al. (1979), however, were unable to find any experimental evidence in support of this hypothesis. Other postsynaptic abnormalities have been proposed, such as a defect in the transverse tubular system (Okumura et al., 1980).

Screening for PSS

A number of methods for the identification of PSS in potential breeding stock have been developed with the hope of eradicating the condition genetically. Identification methods range from those suitable only in experimental work, because of their complexity, to rapid and relatively inexpensive tests intended for commercial use (Sybesma and Eikelenboom, 1978).

The propensity for anaerobic glycolysis exhibited by PSS pigs may result in high levels of glycolytic intermediates (products of glycogenolysis) in muscle. Schmidt et al. (1971, 1972b, 1974) investigated the value of glycolytic intermediates in biopsy samples of muscle for the identification of PSS pigs. Glucose-6-phosphate (G-6-P) was selected as a likely candidate since it had the strongest correlation with meat quality. Another biopsy test is used in human medicine to identify patients with *malignant hyperthermia* or *hyperpyrexia* (Kalow, 1978). After careful removal of a muscle strip, it is exposed to *caffeine* and the development of muscle tension is measured (Kalow et al., 1978). Caffeine injections also trigger the stress syndrome in pigs (Watson et al., 1980), and caffeine methods may have some use in PSS research. People with malignant hyperthermia and pigs with PSS both react to *halothane* anesthetic with uncontrolled and often fatal heat-generating muscle glycolysis. This may have tragic consequences in human medicine but in agriculture it offers a method of screening for PSS.

When inhaled by PSS pigs, halothane causes an increase in body temperature, muscle rigidity, excess lactate production, and disturbances in water and electrolyte metabolism (Berman and Kench, 1971). Both red and white muscles are affected (McLoughlin and Mothershill, 1976). When used for detecting PSS, halothane at concentrations of less than 5 per cent is administered for up to 5 minutes. When a PSS pig is tested, it becomes rigid and may have muscle spasms. If this happens, the test is promptly stopped and severely affected animals may need to be cooled. A number of field trials have been carried out with satisfactory results (Eikelenboom and Minkema, 1974; Webb and Jordan, 1978; Eikelenboom et al., 1978; McGloghlin et al., 1980), particularly in populations with a high incidence of PSS. The method is less satisfactory for detecting a few PSS animals in a population of largely non-PSS animals.

Halothane has also been used to investigate myogenic aspects of PSS. Consistent with the involvement of the sarcoplasmic reticulum, halothane causes increased tension in neurally stimulated muscles (Campion et al., 1979). *Suxamethonium* also triggers the stress response in PSS pigs (Lister, 1973). *Dantrolene* reduces the release of calcium ions by the sarcoplasmic reticulum (Wendt and Barclay, 1980), and protects PSS pigs against suxamethonium, although it may also affect calcium ion movements in activated cell membranes on the muscle fiber surface (Hall et al., 1977). *Althesin* protects PSS pigs against halothane but not against suxamethonium (Harrison, 1973). Methods for dealing with the inadvertent triggering of the stress response in surgical anesthesia are reviewed by Lucke et al. (1977).

There are four types of blood test for PSS: (1) an erythrocyte osmotic response test (King et al., 1976); (2) a blood antigen test; (3) a test for an erythrocyte marker enzyme; and (4) a number of tests for blood enzymes associated with muscle disorders. The antigen test is based on the H system of red cell antigens, and it separates a number of genotypes (Rasmusen and Christian, 1976). The erythrocyte marker is *phosphohexose isomerase*, a protein that shows genetically determined structural variation (Jorgensen et al., 1976). In the last category are tests for enzymes which may have leaked from damaged or abnormal muscle fibers; these include 1,6-diphosphofructoaldolase (*aldolase*), lactate dehydrogenase (LDH), glutamate oxaloacetate transaminase (GOT), creatine phosphokinase (CPK), and malate dehydrogenase (MDH). Even in normal pigs, enzyme levels are variable and may change with age (Collis and Stark, 1977). Variability is worse when screening for PSS, since enzyme levels may change with exercise or with stress (Haase et al., 1972; Addis et al., 1974; Elizondo et al., 1976), so that a standardized stress treatment may be needed before samples are taken in commercial screening (Richter et al., 1973). Tissue damage must be minimized in the collection of blood samples so as to avoid contamination with enzymes from damaged cells. One of the commercial tests for CPK uses a very small blood sample which is dried onto paper. This is then analyzed with a bioluminescent method, the Antonik test (Hwang et al., 1977, 1978).

Webb (1980) considered that a single recessive gene is responsible for sensitivity to halothane. A *locus* is the point at which a gene is located on a chromosome. The locus for the gene which causes sensitivity to halothane is near the locus for the gene that determines the H blood group (Andresen, 1980) and the locus for the gene that determines the structure of phosphohexose isomerase (Andresen, 1979). Because of their proximity, these three genes travel together when they pass from one generation to the next. Thus one member of the group can be used as a marker for the others.

Cardiovascular Physiology

Efficient cardiovascular activity facilitates the aerobic removal of lactate and the dissipation of heat, and thus provides a measure of protection for pigs that are being stressed. In PSS pigs, the metabolic removal of lactate from the blood proceeds at a normal rate (Darrah et al., 1981). In the initial stages of the stress response in PSS pigs, however, heat production is largely aerobic (Hall et al., 1976a).

Normal parameters of cardiovascular physiology in pigs are known in some detail (Wegner et al., 1975; Ackermann et al., 1976; Gluck and Paul, 1977; Liedtke et al., 1978). Quantitative anatomical features of the heart and major blood vessels are heritable traits in pigs (Reetz et al., 1975). Thus it may be no coincidence that Large White pigs, in which there is usually only a low incidence of PSS, have a higher heart-to-muscle weight ratio than do Pietrain pigs, in which PSS is more common (Davies, 1974). Similar findings have been reported in comparisons between other breeds

(Unshelm et al., 1971). The blood flow through muscles is affected by arterial baroreceptor and chemoreceptor reflexes, which are mediated by postganglionic neurons of the autonomic nervous system (Blumberg et al., 1980).

Pigs are rather prone to *nutritional myodegeneration* caused by vitamin E and/or selenium deficiency. Briefly, the pig would like both factors, but a deficiency in one can be offset by an excess of the other. Pigs on a deficient diet may die or they may survive long enough to produce pale meat. Separation of these two causes of pale pork (PSS and nutritional myodegeneration) created some difficulties in early research on PSS. Even now, separation of PSS from nutritional myodegeneration may require some detective work. In both conditions, cardiac muscle is affected. Cardiac muscle damage found after stress (Johansson et al., 1973; Bergmann, 1979) can, however, be separated histologically from that which is caused by nutritional myodegeneration (Johansson and Jonsson, 1977). A major diagnostic feature is the occurrence of long-term signs of muscle degeneration, such as fibers with hyaline degeneration (a translucent appearance with loss of transverse striations) in nutritional myodegeneration.

Back Muscle Necrosis

A number of reports from Europe describe a type of muscle degeneration in pigs which is apparently associated with PSS (Thoonen and Hoorens, 1960; Bickhardt et al., 1967, 1968, 1972; Ilchmann, 1969; Bergmann and Wesemeier, 1973; Bradley et al., 1979). Pigs that survive bouts of severe stress suffer considerable damage to their muscles, either as a consequence of severe contraction or from anoxia. Muscle fibers are often found with greatly enlarged diameters along part of their length (Ilchmann, 1969). Similar giant fibers may be found in PSE pork (Cassens et al., 1969).

Giant fibers may be devoid of transverse striations, mitochondria, and sarcoplasmic reticulum, and may have increased numbers of dark-staining nuclei (Dutson et al., 1978). Since muscle fiber degeneration and giant fibers can also be found in non-PSS pigs (Todorov and Petrov, 1969; Muir, 1970; Bruin, 1971), degenerative changes in PSS and PSE may represent severe cases of regular muscle damage resulting from transport and slaughter stress. Supercontracted fibers occur in both normal and PSS pigs but are more common in the latter (Palmer et al., 1977). Wells and Pinsent (1980) have reported a myopathy of Pietrain pigs which starts at 2 to 3 weeks of age. Its relationship to the stress problems in this breed is unknown at present.

PRESLAUGHTER ENVIRONMENTAL FACTORS

The degree of muscle activity before slaughter affects the amount of stored muscle glycogen at the time of slaughter: Glycogen is preserved in resting animals but is depleted by exercise. In exhausted animals, glycogen may be almost completely depleted so that postmortem glycolysis and ultimate carcass pH values are high (in the high 6 range) relative to normal meat. In high-pH meat, myofibrillar proteins tightly bind the water of the meat, so that the meat lacks juiciness when eaten. Reflectance of light is reduced in high-pH meat and the meat appears dark. Darkness due to incomplete exsanguination can usually be distinguished by the presence of blood-filled capillaries in fat, or by clotted blood in large blood vessels.

Exercise

In horses, fibers with a high aerobic capacity (either fast or slow contraction speed) are depleted at all exercise intensities, but glycogen in anaerobic fibers (fast contracting) is only used for high-speed locomotion or as a reserve for prolonged exercise (Lindholm et al., 1974). In rats that have been trained for exercise, glycogenolysis in all three fiber types is reduced, relative to untrained animals (Baldwin et al., 1975). In pigs, exercise-trained animals have higher initial glycogen levels after slaughter (Lindberg et al., 1973a).

The ideal carcass is one with an initial amount of glycogen just after slaughter that will allow glycolysis to lower the ultimate pH to approximately 5.6. An ideal rate of glycolysis is more difficult to specify. It must be fast enough to allow the meat to set in rigor before it gets too cold, but slow enough to avoid exposing myofibrillar proteins at body heat to a low pH. Exercise not only affects the initial glycogen level at slaughter, but may also activate postmortem glycolysis. Thus, while normal stress or exhaustive exercise may sometimes cause DFD meat, moderate exercise may cause the opposite condition—PSE meat. Exhaustive exercise acts by depleting glycogen, but exercise started just before slaughter may activate the glycolytic system.

Complete rest may preserve stored glycogen, but a short rest following exercise prior to slaughter may introduce another variable. If meat animals resemble laboratory animals, they may show differences between histochemical fiber types in their ability to replace glycogen depleted during exercise. In rats, fast-contracting aerobic fibers more rapidly replace glycogen than do slow-contracting aerobic fibers; anaerobic fast-contracting fibers are intermediate (Terjung et al., 1974). Pigs that are held in abattoir holding pens prior to slaughter sometimes rest and sometimes fight. Thus overnight holding may sometimes reduce the incidence of PSE and increase the incidence of DFD, probably because of aggressive behavior and glycogen depletion (Moss and Robb, 1978).

Stress and Epinephrine

In practical situations in the abattoir, exercise and stress are interrelated. Exercise imposes or is associated with some degree of stress, while stressed animals often increase their level of muscular activity. The mixing of animals from different origins sometimes causes aggressive behavior, particularly among bulls in confined spaces (Martin and Fredeen, 1974). In pigs, fighting is sometimes worse between animals from the same farm which have already established a dominance order. In young bulls, Price and Tennessen (1981) found that agonistic encounters after the mixing of animals had a far greater effect on the incidence of dark-cutting beef than either transport conditions or a strange environment.

In pigs, serum activities of CPK, LDH, GOT, and pyruvate kinase tend to increase progressively during transport and handling from the farm to the abattoir (Moss and McMurray, 1979). Body temperature and heart rate increase after preslaughter stress, with a greater response in PSS pigs (Veum et al., 1979). The cellular composition of blood also changes following preslaughter stress in pigs (Ellersieck et al., 1979). Factors such as walking up a loading ramp, entering a dark space, or being herded with an electric prod cause considerable stress to pigs (Van Putten and Elshof, 1978). Adrenal ascorbic acid depletion has been proposed as a method of monitoring preslaughter stress in pigs (Warriss, 1979).

Epinephrine injections have been used experimentally to deplete muscle glycogen to simulate the effects of stress on meat quality. This treatment causes high PC and ATP levels immediately postmortem, and is followed by a faster than normal rate of ATP turnover (Bendall, 1973). Epinephrine injection causes increased toughness in chicken breast meat due to muscle shortening (Wood and Richards, 1975). In young pigs, however, epinephrine injection causes no change in muscle glycogen levels (Stanton and Mueller, 1974).

After exsanguination, carcasses can be subjected to severe treatment without bruising, since there is no longer any blood to escape from damaged blood vessels. Thus carcass bruises originate prior to exsanguination, usually during transport to the abattoir (Wythes et al., 1979). Bruising is a serious commercial problem, particularly in beef carcasses, where it often necessitates extensive trimming. Bruises in poultry carcasses may be caused when flocks are captured and crated.

Environmental Temperature and Preslaughter Nutrition

High environmental temperatures increase heart and respiration rates, and also increase the incidence of PSE in pigs (Forrest et al., 1965). In PSS pigs, warm conditions may lead to hypoxia (Forrest et al., 1968). Pigs kept on a diet that is high in fat and protein but low in carbohydrates tend to produce meat with a higher pH, while those fed a high-sucrose diet may produce low pH meat (Sayre et al., 1963a). Sucrose feeding lengthens both internal and external glycogen chains (Sayre et al., 1963b). Initial postmortem glycogen levels may be lowered in pigs that have been fasted for 70 hours prior to slaughter (Sayre et al., 1963c) and may be elevated in fasted pigs that have been refed prior to slaughter (Lee et al., 1971). The response of individual meat animals to one or more days of fasting prior to slaughter is difficult to predict, since their general health and nutritional status must be taken into account. For example, culled dairy cows undergo greater glycogen depletion if they are pregnant (Warnock et al., 1978). Diurnal variation in muscle glycogen has been observed in fasted laboratory rats (Funabiki et al., 1975), and should be taken into account when planning experiments with meat animals. When comparing data on pork quality from different abattoirs it is essential to make allowances for the time of day at which animals are slaughtered, since there may be considerable differences between the first and the last pigs slaughtered (Swatland, 1982b).

ENVIRONMENTAL FACTORS DURING AND AFTER SLAUGHTER

Stunning

Slaughter procedures normally start with a stunning method to render animals unconscious prior to the exsanguination which actually kills them. In meat animals, this is primarily for humanitarian reasons and for the safety of personnel. In poultry there is an added advantage from electrical stunning since it increases the efficiency of exsanguination (Kuenzel and Ingling, 1977), and it may improve meat tenderness (Lee et al., 1979). When the Royal Society for the Prevention of Cruelty to Animals was founded in Britain in 1824, it was not uncommon for animals to be skinned while they were still almost alive (Fairholme and Pain, 1924). The requirements for an efficient and safe method of stunning which were proposed in the 1920s were:

1. that it should allow the operator to stand at the side of the animal and enable him to place the instrument exactly on the spot where perforation is required;
2. that it will instantaneously injure the brain to an extent ensuring immediate and lasting insensibility;
3. that it will either itself cause loss of involuntary motion or permit the use of a cane to effect that purpose.

Modern stunning methods have progressed beyond the point of requiring a bamboo cane to be inserted through the cranial cavity and foramen magnum so as to destroy the spinal cord and terminate reflex activity (*pithing*). This is due partly to improvements in abattoir design and partly to modern equipment which enables animals to be hoisted off the ground rapidly and immobilized. Thornton and Gracey (1974), however, describe a case in which blood splashing in beef was eradicated following the introduction of pithing. To the list of features desired in a stunning method, we should now add the potential of the method for automation.

Neuromuscular Response to Stunning

Little is known about the physiological mechanisms involved in animal stunning. It is difficult to use any method other than *electroencephalography* (EEG) to monitor neural activity in the abattoir, and even this method is difficult to use with electrical stunning. EEG electrodes are placed in electrical contact with the skin over the cranium. The electrodes detect very small voltage changes caused by waves of electrical activity in the brain. Waves of activity are created by the sum of coordinated activity in large numbers of brain cells (Klemm, 1977; Bures et al., 1967). Consciousness is dependent on the coordinated activity of brain cells. Thus when there is no EEG activity, the animal is either (1) anesthetized or reversibly unconscious; (2) dead or irreversibly unconscious; (3) the EEG apparatus is not working properly; or (4) there is insufficient amplification of an existing EEG signal. In the first category of situations where EEG activity is absent may be placed a condition known as *curarization* (Croft, 1957). This condition is thought to occur during the recovery process of animals that have been stunned but not exsanguinated, usually because of a delay on the slaughter line. As the animal recovers consciousness and EEG activity again becomes detectable, it may pass through a period during which it is sensitive to pain but lacks the means to express this sensation because efferent motor systems are not yet operative.

Under abattoir conditions, it is extremely difficult to determine how much pain or suffering is associated with stunning techniques. Leach et al. (1980) concluded that electrical stunning is painless since animals do not anticipate pain after receiving a visual stimulus which they should have learned to associate with stunning. However, the stunning of animals during their supposed period of learning (7 days) might well have disturbed their short-term memory and thus have made it impossible for them to associate the stimulus with stunning. Painful or irritating stimuli such as physical contact with the eyeball may be used to determine if an animal is unconscious; but if the eyeball rotates to its extreme dorsal position, even this test can be difficult to assess. When EEG is used to monitor brain activity, it is sometimes difficult to be sure that *electromyographic* (EMG) activity from the jaw muscles or extraocular muscles is not contributing to EEG activity. Reflex activity in the limbs may continue after EEG activity is lost postmortem. Thus cessation of EEG activity does not indicate cessation of all neural activity in the central nervous system. However, loss of EEG activity is a reasonable sign that animals are suffering no pain or distress. Recordings from electrodes placed through the skull and onto or into the cerebral cortex [*electrocorticography* (ECoG)] show that small groups or individual cells may remain active for some while after exsanguination. Reflexes and ECoG activity soon return if electrical stunning is not immediately followed by exsanguination (Croft, 1952). In unconscious pigs, however, both reflexes and ECoG activity may be found after exsanguination (Figure 9-5; Swatland, 1976d).

The central nervous system of a live animal maintains a fine balance between excitatory and inhibitory neural systems. During the necrobiotic activity of commercially slaughtered meat animals, it is difficult to determine whether inhibitory systems maintain or have lost their control over excitatory systems. The occurrence of reflex limb movements and ECoG activity in exsanguinated meat animals might be a consequence of the loss of inhibitory control. Unlike humans, meat animals lack a large motor cortex with neurons that are directly connected to the motor neurons of the spinal cord (Breazile et al., 1966). Under abattoir conditions, it is difficult to get ECoG electrodes onto the motor cortex with any speed or precision, so that even when ECoG activity can be correlated with visible reflex limb movements, it is difficult to be sure that the motor cortex is initiating the activity. The cortex might simply be receiving incoming sensory information on an event that was initiated peripherally. Consequently, it is not surprising that sectioning of the spinal cord in the anterior cervical region of pork carcasses after exsanguination produces no gross changes in meat quality of the longissimus dorsi muscle (Westervelt and Stouffer, 1978).

Stunning methods sometimes cause blood splashes or hemorrhages in meat (Warrington, 1974). These may be small and dark (*punctate* or *petechial hemorrhages*) or diffuse (*ecchmyoses*), and they may be particularly visible in pale meat such as pork. Both conditions are probably caused by the rupture of capillaries by high blood pressure or by violent muscle contraction. They can be reduced by using high-frequency stunning currents, and by exsanguinating animals rapidly (<5 seconds)

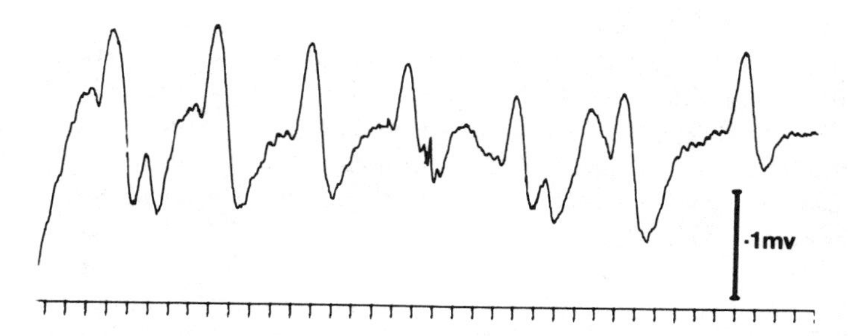

Figure 9-5 Electrocorticographic brain activity associated with reflex muscle contraction after exsanguination. Time scale in seconds starting at approximately 6 minutes from the end of exsanguination. (From Swatland, 1976d.)

after stunning. The incidence of blood splashes in lamb may be reduced with a nonpenetrating percussion method of stunning (Blackmore, 1979).

The factors that cause blood splashes in meat have not yet been completely elucidated, even though blood splashing in meat is a serious commercial problem which may involve up to 10 per cent of carcasses. Blood splashes are rather unattractive in appearance, but they present no known hazard to the consumer. Stunning methods (such as electrical stunning) which cause simultaneous contraction of large groups of muscles appear to cause more of a problem than stunning methods which cause only the sporadic contraction of individual muscles. Individual animals differ considerably in the length of time that their blood takes to clot after exsanguination. Possibly, this has some relationship to the incidence of blood splashes. Petechial hemorrhage may also occur in the layers of fat and connective tissue that occur outside the carcass muscles. In this location, they are due to violent muscle contraction rather than to high blood pressure (Gilbert and Devine, 1982).

Different Stunning Methods

The three stunning methods which can be used for pigs (captive bolt, electrical, and carbon dioxide) have been compared in a number of studies. Captive bolt pistols drive a metal bolt through the frontal bones of the skull to cause brain damage. The bolt is retractable and is usually driven by compressed air or by a small ammunition cartridge. To obtain unconsciousness in sheep, it is necessary to crush the cortex and deeper parts of the brain, either directly or by shockwaves (Lambooy, 1982b). In pigs, brain damage from concussion may cause reflex activity which adversely affects meat quality by lowering pH while carcass muscles are still warm (Klingbiel and Naude, 1972). Dreyer et al. (1972) compared muscle fiber diameters between pigs slaughtered by exsanguination, with or without prior stunning by captive bolt. Captive bolt stunning caused a decrease of fiber diameters, since the low pH led to a loss of intracellular water and to fiber shrinkage. Beecher et al. (1965a) concluded that captive bolt stunning only lowered the pH of white muscles. The adverse effects of electrical and carbon dioxide stunning on meat quality are less than those of captive bolt stunning (Overstreet et al., 1975).

Carbon dioxide stunning systems for pigs became quite common following their technical development by the Hormel company in the United States. Carbon dioxide gas is used in concentrations of 65 to 70 per cent in air, and it causes *hypercapnia* (elevated carbon dioxide concentration in blood) and hypoxia. Carbon dioxide is heavier than air, and is usually trapped in a tunnel or pit through which pigs are transported on a conveyer belt, cage, or Ferris wheel for not less than 45 seconds (Glen, 1971). Most pigs lose consciousness within 60 to 75 seconds. Exsanguination must be initiated within 30 seconds after pigs leave the stunner. Overexposure in the stunner may cause vasodilation and a bluish skin color after scalding. Stunning of pigs by inhalation of a high (68 per cent) concentration of carbon dioxide causes an increase in amplitude and a decrease in frequency of EEG waves (Mullenax and Dougherty, 1963). Hoenderken et al. (1979) have reported that the outcome of carbon dioxide stunning sometimes resembles suffocation. This has been disputed (Lomholt, 1981) on the grounds that the conditions that were examined by Hoenderken et al. (1979) did not adequately represent the methods used industrially. Carbon dioxide (33% to 36%) can also be used to stun poultry, with some improvement in exsanguination (Kotula et al., 1957, 1961).

Exposure to carbon dioxide causes an increase in the free calcium ion concentration of the muscle fiber cytosol (Lea and Ashley, 1978). This might add to the acceleration of glycolysis already initiated by the animal struggling. The effects of different stunning methods on muscle glycolysis can be detected biochemically, even to the point of finding differences in cyclic AMP levels (Ono et al., 1976). Apart from the direct neural activation of muscle, glycolysis may be enhanced by the release into the bloodstream of catecholamines from the adrenal glands (Van der Wal, 1978).

There are many combinations of voltage, duration, and electrode placement that can be used for the electrical stunning of meat animals. EEG data show that only conditions at or exceeding 300 V at 1 A can guarantee the effective stunning of at least 90 per cent of pigs slaughtered (Hoenderken, 1978). Thornton and Gracey (1974) gave the minimum recommended stunning current for pigs as 0.25 A at 75 V for 10 seconds ($0.25 \times 75 \times 10 = 187$ W-s). For sheep and calves, 198 W-s was recommended. For sheep, Lambooy (1982a) suggests that a current of at least 0.5 A (160 V, 50 Hz) should be used. For cattle, 285 W-s was recommended by Thornton and Gracey (1974). However, bulls need 430 W-s (Croft, 1957). Lambooy and Spanjaard (1982) found that a current strength of 0.87 A was needed for the effective stunning of 90 per cent of cattle. Electrical stunning of pigs may lead to elevated muscle temperatures (Monin, 1973). The major problem with electrical stunning is not with the inducement of unconsciousness but with its duration, since animals must be exsanguinated before they recover consciousness. The recommended currents are often inadequate for this purpose.

Electrical stunning of poultry (approximately 5 seconds at < 100 V, 50 to 60 Hz, 0.25 A) initiates an *electoplectic fit* which is characterized by (1) open eyes without corneal reflex; (2) the head arched toward the back; (3) wings with steady downward extension and feathers spread; (4) legs fully extended; (5) tail feathers turned up over the back; and (6) absence of defecation (Scott, 1971). EEG activity is lost within 15 seconds of electrical stunning, but is regained after several minutes if birds are not exsanguinated (Richards and Sykes, 1967). If the comb of a chicken is pinched, the chicken responds by withdrawing and shaking its head. Richards and Sykes (1964) used this reflex to investigate the loss of consciousness following exsanguination, with or without prior electrical stunning. After exsanguination alone, the birds showed a comb reflex for over 2 minutes. Electrically stunned birds showed no reflex. High-voltage (>400 V) stunning of poultry may cause violent muscle contractions which damage the sternum and other bones.

Reflex Activity

Lower reflex activity occurs quite often in meat animals and poultry after exsanguination and loss of consciousness. Since the intrinsic activity of individual muscle fibers is imperceptible in a whole carcass, visible movements of large parts of the body must involve some degree of neural coordination. Modern research on necrobiotic activity in the nervous system is scarce. Rosenthal (1881) considered that death and loss of excitability in the axons of necrobiotic motor neurons followed a proximal to distal sequence preceded by hyperexcitability. The rate of loss of excitability was found to be temperature dependent.

The situation of necrobiotic motor neurons in a commercial carcass is extremely complex. Motor neurons in the ventral horn of the spinal cord receive both inhibitory and excitatory inputs. The descending and ascending tracts which link the brain and spinal cord in meat animals are *polysynaptic* (a relay system of neurons), so that motor neurons have a greater degree of autonomy than in humans (Palmer, 1976). Meat animals have only a small direct motor cortex in the *cerebrum*, and only short motor tracts descend the spinal cord. In meat animals, the sensory receiving areas of the cerebral cortex are dominated by inputs from the lips and snout, and the areas which correspond to the limbs and trunk are small or nonexistent (Adrian, 1943; Woolsey and Fairman, 1946).

The normal physiological definition of a reflex (involuntary activity elicited by stimulation of a sensory receptor) must be considerably enlarged in scope to accommodate what a slaughterer or butcher may call reflex activity. Neural activity may originate intrinsically from motor neurons which have just been released (by cell death or disruption) from their normal inhibitory control. Motor neurons are normally restrained by a variety of inputs, such as those from *Renshaw cells* (Eccles et al., 1954). True reflexes such as the *myotatic* or stretch reflex of muscles might occur briefly in carcasses. Myotatic reflexes normally act via a sensory input from *neuromuscular spindles* so that a muscle will contract to resist a stretching force.

Carcasses are normally suspended on an overhead rail, and some muscles are stretched before exsanguination. However, since severe stretching occurs in some muscles, an opposite response may be created by activation of *Golgi tendon organs*. These sensory receptors are located on tendons and, following extreme muscle stretching, they inhibit appropriate motor neurons and cause sudden muscle relaxation. Since stunned animals are usually shackled by only one hind limb, reflex jerking by the free hind limb may originate from a *crossed extensor* reflex which acts through reflex pathways that are normally associated with locomotion or withdrawal from a painful stimulus. The summation of these events can be seen by electromyography during the time that the carcass is handled after stunning (Figure 9-6). EMG may reveal *isometric* muscle contraction in the shackled limb.

The occurrence of reflex activity after the stunning of pigs is extremely important in the meat industry. In some abattoirs the stunning methods cause massive reflex activity, and it may be difficult and dangerous to exsanguinate the animals. Reflex activity also exacerbates the development of the PSE condition. The neural origins of this violent muscle activity are unknown. As the degree of reflex activity diminishes, however, the animal's respiratory reflexes become evident. Most commonly, the animal coughs and moves its head downwards or ventrally. This can be detected by EMG activity in the ventral muscles of the throat. Less commonly, the pig throws its head backwards or dorsally so that its vertebral column is concave dorsally (Swatland, 1983).

Neural Necrobiosis

The importance of intrinsic neural activity on postmortem muscle metabolism is difficult to evaluate. It may be a major factor which accelerates glycolysis by causing muscle contraction. There is some evidence that the production of PSE pork from PSS

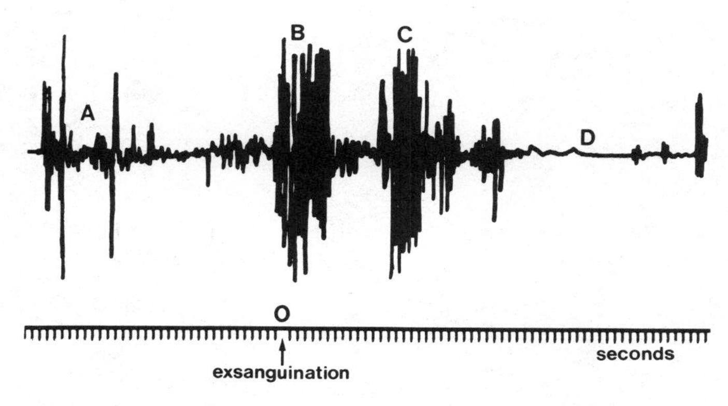

Figure 9-6. Electromyographic muscle activity in a pork carcass before and after exsanguination. Exsanguination of the pig was started at zero time. Large-scale reflex activity occurred in the unconscious pig before (A) and after (B, C) exsanguination. Low amplitude and otherwise hidden activity started after a period of inactivity (D). Source: from Swatland (1976a).

pigs is aggravated by neural activity (Hall et al., 1976b; Somers et al., 1977). Neuromuscular junctions are enlarged in PSS pigs (Swatland and Cassens, 1972). In normal pigs, treatment with curare to block neuromuscular junctions causes a slow and constant rate of postmortem glycolysis (Bendall, 1966). This indicates that neural impulses normally provide an irregular stimulation of postmortem glycolysis during slaughter procedures.

Rosenthal (1881) found that excitability remained in muscles after it had been lost in necrobiotic nerves. In more or less intact pork carcasses, nerves and muscles can still respond to electrical stimulation after the cessation of normally occurring necrobiotic EMG activity (Swatland, 1976a). Stimulation via the spinal cord shows that nerves remain excitable for approximately 15 minutes postmortem, and that muscles can still be stimulated directly even after their nerves are no longer excitable (Swatland, 1975c).

In commercial carcasses, extensive muscle damage occurs during operations such as decapitation and evisceration. Muscle twitching in damaged areas of muscle may continue for a long time, particularly in beef carcasses. Sometimes whole muscles contract irregularly, such as the sternomandibularis muscle which is transected in the neck. *Fasciculation*, the local twitching of muscle fiber bundles, is evident in many exposed muscles. These types of activity originate from nerves that have been cut or damaged, either between or within muscles. EMG activity (Figure 9-7) and the pattern of glycogen depletion in individual muscle fibers (Plate 27; Swatland, 1976b) indicate that individual or small groups of *motor units* of any histochemical type may be activated by damaged axons.

When muscle strips are prepared for experiments, tissue damage is a major source of experimental error (Bendall, 1978), and neural components are probably involved. Surface drying and aerobic exposure facilitate motor unit irritability, but it is inhibited by keeping tissue surfaces moist, anaerobic, or cold (Figure 9-8). The biophysical cause of necrobiotic motor unit activity has not been established, but a reasonable working hypothesis is that surface drying, oxygen, and warmth allow transected axons to be capped by clotted protein. The sealed damaged region may then

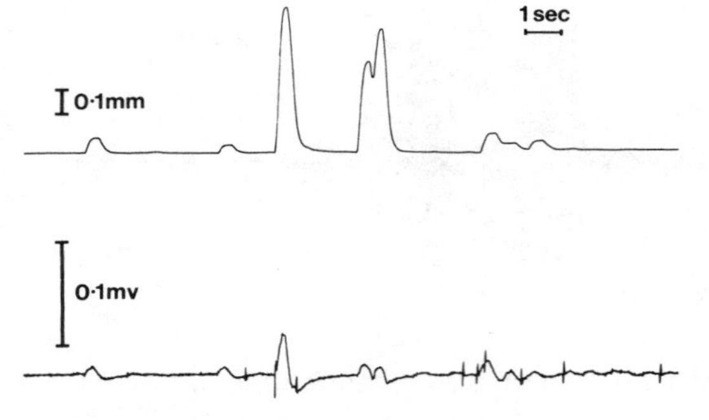

Figure 9-7 Motor unit activity originating from a transected nerve in a sample of meat taken soon after exsanguination of a steer. The electromyographic activity of motor units (bottom trace) corresponds to muscle twitches (top trace). The contraction of individual muscle fibers (spikes on the bottom trace) does not result in detectable movement. (From Swatland, 1976b.)

initiate action potentials in the remaining distal parts of the axon which have been able to restore almost normal resting potentials. Resting potentials can be recorded along transected beef muscle fibers postmortem (Swatland, 1980a). Loss of muscle fiber resting potentials is more rapid in pork, particularly if it is PSE (Schmidt et al., 1972a). Lactate accumulation in living muscle is correlated with decreases in membrane-resting potentials (Jennische et al., 1978).

Despite the universal acceptance that pH is a primary factor which controls postmortem sources of variation in meat quality, the correlation of pH with ultimate meat quality is not particularly reliable in industrial situations (Martin et al., 1981). Carcasses that ultimately develop the most extreme PSE may have had a pH value well within the normal range when measured just after slaughter. Martin and Fredeen (1974) reported that their most extreme PSE carcasses had a mean of pH 6.55 at 40 minutes postmortem, whereas mild PSE carcasses had a pH of 6.28. These exceptional cases are usually regarded as the product of experimental error, but this need not necessarily be true. Two connotations of the phenomenon of *hysteresis* can be invoked to explain such exceptions to the pH rule: (1) hysteresis as the state of a substance that can only be completely described by reference to its previous history; and (2) hysteresis as the lack of coincidence of two otherwise associated phenomena. The first of these connotations is already widely known to apply to the PSE condition; ultimate pH is related to ultimate meat quality, but knowledge of this relationship is greatly enhanced if information is added on the rate and time at which the ultimate pH was attained. It can also be argued that the second connotation of hysteresis is equally applicable to the PSE condition. A possible mechanism for this is shown in Figure 9-9.

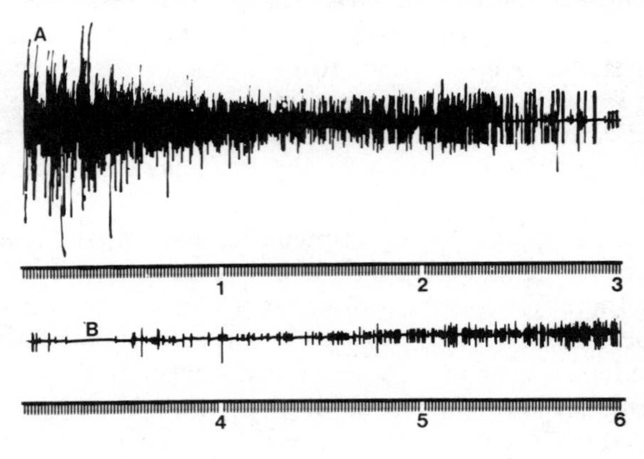

Figure 9-8 Electromyographic activity initiated by a transected intramuscular nerve which has been cooled at point A and rewarmed at point B. Time scale in minutes. (From Swatland, 1976a.)

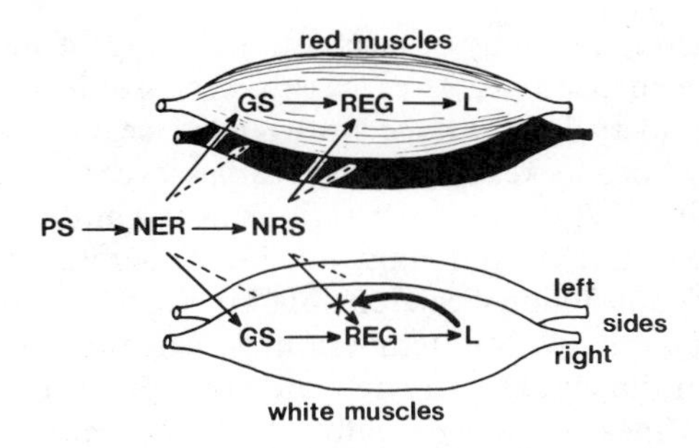

Figure 9-9 Environmental factors that may affect lactate production in red and white muscles on the left and right sides of the carcass, including a possible negative feedback of lactic acid on reflex muscle activity: (PS preslaughter stress; (NER) neural and endocrine response to stress; (NRS) neural response to slaughter; (GS) glycolytic status at the start of slaughter; (REG) rate and extent of glycolysis postmortem; (L) lactic acid.

PSE pork has a greater threshold to electrical excitation than normal pork (Forrest et al., 1966). The release of acetylcholine at the neuromuscular junction is inhibited by a low pH (Landau and Nachshen, 1975). Thus excessive reflex activity immediately after exsanguination may accelerate glycolysis, particularly in white muscles. Excessive heat and acidity may initiate the protein damage that leads to the PSE condition, but it may also terminate any further reflex activity in the muscles involved. In other words, a relatively small amount of lactate which is formed immediately after exsanguination (<3 minutes) might produce as much or more damage than a larger amount produced over a longer time (up to 38 minutes). The times given here are those of the minimum and maximum duration of electromyographic activity after exsanguination (Swatland, 1976a), but other similar values would be equally acceptable in this argument. Left and right sides of the carcass must be considered separately when the animal is shackled by only one hind limb, since this modifies the pattern of reflex activity.

Meat Cooling

Abattoir kill floors are usually quite hot places, and carcasses also generate their own heat during rigor mortis (Morley, 1974). To reduce bacterial spoilage and to reduce the length of time that carcasses remain in the abattoir, it is desirable to cool meat rapidly after slaughter. The proliferation of spoilage microorganisms is slower at low temperatures, and spoilage in deep parts of the carcass such as the hip may cause *bone taint* if deep temperatures remain high. Carcasses are normally moved from the kill floor to a meat cooler at approximately 0° C. Convection currents around hot carcasses and the latent heat of water evaporation from wet surfaces have a slight cooling effect, but air has to be circulated at several meters per second through overhead heat exchangers in order to cool the carcasses at the required rate.

The rate of carcass cooling depends on air temperature and speed, and on carcass exposure. Heavy carcasses have a poor exposure because of an unfavorable surface-to-volume ratio, while fat carcasses may cool more slowly because of their greater insulation. Carcasses lose weight by evaporation, and this is increased by carcass exposure and by a fast dry stream of air. Rapidly chilled carcasses tend to have a lower drip loss when they are broken into retail cuts after cooling. Meat removed from carcasses before the development of rigor mortis (*hot boning*) can be cooled more rapidly, and it has a reduced drip loss when packaged (Follett et al., 1974; Kastner and Russell, 1975). The removal of heat from deep within the carcass depends on the conduction of heat through the meat.

Although advantageous in many respects, the chilling of meat can easily be pushed too far. There is evidence that fat carcasses, because they cool more slowly, have increased autolytic activity and produce more tender meat (Smith et al., 1976). A similar advantage can be gained by keeping normal carcasses at higher temperatures (7 to 8°C; Cliplef and Strain, 1976). The most serious problem due to rapid cooling, however, is *cold shortening* (Locker and Hagyard, 1963). Before muscles set in rigor mortis they are capable of contraction if calcium ions are not removed from the cytosol by the sarcoplasmic reticulum. At low temperatures, calcium ion accumulation is weak, and calcium ions in the cytosol may activate myofibrillar contraction (Davey and Gilbert, 1974). When meat does eventually set in rigor mortis, thick and thin filaments of cold shortened meat may overlap to the maximum extent allowed by skeletal restraint in different muscles (Joseph and Connolly, 1977), and the meat may be unduly tough. Contributing to the cold-induced release of calcium ions from the sarcoplasmic reticulum (Newbold and Tume, 1977), calcium ions may also be released from mitochondria (Buege and Marsh, 1975; Mickelson and Marsh, 1980; Cornforth et al., 1980).

Although sarcomere length and the degree of thick and thin filament overlap are important, other factors may also be involved in the relation between cold shortening and meat tenderness, since cold-shortened sarcomeres which set in rigor mortis after secondary warming may revert to a normal degree of tenderness (Locker and Daines, 1976). Similarly, when extreme cold shortening occurs, ultrastructural damage begins to reduce meat toughness (Marsh et al., 1974). Further complications are (1) that cold shortening does not affect all fibers equally since active fibers are straight and passive fibers are wavy (Voyle, 1969); and (2) that red muscles are more severely affected than white muscles (Bendall, 1975). In fast-contracting fibers, the volume of the sarcoplasmic reticulum is about twice that of slow fibers (Schmalbruch, 1979), and the calcium-uptake capabilities of the sarcoplasmic reticulum are greater in fast-contracting muscles than in slow muscles (Briggs et al., 1977). To avoid cold shortening, meat must not reach 10°C within 10 hours postmortem in lamb carcasses. Beef sides should not be exposed to air below 5°C and faster than 1 m/second within 24 hours after slaughter (Cutting, 1974).

If muscles are accidentally frozen before the onset of rigor mortis, a stronger form of contraction, *thaw contracture*, may occur on thawing. At temperatures just above freezing, the sarcoplasmic reticulum is only weakly active, and calcium ions in the cytosol are free to initiate myofibrillar contraction. The sarcoplasmic reticulum is also damaged by ice crystals while frozen. Thus, as prerigor meat is thawed, it passes through a stage at which myofibrils are capable of contraction, but the sarcoplasmic reticulum is unable to stop them. If higher temperatures are reached after thawing and if ATP is still present in sufficient quantity, the sarcoplasmic reticulum may be able to resume its normal function and muscle relaxation may occur (Bendall, 1973). In chicken muscle, thaw contracture can be elicited several times as muscles are alternately frozen and thawed (Whiting and Richards, 1975). The ultrastructure of thaw contracted muscles is severely disrupted and irregular dark bands across fibers may be seen under the light microscope (Cassens et al., 1963). Sarcomeres are severely shortened (Herring et al., 1964).

The freezing and thawing of beef muscles prior to the complete development of rigor mortis usually lead to an increase in meat toughness. In poultry meat the situation is different. Meat toughness reaches its maximum level as recently slaughtered carcasses are washed, but the meat becomes more tender as carcasses are chilled. Maximum tenderness is reached after blast-freezing and thawing (Grey and Jones, 1977).

Glycolysis in necrobiotic muscle slows down as temperatures are decreased from body temperature down to approximately 5° C (Cassens and Newbold, 1967) but below this there is an increase in glycolytic rate (Cassens and Newbold, 1966). Similar findings, but with a minimum rate at 10° C, were reported by Jeacocke (1977b). Activation of glycolysis at low temperatures is probably due to an accumulation of AMP, which activates phosphorylase b (Newbold and Scopes, 1967). Freezing of necrobiotic muscle does not stop glycolysis or the development of rigor mortis, and in lightly frozen meat, these processes continue in a slow but normal manner (Winger et al., 1979).

Electrical Stimulation of Carcasses

The electrical stimulation of carcass muscles soon after slaughter accelerates their normal decline in pH and may enhance tenderization during conditioning (Harsham and Deatherage, 1951). Commercial interest in electrical stimulation increased after it was shown that stimulation prevented cold shortening in lamb (Carse, 1973) and in beef (Davey et al., 1976). Electrical stimulation of carcasses is used commercially in the United States (Dutson, 1981), New Zealand (Davey and Chrystall, 1981), Sweden (Fabiansson, 1981), Great Britain (Cuthbertson, 1981), and other countries. As well as protecting against cold shortening, electrical stimulation may improve meat tenderness, color, and appearance (Cross, 1979; Bendall, 1980). With the beef grading system used in the United States, electrical stimulation may improve the score for youthfulness and meat color (Savell et al., 1978). However, cooking losses may be increased by electrical stimulation (Bouton et al., 1980a).

A number of different voltages have been used; increments to 32 V (Taylor and Marshall, 1980), 100 V (Savell et al., 1977), increments to 110 V (Shaw and Walker, 1977; Bouton et al., 1978), 250 V (Carse, 1973), 300 V (Will et al., 1979), 440 V (Savell et al., 1978), 550 V (Savell, 1978; Riley et al., 1980), 700 V (Bendall et al., 1976; Bendall and Rhodes, 1976), and 1600 V (Chrystall and Hagyard, 1976; Davey et al., 1976). Amperages, if measured at all, have included 0.5 A (Shaw and Walker, 1977), 1.9 A (Will et al., 1979), 2 A (Chrystall and Hagyard, 1976; Davey et al., 1976; Taylor and Marshall, 1980), 5 A (Riley et al., 1980; Savell et al., 1977, 1978), and 6 A (Bendall et al., 1976).

Types of electrode, if described at all, have included aluminum foil (Carse, 1973), probe or pin types (Savell, 1978; Carse, 1973; Davey et al., 1976), clamps (Savell et al., 1977), rectal probes (Bouton et al., 1978), and multipoint electrodes (Shaw and Walker, 1977; Bouton et al., 1978; Taylor and Marshall, 1980). Hooks and shackles suspending carcasses from overhead rails have frequently been used, either intentionally or unintentionally, as ground or return electrodes. Electrode positions have usually spanned great lengths of carcass from neck to hind limb, usually with electrodes placed in muscles, but sometimes into the spinal cord (Carse, 1973).

The time delay between exsanguination and stimulation has been extended up to 60 minutes, during which time carcasses have received variable treatments such as evisceration and splitting. Impulse frequency and duration of application have included 3 Hz for 30 minutes (Carse, 1973), 3.5 Hz for 70 seconds (Savell, 1978), 3.5 Hz for 1 minute (Riley et al., 1980), 15 Hz for 55 seconds (Chrystall and Hagyard, 1976), 15 Hz for 0.5 to 10 minutes (Davey et al., 1976), 25 Hz for 2 minutes (Bendall et al., 1976; Bendall and Rhodes, 1976), 40 Hz for 4 minutes (Shaw and Walker, 1977; Bouton et al., 1978), 50 Hz for 2 minutes (Savell et al., 1977), 50 to 60 Hz for 25 to 75 one-second periods (Savell et al., 1978), and 400 Hz for 15 minutes (Will et al., 1979). Wave forms of stimulatory impulses have included square waves of interrupted direct current (Carse, 1973; Bouton et al., 1978; Will et al., 1979) and an assortment of unchanged or partly modified sinusoidal waves of alternating current.

The origin of samples used to evaluate stimulatory treatments has varied considerably, sometimes being directly between electrodes and sometimes to one side. There is little point in attempting to correlate treatments with responses in all these different methods. It has been claimed that, apart from guarding against cold shortening, the beneficial effects of electrical stimulation on meat tenderness are probably a simple consequence of muscle fiber fracture (Marsh et al., 1981; Sorinmade et al., 1982). However, electrical stimulation is ineffective on dark-cutting beef and Dutson et al. (1982) argue that postmortem muscle stimulation is of no value without an accelerated decline in pH.

Even in a simplified laboratory model of carcass stimulation, with a muscle strip and a pair of stimulatory electrodes, the response of necrobiotic muscle may be quite complex. Although stimulation may accelerate postmortem metabolism, muscle with an already accelerated rate of metabolism may lose its excitability at a faster than normal rate. Thus with both pork (Forrest et al., 1966) and turkey muscle (Ma et al., 1971), animals with intrinsically fast glycolytic rates can be detected by their reduced electrical excitability. Unless special precautions are taken to the contrary, most muscle strips contain severed intramuscular nerves and neuromuscular junctions among the muscle fibers. Immediately postmortem, all three components may be excitable with their own particular activation thresholds, and as these change postmortem, it is difficult to identify the point between the axon and the muscle fiber which is first to respond to electrical stimulation. The excitability of muscle strips decreases progressively postmortem so that either a higher voltage and/or a longer duration stimulus is needed to obtain a constant response. If neuromuscular junctions are pharmacologically blocked in samples taken shortly after animal exsanguination, excitability is decreased. This suggests that the high excitability of muscle strips at this time is due to intramuscular motor axons and/or their neuromuscular junctions (Swatland, 1977b). Similar results have been found with isolated hind limbs of lambs (Chrystall et al., 1980).

Living muscles or strips taken immediately after animal exsanguination respond to a progressive increase in stimulus frequency by twitching at a correspondingly faster rate, until the twitches merge into a sustained contraction or *tetanus* (Figure 9-10). As the time between animal exsanguination and muscle stimulation is increased, muscle strips become progressively less able to maintain tetanus (Swatland, 1977b; Chrystall and Devine, 1978). Immediately after the excitation of axons and muscle fibers, there follows an *absolute refractory period* of complete inexcitability to a second stimulus since the response to the first stimulus is still in progress. Next comes a *relative refractory period* when if the second stimulus is of sufficient magnitude, excitation can

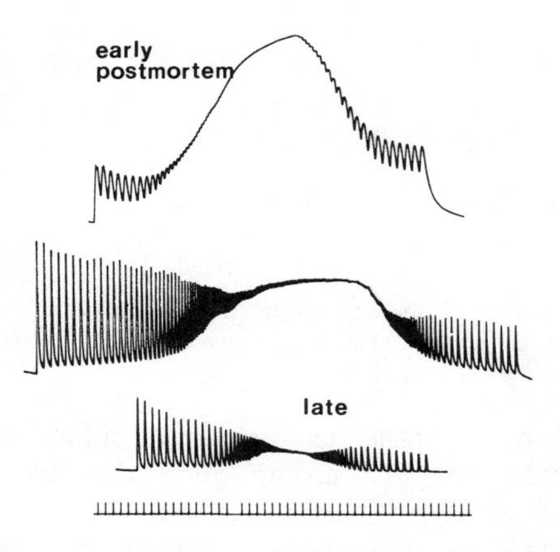

Figure 9-10 Loss of ability to sustain tetanus in necrobiotic muscle. Time scale in seconds. In muscles soon after slaughter, if the muscle is stimulated at an increasing rate, individual twitch responses fuse to form a steady strong contraction (tetanus). Some time later, tetanic contraction is weaker than individual twitch contractions.

be elicited. The duration of the relative refractory period probably increases progressively postmortem. Many of the stimuli delivered at a high frequency may, therefore, arrive during a relative refractory period and elicit no response (Figure 9-11). For this reason, the stimulus frequency for electrical stimulation of carcasses is usually kept low. Muscles may differ in the optimum frequency required for their maximum stimulation. Bouton et al. (1980b) identified 14 Hz for the longissimus dorsi and 40 Hz for the semimembranosus in 6-month-old calves. In living muscle, electrical stimulation results in a rapid drop in pH, a slow rise in pH, and finally, reacidification (Connett, 1978). The pH decline in meat postmortem may render meat less excitable to electrical stimulation via the nervous system since a low pH reduces the amount of acetylcholine released at the neuromuscular junction (Landau and Nachshen, 1975). The accumulation of calcium ions by transverse tubules (Bianchi and Narayan, 1982) might also be involved in the decreased excitability of muscles after electrical stimulation.

Pale muscle areas show a greater response to electrical stimulation than do dark or red areas (Tarrant et al., 1972a), because anaerobic fibers respond more readily (Swatland, 1975a; Houlier et al., 1980). Another feature of postmortem stimulation is that the effects of stimulation are not limited to the actual period of stimulation, but linger on afterward (Bendall, 1966; Tarrant et al., 1972b). This may be due to changes produced in the sarcoplasmic reticulum (Tume, 1979). Electrical stimulation results in swelling of the sarcoplasmic reticulum, transverse tubules, and mitochondria, together with autolytic ultrastructural changes (Will et al., 1980). Increased binding of glycolytic enzymes to actin filaments may also be involved (Clarke et al., 1980).

Current pathway is difficult to assess in whole carcasses. In homogeneous

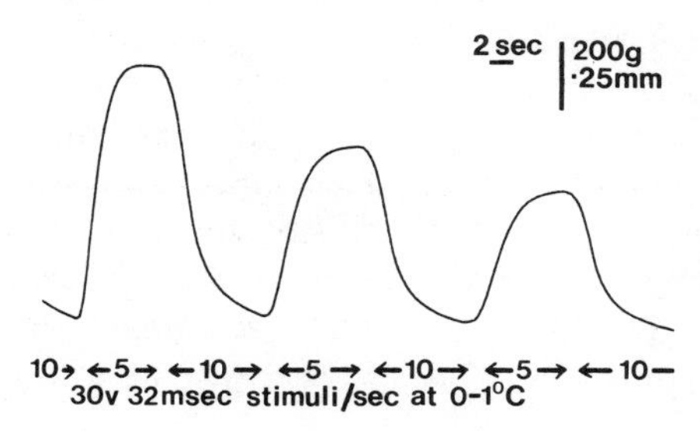

Figure 9-11 Lengthening of a contracted muscle strip as stimuli arrive during the refractory period. (From Swatland, 1977b.)

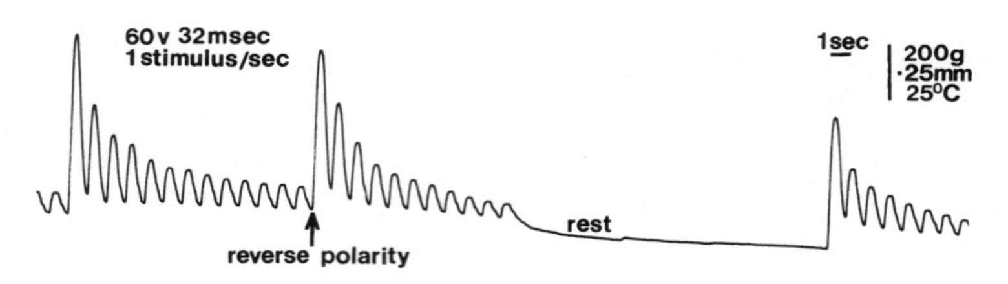

Figure 9-12 Effect of electrical polarization with square-wave stimulation of meat. The contractile response of the meat progressively diminishes but can be briefly restored by reversing the polarity of the square waves.

conductors, *resistance* is proportional to *resistivity* and to the distance between electrodes, and is inversely proportional to conductor cross-sectional area. In homogeneous conductors, resistivity (resistance specific to conductor material) is normally measured between opposite faces of a 1-centimeter cube. However, not only are carcasses interrupted by tracts of fat with a high resistance and by bones with a variable resistance (Geddes and Baker, 1967), but muscles themselves are electrically anisotropic (Burger and van Dongen, 1960). The concept of resistivity is useful in considering current flow through a carcass but heterogeneity and anisotropy render ohm-centimeter values little more than approximations. Resistivity is inversely proportional to meat temperature, it tends to be greater across rather than along muscle fibers, and it may show a transient increase postmortem followed by a progressive decline (Swatland, 1980b). Resistance of a whole carcass is modified by factors such as (1) the time lapse between animal exsanguination and muscle stimulation; (2) the distance between electrodes; (3) electrode surface area in contact with meat or connective tissue and not blocked by fat; (4) whether or not the carcass is whole, eviscerated or split; and (5) electrode location.

Continuous dc currents elicit strong stimulation only when they are first applied, but interrupted dc currents of square-wave impulses can be used to prolong this initial response. However, dc currents of any type soon cause *polarization* at the electrode-tissue interface. Polarization increases resistance and decreases response (Figure 9-12). A muscle that has almost ceased responding to unidirectional square waves may respond with original intensity if polarity is reversed. To avoid polarization, square waves of alternating polarity may be used, but apparatus to produce these at high voltage may become expensive to construct. Thus there are some advantages to using apparatus that supplies modified sine-wave impulses derived from the regular 50- or 60-Hz commercial power supply. However, for optimum response the frequency must be reduced.

Diagnostic Electrical Properties of Meat

For alternating currents, the equivalent of resistance in Ohm's law is *impedance*. Impedance is a combination of resistance and *reactance*. Reactance produces the same end result as resistance but is due to the current surging in and out of *capacitors* and *inductors*. If capacitive components are dominant in an electrical network, the voltage lags behind the current. If inductive components are dominant, the current lags behind the voltage. Meat does not, of course, contain discrete capacitors and inductors like an electric network, but there are components in meat, probably membranes separating electrolytes, which exhibit some of the properties of capacitors. Impedance values for meat carcasses are reported by Bendall (1980), and have been used by Sale (1972) to identify meat that has been frozen and thawed prior to marketing. Low impedance can be used to identify turkeys with rapid glycolysis and rigor development (Aberle et al., 1971).

Electrical capacitance declines in meat postmortem. The rate of decline is faster in PSE pork relative to normal pork (Swatland, 1980d), and it is often lost in conditioned or frozen meat. Despite variation in pH between muscles, even a superficial muscle of the carcass (adductor) can be used for capacitance measurements to assess PSE development in other muscles (Swatland, 1981c, 1982a). Intermuscular variation is the major problem in any method used to assess meat quality in a whole carcass. For the detection of PSE in an intact carcass, the reliability of capacitance measurements is usually slightly better than that of pH measurement (Swatland, 1982c). The major advantage is in being able to dispense with glass pH electrodes.

The dielectric (capacitance) properties of fish can be used to estimate when the fish were caught and to detect if they have been frozen at any time (Kent and Jason, 1975; Jason and Lees, 1971). The scientific literature on membrane capacitance is reviewed by Cole (1970). Resistance measurements have been used during meat processing operations to monitor the penetration of curing ingredients (Callow, 1936; Banfield, 1935; Banfield and Callow, 1935a, 1935b) and to measure emulsifying capacity of meat (Webb et al., 1970). Conductivity measurements can be used to determine the age of a bruise in the carcass (Hamdy et al., 1957).

Postmortem changes in the electrical resistance of muscle are probably due to changes in the distribution of electrolytes between intracellular and extracellular compartments (Aberle et al., 1971; Swatland, 1980b). Intracellular fluid has a higher resistance than extracellular fluid (Schanne, 1969), and the extracellular space increases postmortem (Heffron and Hegarty, 1974). This may explain the eventual decline in resistivity of meat and the low resistivity of low-pH meat with poor myofibrillar water binding (Penny, 1977). The initial transient increase in resistivity which sometimes occurs after exsanguination may be due to a transient uptake of extracellular fluid because of increasing osmotic pressure inside fibers (Conway et al., 1955). Winter and Pope (1981) showed that the osmotic pressure of beef muscle increased from about 300 milliosmolals to about 500 mOs during the occurrence of rigor mortis. At first it was thought that the loss of capacitance in meat postmortem might be due to lactate-induced membrane damage. However, dark-cutting beef with a high pH does not retain the capacitance that it has immediately after slaughter (Swatland et al., 1982). Both PSE pork and dark-cutting beef deplete their ATP soon after slaughter. Perhaps this has some relationship to the rate of loss of capacitance.

Conditioning

A number of changes occur postmortem in the ultrastructure of muscle (Collan and Salmenpera, 1976; MacNaughtan, 1978a, 1978b). Polyribosomes and glycogen granules disappear within a few hours. Mitochondria soon become swollen and lose their inner matrix of granules. They sometimes undergo strange changes in morphology—their membranes may appear like a roll of paper (*myelin figures*) or like crystals. In myofibrils, the Z lines degenerate and disappear. The I bands degenerate, but the A bands appear fairly resistant to postmortem changes. Minor proteins of the myofibril such as desmin and connectin are also broken down postmortem (Young et al., 1980). A number of enzymes are involved in producing these changes, but firm evidence of proteases *within* muscle fibers is limited, at present, to *calcium-activated protease, cathepsin B*, and *cathepsin D* (Bird et al., 1980).

Calcium-activated protease is located in the cytosol, and it removes Z lines by releasing alpha-actinin (Dayton et al., 1976; Suzuki et al., 1975; Reville et al., 1976). Cathepsin B and D may be found in *lysosomes* in two locations (Bird et al., 1980): (1) in lysosomes associated with *Golgi bodies* near muscle fiber nuclei; and (2) in lysosomal parts of the sarcoplasmic reticulum around myofibrils. Acid phosphatase activity can be localized histochemically within muscle fibers, as either a granular or a diffuse reaction product (Stoward and Al-Sarraj, 1981). The rate of autolysis in muscle is influenced by the nervous system since autolysis is increased after axoplasmic transport is halted (Ramirez, 1980; Boegman and Scarth, 1981). Small increases in lysosomal enzyme activity are induced by exercise (Schott and Terjung, 1979).

Beef tenderness and taste are usually improved if carcasses or vacuum-packed cuts are aged or conditioned for a number of days after slaughter. Paradoxically, however, beef is quite tender 2 hours after slaughter, and several days of conditioning postmortem are required to restore this degree of tenderness (Ramsbottom and Strandine, 1949). Gains in tenderness due to conditioning are particularly important in grass-fed beef (Meyer et al., 1960). Conditioning is faster in carcasses maintained near body temperature (Davey and Gilbert, 1976), but commercial beef carcasses are normally kept at temperatures just above freezing point for about 10 days. The enzymology of conditioning is reviewed in detail by Penny (1980).

CELLULAR HETEROGENEITY

The overall biochemical and physiological responses which may be measured during the conversion of muscles to meat are mean values which indicate trends in the varied activity of countless individual muscle fibers. In living muscles, fibers are specialized for different types of activity, and these differences affect the course of postmortem metabolism. For example, differences in postmortem metabolism between red and white muscles are correlated with their ratios of histochemical fiber types (Beecher et al., 1969). A slow-contracting aerobic fiber in a white muscle might behave in the same manner as a slow-contracting aerobic fiber in a red muscle, and differences between white and red muscles might be due largely to differences in their relative numbers of slow-contracting aerobic fibers. However, random sources of variation such as reflex activity and neural irritability are superimposed on the differences between muscles which are due to fiber-type composition. Antemortem glycogen depletion in different fiber types may variably limit the duration of necrobiotic activity. Before reviewing the evidence in support of necrobiotic cellular heterogeneity, there are other viewpoints which deserve attention. Hamm (1977), for example, considered that, soon after death, the permeability of cell membranes is changed, that membrane potentials are abolished, and that even extreme treatments such as grinding or mincing cannot influence the pattern but only the rate of postmortem metabolism.

In pork at 24 hours postmortem, Z lines show more disruption in white than in red fibers (Dutson et al., 1974), and in cooked turkey muscle, I bands are removed from white but not red fibers (Dahlin et al., 1976). Similar results have been obtained with conditioned beef (Gann and Merkel, 1978). These differences may be affected by earlier differences in glycolytic activity between histochemical fiber types. Z lines in slow fibers are more resistant to biochemical extraction procedures than are those of fast fibers (Hikida, 1978).

Three methods can be used to study cellular heterogeneity in necrobiotic muscle, and more are needed. First, conventional biochemical techniques that utilize homogenized muscle extracts can be used for samples which differ in fiber-type composition. The problem with this method is that there are few muscles which reliably contain only a single fiber type. A second method is to use microelectrodes inserted into single fibers (Carter et al., 1967). Different fiber types can be identified from their activation thresholds and contractile properties. The injection of markers into muscle fibers after completion of observations is possible (Eisenberg et al., 1979), and there are a number of techniques for nerve cells that might be adaptable to muscle (Berthold et al., 1979; Zimmermann, 1977; Chowdhury, 1969). The third and most readily available method is the histochemical analysis of frozen muscle sections (Figure 9-13). Although the postmortem disappearance of glycogen from beef muscle fibers was first demonstrated histochemically many years ago (Robertson and Baker, 1933), the analytical possibilities of this method remained unused until recently. The periodic acid–Schiff (PAS) reaction stains glycogen, and the relationships between glycogenolysis and other aspects of postmortem metabolism in muscle are well documented (Bendall, 1973). Thus from the disappearance of glycogen, we can measure the rate and extent of lactate production at the cellular level.

The problems involved in the quantitative assessment of PAS-stained glycogen

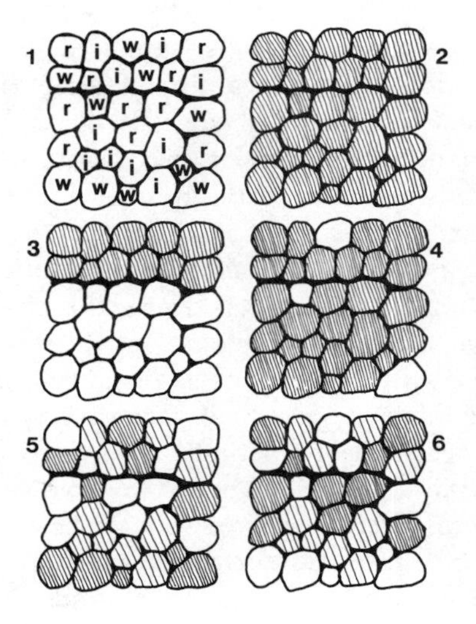

Figure 9-13 Possible glycogen depletion patterns in serial sections of bovine sternomandibularis: (1) fiber types identified by their reactions for ATPase and SDH; (2) control section from a rested animal immediately after slaughter, all fibers are stained by the PAS reaction; (3) effect of fascicular activation from a transected intramuscular nerve; (4) effect of motor unit activation from a transected peripheral nerve; (5) glycogen depletion pattern in a quiescent muscle; (6) glycogen depletion pattern due to electrical stimulation. The depth of shading indicates the extent of glycogen depletion.

in meat are quite complex. The first problem is to keep all, or a constant proportion of the glycogen in situ by histological fixation: different fixatives vary in their effectiveness (Smitherman et al., 1972). The next problem is to oxidize a constant proportion of the fixed glycogen. Even prolonged oxidation may reach only a fraction of the glycogen granule, and the size of this fraction may vary with electrolyte concentration (Ovadia and Stoward, 1971). A similar problem occurs with the variable degree of penetration of the Schiff's stain into the glycogen granules (Puchtler et al., 1974).

The absorbance of PAS-stained glycogen can be measured with a *microscope spectrophotometer* in individual muscle fibers. Glycogen distribution and glycogenolysis appear to be fairly uniform in the sarcoplasm across fiber transverse sections, and problems with differential activity beneath the sarcolemma are not as serious as with the microspectrophotometry of mitochondrial enzymes (Plate 24). However, heavy pigs often have a solid core of glycogen in the axis of their fibers (Plate 25). Longitudinally, glycogen distribution follows the pattern of sarcomere transverse striations (Pette, 1975). Thus not only must section thickness be controlled to comply with the photometric law of Bouguer or Lambert, but sarcomere length must also be regulated to give a constant number of sarcomeres in the light path.

PAS-stained glycogen has an uneven microdistribution since it is restricted to the sarcoplasm. Absorbance values may be decreased by passage of uninterrupted light through myofibrils (an example of *distributional error* in microspectrophotometry). Distributional error can be corrected by a two-wavelength correction (Oostveldt and Boeken, 1977), but many of the commercial dyes that may be used to prepare Schiff's reagent have absorption maxima which are affected by experimental conditions such as disulfite concentration (Weissig, 1978). Aerobic exposure of muscle samples must also be avoided since PAS-stained glycogen may be preserved in superficial fibers because of a reduction in glycolysis (Leet and Locker, 1973). Despite all these problems, Halkjaer-Kristensen and Ingemann-Hansen (1979a, 1979b) consider that glycogen concentration can be measured in individual muscle fibers by absorbance photometry.

It might be wise, however, to view the supposed linear relationship between absorbance and concentration with some caution. Attempts to verify this relationship by correlating biochemical values for glycogen with mean fiber absorbance are inadmissible. In many situations, this correlation is mainly due to the correlation of the relative number of PAS-positive fibers with glycogen concentration. Microchemical analysis of individual fibers from a frozen section may be the only real test of the relationship (Nelson and Tashiro, 1973). This excessive caution is prompted by the fact that, in samples taken shortly after animal death, the absorbance of PAS-stained glycogen increases relative to control samples taken at death (Figure 9-14). Although the possibility of glycogen synthesis cannot be dismissed, it seems more likely that phosphorylase and branching enzyme activity are making external branches of the glycogen granule more accessible to periodic acid and to Schiff's reagent. This uncertainty in early necrobiosis should not be allowed to obscure the fact that the PAS reaction does reveal several later aspects of cellular heterogeneity which are quite compatible with existing biochemical information on overall tissue metabolism.

Investigation of the response of necrobiotic beef muscle to electrical stimulation

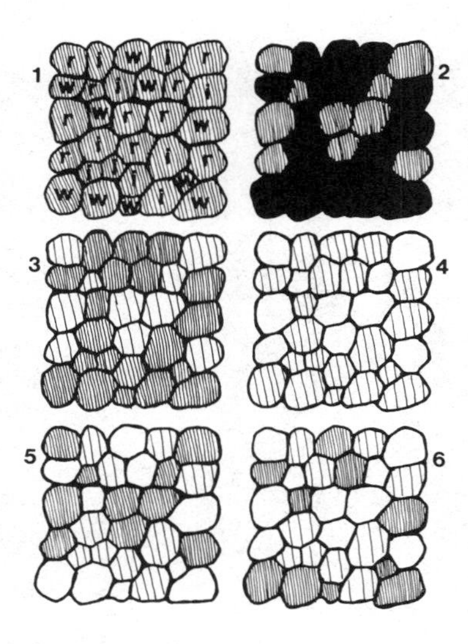

Figure 9-14 Possible glycogen depletion patterns in quiescent bovine sternomandibularis (1 to 4), and differences in initial postmortem glycogen levels due to phasic (5) or tonic (6) muscle activity before slaughter.

shows that glycogenolysis is not accelerated to the same degree in all fibers (Plate 26). Fibers with strong ATPase and weak SDH reactions show the greatest response (Swatland, 1975a, 1981a), but whether this is due to intrinsic biochemical differences between fiber types or to lower neuromuscular activation thresholds is not known at present. Glycogenolysis in fibers with strong ATPase and weak aerobic enzyme activity is also preferentially activated at low temperatures (Swatland, 1979a).

The starting levels of glycogen for postmortem metabolism in different fiber types depend on depletion and repletion patterns which occur during transport and during antemortem holding. Slow-contracting fibers are mainly responsible for shivering (Lupandin and Poleshchuk, 1979), so that animals subject to severe cold might have less glycogen in these fibers. In rats, repletion is more efficient in fast-contracting red muscles and is less efficient in fast-contracting white muscles (Conlee et al., 1978). In market-weight pork carcasses, anaerobic fibers usually contain more glycogen than aerobic fibers at the time of animal exsanguination (Swatland, 1975b), and they also make the greatest contribution to overall glycolysis (Swatland, 1976c). However, this is mainly a consequence of the initial distribution of glycogen, since the maximum rate of glycogen depletion is very similar in the three main fiber types (Swatland, 1977a). Although fibers with strong ATPase and weak SDH reactions respond more readily to electrical stimulation, to cold, and to reflex stimulation, they also show the slowest glycogenolysis in quiescent muscles (Swatland, 1975a), probably as a consequence of having fewer calcium-containing mitochondria and a more efficient sarcoplasmic reticulum to keep calcium levels low in the cytosol (Figure 9-15).

Staining with iodine provides another method of studying glycogen metabolism. Amylose chains longer than 18 units can be stained with iodine, and as chains become longer, the color changes from brown to blue (Bailey and Whelan, 1961). The intrinsic glycogen that is present in a fiber can be used as a primer for the assembly of amylose chains from glucose-1-phosphate by phosphorylase (Takeuchi and Kuriaki, 1955). Branching enzyme activity may redistribute the added glucose units so as to increase the number but decrease the length of outer amylose chains. These shortened outer

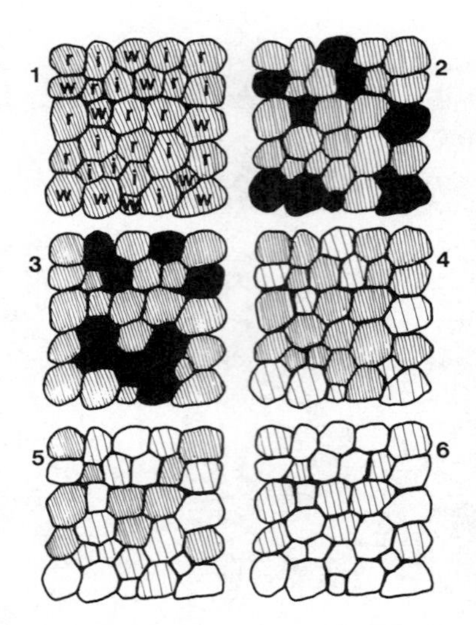

Figure 9-15 Changes in the intensity of the periodic acid–Schiff reaction during glycogen depletion in an electrically stimulated bovine sternomandibularis muscle. (1) through (6) show the sequence of changes from immediately after slaughter to the end of the treatment. Note that (5) fits the pattern described earlier in (5) of Figure 9-14.

chains are stained brown with iodine. On the other hand, the combination of strong phosphorylase and weak branching enzyme activity results in a blue stain (Miyayama, 1977). From the absorbance spectra of beef muscle fibers it appears that aerobic fibers might have stronger branching enzyme activity than that of anaerobic fibers, and that anaerobic fibers are dominated by strong phosphorylase activity (Swatland, 1978a).

RIGOR MORTIS

The conversion of muscles to meat is complete when the muscles have depleted their energy reserves, or when they have lost their ability to utilize any remaining reserves. In living muscles at rest, an ATP molecule binds to each myosin molecule head. In this condition the myosin head is said to be "charged." In resting muscle, further developments between the actin and myosin of thin and thick filaments are prevented by the intrusion of tropomyosin and troponin molecules. Contraction in living muscle is initiated by the release of calcium ions from the sarcoplasmic reticulum, and calcium ions lead to the removal of the tropomyosin-troponin intrusion. As a muscle contracts, charged myosin molecule heads attach to actin molecules, ATP is split to ADP with a release of energy, and the myosin molecule head swivels to cause filament sliding. The myosin molecule head, which is still attached to its site on the actin, can detach itself only if a new ATP molecule is available to be bound. When muscle is converted to meat, myosin molecule heads remain locked to actin and even passive filament sliding is impossible.

Loss of extensibility can be used as a simple physical measure of rigor development. The review by Bendall (1973) is the standard reference, and it should be consulted for details beyond those given here. Loss of extensibility can be measured by apparatus which can gently load the free end of a hanging muscle strip and record subsequent movement of the free end (Figure 9-16). If a suitable load is placed on a muscle strip that still contains ATP, the free end drops as the muscle stretches. When the load is removed, the muscle returns to its original length. When ATP is no longer

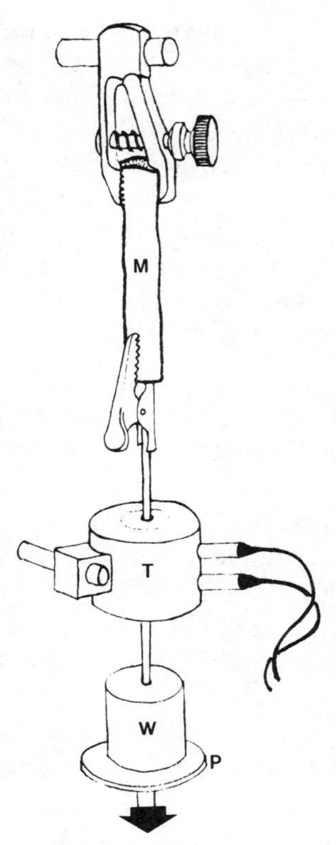

Figure 9-16 Simplified rigorometer. When the platform (P) drops, the weight (W) hangs on the muscle (M). If the muscle stretches in response to this treatment (prerigor condition), the subsequent movement is recorded by a transducer (T). The platform drops and then returns to its original position at regularly timed intervals.

available within the muscle, the muscle is only very slightly extensible, and there is very little drop of the loaded free end (Figure 9-17). Apparatus with this capability may be called a *rigorometer*. Muscle strips are usually maintained in an atmosphere of nitrogen gas to prevent aerobic resynthesis of ATP.

For a short time after exsanguination, ATP can be resynthesized from creatine

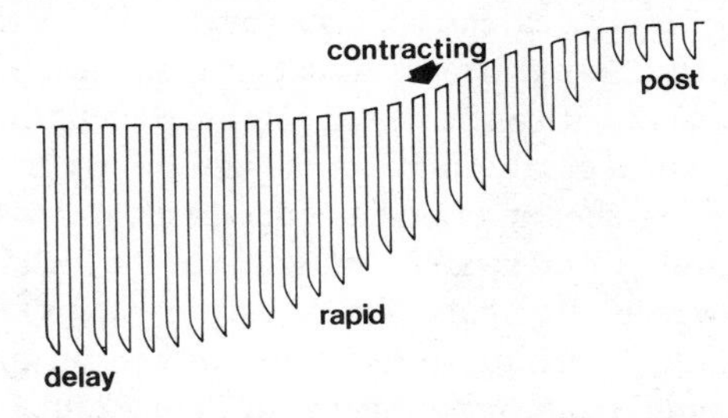

Figure 9-17 Simplified diagram of the trace from a rigorometer. A real trace might extend over many meters of chart paper on which would be marked thousands of muscle movements. During the delay period (often many hours) the muscle extends to a constant length every time it is loaded. As the muscle sets into rigor mortis there is a rapid phase of diminishing extensibility. After rigor mortis is fully developed (postrigor phase), only a small and constant amount of extension occurs when the muscle is loaded. The terminology for the features of the trace follows that used by Bendall (1973). Sometimes the muscle contracts during the development of rigor mortis and the bottom end of the muscle sample is elevated.

phosphate. After the creatine phosphate has been used up, the length of time before the occurrence of rigor mortis depends on the amount of glycogen which is available within the muscle and on the survival of glycolytic enzymes. A number of features appear on a rigorometer trace (Figure 9-17). The *delay period* is when ATP is still being resynthesized, the *rapid phase* is when individual muscle fibers run out of ATP, and the final *postrigor phase* is when a sufficient number of fibers have set in rigor mortis to prevent any further extensibility. Sometimes, the muscle may slowly and weakly contract as it goes into rigor. The behavior of muscle strips in a rigorometer and of whole muscles in a carcass is greatly affected by the condition of the animal at the time of slaughter. These differences may be seen on the rigorometer as follows;

1. An animal that is calm at the time of slaughter gives a long delay, a slow rapid phase, and a decrease in unloaded length at body temperature but not at room temperature.
2. An animal that struggles at the time of slaughter gives a short delay, a short rapid phase, and a decrease in unloaded length at body temperature but not at room temperature.
3. An exhausted animal gives an extremely short or nonexistent delay period, a very short rapid phase, and shortens its unloaded length at both body and room temperature.
4. A starved animal gives a short delay period, a fairly long rapid phase, and a decrease in unloaded length at body temperature but not room temperature.

Certain muscles may attempt to shorten as they set in rigor mortis, depending on animal condition at the time of slaughter and on muscle temperature (superficial muscles will cool faster if not extensively insulated by fat). Some of these muscles will be completely prevented from shortening because they are already stretched by the weight of the hanging carcass. Unrestrained muscles, such as the posterior hindlimb muscles of a hanging carcass, are free to contract. If rigor contraction is stronger than normal, as in PSE pork carcasses (Davis et al., 1978b), the position of the forelimb may be raised against gravity. Unrestrained carcass muscles that contract as they go into rigor produce meat with a shortened sarcomere length, a greater overlap of thick and thin filaments, and a greater number of rigor bonds linking the filaments. Meat in this condition is less tender and often has a decreased water-binding capacity. The extent of rigor contraction of unrestrained muscles can be lessened by hanging carcasses from the pelvis. However, retail cuts of meat may lose their familiar appearance, the normal tenderization by stretching may be lost in certain muscles, and carcasses may be difficult to handle on a normal overhead rail system.

Congratulations to the reader who has made it to the end, even if it has been by leaps and bounds. I hope you will agree that the scientific study of the structure and development of meat animals is a rather intriguing tangle of technology and basic science. As a research field, it offers both the intellectual challenge of solving some complex biological problems together with the satisfaction of undertaking something directly useful to society.

REFERENCES

Aberle, E. D., Stadelman, W. J., Zachariah, G. L., and Haugh, C. G. 1971. *Poult. Sci.* 50:743.

Ackermann, R. D., Hamlin, R. L., and Muir, W. W. 1976. *Am. J. Vet. Res.* 37:715.

Addis, P. B., Nelson, D. A., Ma, R. T.-I., and Burroughs, J. R. 1974. *J. Anim. Sci.* 38:279.

Adrian, E. D. 1943. *Brain* 66:89.

Althen, T. G., Ono, K., and Topel, D. G. 1977. *J. Anim. Sci.* 44:985.

Althen, T. G., Steele, N. C., and Ono, K. 1979. *J. Anim. Sci.* 48:531.

Altrogge, D. M., Topel, D. G., Cooper, M. A., Hallberg, J. W., and Draper, D. D. 1980. *J. Anim. Sci.* 51:74.

Andersen, L. D., Parrish, F. C., and Topel, D. G. 1975. *J. Anim. Sci.* 41:1600.

Andresen, E. 1979. *Acta Agric. Scand.* 29:369.

Andresen, E. 1980. *Livestock Prod. Sci.* 7:155.

Bailey, J. M., and Whelan, W. J. 1961. *J. Biol. Chem.* 236:969.

Baldwin, K. M., Fitts, R. H., Booth, F. W., Winder, W. W., and Holloszy, J. O. 1975. *Pfluegers Arch.* 354:203.

Ball, R. A., Annis, C. L., Topel, D. G., and Christian, L. L. 1973. *Vet. Med. Small Anim. Clin.* 68:1156.

Ball, R. A., Topel, D. G., Marple, D. N., and Annis, C. L. 1974. *Can. J. Comp. Med.* 38:153.

Banfield, F. H. 1935. *J. Soc. Chem. Ind.* 13:411.

Banfield, F. H., and Callow, E. H. 1935a. *J. Soc. Chem. Ind.* 13:411.

Banfield, F. H., and Callow, E. H. 1935b. *J. Soc. Chem. Ind.* 13:414.

Banks, P., Bartley, W., and Birt, L. M. 1976. *The Biochemistry of the Tissues*. John Wiley & Sons, London.

Barnes, E. M., Mead, G. C., and Griffiths, N. M. 1973. *Br. Poult. Sci.* 14:229.

Beecher, G. R., Briskey, E. J., and Hoekstra, W. G. 1965a. *J. Food Sci.* 30:477.

Beecher, G. R., Cassens, R. G., Hoekstra, W. G., and Briskey, E. J. 1965b. *J. Food Sci.* 30:969.

Beecher, G. R., Kastenschmidt, L. L., Hoekstra, W. G., Cassens, R. G., and Briskey, E. J. 1969. *Agric. Food Chem.* 17:29.

Bembers, M., and Satterlee, L. D. 1975. *J. Food Sci.* 40:40.

Bendall, J. R. 1966. *J. Sci. Food Agric.* 17:333.

Bendall, J. R. 1973. Postmortem changes in muscle. In G. H. Bourne (ed.), *The Structure and Function of Muscle*, Vol. 2, Pt. 2, pp. 244–309. Academic Press, New York.

Bendall, J. R. 1975. *J. Sci. Food Agric.* 26:55.

Bendall, J. R. 1978. *Meat Sci.* 2:91.

Bendall, J. R. 1979. *Meat Sci.* 3:143.

Bendall, J. R. 1980. The electrical stimulation of carcasses of meat animals. In R. A. Lawrie (ed.), *Developments in Meat Science 1*, pp. 37–59, Applied Science Publishers, London.

Bendall, J. R., and Rhodes, D. N. 1976. Electrical stimulation of the beef carcass and its practical application. *Proc. 22nd Eur. Meat Res. Workers Congr.*, Kavlinge, Sweden, B2:3.

Bendall, J. R., and Taylor, A. A. 1970. *Biochem. J.* 118:887.

Bendall, J. R., and Wismer-Pedersen, J. 1962. *J. Food Sci.* 27:144.

Bendall, J. R., Ketteridge, C. C., and George, A. R. 1976. *J. Sci. Food Agric.* 27:1123.

Bergmann, V. von. 1979. *Exp. Pathol.* 17:243.

Bergmann, V. von, and Wesemeier, H. 1970. *Arch. Exp. Veterinaermed.* 24:1241.

Bergmann, V. von, and Wesemeier, H. 1973. *Mh. VetMed.* 28:245.

BERMAN, M. C., and KENCH, J. E. 1971. *Proc. 2nd Int. Symp. Condition Meat Quality Pigs*, Zeist, Pudoc, Wageningen, pp. 29–35.

BERTHOLD, C.-H., KELLERTH, J.-O., and Conradi, S. 1979. *J. Comp. Neurol.* 184:709.

BESSMAN, S. P., and GEIGER, P. J. 1981. *Science* 211:448.

BETHLEM, J. 1970. *Muscle Pathology*. North-Holland Publishing Company, Amsterdam.

BIANCHI, C. P., and NARAYAN, S. 1982. *Science* 215:296.

BICKHARDT, K. VON, and CHEVALIER, H. J. 1968. *Dtsch. Tierarztl. Wochenschr*. 75:198.

BICKHARDT, K. VON, and CHEVALIER, H. J. 1969. *Proc. 1st Congr. Int. Pig Vet. Soc*., Cambridge, p. 60.

BICKHARDT, K. VON, UEBERSCHAR, S., and GIESE, W. 1967. *Dtsch. Tierarztl. Wochenschr*. 74:324.

BICKHARDT, K. VON, CHEVALIER, H. J., GIESE, W., and REINHARD, H. J. 1972. *Zentalbl. Veterinaermed. Suppl*. 18.

BIRD, J. W. C., CARTER, J. H., TRIEMER, R. E., BROOKS, R. M., and SPANIER, A. M. 1980. *Fed. Proc*. 39:20.

BIRTH, G. S., DAVIS, C. E., and TOWNSEND, W. E. 1978. *J. Anim. Sci*. 46:639.

BLACK, W. H., HINER, R. L., BURK, L. B., ALEXANDER, L. M., and WILSON, C. V. 1940. *USDA Tech. Bull. 717*.

BLACKMORE, D. K. 1979. *Vet. Rec*. 105:372.

BLUMBERG, H., JANIG, W., RIECKMANN, C., and SZULCZYK, P. 1980. *J. Auton. Nerv. Syst*. 2:223.

BOEGMAN, R. J., and SCARTH, B. 1981. *Exp. Neurol*. 73:37.

BOUTON, P. E., FORD, A. L., HARRIS, P. V., and SHAW, F. D. 1978. *J. Food Sci*. 43:1392.

BOUTON, P. E., FORD, A. L., HARRIS, P. V., and SHAW, F. D. 1980a. *Meat Sci*. 4:145.

BOUTON, P. E., WESTE, R. R., and SHAW, F. D. 1980b. *J. Food Sci.* 45:148.

BRADLEY, R., WELLS, G. A. H., and GRAY, L. J. 1979. *Vet. Rec*. 104:183.

BREAZILE, J. E., SWAFFORD, B. C., and THOMPSON, W. D. 1966. *Am. J. Vet. Res*. 27:1369.

BRIGGS, F. N., POLAND, J. L., and SOLARO, R. J. 1977. *J. Physiol*. 266:587.

BRISKEY, E. J. 1964. *Adv. Food Res*. 13:89.

BRISKEY, E. J., CASSENS, R. G., and TRAUTMAN, J. C. 1966. *The Physiology and Biochemistry of Muscle as a Food.* University of Wisconsin Press, Madison.

BRISKEY, E. J., CASSENS, R. G., and MARSH, B. B. 1970. *The Physiology and Biochemistry of Muscle as a Food, 2*. University of Wisconsin Press, Madison.

BRITT, B. A., KALOW, W., and ENDRENYI, L. 1977. Malignant hyperthermia-pattern of inheritance in swine. In J. A. Aldrete and B. A. Britt (eds.), *The Second International Symposium on Malignant Hyperthermia*, pp. 195–211. Grune & Stratton, New York.

BROOKS, G. A., BRAUNER, K. E., and CASSENS, R. G. 1973. *Am. J. Physiol*. 224:1162.

BRUIN, A. DE. 1971. *Proc. 2nd Int. Symp. Condition Meat Quality Pigs*, Zeist, Pudoc, Wageningen, pp. 86–89.

BUEGE, D. R., and MARSH, B. B. 1975. *Biochem. Biophys. Res. Commun*. 65:478.

BURES, J., PETRAN, M., and ZACHAR, J. 1967. *Electrophysiological Methods in Biological Research*, 3rd ed., pp. 507–546. Academic Press, New York.

BURGER, H. C., and VAN DONGEN, R. 1960. *Phys. Med. Biol.* 5:431.

BYGRAVE, F. L. 1978. *Biol. Rev*. 53:43.

CALLOW, E. H. 1936. *The Electrical Resistance of Muscular Tissue and Its Relation to Curing*. Dept. Sci. Ind. Res. (Lond.), Food Invest. Board, Spec. Rep. 75.

CAMPION, D. R., and TOPEL, D. G. 1975. *J. Anim. Sci*. 41:779.

CAMPION, D. R., CASSENS, R. G., and NAGLE, F. J. 1973. *Growth* 37:257.

CAMPION, D. R., EIKELENBOOM, G., and CASSENS, R. G. 1974. *J. Anim. Sci.* 39:68.

CAMPION, D. R., EIKELENBOOM, G., and CASSENS, R. G. 1979. *Res. Vet. Sci.* 27:116.
CARSE, W. A. 1973. *J. Food Technol.* 8:163.
CARTER, N. W., RECTOR, F. C., CAMPION, D. R., and SELDIN, D. W. 1967. *Fed. Proc.* 26:1322.
CASSENS, R. G., and COOPER, C. C. 1971. *Adv. Food Res.* 19:1.
CASSENS, R. G., and NEWBOLD, R. P. 1966. *J. Sci. Food Agric.* 17:254.
CASSENS, R. G., and NEWBOLD, R. P. 1967. *J. Food Sci.* 32:13.
CASSENS, R. G., BRISKEY, E. J., and HOEKSTRA, W. G. 1963. *Biodynamica* 9:165.
CASSENS, R. G., JUDGE, M. D., SINK, J. D., and BRISKEY, E. J. 1965. *Proc. Soc. Exp. Biol. Med.* 120:854.
CASSENS, R. G., COOPER, C. C., and BRISKEY, E. J. 1969. *Acta Neuropathol.* 12:300.
CHEAH, K. S., and CHEAH, A. M. 1976. *J. Sci. Food Agric.* 27:1137.
CHOWDHURY, T. K. 1969. Techniques of intracellular microinjection. In M. Lavalee, O. F. Schanne, and N. C. Hebert (eds.), *Glass Microelectrodes*, pp. 404–423. John Wiley & Sons, New York.
CHRISTIAN, L. L. 1972. A review of the role of genetics in animal stress susceptibility and meat quality. In R. Cassens, F. Giesler, and Q. Kolb (eds.), *The Proceedings of the Pork Quality Symposium*, pp. 91–115. University of Wisconsin Press, Madison.
CHRYSTALL, B. B., and DEVINE, C. E. 1978. *Meat Sci.* 2:49.
CHRYSTALL, B. B., and HAGYARD, C. J. 1976. *N.Z. J. Agric. Res.* 19:7.
CHRYSTALL, B. B., DEVINE, C. E., and DAVEY, C. L. 1980. *Meat Sci.* 4:69.
CLARKE, F. M., SHAW, F. D., and MORTON, D. J. 1980. *Biochem. J.* 186:105.
CLIPLEF, R. L., and STRAIN, J. H. 1976. *Can. J. Anim. Sci.* 56:417.
COLE, K. S. 1970. Dielectric properties of living membranes. In F. Snell, J. Wolken, G. Iverson, and J. Lam (eds.), *Physical Principles of Biological Membranes*. Gordon and Breach, Science Publishers, New York.
COLLAN, Y., and SALMENPERA, M. 1976. *Acta Neuropathol.* 35:219.
COLLIS, K. A., and STARK, A. J. 1977. *Res. Vet. Sci.* 23:326.
CONLEE, R. K., HICKSON, R. C., WINDER, W. W., HAGBERG, J. M., and HOLLOSZY, J. O. 1978. *Am. J. Physiol.* 235:R145.
CONLEE, R. K., MCLANE, J. A., RENNIE, M. J., WINDER, W. W., and HOLLOSZY, J. O. 1979. *Am. J. Physiol.* 237:R291.
CONNETT, R. J. 1978. *Am. J. Physiol.* 234:C110.
CONNETT, R. J. 1979. *Am. J. Physiol.* 237:C231.
CONWAY, E. J., GEOGHEGAN, H., and MCCORMACK, J. I. 1955. *J. Physiol.* 130:427.
COOPER, C. C., CASSENS, R. G., and BRISKEY, E. J. 1969. *J. Food Sci.* 34:299.
CORNFORTH, D. P., PEARSON, A. M., and MERKEL, R. A. 1980. *Meat Sci.* 4:103.
CROFT, P. G. 1952. *Vet. Rec.* 64:255.
CROFT, P. G. 1957. Electrical stunning. In, *Meat Hygiene*. Monogr. 33, pp. 147–159. World Health Organization, Geneva.
CROSS, H. R. 1979. *J. Food Sci.* 44:509.
CURRIE, R. W., and WOLFE, F. H. 1980. *Meat Sci.* 4:123.
CUTHBERTSON, A. 1981. Commercial application in Britain and the rest of Europe. In *Commercial Application of Electrical Stimulation*. Meat and Livestock Commission, Bletchley, England.
CUTTING, C. L. 1974. *Inst. Meat Bull.* 84:8.
DAHLIN, K. J., ALLEN, C. E., BENSON, E. S., and HEGARTY, P. V. J. 1976. *J. Ultrastruct. Res.* 56:96.
DALRYMPLE, R. H., KASTENSCHMIDT, L. L., and CASSENS, R. G. 1973. *Growth* 37:19.
DARRAH, P. S., BEITZ, D. C., TOPEL, D. G., and CHRISTIAN, L. L. 1981. *J. Anim. Sci.* 53:1000.

Davey, C. L., and Chrystall, B. B. 1981. Commercial application of electrical stimulation in New Zealand. In *Commercial Application of Electrical Stimulation*. Meat and Livestock Commission, Bletchley, England.

Davey, C. L., and Gilbert, K. V. 1974. *J. Food Technol.* 9:51.

Davey, C. L., and Gilbert, K. V. 1976. *J. Sci. Food Agric.* 27:244.

Davey, C. L., Gilbert, K. V., and Carse, W. A. 1976. *N.Z. J. Agric. Res.* 19:13.

Davies, A. S. 1974. *Anim. Prod.* 19:377.

Davis, C. E., Birth, G. S., and Townsend, W. E. 1978a. *J. Anim. Sci.* 46:634.

Davis, C. E., Townsend, W. E., and McCampbell, H. C. 1978b. *J. Anim. Sci.* 46:376.

Dayton, W. R., Goll, D. E., Zeece, M. G., Robson, R. M., and Reville, W. J. 1976. *Biochemistry* 15:2150.

de Fremery, D., and Pool, M. F. 1960. *Food Res.* 25:73.

de la Haba, G., Cooper, G. W., and Elting, V. 1968. *J. Cell. Physiol.* 72:21.

Dildey, D. D., Aberle, E. D., Forrest, J. C., and Judge, M. D. 1970. *J. Anim. Sci.* 31:681.

Dolken, G., Leisner, E., and Pette, D. 1975. *Histochemistry* 43:113.

Dreyer, J. H., Naude, R. T., and Gouws, P. J. 1972. *S. Afr. J. Anim. Sci.* 2:109.

Dumont, B. L., and Schmitt, O. 1970. *Ann. Genet. Sel. Anim.* 2:381.

Dunn, R. B., and Critz, J. B. 1975. *Am. J. Physiol.* 229:255.

Dutson, T. R. 1981. Commercial application of electrical stimulation in the USA. In *Commercial Application of Electrical Stimulation*. Meat and Livestock Commission, Bletchley, England.

Dutson, T. R., Pearson, A. M., and Merkel, R. A. 1974. *J. Food Sci.* 39:32.

Dutson, T. R., Merkel, R. A., Pearson, A. M., and Gann, G. L. 1978. *J. Anim. Sci.* 46:1212.

Dutson, T. R., Savell, J. W., and Smith, G. C. 1982. *Meat Sci.* 6:159.

Dvorak, M. 1973. *Zentralbl. Veterinaermed. A.* 20:370.

Eccles, J. C., Fatt, P., and Koketsu, K. 1954. *J. Physiol.* 126:524.

Eighmy, J. J., Williams, C. H., and Anderson, R. A. 1978. In J. A. Aldrete and B. A. Britt (eds.), *The Second International Symposium on Malignant Hyperthermia,* pp. 161–173. Grune & Stratton, New York.

Eikelenboom, G., and Minkema, D. 1974. *Tijdschr. Diergeneeskd.* 99:(61) 421.

Eikelenboom, G., and Wiess, G. M. 1972. *J. Anim. Sci.* 35:1096.

Eikelenboom, G., Campion, D. R., Kauffman, R. G., and Cassens, R. G. 1974. *J. Anim. Sci.* 39:303.

Eikelenboom, G., Minkema, D., van Eldik, P., and Symbesma, W. 1977. In J. A. Aldrete and B. A. Britt (eds.), *The Second International Symposium on Malignant Hyperthermia*, pp. 141–146. Grune & Stratton, New York.

Eikelenboom, G., Minkema, D., van Eldik, P., and Sybesma, W. 1978. *Livestock Prod. Sci.* 5:277.

Eisenberg, B. R., Mathias, R. T., and Gilai, A. 1979. *Am. J. Physiol.* 237:C50.

Elizondo, G., Addis, P. B., Rempel, W. E., Madero, C., Martin, F. B., Anderson, D. B., and Marple, D. N. 1976. *J. Anim. Sci.* 43:1004.

Ellersieck, M. R., Veum, T. L., Durham, T. L., McVickers, W. R., McWilliams, S. N., and Lasley, J. F. 1979. *J. Anim. Sci.* 48:453.

Ernst, L. J. 1980. *Proc. Reciprocal Meat Conf.* 33:37.

Fabiansson, S. 1981. Commercial application of low voltage stimulation of beef meat in Sweden. In *Commercial Application of Electrical Stimulation*. Meat and Livestock Commission, Bletchley, England.

Fairholme, E. G., and Pain, W. 1924. *A Century of Work for Animals. The History of the R.S.P.C.A., 1824–1924*. John Murray (Publishers), London.

Fischer, C., Hamm, R., and Honikel, K. O. 1979. *Meat Sci.* 3:11.

FITTS, R. H., NAGLE, F. J., and CASSENS, R. G. 1974. *Eur. J. Appl. Physiol.* 33:275.
FOLLETT, M. J., NORMAN, G. A., and RATCLIFF, P. W. 1974. *J. Food Technol.* 9:509.
FORREST, J. C., KASTENSCHMIDT, L. L., BEECHER, G. R., GRUMMER, R. H., HOEKSTRA, W. G., and BRISKEY, E. J. 1965. *J. Food Sci.* 30:492.
FORREST, J. C., JUDGE, M. D., SINK, J. D., HOEKSTRA, W. G., and BRISKEY, E. J. 1966. *J. Food Sci.* 31:13.
FORREST, J. C., WILL, J., SCHMIDT, G. R., JUDGE, M. D., and BRISKEY, E. J. 1968. *J. Appl. Physiol.* 24:33.
FORREST, J. C., ABERLE, E. D., HEDRICK, H. B., JUDGE, M. D., and MERKEL, R. A. 1975. *Principles of Meat Science.* W. H. Freeman and Company, Publishers, San Francisco.
FRASER, D., RITCHIE, J. S. D., and FRASER, A. F. 1975. *Br. Vet. J.* 131:653.
FUNABIKI, R., IDE, S., EBE, N., TAGUCHI, O., and YASUDA, K. 1975. *Agric. Biol. Chem.* 39:791.
GALLANT, E. M. 1980. *Am. J. Vet. Res.* 41:1069.
GALLWEY, W. J., and TARRANT, P. V. 1979. *Acta Agric. Scand. Suppl.* 21:32.
GANN, G. L., and MERKEL, R. A. 1978. *Meat Sci.* 2:129.
GEDDES, L. A., and BAKER, L. E. 1967. *Med. Biol. Eng.* 5:271.
GILBERT, K. V., and DEVINE, C. E. 1982. *Meat Sci.* 7:197.
GLEN, J. B. 1971. The use of carbon dioxide for pre-slaughter anaesthesia. In *Humane Killing and Slaughterhouse Techniques*, pp. 15–23. Universities Federation for Animal Welfare, Potters Bar, Hertfordshire, England.
GLUCK, E., and PAUL, R. J. 1977. *Pfluegers Arch.* 370:9.
GOLDFISCHER, S. 1967. *J. Cell Biol.* 34:398.
GREASER, M. L., CASSENS, R. G., HOEKSTRA, W. G., BRISKEY, E. J., SCHMIDT, G. R., CARR, S. D., and GALLOWAY, D. E. 1969. *J. Anim. Sci.* 28:589.
GREY, T. C., and JONES, J. M. 1977. *Br. Poult. Sci.* 18:671.
GRIFFITHS, N. M. 1975. *Br. Poult. Sci.* 16:83.
GRILLO, T. A. I., and OZONE, K. 1962. *Nature* (*Lond.*) 195:902.
GRONERT, G. A., MILDE, J. H., and THEYE, R. A. 1978. Role of sympathetic activity in porcine malignant hyperthermia. In J. A. Aldrete and B. A. Britt (eds.), *The Second International Symposium on Malignant Hyperthermia*, pp. 159–160. Grune & Stratton, New York.
GRONERT, G. A., HEFFRON, J. J. A., and TAYLOR, S. R. 1979. *Eur. J. Pharmacol.* 58:179.
GRONERT, G. A., MILDE, J. H., and THEYE, R. A. 1978. Role of sympathetic activity in porcine
GROSS, S. R., and MAYER, S. E. 1974. *Life Sci.* 14:401.
GUTIERREZ, M., DE BURGOS, N. M., BURGOS, C., and BLANCO, A. 1974. *Comp. Biochem. Physiol.* 48B:379.
HAASE, S. VON, KALLWEIT, E., STEINHAUF, D., and WENIGER, J. H. 1972. *Z. Tierz. Zuechtungsbiol.* 89:217.
HALKJAER-KRISTENSEN, J., and INGEMANN-HANSEN, T. 1979a. *Histochem. J.* 11:629.
HALKJAER-KRISTENSEN, J., and INGEMANN-HANSEN, T. 1979b. *Histochem. J.* 11:737.
HALL, G. M., BENDALL, J. R., LUCKE, J. N., and LISTER, D. 1976a. *Br. J. Anaesth.* 48:305.
HALL, G. M., LUCKE, J. N., and LISTER, D. 1976b. *Br. J. Anaesth.* 48:1135.
HALL, G. M., LUCKE, J. N., and LISTER, D. 1977. *Anaesthesia* 32:472.
HALL, J. T. 1972. Economic importance of pork quality. In R. Cassens, F. Giesler, and Q. Kolb (eds.), *The Proceedings of the Pork Quality Symposium*, pp. ix–xii. University of Wisconsin Press, Madison.
HAMDY, M. K., KUNKLE, L. E., and DEATHERAGE, F. E. 1957. *J. Anim. Sci.* 16:490.
HAMM, R. 1972. *Kolloidchemie des Fleisches- das Wasserbindungsvermoegen des Muskeleiweisses in Theorie und Praxis.* Verlag Paul Parey, Berlin.
HAMM, R. 1977. *Meat Sci.* 1:15.
HAND, A. R. 1974. *J. Histochem. Cytochem.* 22:207.
HARRISON, G. G. 1973. *Br. J. Anaesth.* 45:1019.

HARSHAM, A. A., and DEATHERAGE, F. E. 1951. U.S. Patent No. 2,544,681.
HEFFRON, J. J. A., and HEGARTY, P. V. J. 1974. *Comp. Biochem. Physiol.* 49A:43.
HEGARTY, P. V. J., DAHLIN, K. J., and BENSON, E. S. 1978. *Experientia* 34:1070.
HERRING, H. K., CASSENS, R. G., and BRISKEY, E. J. 1964. *Biodynamica* 9:257.
HIKIDA, R. S. 1978. *J. Ultrastruct. Res.* 65:266.
HOENDERKEN, R. 1978. Elektrische Bedwelming van Slachtvarken. Thesis, Utrecht.
HOENDERKEN, R., VAN LOGTESTIJN, J. G., SYBESMA, W., and SPANJAARD, W. J. M. 1979. *Fleischwirtschaft* 59:1572.
HOULIER, B., VALIN, C., MONIN, G., and SALE, P. 1980. Is electrical stimulation efficiency muscle dependent? *Proc. 26th Eur. Meat Res. Workers Congr.*, Colorado Springs, Colo.
HUFFMAN, D. L. 1980. *Proc. Reciprocal Meat Conf.* 33:4.
HWANG, P. T., ADDIS, P. B., REMPEL, W. E., and ANTONIK, A. 1977. *J. Anim. Sci.* 45:1015.
HWANG, P. T., MCGRATH, C. J., ADDIS, P. B., REMPEL, W. E., THOMPSON, E. W., and ANTONIK, A. 1978. *J. Anim. Sci.* 47:630.
ILCHMANN, G. VON, 1969. *Mh. VetMed.* 24:769.
IVY, J. L., WITHERS, R. T., VAN HANDEL, P. J., ELGER, D. H., and COSTILL, D. L. 1980. *J. Appl. Physiol.* 48:523.
JAMES, N. T. 1968. *Nature* 219:1174.
JAMES, N. T. 1972. *Comp. Biochem. Physiol.* 41B:457.
JASON, A. C., and LEES, A. 1971. *Estimation of Fish Freshness by Dielectric Measurement.* Report T71/7. Torrey Research Station, Aberdeen, Scotland.
JEACOCKE, R. E. 1977a. *J. Food Technol.* 12:375.
JEACOCKE, R. E. 1977b. *J. Sci. Food Agric.* 28:551.
JENKINS, R. R., and TENGI, J. 1981. *Experientia* 37:67.
JENNISCHE, E., ENGER, E., MEDEGARD, A., APPELGREN, L., and HALJAMAE, H. 1978. *Circ. Shock* 5:251.
JOHANSSON, G., and JONSSON, L. 1977. *J. Comp. Pathol.* 87:67.
JOHANSSON, G., JONSSON, L., LANNEK, N., and LINDBERG, P. 1973. *Acta Vet. Scand.* 14:764.
JORGENSEN, P. F., HYLDGAARD-JENSEN, J., MOUSTGAARD, J., and EIKELENBOOM, G. 1976. *Acta Vet. Scand.* 17:370.
JOSEPH, R. L., and CONNOLLY, J. 1977. *J. Food Technol.* 12:231.
JUDGE, M. D., BRISKEY, E. J, and MEYER, R. K. 1966. *Nature* (*Lond.*) 212:287.
KACIUBA-USCILKO, H., DUDLEY, G. A., and TERJUNG, R. L. 1980. *Am. J. Physiol.* 238:E518.
KAGEN, L. J., and GUREVICH, R. 1967. *J. Histochem. Cytochem.* 15:436.
KALOW, W. 1978. Concluding remarks. In J. A. Aldrete and B. A. Britt (eds.), *The Second International Symposium on Malignant Hyperthermia*, pp. 553–555. Grune & Stratton, New York.
KALOW, W., BRITT, B. A., and PETERS, P. 1978. Rapid simplified techniques for measuring caffeine contraction for patients with malignant hyperthermia. In J. A. Aldrete and B. A. Britt (eds.), *The Second International Symposium on Malignant Hyperthermia*, pp. 339–350. Grune & Stratton, New York.
KANG, C. G., MUGURUMA, M., FUKAZAWA, T., and ITO, T. 1978. *J. Food Sci.* 43:508.
KASTENSCHMIDT, L. L., HOEKSTRA, W. G., and BRISKEY, E. J. 1968. *J. Food Sci.* 33:151.
KASTNER, C. L., and RUSSELL, T. S. 1975. *J. Food Sci.* 40:747.
KAUFFMAN, R. G., WACHHOLZ, D., HENDERSON, D., and LOCHNER, J. V. 1978. *J. Anim. Sci.* 46:1236.
KEMPSTER, A. J. 1979. *Inst. Meat Bull.* 103:12.
KENT, M., and JASON, A. C. 1975. Dielectric properties of foods in relation to interactions between water and the substrate. In R. B. Duckworth (ed.), *Water Relations of Food*, pp. 221–232. Academic Press, New York.
KING, W. A., OLLIVIER, L., and BASRUR, P. K. 1976. *Ann. Genet. Sel. Anim.* 9:537.
KJOLBERG, O., MANNERS, D. J., and LAWRIE, R. A. 1963. *Biochem. J.* 87:351.

KLEMM, W. R. 1977. Alertness, sleep and related states. In M. J. Swenson (ed.), *Dukes' Physiology of Domestic Animals*, 9th ed., pp. 621–639. Comstock Publishing Associates, Ithaca, N.Y.

KLINGBIEL, J. F. G., and NAUDE, R. T. 1972. *S. Afr. J. Anim. Sci.* 2:105.

KOTULA, A. W., DREWNIAK, E. E., and DAVIS, L. L. 1957. *Poult. Sci.* 36:585.

KOTULA, A. W., DREWNIAK, E. E., and DAVIS, L. L. 1961. *Poult. Sci.* 40:213.

KRAELING, R. R., and GERRITS, R. J. 1973. *J. Anim. Sci.* 37:1124.

KRAELING, R. R., and RAMPACEK, G. B. 1977. *J. Anim. Sci.* 45:71.

KRAELING, R. R., ONO, K., DAVIS, B. J., and BARB, C. R. 1975. *J. Anim. Sci.* 40:604.

KROPF, D. H. 1980. *Proc. Reciprocal Meat Conf.* 33:15.

KUENZEL, W. J., and INGLING, A. L. 1977. *Poult. Sci.* 56:2087.

LACOURT, A., and TARRANT, P. V. 1980. *Curr. Top. Vet. Med. Anim. Sci.* 10:440.

LAMBOOY, E. 1982a. *Meat Sci.* 6:123.

LAMBOOY, E. 1982b. *Meat Sci.* 7:51.

LAMBOOY, E. and SPANJAARD, W. 1982. *Meat Sci.* 6:15.

LANDAU, E. M., and NACHSHEN, D. A. 1975. *J. Physiol.* 251:775.

LAWRIE, R. A. 1950. *J. Agric. Sci., Camb.* 40:356.

LAWRIE, R. A. 1953. *Biochem. J.* 55:298.

LAWRIE, R. A. 1974. *Meat Science.* Pergamon Press, Oxford.

LEA, T. J., and ASHLEY, C. C. 1978. *Nature (Lond.)* 275:5677.

LEACH, T. M., WARRINGTON, R., and WOTTON, S. B. 1980. *Meat Sci.* 4:203.

LEE, Y. B., KAUFFMAN, R. G., GRUMMER, R. H., SCHMIDT, G. R., and BRISKEY, E. J. 1971. *J. Anim. Sci.* 32:457.

LEE, Y. B., HARGUS, G. L., WEBB, J. E., RICKANSRUD, D. A., and HAGBERG, E. C. 1979. *J. Food Sci.* 44:1121.

LEET, N. G., and LOCKER, R. H. 1973. *J. Sci. Food Agric.* 24:1181.

LEWIS, P. K., HECK, M. C., and BROWN, C. J. 1961. *J. Anim. Sci.* 20:727.

LEWIS, P. K., BROWN, C. J., and HECK, M. C. 1962a. *J. Anim. Sci.* 21:196.

LEWIS, P. K., BROWN, C. J., and HECK, M. C. 1962b. *J. Anim. Sci.* 21:433.

LEWIS, P. K., BROWN, C. J., and HECK, M. C. 1967. *J. Anim. Sci.* 26:1276.

LIEDTKE, A. J., NELLIS, S., and NEELY, J. R. 1978. *Circ. Res.* 43:652.

LIN, T.-S., and HULTIN, H. O. 1978. *J. Food Biochem.* 2:39.

LINDBERG, P., LANNEK, N., and BLOMGREN, L. 1973a. *Acta Vet. Scand.* 14:359.

LINDBERG, P., LANNEK, N., BLOMGREN, L., JOHANSSON, G., and JONSSON, L. 1973b. *Nord. Veterinaermed.* 25:619.

LINDHOLM, A., BJERNELD, H., and SALTIN, B. 1974. *Acta Physiol. Scand.* 90:475.

LISTER, D. 1973. *Br. Med. J.* 1:208.

LISTER, D. 1976. Hormonal influences on the growth, metabolism and body composition of pigs. In D. Lister, D. N. Rhodes, V. R. Fowler, and M. F. Fuller (eds.), *Meat Animals. Growth and Productivity*, pp. 355–371. Plenum Press, New York.

LISTER, D., HALL, G. M., and LUCKE, J. N. 1976. *Br. J. Anaesth.* 48:831.

LOCKER, R. H., and DAINES, G. J. 1976. *J. Sci. Food Agric.* 27:193.

LOCKER, R. H., and HAGYARD, C. J. 1963. *J. Sci. Food Agric.* 14:787.

LOMHOLT, N. 1981. *Inst. Meat Bull., Lond.* 113:13.

LUCKE, J. N., HALL, G. M., and LISTER, D. 1977. *Vet. Rec.* 100:45.

LUDVIGSEN, J. 1954. *Beret. Forsogslab.*, p. 272.

LUNDSTROM, K., NILSSON, H., and HANSSON, I. 1977. *Swed. J. Agric. Res.* 7:193.

LUPANDIN, Y. V., and POLESHCHUK, N. K. 1979. *Neirofiziologiya* 11:263.

MA, R. T.-I., ADDIS, P. B., and ALLEN, E. 1971. *J. Food Sci.* 36:125.

MACDONALD, V. W., and JOBSIS, F. F. 1976. *J. Gen. Physiol.* 68:179.

MACNAUGHTAN, A. F. 1978a. *J. Anat.* 126:7.

MACNAUGHTAN, A. F. 1978b. *J. Anat.* 126:461.

MAINWOOD, G. W., and WORSLEY-BROWN, P. 1975. *J. Physiol.* 250:1.

MANKTELOW, B. W., and HARTLEY, W. J. 1975. *J. Comp. Pathol.* 85:139.

MANN, I. 1960. *Meat Handling in Underdeveloped Countries.* FAO Agric. Dev. Paper 70.

MARPLE, D. N., and CASSENS, R. G. 1973a. *J. Anim. Sci.* 37:546.

MARPLE, D. N., and CASSENS, R. G. 1973b. *J. Anim. Sci.* 36:1139.

MARPLE, D. N., TOPEL, D. G., and MATSUSHIMA, C. Y. 1969. *J. Anim. Sci.* 29:882.

MARPLE, D. N., JUDGE, M. D., and ABERLE, E. D. 1972. *J. Anim. Sci.* 35:995.

MARPLE, D. N., CASSENS, R. G., TOPEL, D. G., and CHRISTIAN, L. L. 1974. *J. Anim. Sci.* 38:1224.

MARPLE, D. N., NACHREINER, R. F., MCGUIRE, J. A., and SQUIRES, C. D. 1975. *J. Anim. Sci.* 41:799.

MARSH, B. B., LEET, N. G., and DICKSON, M. R. 1974. *J. Food Technol.* 9:141.

MARSH, B. B., LOCHNER, J. V., TAKAHASHI, G., and KRAGNESS, D. D. 1981. *Meat Sci.* 5:479.

MARTIN, A. H., and FREDEEN, H. T. 1974. *Can. J. Anim. Sci.* 54:127.

MARTIN, A. H., FREDEEN, H. T., L'HIRONDELLE, P. J., MURRAY, A. C., and WEISS, G. M. 1981. *Can. J. Anim. Sci.* 61:289.

MCGLOUGHLIN, P., AHERN, C. P., BUTLER, M., and MCLOUGHLIN, J. V. 1980. *Livestock Prod. Sci.* 7:147.

MCLOUGHLIN, J. V., and MOTHERSHILL, C. 1976. *J. Comp. Pathol.* 86:465.

MEAD, G. C., BARNES, E. M., and IMPEY, C. S. 1974. *Br. Poult. Sci.* 15:381.

MERKEL, R. A. 1971. *Proc. 2nd Int. Symp. Condition Meat Quality Pigs*, Zeist, Pudoc, Wageningen, pp. 97–103.

MEYER, B., THOMAS, J., BUCKLEY, R., and COLE, J. W. 1960. *Food Technol.* 14:4.

MICKELSON, J. R., and MARSH, B. B. 1980. *Cell Calcium* 1:119.

MITCHELL, H. H., and HAMILTON, T. S. 1933. *J. Agric. Res.* 46:917.

MIYAYAMA, H. 1977. *Acta Histochem. Cytochem.* 10:218.

MONIN, G. 1973. *Ann. Zootech.* 22:73.

MORITA, S., CASSENS, R. G., and BRISKEY, E. J. 1969. *Stain Technol.* 44:283.

MORITA, S., CASSENS, R. G., and BRISKEY, E. J. 1970a. *J. Histochem. Cytochem.* 18:364.

MORITA, S., COOPER, C. C., CASSENS, R. G., KASTENSCHMIDT, L. L., and BRISKEY, E. J. 1970b. *J. Anim. Sci.* 31:664.

MORLEY, M. J. 1974. *J. Food Technol.* 9:149.

MOSS, B. W. 1978. *Appl. Anim. Ethol.* 4:323.

MOSS, B. W. 1980. *J. Sci. Food Agric.* 31:308.

MOSS, B. W. and MCMURRAY, C. H. 1979. *Res. Vet. Sci.* 26:1.

MOSS, B. W. and ROBB, J. D. 1978. *J. Sci. Food Agric.* 29:689.

MUIR, A. R. 1970. *J. Comp. Pathol.* 80:137.

MULLENAX, C. H., and DOUGHERTY, R. W. 1963. *Am. J. Vet. Res.* 24:329.

MURRAY, D. M. 1970. *Proc. Aust. Soc. Anim. Prod.* 8:226.

MUYLLE, E. VON, VAN DEN HENDE, C., and OYAERT, W. 1968. *Dtsch. Tierarztl. Wochenschr.* 75:29.

NELSON, J. S., and TASHIRO, K. 1973. *J. Neuropathol. Exp. Neurol.* 32:371.

NELSON, T. E., JONES, E. W., HENRICKSON, R. L., FALK, S. N., and KERR, D. D. 1974. *Am. J. Vet. Res.* 35:347.

NEWBOLD, R. P., and SCOPES, R. K. 1967. *Biochem. J.* 105:127.

NEWBOLD, R. P., and TUME, R. K. 1977. *Aust. J. Biol. Sci.* 30:519.

OKUMURA, F., CROCKER, B. D., and DENBOROUGH, M. A. 1980. *Br. J. Anaesth.* 52:377.

OLLIVIER, L. 1977. *Ann. Genet. Sel. Anim.* 9:353.

OLLIVIER, L. 1980a. *Livestock Prod. Sci.* 7:57.

OLLIVIER, L. 1980b. *Ann. Genet. Sel. Anim.* 12:383.

ONO, K., ALTHEN, T. G., and MAHEN, K. W. 1976. *J. Anim. Sci.* 43:1000.

OOSTVELDT, P. VAN, and BOEKEN, G. 1977. *J. Histochem. Cytochem.* 25:1337.

OVADIA, M., and STOWARD, P. J. 1971. *Histochem. J.* 3:233.
OVERSTREET, J. W., MARPLE, D. N., HUFFMAN, D. L., and NACHREINER, R. F. 1975. *J. Anim. Sci.* 41:1014.
PALMER, A. C. 1976. *Introduction to Animal Neurology*, 2nd ed., pp. 21–34. Blackwell Scientific Publications, Oxford.
PALMER, E. G., TOPEL, D. G., and CHRISTIAN, L. L. 1977. *J. Anim. Sci.* 45:1032.
PARK, H. K., MUGURUMA, M., FUKAZAWA, T., and ITO, T. 1975. *Agric. Biol. Chem.* 39:1363.
PASSBACH, F. L., MULLINS, A. M., WIPF, V. K., and PAUL, B. A. 1970. *J. Anim. Sci.* 30:507.
PASTEA, Z., MANOIU, I., and COSTEA, V. 1972. *Lucr. Inst. Cercet. Vet. Bioprep. Pasteur.* 9:85.
PEASE, A. H. R., and SMITH, C. 1965. *Anim. Prod.* 7:273.
PENNY, I. F. 1977. *J. Sci. Food Agric.* 28:329.
PENNY, I. F. 1980. The enzymology of conditioning. In R. A. Lawrie (ed.), *Developments in Meat Science 1*, pp. 115–143. Applied Science Publishers, London.
PETTE, D. 1975. *Acta Histochem. Suppl.* 14:47.
PHAIR, R. D., and SPARKS, H. V. 1979. *Am. J. Physiol.* 237:H1.
POEL, W. E. 1949. *Am. J. Physiol.* 156:44.
PRICE, J. F., and SCHWEIGERT, B. S. 1971. *The Science of Meat and Meat Products*, 2nd ed. W. H. Freeman and Company, Publishers, San Francisco.
PRICE, M. A., and TENNESSEN, T. 1981. *Can. J. Anim. Sci.* 61:205.
PUCHTLER, H., MELOAN, S. N., and BREWTON, B. R. 1974. *Histochemistry* 40:291.
RAMIREZ, B. U. 1980. *Exp. Neurol.* 67:257.
RAMSBOTTOM, J. M., and STRANDINE, E. J. 1949. *J. Anim. Sci.* 8:398.
RANTSIOS, A. 1972. *Vet. Rec.* 90:369.
RASMUSEN, B. A., and CHRISTIAN, L. L. 1976. *Science* 191:947.
REETZ, I., WEGNER, W., and FEDER, H. 1975. *Zentralbl. Veterinaermed. A* 22:741.
REVILLE, W. J., GOLL, D. E., STROMER, M. H., ROBSON, R. M., and DAYTON, W. R. 1976. *J. Cell Biol.* 70:1.
REY, C. R., KRAFT, A. A., TOPEL, D. G., PARRISH, F. C., and HOTCHKISS, D. K. 1976. *J. Food Sci.* 41:111.
RICHARDS, B. B., EDWARDS, J. R., COOK, R. D., and WHITE, R. R. 1977. *Neuropathol. Appl. Neurobiol.* 3:45.
RICHARDS, S. A., and SYKES, A. H. 1964. *Vet. Rec.* 76:835.
RICHARDS, S. A., and SYKES, A. H. 1967. *Res. Vet. Sci.* 8:361.
RICHTER, L. VON, FLOCK, D. K., and BICKHARDT, K. 1973. *Zuchtungskunde* 4:429.
RILEY, R. R., SAVELL, J. W., SMITH, G. C., and SHELTON, M. 1980. *J. Food Sci.* 45:119.
ROBERTSON, D. D., and BAKER, D. D. 1933. *Univ. Mo. Agric. Exp. Stn. Bull. 200.*
ROSENTHAL, I. 1881. *General Physiology of Muscle and Nerves*, pp. 119–124. D. Appleton and Company, New York.
SAIR, R. A., LISTER, D., MOODY, W. G., CASSENS, R. G., HOEKSTRA, W. G., and BRISKEY, E. J. 1970. *Am. J. Physiol.* 218:108.
SAIR, R. A., KASTENSCHMIDT, L. L., CASSENS, R. G., and BRISKEY, E. J. 1972. *J. Food Sci.* 37:659.
SALE, P. 1972. Appareil de detection des viandes decongelees par mesure de conductance electrique. *Bull. Inst. Int. Froid* 265–275.
SAVELL, J. W. 1978. *Anim. Ind. Today* 1:3.
SAVELL, J. W., SMITH, G. C., DUTSON, T. R., CARPENTER, Z. L., and SUTER, D. A. 1977. *J. Food Sci.* 42:702.
SAVELL, J. W., SMITH, G. C., and CARPENTER, Z. L. 1978. *J. Anim. Sci.* 46:1221.
SAYRE, R. N., BRISKEY, E. J., HOEKSTRA, W. G., and BRAY, R. W. 1961. *J. Anim. Sci.* 20:487.
SAYRE, R. N., BRISKEY, E. J., and HOEKSTRA, W. G. 1963a. *J. Food Sci.* 28:292.
SAYRE, R. N., BRISKEY, E. J., and HOEKSTRA, W. G. 1963b. *Proc. Soc. Exp. Biol. Med.* 112:223.
SAYRE, R. N., BRISKEY, E. J., and HOEKSTRA, W. G. 1963c. *J. Food Sci.* 28:472.

SCHANNE, O. F. 1969. Measurement of cytoplasmic resistivity by means of the glass microelectrode. In M. Lavallee, O. F. Scanne, and N. C. Hebert (eds.), *Glass Microelectrodes*, p. 300. John Wiley & Sons, New York.

SCHEPER, J. 1971. *Proc. 2nd Int. Symp. Condition Meat Quality Pigs*, Zeist, Pudoc, Wageningen, pp. 271–277.

SCHMALBRUCH, H. 1979. *Cell Tissue Res.* 204:187.

SCHMIDT, G. R., ZUIDAM, L., and SYBESMA, W. 1971. *Proc. 2nd Int. Symp. Condition Meat Quality Pigs*, Zeist, Pudoc, Wageningen, pp. 73–80.

SCHMIDT, G. R., GOLDSPINK, G., ROBERTS, T., KASTENSCHMIDT, L. L., CASSENS, R. G., and BRISKEY, E. J. 1972a. *J. Anim. Sci.* 34:379.

SCHMIDT, G. R., ZUIDAM, L., and SYBESMA, W. 1972b. *J. Anim. Sci.* 34:25.

SCHMIDT, G. R., CRIST, D. W., and WAX, J. E. 1974. *J. Anim. Sci.* 38:295.

SCHOTT, L. H., and TERJUNG, R. L. 1979. *Eur. J. Appl. Physiol.* 42:175.

SCOTT, W. N. 1971. The use of electrical stunning devices in poultry slaughter. In *Humane Killing and Slaughterhouse Techniques*, pp. 24–33. Universities Federation for Animal Welfare, Potters Bar, Hertfordshire, England.

SCOTT, W. N. 1978. *Anim. Regul. Stud.* 1:227.

SEBRANEK, J. G., MARPLE, D. N., CASSENS, R. G., BRISKEY, E. J., and KASTENSCHMIDT, L. L. 1973. *J. Anim. Sci.* 36:41.

SHAW, F. D., and WALKER, D. J. 1977. *J. Food Sci.* 42:1140.

SHELLEY, H. J. 1961. *Br. Med. Bull.* 17:137.

SINK, J. D., and JUDGE, M. D. 1971. *Growth* 35:349.

SMITH, G. C., DUTSON, T. R., HOSTETLER, R. L., and CARPENTER, Z. L. 1976. *J. Food Sci.* 41:748.

SMITH, R. N., LEMIEUX, M. D., and COUCH, N. P. 1969. *Surg. Gynecol. Obstet.* 128:533.

SMITH, W. C., and LESSER, D. 1979. *Anim. Prod.* 28:442.

SMITHERMAN, M. L., LAZAROW, A., and SORENSON, R. L. 1972. *J. Histochem. Cytochem.* 20:463.

SOMERS, C. J., WILSON, P., AHERN, C. P., and MCLOUGHLIN, J. V. 1977. *J. Comp. Pathol.* 87:177.

SORINMADE, S. O., CROSS, H. R., ONO, K., and WERGIN, W. P. 1982. *Meat Sci.* 6:71.

STANTON, H. C., and MUELLER, R. L. 1974. *Biol. Neonate* 25:307.

STEINHAGEN, C., HIRCHE, H. J., NESTLE, H. W., BOVENKAMP, U., and HOSSELMANN, I. 1976. *Pfluegers Arch.* 367:151.

STEINHARDT, M. VON, LYHS, L., and BUNGER, U. 1974. *Mh. VetMed.* 29:169.

STOWARD, P. J., and AL-SARRAJ, B. 1981. *Histochemistry* 71:405.

SUGITA, H., ISHIURA, S., SUZUKI, K., and IMAHORI, K. 1980. *Muscle Nerve* 3:335.

SUNG, S. K., ITO, T., and FUKAZAWA, T. 1976. *J. Food Sci.* 41:102.

SUZUKI, A., NONAMI, Y., and GOLL, D. E. 1975. *Agric. Biol. Chem.* 39:1461.

SWATLAND, H. J. 1974. *Vet. Bull.* 44:179.

SWATLAND, H. J. 1975a. *Histochem. J.* 7:367.

SWATLAND, H. J. 1975b. *Histochem. J.* 7:459.

SWATLAND, H. J. 1975c. *Can. Inst. Food Sci. Technol. J.* 8:122.

SWATLAND, H. J. 1976a. *J. Anim. Sci.* 42:838.

SWATLAND, H. J. 1976b. *Can. Inst. Food Sci. Technol. J.* 9:177.

SWATLAND, H. J. 1976c. *Histochem. J.* 8:455.

SWATLAND, H. J. 1976d. *J. Anim. Sci.* 43:577.

SWATLAND, H. J. 1977a. *Histochem. J.* 9:163.

SWATLAND, H. J. 1977b. *Can. Inst. Food Sci. Technol. J.* 10:280.

SWATLAND, H. J. 1978a. *J. Anim. Sci.* 46:113.

SWATLAND, H. J. 1978b. *Can. J. Anim. Sci.* 58:427.

SWATLAND, H. J. 1979a. *Histochem. J.* 11:391.

SWATLAND, H. J. 1979b. *Stain Technol.* 54:245.

SWATLAND, H. J. 1980a. *Can. Inst. Food Sci. Technol. J.* 13:45.

SWATLAND, H. J. 1980b. *J. Anim. Sci.* 50:67.

SWATLAND, H. J. 1980c. *Histochem. J.* 12:39.

SWATLAND, H. J. 1980d. *J. Anim. Sci.* 51:1108.

SWATLAND, H. J. 1981a. *Meat Sci.* 5:451.

SWATLAND, H. J. 1981b. *Can. Inst. Food Sci. Technol. J.* 14:147.

SWATLAND, H. J. 1981c. *J. Anim. Sci.* 53:666.

SWATLAND, H. J. 1982a. *J. Anim. Sci.* 54:264.

SWATLAND, H. J. 1982b. *Can. J. Anim. Sci.* 62:725.

SWATLAND, H. J. 1982c. *Can. Inst. Food Sci. Technol. J.* 15:92.

SWATLAND, H. J. 1983. *Can. Inst. Food Sci. Technol. J.* 16:35.

SWATLAND, H. J., and CASSENS, R. G. 1972. *J. Comp. Pathol.* 82:229.

SWATLAND, H. J., and CASSENS, R. G. 1973. *J. Anim. Sci.* 37:885.

SWATLAND, H. J., WARD, P., and TARRANT, P. V. 1982. *J. Food Sci.* 47:686.

SYBESMA, W., and EIKELENBOOM, G. 1978. *Meat Sci.* 2:79.

TAKEUCHI, T., and KURIAKI, H. 1955. *J. Histochem. Cytochem.* 3:153.

TARRANT, P. J. V., HEGARTY, P. V. J., and MCLOUGHLIN, J. V. 1972a. *Proc. R. Ir. Acad.* 72B:229.

TARRANT, P. J. V., MCLOUGHLIN, J. V., and HARRINGTON, M. G. 1972b. *Proc. R. Ir. Acad.* 72B:55.

TARRANT, P. V., and SHERINGTON, J. 1980. *Meat Sci.* 4:287.

TARRANT, P. V., GALLWEY, W. J., and MCGLOUGHLIN, P. 1979. *Ir. J. Agric. Res.* 18:167.

TAYLOR, D. G., and MARSHALL, A. R. 1980. *J. Food Sci.* 45:144.

TE BRAKE, J. H. A. 1971. *Proc. 2nd Int. Symp. Condition Meat Quality Pigs*, Zeist, Pudoc, Wageningen, pp. 53–60.

TERJUNG, R. L., BALDWIN, K. M., WINDER, W. W., and HOLLOSZY, J. O. 1974. *Am. J. Physiol.* 226:1387.

THOMAS, N. W., and JUDGE, M. D. 1970. *J. Agric. Sci., Camb.* 74:241.

THOONEN, J., and HOORENS, J. 1960. *Vlaams Diergeneeskd. Tijdschr.* 29:205.

THORNTON, H., and GRACEY, J. F. 1974. *Textbook of Meat Hygiene*, 6th ed. Ballière Tindall, Publishers, London.

TODOROV, A., and PETROV, J. 1969. *Anat. Anz.* 125:88.

TOMINAGA, S., CURNISH, R. R., BELARDINELLI, L., RUBIO, R., and BERNE, R. M. 1980. *Am. J. Physiol.* 238:H156.

TOPEL, D. G. 1971. A review of animal physiology and the porcine stress syndrome in relation to meat quality. In R. Cassens, F. Giesler, and Q. Kolb (eds.), *The Proceedings of the Pork Quality Symposium*, pp. 26–67. University of Wisconsin Press, Madison.

TOPEL, D. G., BICKNELL, E. J., PRESTON, K. S., CHRISTIAN, L. L., and MATSUSHIMA, C. Y. 1968. *Mod. Vet. Pract.* 49:40.

TUME, R. K. 1979. *Aust. J. Biol. Sci.* 32:163.

UNSHELM, J., HOHNS, H., OLDIGS, B., and RUHL, B. 1971. *Proc. 2nd Int. Symp. Condition Meat Quality Pigs*, Zeist, Pudoc, Wageningen, pp. 208–214.

VAN DER WAL, P. G. 1978. *Meat Sci.* 2:19.

VAN PUTTEN, G., and ELSHOF, W. J. 1978. *Anim. Regul. Stud.* 1:247.

VEUM, T. L., ELLERSIECK, M. R., DURHAM, T. L., MCVICKERS, W. R., MCWILLIAMS, S. N., and LASLEY, J. F. 1979. *J. Anim. Sci.* 48:446.

VOYLE, C. A. 1969. *J. Food Technol.* 4:275.

WANGSNESS, P. J., MARTIN, R. J., and GAHAGAN, J. H. 1977. *Am. J. Physiol.* 233:E104.

WANSON, J. C., and DROCHMANS, P. 1972. *J. Cell Biol.* 54:206.

WARDLE, C. S. 1978. *J. Exp. Biol.* 77:141.

WARNOCK, J. P., CAPLE, I. W., HALPIN, C. G., and MCQUEEN, C. S. 1978. *Aust. Vet. J.* 54:566.

WARRINGTON, R. 1974. *Vet. Bull.* 44:617.
WARRISS, P. D. 1979. *Meat Sci.* 3:281.
WARRISS, P. D., and AKERS, J. M. 1980. *J. Food Technol.* 15:629.
WARRISS, P. D., and LISTER, D. 1982. *Meat Sci.* 7:183.
WASSMUTH, R. VON, and REUTER, H. 1973. *Z. Tierz. Zuechtungsbiol.* 90:56.
WATSON, C. G., TOPEL, D. G., KUHLERS, D. L., and CHRISTIAN, L. L. 1980. *J. Anim. Sci.* 50:442.
WEBB, A. J. 1980. *Vet. Rec.* 106:410.
WEBB, A. J., and JORDAN, C. H. C. 1978. *Anim. Prod.* 26:157.
WEBB, N. B., IVEY, F. J., CRAIG, H. B., JONES, V. A., and MONROE, R. J. 1970. *J. Food Sci.* 35:501.
WEGNER, W., FEDER, H., and REETZ, I. 1975. *Zentralbl. Veterinaermed. A* 22:645.
WEISS, G. M., TOPEL, D. G., SIERS, D. G., and EWAN, R. C. 1974. *J. Anim. Sci.* 38:591.
WEISSIG, K. 1978. *Histochemistry* 58:121.
WELLS, G. A. H., and PINSENT, P. J. N. 1980. *Vet. Rec.* 106:556.
WENDT, I. R., and BARCLAY, J. K. 1980. *Am. J. Physiol.* 238:C56.
WESTERVELT, R. G., and STOUFFER, J. R. 1978. *J. Anim. Sci.* 46:1206.
WHITING, R. C., and RICHARDS, J. F. 1975. *J. Food Sci.* 40:960.
WILL, P. A., HENRICKSON, R. L., MORRISON, R. D., and ODELL, G. V. 1979. *J. Food Sci.* 44:1646.
WILL, P. A., OWNBY, C. L., and HENRICKSON, R. L. 1980. *J. Food Sci.* 45:21.
WILLIAMS, C. H., SHANKLIN, M. D., HEDRICK, H. B., MUHRER, M. E., STUBBS, D. H., KRAUSE, G. F., PAYNE, C. G., BENEDICT, J. D., HUTCHESON, D. P., and LASLEY, J. F. 1977. The fulminant hyperthermia-stress syndrome: Genetic aspects, hemodynamic and metabolic measurements in susceptible and normal pigs. In A. Aldrete and B. A. Britt (eds.), *The Second International Symposium on Malignant Hyperthermia*, pp. 113–140. Grune & Stratton, New York.
WINGER, R. J., and POPE, C. G. 1981. *Meat Sci.* 5:355.
WINGER, R. J., FENNEMA, O., and MARSH, B. B. 1979. *J. Food Sci.* 44:1681.
WOOD, D. F., and RICHARDS, J. F. 1975. *Poult. Sci.* 54:520.
WOOD, J. D., DRANSFIELD, E., and RHODES, D. N. 1979. *J. Sci. Food Agric.* 30:493.
WOOLSEY, C. N., and FAIRMAN, D. 1946. *Surgery* 19:684.
WOOTEN, W. L., and CASCARANO, J. 1980. *J. Bioenerg. Biomembr.* 12:1.
WYTHES, J. R., GANNON, R. H., and HORDER, J. C. 1979. *Vet. Rec.* 104:71.
YOUMANS, W. B. 1967. *Am. J. Phys. Med.* 46:173.
YOUNG, O. A., GRAAFHUIS, A. E., and DAVEY, C. L. 1980. *Meat Sci.* 5:41.
ZIMMERMAN, M. R., and TEDFORD, R. H. 1976. *Science* 194:183.
ZIMMERMANN, P. 1977. *Cell Tissue Res.* 178:239.

Index

B

C

D

E

F

G

H

I

J

K

L

M

N

O

P

Q

R

S

T

U

V

W

Y

Z